Social Studies for the Seventies

In Elementary and Middle Schools

Second Edition

Leonard S. Kenworthy

Formerly of Brooklyn College of
The City University of New York

XEROX

Xerox College Publishing | Lexington, Massachusetts | Toronto

ISB Number: 0-536-00816-7

Library of Congress Catalog Card Number: 72-85445

Printed in the United States of America.

Acknowledgments

The author is grateful to the following companies and school systems for their permission to reprint and their cooperation in making the photographs listed below. He also wishes to thank two photographers, Maurice L. Lehv of Brooklyn, New York and Don Smith of the Hamilton Studios of Wayne, New Jersey, for their cooperation during the on-location assignments. The photograph adapted for use on the cover of this book was supplied by the United Nations.

Page xvi	1.	United Nations
	2.	General Motors Corporation
Page xvii	3.	TWA
	4.	General Motors Corporation
	5.	Westinghouse Electric Corporation
	6.	AT & T Company
Page 46	7.	Bruce Davidson, Magnum Photos, Inc.
	8.	Whitman Junior High School, Brooklyn, N.Y.
Page 47	9.	Whitman Junior High School, Brooklyn, N.Y.
	10.	Brooklyn College, Early Childhood Center, Brooklyn, N.Y.
	11.	Western Electric Company
	12.	The New York Association for the Blind
Page 76	13.	Little Red Schoolhouse, New York, N.Y.
	14.	Bronxville Public Schools, New York, N.Y.
	15.	Fairlawn Public Schools, Fairlawn, N.J.
	16.	P.S. 139, New York, N.Y.
Page 77	17.	Ditmas Junior High School, Brooklyn, N.Y.
	18.	Ditmas Junior High School, Brooklyn, N.Y.
	19.	American Museum of Natural History, New York, N.Y.
Page 116	20.	Fairlawn Public Schools, Fairlawn, N.J.
	21.	Fairlawn Public Schools, Fairlawn, N.J.
Page 117	22.	P.S. 139, New York, N.Y.
	23.	Fairlawn Public Schools, Fairlawn, N.J.
	24.	Fairlawn Public Schools, Fairlawn, N.J.
Page 145	25.	Maurice L. Lehv, Brooklyn, N.Y.
Page 158	26.	P.S. 139, New York, N.Y.
	27.	Whitman Junior High School, Brooklyn, N.Y.

Page 204   28.   Fairlawn Public Schools, Fairlawn, N.J.
            29.   Little Red Schoolhouse, New York, N.Y.
            30.   Northern Illinois University, Dekalb, Ill.

Page 255   31.   P.S. 139, New York, N.Y.
            32.   Brooklyn College, Early Childhood Center, Brooklyn, N.Y.
            33.   Brooklyn College, Early Childhood Center, Brooklyn, N.Y.

Page 270   34.   P.S. 139, New York, N.Y.
            35.   Brooklyn College, Early Childhood Center, Brooklyn, N.Y.
            36.   P.S. 139, New York, N.Y.

Page 285   37.   P.S. 139, New York, N.Y.
            38.   Detroit Public Schools, Detroit, Mich.
            39.   Fairlawn Public Schools, Fairlawn, N.J.

Page 317   40.   Fairlawn Public Schools, Fairlawn, N.J.
            41.   P.S. 139, New York, N.Y.
            42.   Fairlawn Public Schools, Fairlawn, N.J.

Page 344   43.   Whitman Junior High School, Brooklyn, N.Y.
            44.   P.S. 139, New York, N.Y.
            45.   Ditmas Junior High School, Brooklyn, N.Y.

Page 345   46.   Whitman Junior High School, Brooklyn, N.Y.
            47.   P.S. 139, New York, N.Y.
            48.   Whitman Junior High School, Brooklyn, N.Y.

Page 420   49.   Whitman Junior High School, Brooklyn, N.Y.
            50.   Whitman Junior High School, Brooklyn, N.Y.

Page 421   51.   Whitman Junior High School, Brooklyn, N.Y.
            52.   P.S. 139, New York, N.Y.

Page 458   53.   P.S. 139, New York, N.Y.
            54.   Whitman Junior High School, Brooklyn, N.Y.
            55.   Whitman Junior High School, Brooklyn, N.Y.

Preface

Tremendous changes have occurred in the United States as well as in other parts of the world in recent years. Even more drastic changes are probably on their way. Yet the social studies programs of most schools are still preparing boys and girls for the world of 1900, 1925, or possibly 1950. Many colleges and universities are basing their programs for the preparation of social studies teachers in elementary and junior high schools on archaic and outmoded plans.

Fortunately, however, there is real ferment in the social studies field at the present time. Individual schools and school systems are examining their social studies programs in the light of changes in our country and in other parts of the world. Research programs are shedding new light on learning theory and on instructional strategies. Grants from foundations and from the government are making better library facilities possible. Textbook companies and audio-visual firms are producing new materials, many of them in "packages," including texts, pictures, films, filmstrips, and other learning devices.

Social studies instruction in our schools dare not lag behind the times. Teachers now in service need to be brought up-to-date in their thinking and in their teaching. Students preparing for teaching in the next decade need to be educated in the most current thinking about the contemporary world and the best ways to arouse the interest of pupils in the social studies. Courses of study need to be revised and new approaches tried out.

It is the hope of the author of this book that this volume will contribute to the changes which must occur in the social studies field in the next few years. A broad frame of reference for social studies in the seventies and beyond is suggested. New programs are outlined and evaluated. The nature of the various social science disciplines is summarized. New methods or strategies for learning are suggested and many examples of them given. New materials are listed and sources of other materials presented. It is hoped that all of these approaches will prove stimulating to teachers and teachers-to-be.

This volume is intended primarily for the pre-service education of teachers from kindergarten through grade eight. But it should also prove useful to experienced teachers and to principals, supervisors, librarians, and others concerned with the social studies field.

This is a new type of social studies methods book. It is patterned in part after the author's volume on *A Guide to Social Studies Teaching in Secondary Schools* (Wadsworth Publishing Co.). In order to make the best possible use of the available space and also to increase the speed of reading, certain sections are presented in outline form. In order to include the reader as co-author, several sections are left blank and the reader is invited to fill them in. A large number of bibliographies are included so readers may pursue many topics in depth. Lists of books for children are also included to assist teachers in obtaining a wide variety of volumes for their pupils.

It is the author's firm conviction that we know enough now to revise our social studies curricula and methods radically to meet the changing times. Therefore this book represents a distinct point of view. It stresses a broad interpretation of the social studies, drawing upon *all* of the social science disciplines. It pleads for the continued use of all that is known about child growth and development as well as the use of newer research on cognitive learning. It calls for limiting the number of topics to be studied and for greater depth in the examination of those themes and areas selected. It emphasizes new as well as old strategies for learning or methods. It urges schools to introduce boys and girls to the world much earlier than is now done in most places. It calls for a study of the United States today as well as its history. And it proposes a year of attention in the upper grades to contemporary problems.

Because many of the programs suggested are new, a large part of the book is devoted to detailed accounts of studying families in the United States and in other parts of the world, to studying communities in our country as well as in other parts of the world, and to the study of the United States as a nation, followed by the study of a selected number of other nations. It gives a few examples, too, of some contemporary problems which might well be undertaken by pupils in the upper grades of elementary schools.

Because of space limitations, some topics traditionally covered in methods books are covered quickly. But suggestions for further reading are made.

The writer owes debts of gratitude to hundreds of persons in the preparation of this volume. Some have been teachers in times past. Many have been cooperating teachers, student teachers, and pupils in college classes. Others have been participants in workshops in many parts of the United States. Still others have been hardworking members of curriculum committees in school systems where the author has worked. Many are writers of books and articles. Special thanks are due to the officers of Wadsworth Publishing Co. for permission to use some materials which appeared in the author's guide to social studies teaching in secondary schools. Thanks, too, are extended to several colleagues in the Education Department of Brooklyn College who have assisted the author in many, many ways. Richard A. Birdie has helped with many of the details in the preparation of this volume.

In this second edition of *Social Studies for the Seventies* new sections have been added on environment and ecology; on conflicts and conflict resolution, or war and peace; and on studying Africa. At the suggestion of readers, numbers have been added to the chapters and captions to the pictures. Bibliographies have been up-dated and various sections revised. Suggestions for further improvement of this volume are welcomed by the author.

LEONARD S. KENWORTHY

Contents

9.

WHAT GENERAL RESOURCES ARE AVAILABLE FOR SOCIAL STUDIES TEACHING?

10.

WHAT ROLE SHOULD CURRENT AFFAIRS AND SPECIAL OCCASIONS PLAY IN SOCIAL STUDIES PROGRAMS?

11.

HOW, THEN, SHOULD A TEACHER PLAN LEARNING ACTIVITIES IN THE SOCIAL STUDIES?

12.

HOW CAN A TEACHER PROVIDE FOR INDIVIDUAL DIFFERENCES? 195

13.

HOW CAN A TEACHER UTILIZE OTHER SUBJECT FIELDS IN TEACHING THE SOCIAL STUDIES?

22.

STUDYING THE WORLD — GENERAL   459

23.

STUDYING PERSONAL PROBLEMS AND PROBLEMS OF THE LOCAL COMMUNITY, THE UNITED STATES, AND THE REST OF THE WORLD   471

24.

ADDRESSES OF ORGANIZATIONS AND PUBLISHERS   492

**Social
Studies
for the
Seventies**

1. A world of six to seven billion "neighbors"

The World of the Present and the Foreseeable Future

2. With many new satellite cities in various parts of our planet

3. Linked by super-jets

4. Making more use of our oceans

5. With more power supplied by atomic energy

6. And communication by satellites

1

What Are the Purposes
of Social Studies Teaching?

WHAT THE TERM "SOCIAL STUDIES" MEANS TO YOU

What springs into your mind when you hear or see the words "social studies?" Would you be willing to write the words, phrases, or experiences which come to mind at once when you hear or see those two words? Some space is provided below for your response. Don't wait to reflect or select ideas. Merely record what flashes into your mind when you see the words "social studies."

What did you jot down?

Long lists of dates to be memorized — or a parade of fascinating people from many places and many periods of history?

Strange names to remember — or farms, villages, and cities filled with interesting people, sights, and sounds?

The word *history* — or the words *anthropology, economics, geography, history, political science* or *government,* or *sociology*?

A strict chronological approach to the story of the United States — or selected episodes in our history — suffering through a long winter with the Pilgrims, deciding whether to fight England or remain loyal to the motherland, attending Jackson's Inaugural Ball, fighting with the northern or southern forces in the Civil War, trekking west with the pioneers, and a host of similar stories?

Only references to the United States — or exposure to the whole world?

A battered textbook to be plowed through, sentence by sentence—or several textbooks and many trade books, films, filmstrips, trips, and other materials?

Tiny maps cluttered with unpronounceable place-names—or large, colorful maps that reminded you of people at work and at worship, at school and at play, being governed and governing themselves?

Your answers to the question of what springs into your mind when you hear or see the words "social studies" may reveal much about the ways in which you were taught that subject. They may disclose whether you detest or enjoy the social studies. They might indicate how you will teach the social studies, or the way in which you have been teaching for some years.

Perhaps it would be worthwhile to stop and analyze what you wrote in order to determine why those particular items sprang into your mind. Better yet, you might like to analyze what several persons wrote, and why, so far as they can tell.

SOCIAL STUDIES AS INTERPRETED BY TWO TEACHERS

Each person interprets the term social studies in a different way. Let's think about two teachers at almost opposite ends of a continuum. For purposes of identification, let's call them Mrs. Appleby and Mrs. Zelch.

Mrs. Appleby teaches the social studies as a part of her work in a self-contained classroom. She believes that this term means *history and geography* and that the two are separate and seldom, if ever, complement each other. She teaches them in two distinct periods each day. To her the teaching of the social studies begins in the fourth grade because children cannot possibly learn anything about those two subjects until they can read well.

She uses a single textbook for all the pupils and they follow the book sentence by sentence and paragraph by paragraph. "That is why the author wrote it that way," she maintains. Using other textbooks and trade books is "confusing" to children, she says.

Content is of the utmost importance. "All this talk about 'the discovery method' is nonsense," Mrs. Appleby asserts. Her basic philosophy is that teaching is telling and she does a good deal of that. At the end of the day her throat is often sore from telling—and yelling. Dates are highly important. "Of course they are important," she says. "How can you have history without dates?" she asked the principal who tried to get her to downplay dates.

Trips are "excursions" and films and filmstrips are "play" to her. Of course questions are important, but they are asked by the teachers. Hasn't she had years of experience in asking them? How can children know enough to ask questions?

Current events? Not in Mrs. Appleby's class. She feels strongly that children in the fourth grade are much too young to be exposed to what is happening in what she calls "the adult world."

Her bulletin boards are beautiful because she prepares them herself. Some parents point to them as examples of the work of a good teacher. "The children learn from her because she knows so much and makes them toe the line," they say.

Such an interpretation of the social studies makes this subject irrelevant, dull, and boring to her pupils. They learn little that can be applied in their daily lives. There is almost no excitement and almost no joy in the process of learning.

Does this sound like a caricature? Perhaps. But I sat in her class once. The only other teacher who was worse was a specialist in the social studies in the eighth grade. All I can remember of the year I spent with her is the fact that she wore a black dress, had a dirty face, and hated the Germans.

Mrs. Zelch is also a fourth grade teacher. She believes that anything which helps children to become wholesome, worthwhile, contributing members of various groups is a part of the social studies. "I'm not so concerned with the labels as with the results," she says. "Perhaps it is social education. Possibly it is social learnings. Perhaps it is social studies." They are almost the same thing, she maintains.

The social studies bloc takes up a good part of the day in her class. Often that bloc includes music, art, and some reading. Mrs. Zelch has discovered that children learn to read better when they read material of interest to them. Often that material is social studies material.

Mrs. Zelch has made a fairly compact list of concepts or generalizations on which she wants to concentrate and a list of skills and attitudes she wants to develop. Long ago she stopped trying to get definitions of concepts and generalizations. Now she calls them "big ideas" in the social sciences. "People spend too much time on definitions. What is important is to know what you want to stress and to get on with the job," she tells her colleagues.

She believes that the current fourth grade program in her school is archaic and limiting. "A year on the local community is too much. Pupils have already studied it in the first three grades. It is time they broadened their horizons," she told the principal. Between them they decided to inaugurate an experimental program in the class Mrs. Zelch teaches. During the year she and her pupils are studying three selected communities in other parts of the United States and four in other parts of the world. It isn't an ideal program, she feels, but it is a move in the right direction.

"Content? Of course it is important. But the way in which it is acquired is even more important. I want my pupils to discover ideas themselves through a series of related experiences. Then they are more likely to retain it. And they will have learned important skills in the process which will stay with them throughout their lives." So she uses three basic textbooks to provide for different reading levels and for comparisons of content. She also uses many supplementary books. There are occasional trips, films, filmstrips, and other experiences.

Part of the time her class works as a committee-of-the-whole. Much of the time they work in small groups after overviews by all the pupils of the topic they are exploring. Occasionally they work individually. Many of the questions come from the pupils, but not all, by any means. Her questions are usually of the "why" and "how" type.

A few minutes are often spent on current events at the beginning of the day. There is a "Weekly Roundup of the News" on Friday.

Her bulletin boards? They aren't as neat and beautiful as those of Mrs. Appleby. They are prepared by the pupils and contain their work. She uses a wide range of evaluation methods; sometimes the pupils are the evaluators.

Does this sound too ideal? Perhaps. But there are many teachers like Mrs. Zelch. I sat in such a class in second grade, in the fifth grade, and in the seventh grade myself. I owe these three teachers a great deal.

In the second grade we were basically explorers. We explored our home town and the territory around it. We visited the firehouse and the police station, a milk plant and several stores, the telephone exchange, and city hall. We met and talked with older residents who introduced us to history. We read about our community, made experience charts on our trips, drew pictures of what we had seen, and made maps on the floor of the classroom. We learned to observe, to ask questions, to report our findings, and to work individually, in small groups, and as a class. Sometimes we were spurred on by a system of blue, pink, and orange slips for "good behavior." But most of the time we were stimulated by better methods, such as a wide range of exciting experiences. This was real social studies instruction in a small town in southwestern Ohio.

In the fifth grade each pupil was given several hundred dollars to spend on a trip around the world. We studied travel folders, read textbooks and trade books, planned our itineraries and kept track of our funds, exchanging dollars for francs, reichsmarks, rubles, and other kinds of money. We kept individual "logs" of what we saw and in doing so learned how to outline and to write. Above all, we were motivated to learn more about the world. This, too, was a wonderful year in the social studies, with a community in eastern Pennsylvania as our home base and the world as our backyard.

In the seventh grade in Washington, D.C. I had a remarkable social studies teacher. Her chief characteristic was her ability to meet individual differences. When she discovered my interest in government and politics, she urged me to visit the House and Senate on Saturdays and to report to the class on Mondays. I suspect I got "extra credit" for those reports but that kind of "bribe" wasn't necessary. Government became something real and personal during that year.

Have you been fortunate enough to have had several such teachers of the social studies? What was outstanding about them? Here is some space in which to describe the teacher who was best in the field of social studies while you were in elementary school:

The Best Teacher of the Social Studies I Had in Elementary School

THE MEANING AND PURPOSES OF THE SOCIAL STUDIES

What, then, is meant specifically by the term "social studies?" Social Studies is a kind of shorthand for the study of man by pupils in elementary and secondary schools. In reality it should be called anthropology-sociology, economics, geography, history, political science, psychology-social psychology, for it draws upon all these fields. In addition, the social studies are closely related to philosophy and religion, art and music, literature, and science. All of these subject fields need to be utilized to understand man.

In chart form the social studies field might look like this:

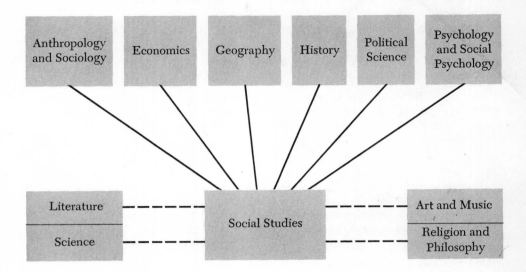

At the college level we usually refer to the social sciences or to history and the social sciences, inasmuch as some historians do not consider their field to be a science. Within recent years some writers and educators have begun to refer to the work in elementary and secondary schools as "the social sciences." For example, Bruce Joyce has written a methods book which he calls *Strategies for Elementary Social Science Education.* The term "social science" is being used by him and others because they feel that it is now exact enough to be called a science and that it will become even more exact as teachers examine the structure of each discipline and try to get pupils to discover the underlying principles through the discovery method.

Two other terms are sometimes used in conjunction with the social studies field. They are social education and social learnings. Both refer to the experiences of children in school which help them to become socialized. To many persons these terms are broader than the social studies field. Yet they are so closely related to the aims of social studies that this writer will include them throughout this book as a part of any effective social studies program.

The field of the social studies is almost limitless. It is as wide as the world and as long as the history of man. For good measure, it includes the millions of years prior to recorded history and also the foreseeable future.

Yet the social studies can be defined with one word. That word is *people.* People live in specific places — which is geography. They live at a specific time — which is history. They earn a living and exchange goods — which is economics. They live in various groups — families, tribes, friendship groups, communities, and interest groups — which is sociology and anthropology. They organize themselves or are organized into political units of many kinds — which is the study of government or political science. All of these important aspects of the study of man need to be included in any effective social studies program, even at the elementary school level.

The Purposes of the Social Studies

Several years ago Robert S. Lynd wrote a provocative book with the title *Knowledge for What?* This was a study of the place of social science in American culture.

His clever caption might be revised for our purposes in this book by asking the question: Social Studies — For What? What are the aims of the social studies? What are its purposes?

Let's listen to several experts give their answers. Then we will be better able to write our own definition.

In her volume on *Curriculum Development for Elementary Schools,* Muriel Crosby writes concerning the aims of the social studies as follows:

> . . . the social studies are identified as those studies which are concerned with how people build a better life for themselves and their fellow men, how people deal with the problems of living together, how people change and are changed by their environment.

Wilhelmina Hill writes in a somewhat different vein, with greater stress on geography and history. She has said:

> This area of the curriculum helps the child understand his social and physical environment. It includes the past but gives emphasis to the present and the foreseeable future. At the same time it has the responsibility of furthering his social development and growth toward good citizenship.

Bruce Joyce has listed the three goals of the social studies in this way:

> *Humanistic education* is the first goal. The social studies should help the child comprehend his experience and find meaning in life.
> *Citizenship education* is the second goal. Each child must be prepared to participate effectively in the dynamic life of his society. Correspondingly, the society needs active, aware citizens who will work devotedly for its improvement.
> *Intellectual education* is the third goal. Each person needs to acquire the analytical ideas and problem-solving tools that are developed by scholars in the social sciences. With increasing maturity the child should learn to ask fruitful questions and examine critical data in social situations.

Writing in *The Instructor* in 1964, Dr. Fannie Shaftel of Stanford University identified the most important strands in the social studies as:

1. The self realization of the child *in* society.
2. The increasing acquaintance of the child with the knowledge about society and the processes of social inquiry that enable us to develop rational solutions to problems.
3. The acquisition of skills for living in groups and engaging in active citizen behavior.
4. The development of a system of democratic values and processes for continually criticizing one's experience and reconstructing one's values.

In a pamphlet on *Social Studies in Transition: Guidelines for Change,* a committee of the National Council for the Social Studies highlighted fifteen themes in the social studies as follows:

1. Recognition of the dignity and worth of the individual.
2. The use of intelligence to improve human living.
3. Recognition and understanding of world interdependence.
4. The understanding of the major world cultures and culture areas.
5. The intelligent uses of the natural environment.
6. The vitalization of our democracy through an intelligent use of our public educational facilities.

7. The intelligent acceptance, by individuals and groups, of responsibility for achieving democratic social action.
8. Increasing the effectiveness of the family as a basic social institution.
9. The effective development of moral and spiritual values.
10. The intelligent and responsible sharing of power in order to attain justice.
11. The intelligent utilization of scarce resources to attain the widest general wellbeing.
12. Achievement of adequate horizons of loyalty.
13. Cooperation in the interest of peace and welfare.
14. Achieving a balance between social stability and social change.
15. Widening and deepening the ability to live more richly.

In 1959 a statewide committee in California summarized months of discussion on the major purposes of the social studies at all levels by listing the following eighteen points:

1. Man's comprehension of the present and his wisdom in planning for the future depend upon his understanding of the events of the past and of the various forces and agencies in society that influence the present.
2. Change is a condition of human society; societies rise and fall; value systems improve or deteriorate; the tempo of change varies with cultures and periods of history.
3. Through all time and in all regions of the world, man has worked to meet common basic needs and to satisfy common human desires and aspirations.
4. People of all races, religions, and cultures have contributed to the cultural heritage. Modern society owes a debt to cultural inventors of other places and times.
5. Interdependence is a constant factor in human relationships. The realization of self develops through contact with others. Social groupings of all kinds develop as a means of group cooperation in meeting individual and societal needs.
6. The culture under which an individual is reared and the social groups to which he belongs exert great influence on his ways of perceiving, thinking, feeling, and acting.
7. In the United States, democracy is dependent on the process of free inquiry; this process provides for defining the problem, seeking data, using the scientific method in collecting evidence, restating the problem in terms of its interrelationships, arriving at a principle that is applicable, and applying the principle in the solution of the problem.
8. The basic substance of a society is rooted in its values; the nature of the values is the most persistent and important problem faced by human beings.
9. Man must make choices based on economic knowledge, scientific comparisons, analytic judgment, and his value system concerning how he will use the resources of the world.
10. The work of society is carried out through organized groups; group membership involves opportunities, responsibilities, and the development of leadership.
11. Organized group life of all types must act in accordance with established rules of social relationships and a system of special controls.
12. All nations of the modern world are part of a global, interdependent system of economic, social, cultural, and political life.
13. Democracy is based on such beliefs as the integrity of man, the dignity of the individual, equality of opportunity, man's rationality, man's morality, man's ability to govern himself and to solve problems cooperatively.

14. Many people believe that physically man is the product of the same biological evolution as the rest of the animal kingdom. Man is in many ways similar to other animals, but a most important difference exists as a result of man's rationality and in the body of knowledge, beliefs, and values that constitute man's culture.

15. All human beings are of one biological species within which occur negligible variations.

16. Environment affects man's ways of living, and man, in turn, modifies the environment.

17. One of the factors affecting man's mode of life is his natural environment. Weather and climate and regional differences in land forms, soils, drainage, and natural vegetation largely influence the relative density of population in the various regions of the world.

18. Because man must use natural resources to survive, the distribution and use of these resources affect where he lives on the earth's surface and to some extent how he lives. The level of his technology affects how he produces, exchanges, transports, and consumes goods.

Your author has already interjected his own thinking about the role of the social studies, stressing the centrality of people. Here is his definition of the task of the teacher of social studies:

To discover and develop the abilities of every child so that he or she may comprehend himself or herself and other human beings better, cope with life more effectively, contribute to society in his or her own ways, help to change society, enjoy it, and share in its benefits.

There are many corollaries of this proposition. Here are some of them:

1. The social studies are a basic part of the education of every pupil, including the poor reader, the underachiever, and the slow and the retarded, as well as the average and the gifted.

2. The goal of all social studies learnings is improved behavior.

3. Knowledge is important, but most knowledge gained should be pertinent in the lives of children now, as well as in the foreseeable future.

4. The experiences of children should be organized to promote the discovery by them of concepts, generalizations, or "big ideas."

5. Social studies teaching should include feelings as well as facts.

6. The formation, change, and improvement of attitudes is even more important than the acquisition of knowledge.

7. The acquisition of skills and the understanding of processes is essential to any effective social studies program.

8. Children should be assisted in examining values and eventually in becoming committed to those of our democratic society, which are rooted in large part in the Judaic–Christian tradition.

9. Children should be helped in identifying with the United States and in developing a refined sense of patriotism.

10. Understanding and accepting oneself, with all one's strengths and limitations, is a first step toward accepting others.

11. Learning to respect the wide variety of persons in our nation and on our planet should flow from self-respect. Pupils should grow toward associating strangeness with friendliness rather than hostility.

12. Children should begin quite early to learn about the groups and institutions people have developed for their own well-being, locally, regionally, nationally, and internationally.
13. Children should be assisted in coping with a variety of problems — personal, economic, social, civic, and cultural.
14. Children should learn very early about rules and authority and responsibilities as well as rights, becoming concerned about justice for all people.
15. A variety of experiences are necessary to promote such learnings, with high priority given to the discovery or inquiry method.

You have been subjected to several pages on the role and purposes of the social studies from kindergarten through the eighth grade — and beyond. By now you are probably ready to frame your own statement about the place of social studies in elementary schools. The space below has been reserved for such a statement by you.

Several months later you may want to write a second statement of aims. Then you can compare your thinking over that period of time and explain to yourself the reasons for changes in your thinking.

My Comments on the Aims and Purposes of the Social Studies

Written on _____

My Comments on the Aims and Purposes of the Social Studies

A Later Version

Written on _____

THE TRADITIONAL PATTERN OF COURSES
OF STUDY IN THE SOCIAL STUDIES

Because of the decentralized nature of schools in the United States, there are many different courses of study in the social studies. Sometimes there is a statewide course of study, but adaptations are usually permitted and sometimes encouraged. Some states do not even suggest what the general pattern should be, leaving this to individual schools and school systems.

However, a fairly common pattern in the social studies has emerged over the years, based upon the expanding-horizons theory of curricular organization. Thus children study first about their homes and their neighborhoods, then about their community and state. Next they study the regions of the United States and our national history. Then they plunge into Old World Backgrounds and/or world geography, approached by continents. Below is a listing of the most common pattern, with some variations — Grade One through Grade Twelve.

Grade One.	The Home, the School, and the Neighborhood.
Grade Two.	The Neighborhood and the Community: Community Helpers.
Grade Three.	The Community and/or Food, Clothing, Shelter, and Transportation.
Grade Four.	Regional Geography of the United States or Regions of the World (by types of physical environment). State History.
Grade Five.	U.S. Geography and/or History.
Grade Six.	World Geography: usually North and South America; sometimes Old World Backgrounds of the U.S.A.
Grade Seven.	World Geography: Europe, Asia, the Middle East, and Africa.
Grade Eight.	United States History.
Grade Nine.	Civics or occasionally Ancient and Medieval History.
Grade Ten.	Modern History or World History.
Grade Eleven.	United States History.
Grade Twelve.	Problems of Democracy and/or one semester of Economics, Sociology, etc.

Most schools do not require twelve years of social studies. However, most states require some study of the state and a year of United States history.

With the exception of the twelfth grade courses, this general pattern has been in existence in the United States for more than 50 years. The course in Problems of Democracy was added in the 1930s, largely as a result of the Depression; the one-semester courses are even more recent.

In the last few years, however, there have been many criticisms of this expanding-horizons approach, especially in the elementary school grades. People have pointed out that such a program postpones until at least the sixth grade any study of people outside the United States. They have been disturbed by the undue emphasis upon geography and history, to the exclusion of the other social sciences. Many persons have criticized the amount of time spent in studying the local community.

What other criticisms could you make of this traditional pattern at the elementary level? At the secondary level?

How Can I Begin
My Preparation for
Teaching the Social Studies?

2

A CHECKLIST OF SUGGESTIONS

If you have not taught yet, you may be eager to get started. You are tired of reading theories about teaching; you want to teach and learn by firsthand experience. Perhaps your apprenticeship must still be postponed a short time. In the meantime there are many things you can "do." Some of them are listed on this page and on the next page. Even if you have already taught for some time, you may want to check the items suggested here. Don't plan to work on all of them. For many readers it is important to think of other subjects besides the social studies and to prepare for teaching them, too.

	Have already started	Should begin at once	This can be postponed
1. Read a good newspaper every day.			
2. Read at least two newsmagazines with different points of view on current affairs. (For example, *Time* or *Newsweek* or *U.S. News and World Report* and *The Nation* or *The New Republic*, or *The Progressive*.			
3. Send for free and inexpensive materials for use in your social studies teaching. You might like to use the Peabody College booklet on "Free and Inexpensive Learning Materials" (Peabody College, Nashville, Tennessee 37203).			
4. Start a small collection of folders on the topics you will probably teach, placing all material on a subject, topic or country in a folder.			
5. Start a collection of pictures and maps.			
6. Visit social studies classes and talk with teachers about their work.			

	Have already started	Should begin at once	This can be postponed
7. Visit one place of interest in your local community which you may want to visit later with a group of pupils.			
8. Visit some part of the state in which you live or some other part of the country which you have never visited.			
9. Join a professional organization in the social studies.			
10. Attend an educational conference, preferably one highlighting teaching about the social studies.			
11. Prepare some visual device for use in your teaching of the social studies.			
12. Examine several books for children related to the social studies. Make file cards, with annotations on the back.			
13. Learn to use some of the equipment you will need in your teaching, such as: (a) Film projector (b) Filmstrip projector (c) Overhead projector (d) Opaque projector (e) Tape recorder (f) Mimeograph machine (g) Rexograph machine			
14. Do some reading in books written about the teaching of the social studies. Take your notes by topics so that they can be dropped into folders and not have to be copied at some future date.			

Other things I would like to do:

15.			
16.			
17.			
18.			
19.			
20.			

BIBLIOGRAPHY ON SOCIAL STUDIES IN ELEMENTARY AND MIDDLE SCHOOLS

Books of readings

1. HERMAN, WAYNE L., JR. *Current Research in Elementary School Social Studies.* New York: Macmillan (1969), 468 pp.
2. HOWES, VIRGIL M. *Individualizing Instruction in Reading and Social Studies: Selected Readings on Programs and Practices.* New York: Macmillan (1970), 236 pp.
3. JAROLIMEK, JOHN and HUBER M. WALSH. *Readings for Social Studies in Elementary Education.* New York: Macmillan (1969), 485 pp. Second edition.
4. LEE, JOHN R. and JONATHON C. MCLENDON. *Readings on Elementary Social Studies: Prologue to Change.* Boston: Allyn and Bacon (1965), 447 pp.
5. MCLENDON, JONATHON C., WILLIAM W. JOYCE, and JOHN R. LEE. *Readings on Elementary Social Studies: Emerging Changes.* Boston: Allyn and Bacon (1970), 592 pp. Second edition.
6. MASSIALAS, BYRON G. and ADREAS M. KAZAMIAS. *Crucial Issues in the Teaching of Social Studies: A Book of Readings.* Englewood Cliffs, New Jersey: Prentice-Hall (1964), 278 pp. Includes some readings on the elementary school level.

Textbooks on methods of teaching the social studies

1. BARNES, DONALD L. and ARLENE B. BURGDORF. *New Approaches to Teaching Elementary Social Studies: With Illustrative Units.* Minneapolis: Burgess (1968), 280 pp.
2. CLEMENTS, H. MILLARD, WILLIAM R. FIEDLER and B. ROBERT TABACHNIK. *Social Study in the Elementary School.* Indianapolis: Bobbs-Merrill (1963), 402 pp.
3. DOUGLASS, MALCOLM. *Social Studies in the Elementary School.* Philadelphia: Lippincott (1967), 600 pp.
4. ESTVAN, FRANK S. *Social Studies in a Changing World.* New York: Harcourt Brace Jovanovich (1968), 532 pp.
5. JAROLIMEK, JOHN. *Social Studies in Elementary Education.* New York: Macmillan (1971), 480 pp. Fourth edition.
6. JOYCE, BRUCE R. *Strategies for Elementary Social Science Education.* Chicago: Science Research Associates (1972), 434 pp. Accents philosophy and inquiry.
7. KALTSOUNIS, THEODORE. *Teaching Elementary Social Studies.* West Nyack, N.Y.: Parker Publishing Company (1969), 197 pp.
8. KENWORTHY, LEONARD S. *Social Studies for the Seventies: In Elementary and Middle Schools.* Lexington, Mass.: Xerox College Publishing (1972), 530 pp.
9. MICHAELIS, JOHN U. *Social Studies for Children in a Democracy: Recent Trends and Development.* Englewood Cliffs, New Jersey: Prentice-Hall (1963), 624 pp.
10. PLOGHOFT, MILTON E. and ALBERT H. SHUSTER. *Social Science Education in the Elementary School.* Columbus, Ohio: Merrill (1971), 415 pp.
11. PRESTON, RALPH C. *Teaching Social Studies in the Elementary School.* New York: Holt, Rinehart and Winston (1968), 370 pp. Third edition.
12. RAGAN, WILLIAM B. and JOHN D. MCAULAY. *Social Studies for Today's Children.* New York: Appleton-Century-Crofts (1970), 281 pp.
13. SERVEY, RICHARD E. *Social Studies Instruction in the Elementary School.* San Francisco: Chandler (1967), 565 pp.
14. SKEEL, DOROTHY J. *The Challenge of Teaching Social Studies in the Elementary School.* Pacific Palisades, Cal.: Goodyear Publishers (1970), 196 pp.
15. SMITH, JAMES A. *Creative Teaching of the Social Studies in the Elementary School.* Boston: Allyn and Bacon (1967), 281 pp.
16. WESLEY, EDGAR B. and WILLIAM H. CARTWRIGHT. *Teaching Social Studies in Elementary Schools.* Boston: Heath (1968), 354 pp.

Books on specialized aspects of the social studies, and related fields

1. ALEXANDER, WILLIAM N. and others. *The Emergent Middle School.* New York: Holt, Rinehart and Winston (1968), 191 pp. An outstanding book in this general field.

2. AMBROSE, EDNA and ALICE MIEL. *Children's Social Learnings*. Washington: Association for Supervision and Curriculum Development (1958), 120 pp. What research suggests for social learnings.

3. ASHTON-WARNER, SYLVIA. *Teacher*. New York: Simon and Schuster (1963), 224 pp. The creative work of a teacher of Maoris in New Zealand, with implications for work in the primary grades in the U.S.A. Also available as a paperback.

4. BEYER, BARRY K. and ANTHONY N. PENNA (Editors). *Concepts in the Social Studies*. Washington, National Council for the Social Studies (1971), 95 pp.

5. BRAITHWAITE, EDWARD R. *To Sir, With Love*. Englewood Cliffs, New Jersey: Prentice-Hall (1959), 216 pp. The work of a West Indian Negro with teen-agers in the East Side of London. Also available as a paperback.

6. CARPENTER, HELEN (Editor). *Skill Development in Social Studies*. Washington: National Council for the Social Studies (1963), 332 pp. Highly recommended.

7. CROSBY, MURIEL. *An Adventure in Human Relations*. Chicago: Follett (1965), 396 pp. Work with teachers in Wilmington, Delaware, with implications for work with inner city children anywhere.

8. CROSBY, MURIEL. *Curriculum Development for Elementary Schools in a Changing Society* (1964), 409 pp. Especially Chapter 10.

9. DARROW, HELEN F. *Social Studies for Understanding*. New York: Bureau of Publications — Teachers College, Columbia University (1964), 91 pp.

10. DENNISON, GEORGE. *The Lives of Children*. New York: Random House (1969), 308 pp. The story of the famous First Street School in New York City.

11. DEPARTMENT OF ELEMENTARY SCHOOL PRINCIPALS. *Focus on the Social Studies*. Washington: Department of Elementary School Principals, N.E.A. (1965), 79 pp. Report of a convention highlighting the social studies.

12. DUNFEE, MAXINE. *Elementary School Social Studies: A Guide to Current Research*. Washington: Association for Supervision and Curriculum Development (1971), 125 pp.

13. DUNFEE, MAXINE and HELEN SAGL. *Social Studies Through Problem Solving*. New York: Holt, Rinehart and Winston (1966), 386 pp.

14. FAY, LEO, THOMAS HORN, and CONSTANCE MCCULLOUGH. *Improving Reading in the Elementary Social Studies*. Washington, D.C.: National Council for the Social Studies (1961), 72 pp.

15. FENTON, EDWIN. *The New Social Studies*. New York: Holt, Rinehart and Winston (1967), 144 pp. An overall account of innovations in the social studies field.

16. GOLDMARK, BERNICE. *Social Studies: A Method of Inquiry*. Belmont, California: Wadsworth (1968), 237 pp.

17. HAMMOND, SARAH L., RUTH J. DALES, DORA S. SHIPPER, and RALPH WITHERSPOON. *Good Schools for Young Children: A Guide for Working with Three, Four, and Five Year Olds*. New York: Macmillan (1963), 397 pp. Chapter 10 on the social studies.

18. HENTOFF, NAT. *Our Children Are Dying*. New York: Viking (1966), 141 pp.

19. HOPKINS, LEE B. and MISHA ARENSTEIN. *Partners in Learning: A Child-Centered Approach to Teaching the Social Studies*. New York: Citation Press (1971), 237 pp.

20. HUDGINS, BRYCE B. *Problem Solving in the Classroom*. New York: Macmillan (1966), 74 pp.

21. HUNNICUTT, CLARENCE W. (Editor). *Social Studies for the Middle Grades: Answering Teachers' Questions*. Washington: National Council for the Social Studies (1960), 122 pp.

22. JAROLIMEK, JOHN (Editor). *Social Studies Education: The Elementary School*. Washington: National Council for the Social Studies (1967), 68 pp. Reprints of articles from *Social Education* magazine.

23. KOZOL, JONATHAN. *Death At An Early Age*. New York: Bantam (1967), 242 pp. A teacher's account of Negro children in a Boston public school.

24. LANGDON, GRACE and IRVING W. STOUT. *Teaching in the Primary Grades*. New York: Macmillan (1965), 359 pp.

25. LEEPER, ROBERT R. and MARY ALBERT O'NEIL (Editors). *Hunters Point Redeveloped: A*

Sixth Grade Venture. Washington: Association for Supervision and Curriculum Development (1970), 64 pp. Sixth graders plan the redevelopment of a depressed area in San Francisco.

26. LOUNSBURY, JOHN H. and JEAN V. MARANI. *The Junior High School We Saw: One Day in the Eighth Grade.* Washington: Association for Supervision and Curriculum Development (1964), 78 pp. Several persons report what they saw on that one day in different schools in the U.S.A.

27. MASSIALAS, BYRON G. and C. BENJAMIN COX. *Social Studies in the United States: A Critical Appraisal.* New York: Harcourt Brace Jovanovich (1967), 355 pp. Some material on elementary and junior high schools.

28. MICHAELIS, JOHN U. (Editor). *Social Studies in Elementary Schools.* Washington: National Council for the Social Studies (1962), 334 pp. A yearbook.

29. MIEL, ALICE. *The Shortchanged Children of Suburbia.* New York: American Jewish Committee (1967), 68 pp. On the lack of interracial experiences.

30. MIEL, ALICE and PEGGY BROGAN. *More Than Social Studies.* Englewood Cliffs, New Jersey: Prentice-Hall (1957), 452 pp. On social learnings.

31. MORRISSETT, IRVING and W. W. STEVENS, JR. *Social Science in the Schools: A Search for Rationale.* New York: Holt, Rinehart and Winston (1971), 204 pp. On major concepts.

32. NELSON, JACK L. *Teaching Elementary Social Studies Through Inquiry.* Highland Park, New Jersey: Dreier Educational Systems (1970), 93 pp.

33. NOAR, GERTRUDE. *Teaching and Learning the Democratic Way.* Englewood Cliffs, New Jersey: Prentice-Hall (1963), 234 pp. Junior high schools.

34. PRESTON, RALPH C. (Editor). *A New Look at Reading in the Social Studies.* Newark, Delaware: International Reading Association (1969), 69 pp.

35. ROBINSON, HELEN and BERNARD SPODEK. *New Directions in the Kindergarten.* New York: Teachers College Press (1965). Pp. 68–93 on social studies.

36. SCHULMAN, LEE S. and KEISLAR, EVANS (Editors). *Learning By Discovery: A Critical Appraisal.* New York: Rand McNally (1966).

37. SHAFTEL, FANNIE and GEORGE. *Role Playing for Social Values: Decision Making in the Social Studies.* Englewood Cliffs, New Jersey: Prentice-Hall (1967), 431 pp. In reality a book about the social studies in general.

38. TABA, HILDA. *Curriculum Development: Theory and Practice.* New York: Harcourt Brace Jovanovich (1962), 529 pp. Many applications to the social studies field.

39. TODD, VIVIAN and HELEN HEFFERNAN. *The Years Before School: Guiding Pre-School Children.* New York: Macmillan (1964), 658 pp.

40. WANN, KENNETH D., MIRIAM S. DORN, and ELIZABETH A. LIDDLE. *Fostering Intellectual Development in Young Children.* New York: Teachers College Press (1962), 140 pp.

41. WEBER, EVELYN. *Intermediate Education: Changing Dimensions.* Washington: Association for Childhood Education (1965), 80 pp.

How Are Courses of Study
in the Social Studies
3 # Determined?

Perhaps you wondered as you read the page on "The Traditional Pattern of Courses of Study in the Social Studies" how such programs are determined. You may have wondered who made the decisions and what criteria they used.

In most nations education is a national enterprise, centralized in the Ministry of Education. Courses of study are made by officials in that Ministry and turned over to officials throughout the country for implementation. There is little or no curriculum-making at the lower levels of administration.

In the United States, education is not a national enterprise. Legally it is reserved to each of the fifty states. This arrangement grew out of our fear of centralized government and out of our belief that democracy works best when the people are closely involved in decision-making. There are drawbacks in this arrangement. All the states do not have equal resources for their schools, and inequality of education results. Similar standards are not always achieved because of this approach. However, there is a flexibility built into our system which many people feel is extremely important. Experimentation is more likely to occur under such a plan, and courses of study can be adapted to local needs better than with a national course of study.

In the social studies field many states have developed courses of study for all their schools. Very few states demand that all schools follow rigidly such a state course; most states permit considerable leeway on the part of local systems provided that the variations seem feasible to the state authorities. Many states limit their authority in constructing courses of study to a few regulations. Most of them require some study of the state and at least one year of United States history. + Government

Because of their comparative freedom, many large cities, many counties, and hundreds of local school systems and schools develop their own courses of study in the social studies. Such programs are usually drafted by a committee of teachers, in consultation with other instructors. Often consultants from nearby colleges are used in the preparation of such courses of study. There are today perhaps forty or fifty college professors who are used widely in such a capacity. Occasionally people from a distance are brought in for consultation. When the courses of study are completed, they are often printed for use by classroom teachers.

But there are other forces involved in curriculum-making. Textbook publishers wield an enormous influence in this regard. Schools often adopt a set of textbooks and these set the pattern for social studies teaching locally. For the most part textbook publishers are alert to changes in thinking about the social studies. Occasionally they are ahead of most school systems in the materials they produce. Increasingly, textbook publishers have a panel of experts from the social science disciplines as consultants. The senior author of a textbook series is usually a recognized expert in the field.

Citizens' groups are often influential, too, in determining courses of study in the social studies. For example, patriotic groups may call for more study of United States history or exert pressure for certain approaches in the teaching of our national history. Local and state historical groups often bring pressure to bear on curriculum committees to include considerable attention to local and state history. Minority groups often call for better treatment of their role in society than has been accorded them in the past.

In a similar way parents may help to determine curriculum. Often they exert influence through the parent-teacher organization.

Teachers in secondary schools also play a part in determining courses of study in elementary schools, working cooperatively with elementary school people or making demands upon curriculum committees.

Professional organizations in the social studies have some influence, too, through their publications and through surveys of newer practices in the social studies.

In larger school systems there are often curriculum bureaus which have a strong influence upon courses of study. Sometimes they include specialists in the social studies who work with local teachers in developing courses of study and then in implementing them.

Boards of Education usually do not take any part in the planning of new courses of study, but they are consulted in many places and usually vote on the proposals of local committees.

A few schools or school systems play an important role nationally in the construction of social studies courses. When they produce a new course of study for their school or school system, other people examine what they have done and often follow a similar pattern. Such schools are likely to be in the suburbs. From time to time a statewide revision of a course of study will attract national attention and parts of it will be adopted by others.

In the mid 1960s, a wave of reform hit the social studies field in the United States in the same way in which reform had already left its mark upon such fields as science, mathematics, and modern languages. Many new projects were launched. Some of them were financed by the federal government. Others were underwritten by private foundations and institutions of higher learning. Although most of them concentrated upon secondary schools, some included elementary schools or dealt solely with them. In the late 1960s and early 1970s, publishers began to utilize some of those projects, thus affecting some changes in the social studies field.

No matter what the course of study says should be taught, it is the classroom teacher who, in the final analysis, determines what is taught. Once the door of a classroom is closed, the teacher becomes the final authority on curriculum.

In planning curricular experiences for children and in planning the broader frame of reference of courses of study, there are at least four major determinants. They are (1) the nature of the social science disciplines, (2) the nature of learning, (3) child growth and development, and (4) the nature of society — locally, nationally, and internationally. In the pages that follow we shall attempt to spell out in some detail all four of these factors. Additional references will be cited so that you can pursue in more depth these determinants. In constructing courses of study all of these factors need to be kept in balance, rather than permitting one to dominate the proposal developed by a curriculum committee.

THE SOCIAL SCIENCE DISCIPLINES AS DETERMINANTS OF CURRICULA

Until quite recently the social studies curricula of elementary schools were based almost exclusively upon geography and history. By the junior high school years civics, government, or political science was added, usually at the seventh or ninth grade level. The family, neighborhood, community, and state were included as focal points but little attention was paid to the methodology of anthropology or sociology in those studies. The geography that was taught was largely physical geography. Economics was almost ignored.

In the last few years social scientists and social studies experts have begun to work more closely together than they have ever worked. As a result, many individuals have begun to ask what place economics should have in the curricula of elementary and junior high schools and they have begun to experiment with economics even in the early grades. Many persons concerned with curricula have begun to realize the dominant place that anthropology and sociology have in social studies programs and the methodology of those fields is beginning to be utilized in programs of social studies for grades K–8. Civics as a separate subject is being reexamined and attempts are now being made to bring such teaching more into line with the findings of political science. The emphasis in the teaching of geography is gradually shifting to human geography. Curriculum planners are raising questions again as to the place of history in the social studies curricula of elementary and junior high schools.

All of these changes mean that there is tremendous ferment in the social studies. Some persons are asking for separate programs in each of these fields in every grade. Obviously that is ridiculous. There is just not enough time for separate programs on each grade level for all of the social science disciplines. Furthermore, what seems to many experts to be needed is the integration of the various social science disciplines, at least in the earlier years of school.

Consequently, any person who is going to teach social studies today in elementary or junior high school needs to be conversant with the latest thinking about anthropology and sociology and social psychology, economics, government or political science, geography and history. In the pages that follow, very brief accounts are given of these separate but related social science disciplines. Ample bibliographies are included in order to encourage you to pursue these subjects in greater depth.

Very few people have a strong background in each of the various social science disciplines. But that need not bother you. You can determine now to explore the field or fields in which you feel least competent. An excellent way to begin would be to purchase and study that pamphlet in the Merrill Social Science Seminar Series, edited by Raymond Muessig and Vincent Rogers, which deals with the discipline in which you feel inadequate.

Readers will find that the lines between these disciplines are somewhat blurred today. Political science is being affected by social psychology. Geography is being affected by anthropology. The other disciplines are being changed in many ways by their kindred subjects. Many changes in the social studies teaching should result from the new insights into the various social sciences. This should be an exciting time to teach the social studies in elementary and junior high schools, with all this ferment in that broad field.

Anthropology–Sociology

Anthropologists and sociologists are interested in the study of man, with particular reference to group life. They observe and analyze the values of groups, the patterns of behavior of people, the structure of groups, and the interrelationships among the members of groups as they play different roles. They are interested in how people pass on a culture and how they change it. The groups they study are many and varied, including families, play groups, peer groups, minority groups, labor unions, communities, and even countries.

It is almost impossible to draw strict lines of demarcation between these two fields or among the three fields of anthropology, sociology, and social psychology. The lines today are blurred as they borrow methods from each other.

Anthropology has one foot in the field of science and the other in the domain of the social sciences. Historically most anthropologists were interested in physical anthropology, studying the physical features of human beings. For a long time anthropologists studied small, contemporaneous societies with pre-literate and/or pre-industrial forms. But in more recent years cultural anthropology has come to the fore and anthropologists today do not limit themselves to Indian tribes or the peoples of the South Pacific. They are just as likely to be interested in observing and analyzing city life or the life of a nation.

Caroline Rose said that "Culture consists of the shared meanings and values that the members of any group hold in common." Clyde Kluckhohn said that "Anthropology holds up a great mirror to man and lets him look at himself in his infinite variety." These and other eminent anthropologists are interested in the totality of a group's life. They do not believe that a group can be viewed correctly in any other way.

The methods of the anthropologists are likely to stress fieldwork, with on-the-spot observation, usually through informants of a culture. But anthropologists are increasingly using other means of analysis, including survey questionnaires, census materials, and psychological tests.

Sociologists tend to look at smaller units of society, such as families, voluntary groups, minorities, and labor groups, but they do not limit themselves to such units. They may also look at communities, regions, and countries. In recent years they have turned much of their attention to population studies or demographic data.

Sociologists may observe on-the-spot, but they also may work from census figures and other statistical data. Often they make content analyses, use questionnaires, conduct in-depth interviews, utilize projective tests, and carry on experiments. Structured interviews and statistical analyses are two other commonly used methods of sociologists.

A third closely allied field is social psychology. Its practitioners emphasize the relation of individuals to groups. Since World War II social psychologists have been increasingly interested in social motivation and social perception. They have added a great deal to our knowledge of group dynamics. Much study has been devoted, too, to the measurement of beliefs and attitudes, particularly in the field of racial prejudice. Group leadership and group morale is another field which has been studied in depth in recent years. Two other fields explored by social psychologists have been industrial conflict and international tensions.

All three of these fields stress certain big ideas. Among them are the following:

1. Human beings are alike and different, with similar basic needs but different ways of meeting them.

2. Every society has kinship patterns. The family is the basic unit in which children grow up and acquire their outlook on life.
3. Culture is the common way of life, the attitudes and feelings shared by a large group of people. Culture is derived from the past but is often adapted to the circumstances and needs of the present.
4. People learn the ways of their own culture primarily in families, but supported by other groups.
5. There are ordinarily a wide variety of voluntary groups in which people are banded together to meet their needs or to pursue their common interests.
6. In all groups there is some kind of organization, with differentiation of roles for the members.
7. People also band together in governmental units, from communities to countries.
8. In most groups there are minorities.
9. Groups are interdependent and from their interaction come conflict and cooperation in varying degrees.
10. Continuity and change are characteristics of human groups; change is a predominant characteristic of societies today.
11. Groups have a variety of creative activities as well as maintaining order.

Teachers and curriculum workers should lean heavily upon the two or three disciplines mentioned in the foregoing paragraphs as they plan curricula and introduce pupils to man in different places and in different periods of history. These fields are interested in people as individuals and as members of various groups. They concentrate on such important aspects of society as families, communities and countries. Therefore teachers and other educators need to turn to them to understand how to analyze and interpret human beings. In the primary grades anthropology and sociology should be of great importance, coupled with psychology. In the middle and upper grades they should yield rich rewards to those who study these two fields. In many ways anthropology and sociology are the central foci of social studies in elementary and junior high schools. This is especially true of anthropology, inasmuch as it sees the total life of groups, including their geography, economics, value system, history, politics, and creative expressions.

As leads to further study of these fields, here are some references for teachers and a few books interpreting anthropology to pupils:

Articles, pamphlets, and books on anthropology and sociology

1. ADAMS, E. MERLE, JR. "New Viewpoints in Sociology" in *New Viewpoints in the Social Sciences.* Washington: National Council for the Social Studies (1958), pp. 97–114.
2. ARENSBERG, C. M. and S. T. KIMBALL. *Culture and Community.* New York: Harcourt Brace Jovanovich (1965), 349 pp.
3. ELLISON, JACK L. "Anthropology Brings Human Nature into the Classroom" in *Social Education* (November, 1960), pp. 313–16, 328.
4. FRANCELLO, JOSEPH A. "Anthropology for Public Schools: Profits and Pitfalls" in *Social Studies* (December, 1965), pp. 272–74.
5. GROSS, EDWARD. "Sociology's Supporting Role" in *Crucial Issues in the Teaching of the Social Studies.* Englewood Cliffs, New Jersey: Prentice-Hall (1964), pp. 105–108.
6. HERTZBERG, HAZEL. "Grasping the Drama of a Culture: An Anthropological Approach to History in Junior High" in *N. E. A. Journal* (February, 1963), pp. 44–46.
7. HINKLE, ROSCOE C. and J. GISELA. *The Development of Modern Sociology.* New York: Random House (1962), 74 pp.
8. INKELES, ALEX. *What Is Sociology? — An Introduction to the Discipline and the Profession.* Englewood Cliffs, New Jersey: Prentice-Hall (1964), 120 pp.

9. KNELLER, GEORGE F. *Educational Anthropology: An Introduction.* New York: Wiley (1965), 171 pp.
10. MEAD, MARGARET. *The Family.* New York: Macmillan (1965), 208 pp. With illustrations.
11. MONTAGU, ASHLEY. "What Anthropology Is" and WILFRED C. BAILEY and FRANCIS J. CLUNE, JR. "Anthropology in Elementary Social Studies" in *The Instructor* (November, 1965), pp. 48–50.
12. NIMKOFF, MEYER F. "Anthropology, Sociology and Social Psychology" in *High School Social Studies Perspectives.* Boston: Houghton Mifflin (1962), pp. 29–52.
13. OLIVER, DOUGLAS. "Cultural Anthropology" in *The Social Studies and the Social Sciences.* New York: American Council of Learned Societies, and the National Council for the Social Studies (1962), pp. 135–155.
14. PELTO, PERTTI. *The Study of Anthropology.* Columbus, Ohio: Merrill (1965), 118 pp.
15. PERRUCCI, ROBERT. "What Sociology Is" and SISTER M. MERCEDES. "Sociology in Elementary Schools" in *The Instructor* (March, 1966), pp. 46–47.
16. ROSE, CAROLINE B. *Sociology: The Study of Man in Society.* Columbus, Ohio: Merrill (1965), 117 pp.
17. SPINDLER, GEORGE B. "New Trends and Applications in Anthropology" in *New Viewpoints in the Social Sciences.* Washington: National Council for the Social Studies (1958), pp. 115–143.
18. SYKES, GRESHAM M. "Sociology" in *The Social Studies and the Social Sciences.* New York: American Council of Learned Societies and the National Council for the Social Studies. Harcourt Brace Jovanovich (1962), pp. 156–170.
19. THOMPSON, LAURA. *The Secret of Culture: Nine Community Studies.* New York: Random House (1969), 394 pp. A paperback book. See especially Chapter 20 on "Culture As a Problem-Solving Device."
20. WOLF, ERIC ROBERT. *Anthropology.* Englewood Cliffs, New Jersey: Prentice-Hall (1966), 113 pp.

Economics

In recent years economics has found a place in the curricula of many elementary and junior high schools. Usually this has meant the incorporation of much material on economics into the social studies programs of schools rather than treating economics as a separate subject.

Particularly noteworthy has been the teaching of economics in the primary grades. This has been largely due to the efforts of the various Joint Councils for Economic Education and the pioneer work of Professor Lawrence Senesh of Purdue University.

Perhaps the idea of teaching economics frightens you. It does scare many teachers who have come to think that they have little or no background in economics and are put off by the vocabulary of that field, with such words as *production, distribution, growth, gross national product,* and a host of similar words and phrases.

But teachers need not be frightened by the teaching of economics if they think in terms of the central concepts of that important discipline. Probably the central idea is that of *unlimited wants* and *limited resources.* From this idea all other concepts emanate.

Phrased in questions, there are three big ideas in economics which pupils need to explore. They are:

1. WHAT shall we produce with our limited resources?
2. HOW MUCH can we produce and HOW FAST can our economy grow?
3. WHO shall get the goods and services produced?

These three basic questions can be asked of any group of people anywhere in the world and at any point in history. They can be applied to families — nuclear or extended — to tribes, to communities, to nations, and to the international community.

These questions can be explored by boys and girls in the primary grades in relation to their own families and the families of others. The skills they develop can then be utilized to help them think in terms of other families they study in other parts of the United States and the world. Concepts and generalizations can be tested in many situations until teachers are certain that pupils really understand them.

Children can and should learn that men have developed specialization because they wanted to obtain more goods and/or services. Such specialization has led to the exchange of goods and thus to interdependence. This in turn has led to markets where the goods could be exchanged. In order to exchange items better, means of exchange have been developed. In most societies today the means of exchange is money. This, in turn, has led to banks and credit. Students can learn, too, how various groups have determined who should share in the distribution of the production of men and women. In some societies the government has controlled tightly the distribution system. In others there has been what we call "free enterprise," as in the United States. This term is used for our economy despite the fact that it is far from "free," considering the many economic controls now carried on by our government units. Actually we should think of our economy as "mixed."

If teachers can help boys and girls to discover these general principles and see their application in a host of different situations, then pupils will become economically literate. They will become better producers, better consumers, and better citizens. Thus economics can play its part in the development of people who understand the basic economic structure of their own society and of other societies and are able to help bring about necessary change in it.

For teachers who feel hesitant about teaching the economic aspects of social studies, there is much help available. The state and national offices of the Joint Council for Economic Education can help such persons in many ways. Another source of help is the Industrial Relations Center of the University of Chicago, which has developed a number of fine materials for the middle grades. A third is Science Research Associates, with their filmstrips, books, and records based on the experimental work of Professor Senesh in the primary grades in particular. Several other possibilities are opened up by the materials listed in the bibliography on the next page.

For pupils there are many recent books on topics in economics. They range from a simple volume like Patricia Miles Martin's *Benjie Goes Into Business* (Putnam) or Elaine Hoffman and Jane Hefflefinger's *About Helpers Who Work at Night* (Melmont), to the volumes in the *Men at Work* series (Putnam), cited in the chapter "Studying the United States Today."

In all their work in economics, teachers need to come back to a list of basic ideas, such as the one below, framed by the writer, in conjunction with several experts in the social studies. Here are eleven generalizations central in the field of economics. They are arranged in order to show the most simple aspects first. Even children in the primary grades can comprehend the first few of these generalizations if the pupils work on them with specific, concrete examples. The later generalizations are more difficult and should be handled with older pupils.

1. People need goods and services to survive.
2. The conflict between unlimited needs and wants and limited resources is man's basic economic problem: scarcity.

3. Production is organized on the bases of such factors as basic needs and wants, labor and skills, capital, values and efficiency.
4. Every society has to determine how much goods and services will be produced — and how they will be produced. The answers vary from society to society.
5. Every society must determine to whom and in what ways goods and services are distributed.
6. Specialization has developed in all societies.
7. Specialization has led to greater interdependence and trade.
8. Various societies have developed media of exchange; money is the chief medium in most societies today.
9. Public policy is a strong determinant in economic affairs, but varies from society to society.
10. The United States has a modified private or free enterprise economic system.
11. Economic changes in the world today have brought more trade and a greater need for international controls.

Articles, pamphlets, and books on economics

1. ASSOCIATION FOR SUPERVISION AND CURRICULUM DEVELOPMENT. *Educating for Economic Competence.* Washington, D.C.: A. S. C. D. (1960), 78 pp. See especially the chapters on "Economic Education in the Primary Grades," "Economic Education in the Intermediate Grades," and "Economic Education in the Junior High School."
2. COLEMAN, JOHN R. *Comparative Economic Systems: An Inquiry Approach.* New York: Holt, Rinehart and Winston (1968), 226 pp. A superior economics text for secondary school pupils.
3. COON, ANNE. "Introducing the Economic World to Primary-Grade Pupils" in *Social Education* (April, 1966), pp. 253–56.
4. GRIBLING, ROBERT. "You Can Teach More Economics Than You Think" in *Grade Teacher* (February, 1966), pp. 82–84.
5. GRUENBERG, SIDONIE M. "Your Child and Money." New York: Public Affairs Committee (1965), 28 pp.
6. HAEFNER, JOHN H. and GERRY R. MOORE. "Economic Forces in American History." Chicago: Scott, Foresman (1964), 88 pp. To accompany an excellent series of booklets on each period of United States history.
7. HARRIS, SEYMOUR E. "Economics" in Erling Hunt and others. *High School Social Studies Perspectives.* Boston: Houghton Mifflin (1962), pp. 53–80.
8. HEILBRONER, ROBERT L. "The World of Economics." New York: Public Affairs Committee (1963), 33 pp. A splendid booklet.
9. JACOBS, A. D. "Economics Through Children's Books" in *Elementary English Journal* (January, 1961), pp. 15–21. An annotated bibliography of trade books dealing with economics for children.
10. JOINT COUNCIL ON ECONOMIC EDUCATION. Write for free materials and a bibliography of materials for pupils on economics.
11. LEWIS, BEN W. "Economics" in *The Social Studies and the Social Sciences.* New York: Harcourt Brace Jovanovich (1962), pp. 106–134.
12. MARTIN, RICHARD S. and REUBEN G. MILLER. *Economics and Its Significance.* Columbus, Ohio: Merrill (1965), 165 pp. Includes an analysis of the field and a section on teaching methods.
13. NOURSE, EDWIN G. "New Viewpoints in the Economic Area" in *New Viewpoints in the Social Sciences.* Washington: National Council for the Social Studies (1958), pp. 87–96. A yearbook.
14. PATTERSON, FRANKLIN (Editor). *Citizenship and a Free Society.* Washington, D.C.: National Council for the Social Studies (1960). A yearbook. See especially chapters VI and VII.

15. PETERSON, ESTHER. "The Child Buyer and Consumer" in *Childhood Education* (May, 1965), pp. 461–465.
16. SCHEER, LORRAINE H. and VINCENT PATRICK. "Guidelines for Incorporating Economics in Intermediate Grades" in *Social Education* (April, 1966), pp. 256–58.
17. SENASH, LAWRENCE. *Economics*. Denver, Colorado: Social Science Education Consortium (1966), 16 pp.
18. SOBOLLAY, JOHN F. "The Junior High School Develops Economic Understanding" in *Social Education* (April, 1966), pp. 259–60, 273.
19. "Today's Economics: Case Studies for Student Understanding." Columbus, Ohio: Merrill (1965), 63 pp. Excellent material for junior high pupils.

Geography

Fortunately the study of geography is being revived in the schools of the United States today. That revival is due to many factors. One of them is the realization that as a nation, we are geographically "illiterate." We know too little about the world in which we live at a time when it is essential to know a great deal about the people of other parts of the world and places where they live. Hence there is concern that the schools teach more geography to children. Furthermore, many people see the relevance of geography to such a topic as urbanization.

But there is a danger that the return of geography to the curriculum will mean a return to the horrible teaching which caused it to be removed or minimized a few years ago. The criticism then was that geography concentrated on the memorization of the names of the capitals of the states, the leading products of the places, the major rivers and their lengths, and similar minutiae, without regard to the more important topic of relationships of people and places. In those days it too often asked questions about "What?" rather than about "Where?" "Why?" and "How?"

In recent times geographers have begun moving away from an emphasis upon physical geography and have begun to emphasize human and cultural geography. That is a great gain. With more emphasis upon human beings, geography has high relevance to a people-centered social studies curriculum. It can help children to understand where people live, why they live there, what they do and why, and how they are interdependent.

There are many definitions of the broad field of geography. Richard Hartshorne defined the field as briefly as anyone in an article in the *Journal of Geography* in which he wrote that "Stated simply, geography is the study of the earth as the home of man." George Cressey has written that "Geography is the study of space relations and is concerned with the over-all personality of place." Preston E. James phrased his definition in these terms: "Geography is that field of learning in which the characteristics of a particular place on the earth's surface are examined. It is concerned with the arrangement of things and with the associations of things that distinguish one area from another. It is concerned with the connections and movement between areas."

What are some of its emphases and applications today? One is on "areal association" or the several factors which are characteristic of one spot or site on the earth's surface. These include such factors as rainfall, soil and/or resources, climate, and the density of population. Another is on "regions." These are large areas of the earth that have enough aspects in common to be grouped together. A third is upon the distribution of population or the reason why people live where they do. A fourth is upon the interdependence of the peoples of the earth. A fifth is upon cultural areas. For example, Preston James, a geographer, wrote not long ago in his volume on *One World Perspective* about the 11 major areas of the world. He selected these more from a cultural point of view than from a geographical point of view. A sixth major emphasis of geographers today is on the field of urban geography. This is a rapidly growing division of geography, with its intensive study of what transpires in the many urban centers on our planet.

These and other key ideas, such as those in the following list, are of the utmost importance to teachers as they help children to understand the people of their own community, of their state, of their region, of their country, and of other parts of the world. They need to help children discover how people are affected by the land where they live and how people have changed the land. They need to direct boys and girls so that they learn how men have turned the riches of the earth into usable products, called resources. They need to set the stage so that pupils can learn how interdependence has developed.

Viewed in this light, even primary grade children can acquire many geographic learnings. They can explore the local environment on trips, and map it. They can make dioramas of the immediate community and see the interrelationships in it. They can be "given" pieces of land and discover what can be done with them — whether they are homesites, factory sites, or farms.

What the calendar is to the historians, the globe and maps are to geographers. These are the tools with which they work. Boys and girls need to have a grade-by-grade introduction to the various skills needed to interpret these tools of geography — the globe and maps. They need to make many kinds of maps (see pp. 137–138) in many media.

Here are the major generalizations in the discipline of geography:

1. People live on the planet Earth, a satellite of the sun and a part of the larger universe.
2. Each place on Earth is unique in respect to certain geographic features; each place is related to every other place in a variety of ways, including size, direction, distance, and time.
3. People everywhere have certain needs in common: food, clothing, shelter. All human societies therefore are forced to establish workable connections with their geographic environment in order to survive.
4. In obtaining and using resources to meet his needs, man is an active agent in reshaping his environment.
5. The environment, in turn, places certain limitations or restrictions on man.
6. The significance to man of the physical features of the earth depend upon his own ingenuity, objectives, attitudes, and technical skills. How man meets his needs depends therefore on both (1) the resources available and (2) the cultural values, history, and technology of his society.
7. The uneven distribution of natural resources is one factor which leads to regional specialization, to interaction, and to interdependence.
8. The change from a cluster of isolated, self-sufficient communities to an interdependent world society increases trade, encourages migration, diffuses ideas and practices, and attaches even greater significance to man's location.
9. Earlier occupancy of a region influences later periods of occupancy. However, over time, geographic patterns change.
10. In the period since 1750, advances by science have so changed worldwide patterns of life expectation as to affect radically the environment for human habitation on the earth.

11. The concept of a region is useful when we attempt to organize knowledge about the earth and its people. A region is any area of any size which is homogeneous in terms of specific criteria.

Articles, pamphlets, and books on geography for teachers

1. BACON, PHILIP (Editor). *Focus on Geography: Key Concepts and Teaching Strategies.* Washington: National Council for the Social Studies (1970), 437 pp. The 1970 yearbook of the N.C.S.S.
2. BACON, PHILIP. "North: An Easier Way to Teach Directions" in *Grade Teacher* (December, 1964), pp. 38–39, 65.
3. BROEK, JAN O. M. *Geography: Its Scope and Spirit.* Columbus, Ohio: Merrill (1965), 114 pp.
4. CHACE, HARRIETT. "Developing Map Skills in Elementary Schools" in *Social Education* (November, 1955), pp. 309–310, 312.
5. CHACE, HARRIETT. "Map Skills in the First Grade" in *Social Education* (December, 1955), pp. 360–361.
6. CHACE, HARRIETT. "Map Skills at the Third Grade Level," in *Social Education* (January, 1956), pp. 13–14.
7. CHACE, HARRIETT. "Map Skills Developed in Grade Six" in *Social Education* (February, 1956), pp. 60–62.
8. CRESSEY, GEORGE B. "Geography" in *High School Social Studies Perspectives.* Boston: Houghton Mifflin (1962), pp. 81–97.
9. HARRIS, RUBY M. *The Rand McNally Handbook of Map and Globe Usage.* Chicago: Rand McNally (1960), 390 pp.
10. HARTSHORNE, RICHARD. *Perspective on the Nature of Geography.* Chicago: Rand McNally (1962), 193 pp.
11. HILL, WILHELMINA (Editor). *Curriculum Guide for Geographic Education.* Norman, Oklahoma: National Council for Geographic Education (1963), 162 pp.
12. JAMES, LINNIE B. and LA MONTE CRAPE. *Geography for Today's Children.* New York: Appleton-Century-Crofts (1968); 310 pp.
13. JAMES, PRESTON E. "Geography" in *The Social Studies and the Social Sciences.* New York: American Council of Learned Societies and the National Council for the Social Studies (1962), pp. 42–87.
14. JAMES, PRESTON E. "New Viewpoints in Geography" in *New Viewpoints in Geography.* Washington: National Council for the Social Studies (1959), pp. 112–143.
15. KOLEVZON, EDWARD R. and RUBIN MALOFF. *Vitalizing Geography in the Classroom.* Englewood Cliffs, New Jersey: Prentice-Hall (1964), 64 pp.
16. MCAULAY, J. D. "Geography Understanding of the Primary Child" in *Journal of Geography* (April, 1966), pp. 170–176.
17. MCAULAY, J. D. "The Place of Programmed Learning in Elementary School Geography" in *Journal of Geography* (May, 1962), pp. 215–221.
18. MITCHELL, LUCY SPRAGUE. *Young Geographers.* New York: Basic Books (1963), 102 pp.
19. MORRIS, JOHN W. and LORRIN KENNAMER. "Geography" in *The Instructor* (April, 1966), pp. 34–35.
20. SABAROFF, ROSE. "Map-Making in the Primary Grades" in *Social Education* (January, 1960), pp. 19–20.
21. SAVELAND, ROBERT. "Whatever Happened to Geography?" in *Saturday Review* (November 17, 1962), pp. 56–57, 77.
22. SORENSON, FRANK E. "Geographical Understandings" in *Social Education of Young Children.* Washington: National Council for the Social Studies (1956), pp. 56–59.
23. WHIPPLE, GERTRUDE. "Geography in the Elementary Social Studies Program" in *New Viewpoints in Geography.* Washington: National Council for the Social Studies (1959), pp. 112–143.

History

Of the writing of definitions of history there seems to be no end. Hundreds of them have been written and more will undoubtedly appear in the years ahead. And what a range of viewpoints they represent!

Samuel Eliot Morison says that "History is the Story of Men." Arnold Toynbee writes of history in a more philosophical vein, maintaining that "History is a search for light on the nature and destiny of man." Bury claims that "History is a science; no less, and no more," but Henry Adams challenged this view. To him "History is incoherent and immoral." Paul Valéry went even further, asserting that "History is the most dangerous product ever concocted by the chemistry of the intellect."

Why all these variations? Probably because historians have to be selective in presenting the past. They have limited sources in most instances. Then the material which they have, has to be screened by their minds. Their own personal values and the values of their culture and their time affect them.

Little wonder, then, that we have so many interpretations of past events. Perhaps Jacob Burckhardt came closer to a reasonable definition of history when he said that it is "what one age finds worthy of note in another." Henry Steele Commager phrased the same idea in this way: "History is organized memory and the organization is all-important." Elaborating upon this idea he maintained that "writing is painting a picture, not taking a photograph."

Teachers need to realize that there is no such thing as "the" interpretation of a past event. There are many such interpretations. And each age looks at the past in the light of its present. Thus, in United States history, we have had political interpretations, economic interpretations, the frontier hypothesis, the melting pot theory, and the history of ideas. Now we are being introduced to the urban interpretation of our national history.

Children can grasp some of these variations if they are asked to tell the "history" of what happened on the playground during recess or on a trip taken last week. There will be many interpretations as each person tells how he perceived the situation he experienced.

With the whole range of human history to select from, curriculum workers and teachers have a real problem on their hands. It is complicated by the fact that time and space are probably the two most difficult concepts for children to grasp. Even yesterday and tomorrow are beyond the comprehension of some primary grade children, let alone the idea of decade or century or Ancient Greece or Rome.

What place, then, should history have in the social studies programs of our schools? Authorities differ widely. In the past, history has had a central position in the middle grades, with a year of United States history in the fifth grade and often a year of Old World Backgrounds in the sixth grade.

Some schools even introduced history as a central approach in the third or fourth grades. Yet psychologists and even historians have questioned whether children can develop much "time sense" before the fifth, sixth, or seventh grades. For example, Carl Gustavson, the eminent historian, has written that "A profound development of the sense of time occurs at about the age of ten or eleven." Some children begin to grasp the sense of time before that, but many pupils do not develop much time sense until the junior high school years.

Even very young children can begin to understand the concept of history. To them it may be the events of yesterday or last week or last summer or "before I went to school." But that is beginning. Gradually their sense of history can be built through celebration of holidays, stories of famous people and events, and trips to places where they can see houses or communities as they existed "a long time ago." Grandparents, parents, and other adults can relate stories about the time when they were children, thus fortifying the sense of history which boys and girls are acquiring. Stories of families can include the historical dimension. For example, Sonia Gidal tells in *Families Live Everywhere* (Ginn, 1972) about the old wall in Nuremberg and how it was used. She also shows a simple map of a house when it was a barn and tells how it was transformed into a house.

In the study of communities in grades three and four, children can expand their understanding of history. The historical dimension of some of the communities they learn about can extend their knowledge of what life was like 50 or 100 or more years ago.

Instead of an entire year devoted exclusively to the study of United States history, this writer has urged schools to study the U.S.A. Today in the first half of the year and the U.S.A. Yesterday in the second semester. He has likewise urged teachers to "posthole" the history that is studied, concentrating on a few selected decades rather than trying to cover every presidential administration.

In the study of other nations, the emphasis should be largely on the present. However, history can be highlighted in the story of some countries. In their book on *Eleven Nations* (Ginn, 1972) Bani Shorter and Nancy Starr select Israel, England, and China as nations in which the historical dimension is stressed.

By the time boys and girls reach the eighth grade, their sense of time should be fairly well developed. This writer would therefore use history as the focus or "carrier" at that point. This would be the first time that history becomes the major discipline to be stressed.

Among the key concepts from history which teachers should bear in mind are the following:

1. History is man's selective record of the past; historians use a variety of tools or methods to interpret the past.
2. History is the record of individuals and groups of people.
3. History is the record of man on the move, his adjustment to his environment, and the changes he has made in it.
4. History is the record of continuity and change.
5. History is the record of man's economic activities and institutions.
6. History is the record of man's search for values.
7. History is the record of man's political arrangements, including the formation of nation-states in recent times.
8. History is the record of man's problems and how he has faced them, including cooperation and conflict.
9. History is the record of man's ideas and inventions.
10. History is the record of man's creative expression.
11. History is the record of increasing industrialization and urbanization.
12. History is the record of increasing interdependence.

Articles, pamphlets, and books on history

1. ALILUNAS, LEO J. "The Problem of Children's Historical Mindedness" in *Social Studies*. (December, 1965), pp. 251–254.
2. ALLEN, JACK. "Social Studies for America's Children" in *Phi Delta Kappan* (April, 1959), pp. 277–280.
3. BUGGEY, JOANNE and MARIO D. RABOZZI. "History in the Middle Grades" in *Social Education* (May, 1968), pp. 469–471.

4. CARSON, GEORGE BARR, JR. "New Viewpoints in History" in *New Viewpoints in the Social Sciences*. Washington, National Council for the Social Studies (1958), pp. 20–38.
5. CHASE, LINWOOD. "American History in the Middle Grades" in Cartwright, William H. and Richard L. Watson, Jr. *Interpreting and Teaching American History*. Washington: National Council for the Social Studies (1961), 430 pp.
6. CLEGG, AMBROSE A., JR. and CARL E. SCHOMBURG. "The Dilemma of History in the Elementary School: Product or Process?" in *Social Education* (May, 1968), pp. 454–457.
7. COMMAGER, HENRY STEELE. *The Nature and Study of History*. Columbus, Ohio: Merrill (1965), 160 pp. Includes one section on history in elementary schools.
8. DOHERTY, JOAN. "The History Component of the Elementary Social Studies Curriculum" in *Social Education* (May, 1968), pp. 465–468.
9. FRIEDMAN, KOPPLE. "How to Develop Time and Chronological Concepts." Washington: National Council for the Social Studies (1960), 8 pp.
10. GILL, CLARK C. "Interpretations of Indefinite Expressions of Time" in *Social Education* (December, 1962), pp. 454–456.
11. HIGHAM, JOHN (Editor). *The Reconstruction of American History*. New York: Harper (1962), 244 pp. Chapters on interpretations of various periods in our national history.
12. JOYCE, BRUCE. *Strategies for Social Science Education*. Chicago: Science Research Associates (1965), 302 pp. See pp. 62–65 on history as a discipline.
13. MCAULAY, JOHN D. "What Understandings Do Second Grade Children Have of Time Relationships?" in *Journal of Educational Research* (April, 1961), pp. 312–314.
14. MORISON, SAMUEL ELIOT. *An Hour of American History*. Boston: Beacon (1960), 87 pp. An eminent historian traces our history in broad strokes. A paperback.
15. NICHOLS, ROY F. and ARTHUR C. BINING. "The Role of History" in *Education for Civic Responsibilities*. Princeton, New Jersey: Princeton University Press (1942), pp. 62–69.
16. OJEMANN, RALPH H. "Social Studies in Light of Knowledge About Children." Section on "How Do Concepts Develop?" in *Social Studies in the Elementary School*. Chicago: National Society for the Study of Education (1957), pp. 76–119.
17. PRESTON, RALPH. *Teaching Social Studies in the Elementary School*. New York: Holt, Rinehart and Winston (1968), 370 pp. See Chapter Six on "Teaching Historical Concepts."
18. QUILLEN, I. JAMES. "American History in the Upper Grades and Junior High School" in *Interpreting and Teaching American History*. Washington: National Council for the Social Studies (1961), pp. 344–361.
19. SPIESEKE, ALICE W. "Developing a Sense of Time and Chronology" in *Skill Development in Social Studies*. Washington: National Council for the Social Studies (1963), pp. 171–200.
20. WILKINSON, FOSTER F. "General Media and the Battle of History at Elementaria" in *Social Education* (May, 1968), pp. 462–464.

Political Science

Governments are established to accomplish the things which the people (or their rulers) think can be carried on better by groups than by individuals. Two basic questions arise from this assumption. They are (1) what power or powers should be given to the government or the various governmental units and (2) how shall the rulers or representatives of the people be chosen? Other questions also arise, such as how the costs of government shall be borne and how the services of government shall be distributed. In brief this is the domain of political science, or what is often called civics in school.

In the past, political scientists have been concerned chiefly or solely with the formal institutions of government and with the structure of government. In recent years, however, political scientists have been affected by other fields, primarily psychology and social psychology. Today many political scientists are interested in the informal as well as the formal institutions of society. They are interested in the operation of government as well as its structure. They are studying individual and group psychology as it affects political behavior.

Does all of this have anything to do with the teaching of social studies in elementary and junior high schools? Definitely so — and in many ways. Children have already developed attitudes toward authority and toward rules, regulations, and some laws, before they come to school. They have lived in the family group and in play groups and have developed attitudes and skills related to group living long before they come to school.

Yet there is ample room for the schools to influence boys and girls as they develop as participants in various groups and become citizens. But the schools need to help children very early in their elementary school years if they are to affect behavior deeply. As long ago as 1931 Professor Charles Merriam of the University of Chicago pointed out that "Social and political attitudes are determined far earlier than is commonly supposed — many of them in fact in the pre-school years." More recently the studies of David Easton and Robert D. Hess of the University of Chicago have indicated that the crucial years for the development of basic attitudes and skills are from five or six to thirteen or fourteen. After that period, changes can be made only with great skill and great depth. The study of Fred T. Greenstein, cited in the bibliography on page 33, indicated that children identify with the policeman and with the President of the United States by grade three and many children have even begun to identify emotionally with a political party, even though this reflects usually the attitude of their parents.

All of this means that teachers play an important role as political scientists whether they are aware of that role or not. The way in which children work in groups or play in groups is determining how they will act as citizens in later years. Whether they are taught to examine problems and reason out solutions helps to decide how they will approach political problems in later life. How they perceive politics and political leaders today will have a profound influence on how they select leaders in their adult years. Their attitudes today toward rules, regulations, and laws will be reflected in the future. Their feeling of attachment to our nation in the future will also be determined in large part by their early experiences regarding nationhood or patriotism.

In short, the classroom, the playground, and other parts of the school are laboratories for learning the ways of democracy—or the ways of absolutism and dictatorship. That places a heavy burden on teachers to examine every phase of their work with boys and girls to see how human relations and political relations can be improved. Teachers need to ask themselves whether respect for all individuals is fostered, whether opportunities for all are provided, whether children help to determine rules and regulations in a democratic way, whether consensus and voting are introduced very early in the making of decisions, whether interest in current events is stimulated and a rational approach to problems is encouraged, and whether attitudes of loyalty to our nation and concern for its improvements are developed.

A provocative question to ask yourself is, "In what ways is my classroom different from a classroom in a totalitarian nation?" Such a question should help you to determine the extent to which you are preparing boys and girls for a democracy rather than for a dictatorship.

In the early years in school, the study of government should focus on the concepts of authority, rules and regulation, rights and responsibilities in homes and schools. Some attention should also be given to current happenings in the local community.

In grades three and four, pupils can learn a great deal about governments in communities in the United States and in other parts of the world. They should also begin to delve into community problems at an elementary level of learning.

By grade five, a rather full study of government in the United States can be undertaken. An example of such a unit occurs in the author's book on *One Nation — The U.S.A.* (Ginn, 1972). In the study of nations in other parts of the world, pupils can learn about other ways of governing than ours.

In the middle school or junior high school, pupils should wrestle in depth with some of the problems of government, especially in our own nation.

Here are some of the major concepts and generalizations of political science teachers need to use:

1. People live in groups and establish goals and rules for those groups, written and unwritten.
2. Communities and larger units are organized as governments.
3. The world is composed of many nations with different forms of government.
4. All governmental units are interrelated.
5. Governments have some common functions and characteristics, such as symbols, leadership, courts, costs, and services.
6. In every unit of government persons have some responsibilities and some rights; these differ in different types of government.
7. The United States is a democracy and a republic.
8. Decision-making is an important aspect of government but the methods differ radically; decision-making in a democracy is closer to the people and their representatives.
9. All governments have problems.
10. Nations of the world are increasingly interdependent; regional and international organizations are developing.

Articles, pamphlets, and books on political science

1. ALMOND, GABRIEL A. and SIDNEY VERBA. *The Civic Culture: Political Attitudes and Democracy in Five Nations.* Boston: Little, Brown (1963), 379 pp.
2. ASSOCIATION FOR CHILDHOOD EDUCATION, INTERNATIONAL. "Children As Responsible Citizens" in special issue *Childhood Education.* Washington: A.C.E.I. (April, 1961), 50 pp.
3. ASSOCIATION FOR SUPERVISION AND CURRICULUM DEVELOPMENT. "Politics and Education" in special issue of *Educational Leadership.* Washington, D.C.: A.S.C.D. (November, 1964), 71 pp.
4. CAMMAROTA, GLORIA. "Children, Politics and Elementary Social Studies" in *Social Education* (April, 1963), pp. 205–207, 211.
5. COLLIER, DAVID. *Political Science.* West Lafayette, Indiana: Purdue University Social Science Consortium (1966), 12 pp.
6. EASTON, DAVID. *A Framework for Political Analysis.* Englewood Cliffs, New Jersey: Prentice-Hall (1965), 143 pp.
7. EASTON, DAVID. *A Systems Analysis of Political Life.* New York: Wiley (1965), 143 pp.
8. EASTON, DAVID. *A Systems Approach to Political Life.* West Lafayette, Indiana: Purdue University Social Science Consortium (1966), 22 pp.
9. EASTON, DAVID and JACK DENNIS. "The Child's Image of Government" in *Annals of the American Academy of Political and Social Science* (September, 1965), pp. 40–57.

10. GREENSTEIN, FRED T. *Children and Politics*. New Haven, Connecticut: Yale University Press (1965), 199 pp. A study of children in grades four to eight in New Haven schools and what they knew about government.

11. HESS, ROBERT D. and JUDITH V. TORNEY. *The Development of Political Attitudes in Children*. Chicago: Aldine (1967), 288 pp.

12. KENWORTHY, LEONARD S. *Introducing Children to the World: Kindergarten Through Grade Eight*. New York: Harper (1956), 268 pp. Chapter 11 on "A World of Many Forms of Government."

13. KIRKPATRICK, EVRON M. and JEANE J. "Political Science" in Erling Hunt et al. *High School Social Studies Perspectives*. Boston: Houghton Mifflin (1962), pp. 99–126.

14. LINDBERG, LUCILLE. *The Democratic Classroom*. New York: Bureau of Publications Teachers College — Columbia University (1954), 115 pp.

15. LONG, NORTON E. "Political Science" in *Social Studies and the Social Sciences*. New York: Harcourt Brace Jovanovich (1962), pp. 88–106.

16. PATTERSON, FRANKLIN (Editor). *Citizenship and a Free Society: Education for the Future*. Washington: National Council for the Social Studies (1960), 292 pp. A yearbook.

17. PATTERSON, FRANKLIN K. *Man and Politics*. Cambridge, Massachusetts: Educational Services, Inc. (1965), 61 pp. A description of a three year junior high school sequence on "Man As a Political Being."

18. PATTERSON, FRANKLIN K. "Political Reality in Childhood: Dimensions of Education for Citizenship" in *The National Elementary Principal* (May, 1963), pp. 18–23.

19. PRESTON, RALPH C. *Teaching Social Studies in the Elementary School*. New York: Holt, Rinehart and Winston (1968), 370 pp. Chapter 8.

20. WOLL, PETER. "Recent Developments in Political Science" in *Social Education* (March, 1966), pp. 168–72.

APPLICATION OF CONCEPTS IN THE SOCIAL SCIENCES TO A WEEK IN SOCIAL STUDIES

Now that you have read about the various social science disciplines and their major ideas or concepts, perhaps you would like to think of how you can apply these big ideas in a week of work in your class. Space is provided below for you to do this. Select a topic on which you are working now or might work. Then think in terms of the four or five big ideas you might concentrate upon in four or five lessons in the social studies.

Topic for the week:

Major concepts or generalizations to be stressed:

(List the concepts or generalizations and indicate from which social science discipline they are taken.)

1.

2.

3.

4.

5.

4 What Are the Chief Determinants of Curricula?

Along with the nature of the social science disciplines and the factors of growth and development, learning theories are a third factor in determining curricula. Certainly there is nothing more central to teaching than knowledge about how people, and especially children, learn. As teachers we need to know how boys and girls learn a skill, develop an attitude, or grasp a significant concept. We need to know about the role of "readiness" in learning. We need to know how the concept of self affects learning. We need to know how beliefs and value systems in the culture influence the learner and learning. We need to know what part "expectations" play.

Does all this sound complex, involved, and controversial? It should. Learning is a very difficult process. Teaching would be easier if there were a unified theory of learning which we could study and apply in our classrooms. But there isn't any such set of principles upon which everyone agrees.

Therefore teachers need to think long, hard, and often about the learning process and to select from a wide range of views the principles they will try to use to foster learning.

There are many different theories of learning now extant. Each has its outspoken advocates, its own literature, and its practitioners— whether they know it or not. At the present time there are a host of experiments underway and a battle royal among psychologists as to theories of learning. If you are still a college student, you have probably surveyed in considerable detail several of the current theories about learning as well as some from the past. You should bring much knowledge and understanding to the next few pages as a result of your studies. If you are a teacher in service you may want to review learning theories as set forth in the next few pages and reflect on their relationships to your teaching.

Many people maintain that theories about learning are merely philosophical in nature. They depend upon one's philosophy of life and of teaching. Such considerations enter into the development of learning theories and practices, but there is also a wealth of experimentation upon which ideas about learning are based. Teachers of the social studies as well as teachers of other subjects need to know about these experiments and to reflect on their meaning as they relate to classroom situations.

For centuries man has been intrigued by the learning process. And for approximately a hundred years men have carefully studied this important and complicated process. The names of William James and John Dewey spring to mind immediately as two great thinkers who have added much to our fund of knowledge about this topic.

Although written in 1910, John Dewey's *How We Think* is still pertinent to thinking today.

Faculty Psychology

In her volume on *Curriculum Development*, Hilda Taba points out that we can find in our schools "the fossilized remains of almost any learning theory that ever existed, no matter how outdated or discredited" it is.

Faculty psychology is one of those "fossilized remains" in many schools today. It dates back to the eighteenth century and has long been discredited. Yet it is the basis of far too much learning in the social studies as in other fields today.

Basically it considered the world and man evil. A strong will was therefore needed to resist the temptations of life. The building of such willpower was best accomplished by tackling difficult and often distasteful tasks. The more difficult schoolwork was, the better it trained the will of children. Dullness was almost a virtue. Fear of severe punishment was accepted and even welcomed. To persons who held this belief, knowledge of immediate, practical use was distasteful and undesirable. Abstract reasoning and "pure theory" were commendable.

To Christian Wolff and his disciples man's general faculties included knowing, feeling, and willing — and the greatest of these was "willing." To them it was impossible to make education a scientific study.

When placed on the defensive by later thinkers, the proponents of faculty psychology became even more dogmatic, rigid, and impervious to change than they had been before the attacks.

Today this general approach to learning is discredited by almost every scholar of learning. Yet there are many people who operate within that framework, although they usually would not be able to identify their position with faculty psychology. Perhaps you knew or have studied under such a person.

Associationist or Behaviorist Psychology

Around the turn of the century in the United States Edward L. Thorndike and others conducted experiments which proved to their satisfaction that the idea of mental discipline was scientifically untenable.

Eventually a school of thought developed and became known as the behaviorist or associationist theory. Some of the people linked with that movement included Tichener, J. B. Watson, Thorndike, and Hall. They disagreed upon many points and stressed different approaches. Nevertheless they are often grouped together under this general umbrella.

In general they did not consider man or the world evil. They tended to think of man as neutral or passive. Man was given such qualities or powers as reasoning, attention, and memory. These could be developed through attention to specific stimuli. Much of their experimentation was with animals and upheld the stimulus-response concept of learning. Learning to them started with irreducible elements which could be combined in various ways.

Training was at the heart of their thinking. Since people, like animals, could be taught to respond to specific stimuli, teachers could manipulate learning to their will. Education was largely a matter of conditioning and trial and error. Motives could be controlled by external rewards and punishments. Drill was important. To them the whole of learning was seen in terms of the separate parts. Transfer was limited.

Since they were attached to the idea that only those things which were observable were "scientific," they were not concerned with such factors in learning as purpose, thought, and insight.

Subject matter in any subject would therefore be carefully selected by qualified adults and situations set up in which learners would acquire these essential facts. To them the past was infinitely more important than the present and the future. Coverage of many items was important since there was so little transfer in methods of learning. To them the right answers or the products were much more important than the processes by which they were obtained.

Little attention was paid to such factors in the learning process as motivation, the needs of pupils, or problem-solving. Central to their concept of learning was conditioning from without. Impression rather than expression was to be encouraged.

There is even more associationist or behaviorist psychology in vogue today than faculty psychology. In fact it has staged a comeback in the last few years in United States educational circles. In the social studies field the emphasis nowadays on the structure of the disciplines, with its stress on an organized, logical sequence in learning, has been utilized by proponents of this branch of psychology to advance its approaches. Teaching machines and programmed instruction have also given such psychologists beachheads in education, including the social studies field. To them programmed instruction demands an organized, logical sequence. Furthermore, this form of learning requires overt responses from pupils. In the third place the machine provides feedback immediately. It rewards the pupil if he obtains the correct answer and thus encourages him to learn more. In this way pupils are conditioned. Thus, the teaching machine is not simply another audiovisual aid. As one psychologist has written, "It represents the first practical application of laboratory techniques to education."

Critics of this general approach express a good deal of dismay over the fact that people are to be trained in the same way that animals are trained. They also express fear lest the general approach of the behaviorists lead to automated and even manipulated people rather than to the development of free human beings.

Gestalt, Field, or Organismic Psychology

The third general approach to learning is often called gestalt, field, or organismic psychology. Starting in Germany in the early twentieth century, it has spread to other parts of the world, especially to the United States. Some of the persons whose names are most closely associated with this school of thought are Kohler, Werthei-mer, Koffka, Tolman, Lewin, Bruner and even Freud. These and other proponents of this general school of learning have not always agreed with each other. In fact, they have fought each other bitterly upon occasion. Nevertheless they have enough in common to group them together.

According to them man has the potential for good or evil and has considerable choice in what he becomes. He is an active agent and therefore can determine, at least in part, his future. Their views in general have been more humanistic and more relativistic than those of the members of the associationist or behaviorist school.

Their approach to learning is far broader than that which characterized the work of Watson, Thorndike and the others we have named. They have not limited themselves to scientific experiments in laboratories or confined themselves primarily to experiments with animals. They have been much more people-centered. They have drawn heavily upon the field of cultural anthropology with its emphasis upon the totality of human groups. They have borrowed from sociology, from social psychology, from psychiatry and other fields.

Their field of exploration therefore has been broad. They have been and continue to be interested in the cognitive processes, in ideas, in insights, in intelligence, and in organization. They have even been receptive to hunches, intuitive perceptions, shrewd guesses, and intelligent conjectures.

Instead of seeing learning as a series of discrete, separate acts, they have looked upon it as the ability of people to select and organize data, to discover relationships and general principles, concepts, and generalizations.

As they view learning, the whole is greater than the sum of the parts. They are interested in the totality, in the field, in the "gestalt," or total view. Hence the name of this broad school of thought.

Moreover, they have recognized that learning is selective. Each person is affected by his values. Some of these men and women have been interested in the differences arising from socio-economic backgrounds, ethnic or minority groups, or differing religious viewpoints. All these facets of learning are a part of their emphasis upon perception in learning.

Motivation is therefore central. Transfer is not automatic as the proponents of faculty psychology felt. Nevertheless if a person is taught to inquire, discover, and experiment, he will be better able to attack new problems with reason. Learning is therefore cumulative.

To the field theory people, learning represents changes or even gains in thought patterns, in outlooks, and insights.

There are various implications of this thinking for curriculum in all fields, including the social studies. There should certainly be a wide range of experiences, with emphasis upon problem-solving situations. Open-ended materials are desirable. The role of the child as a learner is as a discoverer, an inquirer, an experimenter, an innovator. The background of children is important for it helps teachers to understand the "glasses" learners wear when they undergo new experiences. Since interaction of human beings is so important, the study of group life is of utmost importance.

In such learning situations the role of the teacher changes, too. She is no longer the instructor but the guide, helping children to effect changes in themselves. She is not constantly trying to impress; she is more interested in children expressing and examining ideas.

In some ways this movement bears considerable resemblance to the Progressive Education movement of the 1920s and 1930s but it is a much more sophisticated approach than any educational movement could have been in those years.

Of course it is possible that we will learn in the not-too-distant future that there are areas in the social studies which can be learned best by the use of behaviorist psychology and other areas in which it should not be used. Thus an eclectic philosophy of learning would emerge.

Forming, Reinforcing, and Changing Attitudes

Since you will be working constantly on the formation, reinforcement, and change of attitudes, here are some findings of psychologists, social psychologists, and other experts on this broad and important field. Perhaps you will want to take one important attitude you want to develop and apply these comments to your work.

1. Attitudes are formed, reinforced, or changed best when a person is secure and can accept changes.
2. Most basic attitudes are formed very early in life.
3. Attitudes, however, can be changed at any age.
4. Times of personal and societal crises are conducive to attitude changes.
5. Changes come best when an entire group is touched. This gives individuals security because others are changing, too.
6. Mass media can be a potent influence in affecting changes.
7. Attitudes are changed more easily when people have opportunities to act upon their new beliefs.
8. Membership in new groups helps to reinforce attitude changes.
9. The testimony of prestige persons often encourages others to change. Such statements must be internalized, however, by the less prominent person.
10. Symbols and slogans have some effect in bringing about change.
11. Mass meetings and other emotion-charged situations sometimes help in bringing about changes. They may be momentary, however, in their effect.
12. Appeals to pride or practical necessity can be helpful in attitude formation or change.
13. Information from reliable sources, especially if discovered by the person who should change, may bring about changes.
14. Changes come best when old views are accepted with equanimity so that the change is not threatened.
15. Shock can effect change but is dangerous in the hands of amateurs.

A Brief Bibliography on Attitudes

1. ALLPORT, GORDON W. *The Nature of Prejudice*. Garden City, New York: Doubleday (1958), 496 pp. A paperback.
2. BENNIS, WARREN G., and others. *The Planning of Change*. New York: Holt, Rinehart and Winston (1961), 781 pp.
3. BERELSON, BERNARD and GARY A. STEINER. *Human Behavior — An Inventory of Scientific Findings*. New York: Harcourt Brace Jovanovich (1964), 712 pp.
4. CLARK, KENNETH B. *Prejudice and Your Child*. Boston: Beacon (1955), 149 pp.
5. GLOCK, CHARLES Y. and ELLEN SIEGELMAN (Editors). *Prejudice: U.S.A.* New York: Praeger (1969), 194 pp.
6. GOODMAN, MARY ELLEN. *Race Awareness in Young Children*. New York: Collier-Macmillan (1964), 351 pp. A paperback. A cultural-anthropological study.
7. MACK, RAYMOND W. (Editor). *Prejudice and Race Relations*. Chicago: Quadrangle (1970), 271 pp.
8. STENDLER, CELIA and WALTER MARTIN. *Intergroup Education in Kindergarten*. New York: Macmillan (1958), 151 pp. Includes bibliographies.

Some General Principles of Learning

Here are some general principles of learning which most contemporary psychologists would support. Yet they should not be viewed as laws of learning. Some implications for classrooms are indicated for the first two. Perhaps you would like to try your hand at filling in some implications for the others.

1. *Pupils learn best when they are physically and emotionally comfortable, yet alert. In social studies teaching this implies that:*

    (a) Room conditions should be conducive to learning — comfortable seats, good lighting, proper ventilation, and colorful, challenging surroundings.

   (b) Teachers should know a great deal about each pupil and convey a feeling
       of respect for each pupil.
   (c) Teachers should avoid sarcasm and cutting humor at the expense of pupils.
   (d) Work should be challenging but adjusted to the needs of different pupils,
       so far as that is possible.
   (e) Pupils should be encouraged to ask questions without feeling guilty about
       revealing their ignorance.
   (f) Pupils should be encouraged frequently to help each other.
   (g) Praise should be given frequently, and sincerely.
   (h) Teachers should have high expectations of their pupils, yet expectations
       that can be achieved reasonably well.

2. *Pupils learn best when they select or help select problems and goals of real
   interest to them. In social studies teaching this implies that:*
   (a) Younger children should be given occasional opportunities to select between
       alternatives mentioned by them and/or the teacher.
   (b) Teachers should encourage frequent discussions based on "What was the
       most difficult part of your work for today?"
   (c) Teachers should stop frequently to give pupils a chance to raise questions
       and problems.
   (d) Teachers should ask pupils occasionally to state the order in which they
       would like to study a list of topics.
   (e) Teachers should offer pupils many opportunities to find answers to problems
       they raise. They may report to the group, but this is not always necessary.
   (f) Pupils should be given an opportunity to do reading on problems they
       select.
   (g) Pupils should be given opportunities to work on problems of their own choice
       in groups or committees.
   (h) Some work should be developed around problems the class suggests as
       important.
   (i) Occasional time should be allotted for free-wheeling discussion of problems
       mentioned.

3. *Pupils learn best through concrete, realistic, and predominately firsthand
   experiences. In social studies teaching this implies that:*

4. *Pupils learn best when they are challenged within the range of their abilities.
   In social studies teaching this implies that:*

5. *Pupils learn best when they are stimulated emotionally as well as intellectually. In social studies teaching this implies that:*

6. *Pupils learn best when they are involved in a variety of related activities. In social studies teaching this implies that:*

7. *Pupils learn best when a new learning is related to an older learning. In social studies teaching this implies that:*

8. *Pupils learn best when they have reflected on the meaning of their experiences and have participated in the evaluation of their experiences. In social studies teaching this implies that:*

9. *Pupils learn best when the teacher, and to some extent the pupils, know the different learning "styles" of individuals. In social studies teaching this implies that:*

10. *Pupils learn best when they are engaged in comparing and in contrasting. In social studies teaching this implies that:*

11. *Pupils learn best when they work often on their own, but with access to competent guidance. In social studies teaching this implies that:*

12. *Pupils learn best when learning is reinforced by meaningful repetition. In social studies teaching this implies that:*

13. *Pupils learn best when there is an element of novelty and/or vividness. In social studies teaching this implies that:*

14. *Pupils learn best when their knowledge leads to some action related to it. In social studies teaching this implies that:*

15. *Pupils learn best when they have a sense of personal and/or group achievement. In social studies teaching this implies that:*

BIBLIOGRAPHY ON LEARNING

1. ALMY, MILLIE and others. *Young Children's Thinking*. New York: Teachers College Press (1966), 153 pp.
2. ASSOCIATION FOR SUPERVISION AND CURRICULUM DEVELOPMENT. *Learning and the Teacher*. Washington: A.S.C.D. (1959), 222 pp.
3. BENJAMIN, WILLIAM F. "The Teacher and Learning in the Social Studies" in *National Elementary Principal*. Washington: National Education Association (May, 1963), pp. 35–39.
4. BIGGE, MORRIS L. *Learning Theories for Teachers*. New York: Harper (1964), 366 pp.
5. BURTON, WILLIAM H. *The Guidance of Learning Activities*. New York: Appleton (1962), 581 pp.
6. CANTOR, NATHANIEL F. *Dynamics of Learning*. New York: Stewart (1956), 296 pp.
7. FLAVELL, JOHN H. *The Developmental Psychology of Jean Piaget*. Princeton, New Jersey: Van Nostrand (1963), 472 pp.
8. FRAZIER, ALEXANDER (Editor). *Freeing Capacity to Learn*. Washington: Association for Supervision and Curriculum Development (1960), 97 pp.
9. GAGNE, ROBERT M. *The Conditions of Learning*. New York: Holt, Rinehart and Winston (1965), 308 pp.
10. GREEN, EDWARD J. *The Learning Process and Programmed Instruction*. New York: Holt, Rinehart and Winston (1962), 228 pp.
11. HARRIS, THEODORE L. and WILSON E. SCHWAHN. *Selected Readings on the Learning Process*. New York: Oxford (1961), 428 pp.

12. HILL, WINFRED F. *Learning: A Survey of Psychological Interpretations.* San Francisco: Chandler (1963), 227 pp.
13. HOLT, JOHN. *How Children Fail.* New York: Pitman (1964), 181 pp.
14. HOLT, JOHN. *How Children Learn.* New York: Pitman Publishing Corporation (1967), 189 pp.
15. HUNT, J. McV. *Intelligence and Experience.* New York: Ronald (1961), 416 pp.
16. INHELDER B. and J. PIAGET. *The Early Growth of Logic in the Child.* New York: Harper (1964), 302 pp.
17. JOYCE, BRUCE R. *Strategies for Social Science Education.* Chicago: Science Research Associates (1965), Chapter 7, "The Capacity to Learn," pp. 100–114.
18. KELLER, FRED S. *Learning: Reinforcement Theory.* New York: Random House (1969), 82 pp. Second edition.
19. "A Look at Learning." Special issue of *Childhood Education* (February, 1965).
20. MACDONALD, JAMES B. *Theories of Instruction.* Washington: Association for Supervision and Curriculum Development (1965), 118 pp.
21. SIGEL, IRVING. *A Teaching Strategy Devised from Some Piagetian Concepts.* Bloomington, Indiana: Indiana University Social Science Consortium (1966), 20 pp.
22. TABA, HILDA. *Curriculum Development: Theory and Practice.* New York: Harcourt Brace Jovanovich (1962), 529 pp. Chapter 6 particularly.
23. WAETJEN, WALTER B. *Human Variability and Learning.* Washington: Association for Supervision and Curriculum Development (1961), 88 pp.
24. WAETJEN, WALTER B. *Learning and Mental Health in the School.* Washington: Association for Supervision and Curriculum Development (1966), 174 pp. A yearbook.

THESE ARE
OUR PUPILS

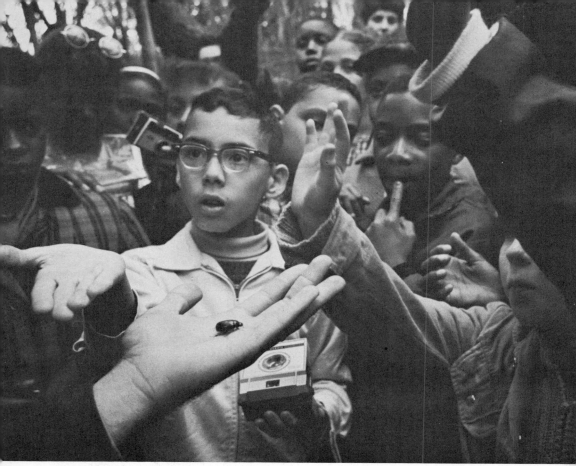

7. Students learning about the outdoors

8. Older girls relaxing
   at recess time

9.   And older boys and girls

These Are Our Pupils

1.   A home-bound pupil

10.   Young girls "trying on" adult roles

12.   And a blind girl

CHILD GROWTH AND DEVELOPMENT AS A DETERMINANT OF CURRICULA

In the past much attention has been given to the growth and development of children as a major factor in determining goals in the social studies and concomitant experiences for achieving such goals.

Possibly this factor has been overemphasized in the past. But the danger today is that it will be forgotten or minimized in the current emphasis upon the structure of the disciplines and upon cognitive learning. As educators we Americans are "pendulum prone" and the pendulum now seems to be swinging away from child growth and development.

Curriculum planners should welcome the new interest of scholars in the schools and in designs for social studies curricula. They have much to offer. At the same time curriculum workers and teachers need to be on guard against scholars whose area of specialty is a single social discipline and whose knowledge of children is often nonexistent or limited to their own bright sons or daughters in the stimulating environment of their middle-class homes. Similar wariness needs to be exercised in listening to the suggestions or demands of aspiring parents who want to get into the act. Curriculum workers and teachers need to listen with open minds and try out some of the ideas presented. At the same time they need to be firm in their insistence that every new idea must be tested to determine whether it fits in with our knowledge of child growth and development.

One example should suffice to make clear what is meant by this warning. A few years ago a state curriculum consultant in California proposed a history sequence for elementary schools which was utterly absurd. He suggested that history be taught in the first grade through the biographies of great personalities. The biographies were to include such men and women as Alexander the Great, Adam Bede, Caesar and Cheops, Drake, Elizabeth I, Plato, Socrates, Washington, Franklin, Lincoln — and the current President of the United States. In the second grade children were to explore the American heritage, with emphasis upon the Boston Tea Party and the Bill of Rights. In the third grade boys and girls were to study world history. A parallel course of study was proposed in geography as a separate subject. In grade one the emphasis was to be on the major land forms of the world. In grade two children were to specialize upon a major region of the world. In grade three there was to be a more detailed study of land forms. To anyone who knows about children or knows the literature of the field, these are incredible ideas. They do not take into account what we know about the difficulties in coping with concepts of time and space, the limitations upon the vocabularies of children, or the ability of young boys and girls to handle the concepts implicit in these schemes.

At our command today are the results of at least a half-century of research on how children grow and develop. Some idea of the extent of such studies is given by Lois Murphy and her collaborators in their volume on *The Widening World of Children: Paths Toward Mastery* (Basic Books, 1962). In one section of that book Dr. Murphy refers to some of the scholars who have been exploring the ways of helping children to cope with life's challenges. In capsule form she writes about several of them in this fashion:

> Murray's approach to needs (leading to action) led to studies by McClelland et al., of the achievement drive, and White's studies of competence. Piaget has contributed a new vocabulary of the cognitive process of adaptation in the infant and child. In psychoanalysis, Gartmann, Rapaport, and others following them have been explicitly concerned with adaptive functions and the pro-

cess of coming to terms with the culture. Karl Menninger has discussed the healing contribution of hope. Erikson has most comprehensively outlined the positive residues from successive phases of psychosexual development available for active response to the environment. Anna Freud has outlined positive lines of development in the young child, and long before had given many illustrations of children's active ways of dealing with stress.

It would be tragic if all this data, and more to come, were ignored or shunted to the sidelines. We know a great deal about the various stages through which children pass on their way to adulthood and maturity, as well as the places where so many of them stop on that long, exciting, yet demanding and hazardous journey. Such data need to be used to the fullest extent possible in planning social studies programs in elementary schools. Let us see what some of those specifics are.

Growth and Development of Children in Kindergarten, Grade One, and Grade Two, and Some Implications for the Social Studies

Here they come! Boys and girls. The fat ones and the thin ones, the dark-skinned and the light-skinned, the sad and the happy, the timid and the belligerent, the advantaged and the disadvantaged.

What do we know about these children entering our kindergartens or first grades? How can we encourage them in understanding themselves and other human beings; in comprehending the big, wide world; in coping with problems; and in capitalizing upon their creativity?

As we have already indicated, we know that they differ. But do we realize how widely they differ? Some of them have already learned that life is hard. They have been taught that they are "no good." Some come confidently. They have been surrounded with love and security from morning until night. They are open to the world.

We also know that they come to school with attitudes toward other people as well as toward themselves. As early as 1929 Bruno Lasker published an extensive collection of anecdotal records showing the awareness by young children of group differences. Those results appeared in his volume on *Race Attitudes in Children* (New York: Henry Holt and Company).

In 1952 Helen G. Trager and Marian Radke Yarrow told the story of an important study in Philadelphia of children in kindergarten, first, and second grade. They called their book *They Learn What They Live: Prejudice in Young Children* (New York: Harper). For example, a high percentage of the children with whom they worked identified groups like the Chinese and Negroes by placing laundryman and work clothing on the cutout figures provided them. In this manner, as well as in other ways, these children demonstrated that they had already developed stereotypes of other people, placing them in categories according to race and religion. They come with patterns of reaction to authority and rules. Although the idea persists that children entering school are like putty or clay, to be molded by competent teachers, this just is not true. They come, bringing with them five or six years of experiences and with emerging personality patterns.

In the earlier years these children were dependent largely upon adults for support. But in the primary grades they begin to reach out to their classmates and peers. More and more they test ideas and actions against the judgment of this new "jury."

These pupils have had limited contacts with other children and adults. Many of

them will have lived in restricted environments. One task of the school is to introduce them to a wide variety of other children so that they can learn to live with
differences as well as with similarities. In this respect the classroom and the school
should be laboratories in democratic education, human rights, and cultural pluralism. But the concepts in these phrases are not attained automatically. They are
learned only when children can feel secure enough about themselves that they can
accept others from different backgrounds. Children need to be exposed to a wide
range of warm, supporting adults, too. Many of them especially need contacts with
men, inasmuch as their experiences in this regard have been limited.

Most children in the primary grades are physically active. But they can tire
easily. Their rate of growth is not as great as in their earlier years, but they are
certainly growing. And this slow, steady growth will continue throughout the primary grades.

Therefore they need a great deal of physical activity. They also need alternate
periods of activity and rest. In the social studies many activities fit into this pattern. They include such experiences as building, spontaneous dramatics, drawing,
painting, games, and dances.

Children learn much through play activities, recreating and reconstructing in
their own way, what they have seen and heard, trying new situations on for size.
As Robison and Mukerji have pointed out in a recent study· "Alternating structured with less structured activities gave the children freedom to practice and
incorporate into their play those more complex ideas and more advanced skills
which had meaning for them." Play is not just play; it is learning.

Most children come to school full of curiosity. We need to capitalize upon this
characteristic and provide a wide range of experiences in which children can get
answers to their questions and be spurred on to ask more questions — and then
seek further answers. Yet we need to be aware, too, of the timid and shy children
— who have been rebuffed in their efforts to explore and discover — and of those
whose range of experiences has been extremely limited. Many young teachers, for
example, are shocked to learn that children can enter school without knowing their
names or having handled a book or used a crayon.

Watch a group of kindergarten or even first grade children and you will realize
that many of them are still in the stage of working and playing *parallel* with other
children but not actually *with* them, or in the stage of working and playing in
teams. As they gain security, they can work in small groups and in the total class
group. All the plans for these grades, including plans for social studies, should provide for a variety of group experiences, often with varied activities going on simulta-
neously.

The ability of most entering children to handle language and language symbols
is limited. With some it is almost nonexistent. The firsthand experiences of children in the school and community, the films and filmstrips and pictures, and other
activities should accelerate their language abilities in all fields, including the
social studies. Experience charts and reading will enhance their learnings in language, too. Throughout the primary grades, children should hear many things read
to them, too. Then they should have opportunities to discuss what they have heard
and to act out the parts that appeal to them.

Some people contend that the attention span of little children is very brief. This
is not always true. If they become intensely involved in an activity, they may
pursue it for a seemingly long time. By and large, however, teachers need to think
in terms of changing activities several times in a day, leaving some large blocks of
time open for free play and activities in small groups.

Time and space are two concepts central to the social studies. Both of them are difficult for children to develop in their early years at school. Nevertheless they are ideas on which teachers need to work. The celebration of birthdays and special events of historic interest, such as national holidays, help. Visits to old houses in the community, if they exist, can help. As children learn to tell time, this adds to their understanding of the time concept. Flannelboards, with pictures arranged in sequences of time, can enhance learning of this concept. Contacts with grandparents, where that is possible, often yield rich dividends in developing the idea of time. Most work in the primary grades should be in the here and now.

Among the many devices for developing concepts of space, there is nothing so productive as trips to other parts of a community — or beyond. Construction activities can help children, too, in gaining ideas of space. Television has undoubtedly helped children to think earlier than in the past about faraway places.

In these crucial years in the development of attitudes toward themselves and toward others, children from minority groups need especially to identify with successful models. Hence the need for visitors from their groups who can help to provide such models. Hence the importance of stories of boys and girls and of families from their groups. And hence the importance of visual materials, ranging from flat pictures to films, filmstrips, and television programs, which show people from the minority groups which boys and girls in the class represent. In these and other ways self-respect and group respect can be enhanced.

By the second grade many of us feel that children can be introduced to the relatively simple unit of the family in other parts of the world. We will devote attention to this topic in a later chapter.

As a part of democratic living, children need to take part in decision-making. At first this may be discussing and even voting on alternatives suggested by them or by the teacher. Later the range of decision-making can be widened. They need, also, to take part early in the making of some rules and regulations and to develop respect for the property rights of others.

Are all these social studies activities? Yes, in the broadest interpretation of that term.

BIBLIOGRAPHY ON SOCIAL STUDIES IN THE PRIMARY GRADES WITH SPECIAL REFERENCE TO CHILD GROWTH AND DEVELOPMENT

1. BARTON, THOMAS F. "Teaching Space in the Primary Grades" in *Journal of Geography*. (March, 1963), pp. 122–129.
2. CAMMAROTA, GLORIA. "Children, Politics, and Elementary Social Studies" in *Social Education*. (April, 1963), pp. 205–211.
3. CAMMAROTA, GLORIA. "New Emphases in Social Studies for the Primary Grades" in *Social Education*. (February, 1963), pp. 77–80.
4. CROSBY, MURIEL. *An Adventure in Human Relations*. Chicago: Follett (1965), 396 pp. Work in the Wilmington, Delaware schools.
5. DE SART, HELEN. "Geographic Readiness in the Kindergarten" in *Journal of Geography*. (October, 1961), pp. 331–335.
6. ELKIND, DAVID. *A Sympathetic Understanding of the Child Six to Sixteen*. Boston: Allyn and Bacon (1971), 154 pp.
7. FRAZIER, ALEXANDER. "Lifting Our Sights in Primary Social Studies" in *Social Education*. (November, 1959), pp. 337–340.
8. HEFFERNAN, HELEN (Editor). *Guiding the Young Child: Kindergarten to Grade Three*. Boston: Heath (1959), 362 pp. Especially good on the characteristics of children and their implications for the social studies.

9. IMHOFF, MYRTLE M. "Better Primary Social Studies" in *Grade Teacher*. (April, 1963), pp. 52, 98, 100, and 102.

10. IMHOFF, MYRTLE M. "They're Never Too Young for Social Studies" in *Grade Teacher*. (October, 1965), pp. 112, 114, 116.

11. ISAACS, SUSAN. *The Children We Teach: Seven to Eleven Years*. New York: Schocken Books (1971 edition), 160 pp.

12. JOYCE, WILLIAM W., ROBERT G. OANA, and W. ROBERT HOUSTON. *Elementary Education in the Seventies: Implications for Theory and Practice*. New York: Holt, Rinehart and Winston (1970), 579 pp. Chapter 3 on Social Studies.

13. MCAULAY, J. D. "Social Studies Interests of the Primary-Grade Child" in *Social Education*. (April, 1962), pp. 199–201.

14. MCAULAY, J. D. "World Understanding Begins Early" in *Grade Teacher*. (April, 1963), pp. 59, 125–128.

15. MICHAELIS, JOHN U. (Editor). *Social Studies in Elementary Schools*. Washington: National Council for the Social Studies (1962), 334 pp. See especially the section on "Social and Psychological Foundations."

16. MIEL, ALICE and PEGGY BROGAN. *More Than Social Studies: A View of Social Learnings in the Elementary School*. Englewood Cliffs, New Jersey: Prentice-Hall (1957), 352 pp. Excellent on social education.

17. OJEMANN, RALPH H. "Social Studies in the Light of Knowledge of Children" in the National Society for the Study of Education yearbook on *Social Studies in the Elementary School*. Chicago: University of Chicago Press (1957), 320 pp.

18. ROBISON, HELEN and ROSE MUKERJI. "Cultural Pluralism in the Primary Grades" in *The Instructor*. (February, 1967), pp. 34–35, 159.

19. ROBISON, HELEN and BERNARD SPODEK. *New Directions in the Kindergarten*. New York: Teachers College Press (1965), 214 pp.

20. SABAROFF, ROSE. "Map Making in the Primary Grades" in *Social Education*. (January, 1960), pp. 19–20.

21. SPODEK, BERNARD. "Developing Social Science Concepts in the Kindergarten" in *Social Education*. (May, 1963), pp. 253–256.

Growth and Development of Children in Grades Three, Four, and Five, and Some Implications for the Social Studies

What are children like in grades three, four, and five? They differ greatly, of course, but there are certain general characteristics which mark these children.

Most of them are sturdy little creatures with better muscular coordination than before. They are still growing steadily rather than by spurts. They have better control of their small muscles and can manipulate easier than before. And they are active, with less sign of fatigue than in their primary grade years. Often, however, they do not get enough sleep. Teachers need to provide for some activity and for some periods of rest.

As a part of growing up and establishing their own identity, these children are likely to try to win considerable independence from adults. Many of them love to "test" teachers with innocent questions and riddles and with unorthodox behavior.

Friends among children are more likely to be the pacesetters than adults. Above all they like one "best friend," of the same sex. But they also like to operate in groups of children. This is the gang and clique age. Rules, procedures, codes, and symbols of their groups interest them.

For the social studies there are several implications in these friendship characteristics. Much teamwork and some committee work can be carried on. Often the children should select their partners and their fellow committee members. Pupils can do much planning with the teacher. They can help in making rules for the classroom and in enforcing their own rules. In content there is a rare opportunity to see how human groups are organized and how they function and the part that rules and laws play in society. The study of local government can be carried on, with the pupils electing their own classroom officials.

This is the period par excellence of coin, stamp, postcard, and doll collections. Often this interest can be utilized by tying in collecting interests with social studies. For example, a pupil with a strong interest in stamps can exhibit his collection of United States issues and relate them to the history of our country.

As a part of developing their own identities, many children have heroes in this period of their lives. This can lead to identification with heroes past and present through books. Biography should be stressed in the social studies. But a whole year need not be devoted exclusively to this approach as is being done in some schools and even in some state curricula at the present time. Children may want to interview older persons or to invite them to their classrooms for interviews.

Self-esteem and group identity need to be fortified, especially in boys and girls from minority groups. Many a pupil who seems "tough" on the outside is giving teachers a warning that he or she actually lacks self-confidence inside.

Special attention needs to be devoted at this stage to the development of skills. All too often such elementary skills as grouping, classifying, comparing, and contrasting are assumed by teachers to have been acquired, when in reality pupils have not reached this stage in their intellectual development.

Because they are older and stronger, children in the upper primary grades can explore the wider community and visit many places. This means that trips can play a large part in social studies instruction in these three years. Studies of the local community and the larger metropolitan area are a "natural" for this age group.

By the time children have reached the third, fourth, and fifth grades, many of them have acquired considerable skill in reading. This is a much coveted occupation for such children and should be encouraged. There should be many books in the classroom relating to the social studies and much use of books in the school and local library.

But many children will not have acquired the skills of reading. They need special help in the way of remedial reading expertise. They also need to find books which appeal to them in subject matter so they will be encouraged to read because of their high interest in a topic. This means acquaintance on the part of teachers with many books for individualized reading. It means group work, too, in which the reading is pitched at various levels, and other types of activity are included so that the poor readers or nonreaders are not excluded from social studies learning.

Because pupils can now manipulate a wide variety of media, all kinds of hand work can be included in social studies experiences. They can work on maps, charts and graphs, do murals, construct models, and prepare simple stage settings for plays.

In these years children begin to think about events around them more than before and want to identify more with adults. This means that some attention should be given to current events work. This will be made easier and more productive by the use of current events papers but some attention should be given, also, to local newspapers, especially by the fifth grade year.

Having studied their own community and other communities of the U.S.A., children should then be able to study a variety of communities in other parts of the world, utilizing the skills they acquired in studying their locality.

The concept of country should be much easier for children to grasp by the fourth or fifth grade and a study of our own country is recommended at this time. Since children are still living primarily in the here and now, however, the first approach might well be a study of the contemporary scene.

By the fourth or fifth grades, the time sense should have been developed much more than in the earlier years. Constant attention to this basic concept should enable children to delve more deeply now into history. The writer believes that a part of the fifth grade year should be devoted to a quick survey of our own national history, with selected periods studied in depth. That postpones the use of strict chronology a little later than is usual in elementary schools, which means that children should be able to grasp such events better than has heretofore been possible. By limiting the amount of detail, pupils should also be able to develop a sense of major periods and the sweep of history in our own country. In line with the gestalt theory of learning, they should be able to see the various aspects of a given period, drawing upon all the social science disciplines for help in explaining the events of a given period.

Some time should be reserved by teachers, however, to merely sit and talk with children about their personal problems and their progress in this critical period of their development. Growth comes in these quiet periods as well as in the rush of activities.

Growth and Development of Children in Grades Six, Seven, and Eight, and Some Implications for the Social Studies

Walk down the corridor of a middle school or junior high school or observe a group of sixth, seventh, and eighth grade boys and girls in an assembly, a cafeteria, or on the playground and you will see some of the ways in which they differ from younger children.

You will note quickly that the girls are taller and more fully developed physically than the boys. You will see these youngsters in separate groups or clusters. Sometimes they will be downright cruel to "outsiders." You may see some evidence that the emotions of the girls are often quite close to the surface.

This may not be the "storm and stress" period described by G. Stanley Hall many years ago, but it is certainly a transitional period of major proportions. It has been made more difficult in our culture because the lines between stages of development are blurred and the rites of transition few.

The physical changes for girls may start as early as the fourth, fifth, or sixth grade. For boys they start later, usually in the seventh or eighth grades. Eventually the boys will catch up with the girls, but the period before that happens will be rough on both groups. The girls may reach out to the boys and be repulsed. Consequently each sex seems to find comfort and security in its own group. Yet even this pattern has been changing in recent years. Encouraged by parents and the mass media, boys and girls have started dating earlier than in former times. Again, considerable confusion as to their roles may result.

This period for boys and girls is often one of tense relationships with adults. This is particularly true in regard to parents, as the chief authority figures. But it also may be true of teachers and other adults. Boys and girls want an increasing degree of independence, before many parents are willing to grant such freedom. Tensions mount and conflicts arise. At the same time these pre-adolescents may want a great deal of support from adults, even to the point of being told not to do certain things. Feelings are often camouflaged.

In all this, boys and girls are seeking their own self-identity. "Who am I?" is a basic question, often considered but not often verbalized. "Am I a child or an adult?" they seem to be asking themselves and others.

During this period many boys and girls are extremely sensitive to criticism. Every comment about them of a derogatory nature is considered a personal attack and even suggestions are often considered insults.

For those who have not matured as fast as their peers, there are worry and doubt and misgivings about themselves.

Yet some boys and girls have been independent a long time. They may have tasted of the fruit of the tree of life and found it bitter or sour. Many a young teacher from a highly protected middle class home may be amazed at the amount of sexual experience some pupils at this age have had. Your author recalls many of the shocks his student teachers at these grade levels have had, such as the young teacher who received a note on a Friday afternoon from her favorite pupil, saying "I won't be back on Monday. I'm going to have a baby. Sorry." What a story that note told!

To young adolescents schools are often dull, boring, childish places, filled with hypocritical teachers and repressive administrators.

Do our social studies classes help pupils through this period in their lives? Seldom, I fear. We travel the ancient roads of Greece and Rome with our pupils in Old World Backgrounds. We start the merry-go-round of American History which they rode on in the fifth grade and spin through our national development in the same dreary fashion as we did once before. Or we subject our pupils to dissections of the Constitution and bore them with the balance of power theory as a part of civics courses.

The social studies, more than any other subject field, should be helpful to boys and girls in this critical stage of their development. Social studies should introduce them to individual and group psychology and help them understand the motivations of people and the influences at work on them. It should help them with their sex roles in society. It should introduce them to the world of work and help them to examine their own talents and shortcomings in relation to possible jobs. It should share with them some of the problems of our industrialized urban society and let them explore the community and metropolitan area to see what is being done or could be done to alleviate some of these pressing problems. It should expose them to the various philosophies of adults and their value systems and encourage them to think about their own philosophies of life. And it should introduce them to the peoples of the world in all of their variety, and yet with similar basic needs to fill.

Most boys and girls are ready for such a broad-based program of social studies in grades six, seven, and eight. They can read better now. They can understand time and space concepts better. They can organize materials. They can move around more easily in the community and region.

They should be encouraged to work individually and in small groups in solving problems of real importance to them. They should see that most problems have local, national, and international dimensions and that solutions are not easy to find or carry out. As much as possible, they should be left to themselves to work out problems, with encouragement and help from teachers when it is needed. This will help them to meet their desire to be treated as adults, yet be supported by sensitive, sympathetic teachers.

In their seventh grade social studies classes, early adolescents could well study a few carefully selected problems — personal, civic, national, and international.

Such studies should have relevance to them and help them to identify with the adult world, which so many of them want to do so badly.

Their study of American History, required by many state legislatures, can also be a problem-centered course, with twenty-five to thirty great decisions to be made, ranging from the best way of life for the early colonists to ways and means today of attacking air pollution, race prejudice, and conflicts between nations.

Some schools are moving in these directions. But too few are doing that yet. Yes, child growth and development do give us many leads for the social studies curriculum if we are open to these ideas. And they need not be in conflict with the current emphasis upon the structure of the disciplines. But more of this later. In the meantime you may want to read and think more about this facet of curriculum planning. On these two pages are several references to help you in this task. Some of them are on the middle grades; most of them are on children and social studies in grades six, seven, and eight.

BIBLIOGRAPHY ON SOCIAL STUDIES IN THE MIDDLE AND UPPER GRADES WITH SPECIAL REFERENCE TO CHILD GROWTH AND DEVELOPMENT

1. ALILUNAS, LEO J. "An Analysis of Social Studies Content in the Middle Grades" in *Social Studies*. (November, 1961), pp. 210–218.
2. BROWN, RALPH ADAMS. "Improving Instruction in Junior High School Social Studies" in *Social Education*. (March, 1961), pp. 139–142.
3. ELKIND, DAVID. *A Sympathetic Understanding of the Child Six to Sixteen*. Boston: Allyn and Bacon (1971), 154 pp.
4. FAUNCE, ROLAND C. and MORREL J. CLUTE. *Teaching and Learning in the Junior High School*. San Francisco: Wadsworth (1961), 367 pp.
5. GROSS, HERBERT H. "Stronger Social Studies in the Intermediate Grades" in *Grade Teacher*. (April, 1963), pp. 53, 102, 104.
6. HUNNICUTT, CLARENCE W. *Social Studies for the Middle Grades: Answering Teachers' Questions*. Washington: National Council for the Social Studies (1960), 122 pp.
7. *Intermediate Education: Changing Dimensions*. Washington: Association for Childhood Education International (1965), 80 pp. Especially "Social Studies for the Middle Grades" by Fannie R. Shaftel.
8. JOHNSON, ERIC. *How to Live Through Junior High School*. Philadelphia: Lippincott (1959), 288 pp. For parents and teachers; reflects a real understanding of junior high school boys and girls.
9. JOYCE, WILLIAM W., ROBERT G. OANA, and W. ROBERT HOUSTON. *Elementary Education in the Seventies: Implications for Theory and Practice*. New York: Holt, Rinehart and Winston (1970), 579 pp. Chapter Three on Social Studies.
10. KNAPP, ROYCE H. "The Space Age in Junior High School Social Studies Programs" in *Social Education*. (December, 1960), pp. 260–264.
11. KRAVITZ, BERNARD. "Factors Related to Knowledge of Current Affairs in Grades 7 and 8" in *Social Education*. (March, 1962), pp. 143–145.
12. MCAULAY, J. D. "Interests of Elementary School Children" in *Social Education*. (December, 1961), pp. 407–409.
13. MCAULAY, J. D. "Mass Media and Third-Grade Social Studies" in *Childhood Education*. (November, 1964), pp. 120–122.
14. MCAULAY, J. D. "Social Studies Interests of the Intermediate Grades" in *Social Education*. (May, 1962), pp. 247–248.
15. MICHAELIS, JOHN U. *Social Studies for Children in a Democracy*. Englewood Cliffs, New Jersey: Prentice-Hall (1963), 624 pp. Chapter Three on "Child Development and Learning."

16. MILOR, JOHN H. "A Superintendent Looks at Fifth-Grade Social Studies" in *Childhood Education*. (November, 1964), pp. 115–118.

17. NOAR, GERTRUDE. *Teaching and Learning the Democratic Way*. Englewood Cliffs, New Jersey: Prentice-Hall (1963), 244 pp. On junior high school social studies work, stressing units.

18. PLUMMER, ROBERT H. and CLYDE BLOCKER. "A Unit on Metropolitan Problems" in *Social Education*. (May, 1963), pp. 257–258.

19. RAGAN, WILLIAM B. and J. D. McAULAY. *Social Studies for Today's Children*. New York: Appleton (1964), Chapter Two on "The Child and the Social Studies Program."

20. WHITTEMORE, KATHERYNE T. "A Program for Geographic Education for Grades 4 Through 6" in *Journal of Geography*. (December, 1960), pp. 423–428.

A CHANGING UNITED STATES AND A CHANGING WORLD AS DETERMINANTS OF CURRICULA

Have you ever stopped to think about the period of history in which today's boys and girls will be living out their lives? Most of them in our elementary schools today will be living well into the twenty-first century — until 2030 or 2035, or even beyond that point if science prolongs their lives, as it is likely to do.

It is our claim as social science specialists that we are preparing pupils to live more effectively or more wisely in the future, as well as helping them to live creatively today. If that is the avowed purpose of the social studies, then we need to know as much as possible about the world of the future. Only in that way can we help to prepare pupils to live in it.

No one can predict the future with any degree of accuracy. There are too many unknown factors in it. But we can sketch in thin pencil lines some characteristics of the foreseeable future because we already know some of the factors which are shaping it. From what we assume will happen, we should be able to obtain some important clues to help us determine directions for social studies curricula now and in the immediate future.

Some Possible Characteristics of the World in the Foreseeable Future. It is likely that the world by 2000 will have between six and seven billion persons in it. We are going to have to produce food for all of them, in greater quantity and hopefully in better quality than is possible today. We are going to have to learn to live together peacefully — or face annihilation.

Tomorrow's world is certain to be one of vastly changed transportation, with airplanes, supersonic jets, space ships, and — who knows what else. One leading airplane executive has stated that by 1990 we will probably be traveling at a height of seventy miles and at a speed of 17,000 miles per hour. That would mean that any place on our planet could be reached in approximately two hours.

Now that men have reached the moon and explored parts of it, it is almost inevitable that we shall soon be exploring other parts of the solar system. We may soon have space platforms with hotels and colonies on the moon. No?

We have already had a preview of the future of communications with Telstar. Global telephones are on the way. David Sarnoff predicted that long before 2000 we will be able to communicate instantaneously in sound and sight with "anyone, anywhere."

We may well move soon into a world-wide system of weights and measures, and into a single global currency.

New applications will certainly be found for lasers. And new and more varied drugs may be utilized to control fatigue, heighten perceptions, and improve health.

Soon we will be utilizing many new types of power — the wind, natural steam, atomic energy, the tides of the sea, and the sun.

In the next few years several more new nations will probably be formed. Existing regional organizations probably will be strengthened and new ones formed. Worldwide planning undoubtedly will increase. Several nations will become increasingly important in the next twenty-five years. Which ones do you think they will be? China? India? Japan? Canada? Indonesia? Australia? Which ones would you name?

Men will explore further the resources of our oceans in the near future, and the question of ownership of these vast storehouses will arise.

What else of importance do you see in the foreseeable future?

Some Possible Characteristics of a Changing United States.   Tremendous changes are also underway in our own nation. Population experts predict that there will be 300 million of us by the year 2000 and that nine out of every ten persons will be living in super-cities and in suburbs. Just what form cities of the future will take is problematical. They may be satellite cities with self-contained skyscrapers. They may be covered with plastic domes like the Astrodome in Houston. Some predict that people will live underground or that some people will live under the water. By 2000 most experts agree that 170 million or so of our populace will be living in five megalopoli: Boston–Washington, Buffalo–Chicago, San Francisco–Los Angeles, San Antonio–Dallas–Houston, and Jacksonville–Miami.

Certainly there will be changes in our sources of power, with a vast increase in the use of atomic energy.

In transportation fantastic changes also are inevitable. Perhaps there will be electric automobiles. There well may be super-trains, traveling at the rate of 300–350 miles per hour. There may also be super-helicopters and large submarines. Some people predict the use of containers to carry goods, with these large plastic bags deflated or destroyed at the end of a journey. Others predict rooftop roadways.

Tremendous increase in the use of computers seems inevitable, freeing most people from routine drudgery.

There may well be considerable control, too, of weather and of climate in the years ahead.

Our lives may also be prolonged with fabulous changes in medicine, including the use of mechanical aids or substitutes for human organs and the use of chemical or biological treatment of mental illnesses.

Someone once asked Carl Sandburg the meaning of the inscription on the National Archives building in Washington, "What is past is prologue." After a moment, he is said to have replied, "It means, 'you ain't seen nothin' yet.' "

Some Effects on Social Studies Curricula.   Is this the kind of U.S.A. and world for which we are preparing boys and girls today? If not, there is something wrong with our courses of study and our teaching.

Surely we are going to have to learn much more about the people of the world than we do today. Certainly we need to help our pupils to learn much more about the science and art of human relations in a tightly knit national and international community than we are now doing. Obviously we are going to have to learn about several nations whose importance is increasing. And we are going to have to learn more about interdependence, space, regional and international organizations, and the causes of conflicts.

In a similar way we need to help today's pupils to understand the changes in our own country and to help them to learn to wrestle with some of the problems of the foreseeable future intelligently.

These are some of the implications for social studies curricula of a changing U.S.A. and a changing world.

How Are Changes
in Courses of Study
and Curricula Brought About? **5**

In the rapidly changing world in which we now live and will probably live in the foreseeable future, radical changes in courses of study and in teaching methods are imperative. The children in our schools today will live half or nearly half of their lives in the twenty-first century and our curricula need to be commensurate with the kind of world in which they will live.

Recognition of the implications of that statement has already altered several fields in elementary schools in recent years. The new math has been introduced. Radical changes have been made in the teaching of science in many elementary schools. Foreign languages have been introduced in many places in the elementary school sequence.

It is in the social studies, however, that the most far-reaching changes still need to be made. Most curricula are preparing boys and girls to live in the early or middle years of the nineteenth century rather than in the latter part of the twentieth century and the early part of the twenty-first century.

Minor changes are made in many school systems from time to time in the social studies field. A few schools provide for periodic examinations of their offerings and of their methods of teaching.

Radical changes are less frequent. They involve a great deal of time and effort on the part of many people. Usually they call for increased expenditure of funds for the further education of teachers and for new equipment and supplies. Frequently they meet with resistance on the part of administrators, of teachers, of parents and other adults, or of members of Boards of Education.

Some Reasons for Resistance to Change. People are sometimes baffled when they find resistance to changes. This is especially true of young teachers, going into school, although they are seldom the persons who initiate change.

Novitiates are sometimes cornered in the corridors, in the teachers' room, in the lunchroom, or elsewhere, and they are urged not to try out too many new ideas or to urge changes. They cannot understand why this is done. Perhaps some explanation is in order for young teachers who enter our schools fresh from courses in college where new ideas often are encouraged. The same explanations may be helpful to older teachers who are ready and eager for changes.

Basically changes in teaching involve changes in people. These are not easily made. Some teachers are tired. They may have taught for years. Some of them have been disappointed in not achieving their goals in life. Some have raised families while teaching and have found their dual roles complicated and demanding. They do not want added burdens.

Some teachers and some administrators have a vested interest in the present status. They fear others might replace them if too many changes were instituted.

Some teachers and administrators are rigid in their thinking. Change constitutes a personal threat to them. Therefore they resist it.

Some teachers and administrators are not aware of newer practices and proposals. They have not attended professional meetings where innovations have been discussed. They have not taken courses recently. They have not read widely in professional books and periodicals.

Some persons believe that what they are doing now constitutes the best program for boys and girls. Perhaps they have taken part in changes which did not represent real gains in teaching and are wary of new proposals. They may be cautious, although ready to be shown.

Or they may have taken part in so many changes that they would like to be left alone for a while. They dread to think of the long hours involved in still another change in courses of study and classroom practices.

The Best Times for Change. All this indicates how complicated the process of change really is. Courses of study can be prepared and handed out to teachers, but that will not necessarily bring about the hoped-for changes. Such a procedure can make teachers so insecure that they do not do as good a job as they have done before.

We need to turn to the psychologists and the social psychologists to know how best to bring about changes. Much that we have said in the previous section on learning applies at this point. People need to want to change. They need to be involved in the changes. They need to see some parts of what they have done retained, for their own personal security. They need to see examples of how the changes will work in actual practice. They need administrative support and encouragement. And they need resources with which to implement the suggested changes. All these factors, plus others, should enter into any proposals for change.

Perhaps the most important factor, however, is the timing. Social psychologists are certain that change comes best in times of crisis. At such times, people are aware that something must be altered and are more receptive to modifications.

A variety of local crises may bring about the climate for change. A flood of dropouts in the upper grades may cause a school system to consider what can be done to make the social studies — and other subjects — more meaningful to pupils. This can affect the curriculum all up and down the line.

The arrival of new pupils in a school may provide the impetus for changes. There may be consternation at first — and then a readiness for change. For example, this may be true of a school where many new pupils from Appalachia, from Mexico, or from Puerto Rico are being enrolled. The writer was involved recently in a school where a new child was enrolling almost every day from Hong Kong and the teachers were ready for a discussion of new content and new methods to meet the needs of these new pupils. The author of this book has also been involved in changing the curricula in schools where many children were being "bused in" as a part of a city-wide move for better integration. The teachers in those schools were bewildered at first. Then they were ready to consider changes in courses of study and in methods to meet changing conditions. In these situations just cited, people were confronted with difficulties. Many of them were therefore ready for any changes which promised to alleviate these difficulties.

Poor results on a national standardized test sometimes create a situation in which teachers are ready and willing to examine needed shifts in the social studies curriculum.

One of the best times for change is at the time of appointment of a new superintendent of schools, a new principal, or a new supervisor, provided there is general approval of the appointment. At such a time people expect some changes and are ready for them. This is a little like the President of the United States pushing through legislation in the first few weeks of his term when people expect changes and he has general public support.

Attendance at a conference or workshop where newer methods have been discussed and/or demonstrated, or new courses of study presented, may bring about some readiness for change.

Often change comes about when a school that is generally considered a leading institution in an area makes a shift in its course of study and/or its methods of teaching. Teachers in other schools are anxious to keep up with the "prestige institution" and therefore are ready to consider at least some of the changes it has made. This may not be the best motivation, but it is one means of obtaining change.

Sometimes a pressure group provokes a study of content or methods. This was true in the 1930s, for example, when various Jewish groups demanded a fair deal for Jews in textbooks. A similar movement, to tell the story of American Negroes, has been underway now for several years and has brought many changes in textbooks and some shifts in courses of study in many schools and school systems.

Who Are the Agents of Change?   There are many change agents affecting improvements in teaching methods and in courses of study. Most of them have been indicated in the foregoing paragraphs.

Sometimes administrative officers play that role. So do supervisors.

Pressure groups may serve as the change agents.

Consultants and college professors are often responsible for changes.

Textbook publishers often have a strong influence in bringing about change. Since so many teachers rely heavily on textbooks, a shift in them is likely to force a change in thousands of classrooms.

In recent years organizations like the Joint Council for Economic Education, the Asia Society, and other groups, have been instrumental in bringing about alterations in social studies programs in our schools.

Some of the projects in social studies research sponsored by the U.S. Office of Education have also been innovative. Their recommendations have been followed by many schools and school systems.

Occasionally college requirements and pressures seep down to the junior high school level and even affect work in elementary schools. This is seldom desirable; nevertheless it happens.

Many changes, however, take place because of the initiative of classroom teachers. A teacher reads a book and gets a new idea which he puts into practice in his classroom. Or he attends a meeting and sees a demonstration lesson of a new method which he then tries out himself. In the final analysis, most changes that are lasting must be made by classroom teachers.

Changes at the Classroom Level.   Most of the effective changes in social studies teaching come through teachers. Such changes may be accomplished in many ways. Among them are the following:

1. *Through reading.* The reading of an article, a pamphlet, or a book may provide a teacher with new background on content or methods or new insights into children and learning.
2. *Through new resource materials.* Providing a teacher with new resources for learning or encouraging a teacher to search for such materials may bring about needed change.
3. *Through new equipment.* Providing a teacher with an overhead or opaque projector, a teaching machine, or other new equipment, may stimulate minor changes in teaching. But be sure they get help on using the equipment.
4. *Through visits and demonstrations.* Teachers need to see new ideas and methods in actual situations. Visiting other teachers within a school or persons in other schools is often extremely helpful in accelerating change.
5. *Through films and recordings.* A good film of teaching, a recording, or a tape, can be discussed by teachers as still another method for initiating changes.
6. *Through a collection of promising practices.* Teachers need to be encouraged to collect examples of promising practices in social studies teaching and to share them with their colleagues. On a statewide and a nationwide basis this can also be useful in stimulating changes.
7. *Through studies of children and learning.* Various ways of studying children and the learning process can involve teachers in examining their present practices and often lead to improvements in teaching the social studies.
8. *Through the study of problem situations.* Groups of teachers in conference or workshops can analyze problem situations which are plaguing them and may often evolve improved ways of dealing with them.
9. *Through courses.* Sometimes teachers need to be encouraged to take courses in nearby colleges or in-service courses in the school itself in order to expose them to new ideas, practices, and programs or to give them needed background in content.
10. *Through conferences.* Often teachers are stimulated by attending conferences where they can hear experts, talk with colleagues, examine new materials and gain new background and insights.
11. *Through examination of courses of study.* Examination of new courses of study in other schools often produces good results in improving classroom teaching.
12. *Through the preparation of new units or lessons.* The work involved in producing new materials for classroom use can often serve as a means of curriculum improvement.
13. *Through travel.* Travel to new places, even in one's own locality as well as to distant places, often improves the background of teaching and alters methods in the classroom.

Some Current Conditions Which Are Generating Radical Changes in Social Studies Curricula

In the opening paragraphs of this chapter the author indicated that many people feel that today's social studies are not meeting the needs of boys and girls in a rapidly changing world. Many of us feel that the majority of courses of study are archaic, obsolete, and even detrimental to children.

Despite the many forms of resistance to change, some progress has been made in recent years. However, most of the changes have been only at the classroom level and in the classes of a small minority of teachers.

These times call for radical changes in the social studies. Resistance needs to be recognized and minimized and opportunities for change maximized. Slight changes in a few classrooms will not suffice. What is needed are far-reaching, extended studies of social studies courses in the light of the changing conditions, with radical revisions as the end results of such studies.

Fortunately such revisions seem possible today. The times seem ripe for change. Studies, revisions, and experiments are underway in many places. Teachers in other places are ready for some changes. Teachers and administrators in many more places need to be encouraged to explore the need for changes.

At no time since the Great Depression of the 1920s and 1930s has there been such an opportunity to overhaul our creaking curriculum or to build new models.

In some respects the late 60s and the early 70s resemble the United States in the late 20s and 30s. In a variety of ways people today are expressing their disillusionment with many aspects of our society. Minority groups are pressing for equal rights and equal opportunities. College students and high school pupils are calling for more relevance in their education. Large numbers of people are disturbed by the decay of our inner cities and by increases in crime. It is in such periods of crises that changes are most likely to be made. This applies to the schools as well as to many other institutions in our society.

What are some of the factors which make this writer so optimistic about the possibilities for change? There are many indications; nine will be singled out for discussion or mentioned here.

1. The Aftermath of Sputnik.   In many ways the launching by the Russians of Sputnik in 1957 was the beginning of a new era in American education. For a time we were paralyzed by this feat of a people whom many Americans had considered stupid. There were people so impressed by this victory of the Russians in space that they wanted us to revise our educational system and pattern it after that of the Russians.

But sanity eventually prevailed and we decided that we need not or dare not become "Russians" to match or overtake them in certain areas of life. But many of our schools did strengthen their curricula in science. From that we moved on to the field of mathematics and the area of foreign languages.

Today many people are asking if it is not time that we took as hard a look at the social studies. Science is not likely to save us from a global catastrophe; the social sciences and the humanities are more likely to help us to live in close proximity with other human beings in peace. So the times are ripe for a new look at the social studies up and down the line, from kindergarten through high school or beyond.

2. Our "Discovery" of the World.   In the past, most Americans were educated to live only in the Western World. They studied European history, European languages, European art and music, and European literature. Such an education was probably sufficient in the days when most of today's adults went to school.

But this type of Europe-centered education hardly suffices today. Whether we like it or not, we must learn to live with upwards of four billion neighbors in all parts of our planet. All Americans need to learn that men everywhere have the same or similar needs and that they meet them in similar and in different ways. We need to incorporate much of the methodology of anthropology in our teaching. We need to update our geography work. We need to introduce many aspects of international relations in simple form. Our windows need to be widened to include the whole world, not just the western parts of it.

Almost all of the subject fields in elementary schools need to be utilized to add this international dimension to education. But it is the social studies field that should bear most of the responsibility in this regard. That means that the old concentric circles concept of the curriculum must be abandoned and a new plan evolved to introduce children very early in their school years to selected aspects of the world. Some suggestions along this line will be made later in this volume.

Fortunately there are many Americans who see this as a new and essential part of education. They have discovered the world and realize that our children must do this, too. This is a second factor in the current scene which is generating change in the social studies field.

3. *A Changing U.S.A.*   Profound changes are also underway in our own nation. By the year 2000 we will have somewhere between 300 and 400 million persons. That vast increase in population will alter our nation radically.

Most of those people will be living in cities and in the larger metropolitan areas. A large percentage will be living in five great metropolises or megalopolises — one running from Virginia to Maine, another from Pittsburgh to Milwaukee, a third along the West Coast, a fourth along the Gulf of Mexico, and a fifth in Florida. Many people will probably be living in satellite cities, too, like Reston, Virginia and Columbia, Maryland.

Our population will be even more mobile than it is today. Already, within a given year, one family in five moves. In the future the percentage may well be much greater. How will this affect the study of the local community? How will it affect the study of state history? These are questions which we need to face.

With the growth of industrialization, regional differences will undoubtedly be less sharp. An approach to the study of the United States by regions would be less effective than it is today.

Changes in the status of our minority groups, long overdue, will certainly come in the foreseeable future. Minorities will play a much more significant role in our national life than they do at present. More attention, then, needs to be paid to minorities in our social studies programs. The children of each minority group need to find themselves in stories of our nation's history and be able to identify with the people of their group who have helped to make our nation. All children need to learn about the advantages of our pluralistic society.

With the increase of automation, leisure will be much greater for most of us than in the past. The thirty-five-hour week and the thirty-hour week are likely. Some kind of guaranteed minimum income will probably be granted every family. Perhaps the question of leisure needs to be upgraded in our elementary school social studies programs.

Old problems will still be with us and new problems will certainly emerge. Our pupils need experience in problem-solving, the handling of controversial issues, and critical thinking if they are to become effective citizens of a democratic society. Perhaps the study of water, air pollution, transportation, and crime need to receive attention in our elementary schools or at least in our junior high schools rather than being left for a Problems of Democracy course in the last year of high school.

4. *New Ideas from All the Social Sciences.*   In the past there was a heavy concentration upon geography in our elementary schools, with an accent upon physical geography, highlighting places rather than people. In the fifth grade most schools introduced history. In the junior high schools, civics was emphasized, with a concentration upon the structure of government.

Now we are beginning to realize that all the social sciences have a contribution to make to any social studies program. And we have begun to understand that aspects of all the social sciences can be utilized at every grade level and in every type of study if properly taught.

The work of the Joint Council on Economic Education and that of Lawrence Senesh have been pioneering efforts in that area in elementary schools. Recent research has indicated that children learn their attitudes toward authority and rules or laws very early, and some experimentation has been conducted with introducing aspects of political science even in the primary grades. Certainly we need to draw upon the rich resources of anthropology in our studies of people and communities. Our geography needs to be updated, with emphasis upon human geography. Of all the social science disciplines, only history seems difficult to use extensively with primary grade children.

5. *New Data on How People Learn.*   As we have pointed out on previous pages, there has been much research in recent years on how people learn. We are certainly clear that people learn best when they are motivated, when they are involved, when new learnings are related to previous experiences, when experiences are raised to the level of the learner's consciousness, and when pupils are stimulated emotionally as well as intellectually. One might add many more factors to this short list.

Such findings call for new methods of learning in the social studies in which the teacher becomes more the dramatics coach than the star actor or actress, the guide rather than the demonstrator, the conductor of a symphony orchestra rather than the soloist. It is tragic that we have not yet translated the findings of research on learning into action in our social studies classrooms.

It is the contention of this writer that if teachers really translated the rules of learning which are listed on pages 39–43 into classroom practice, there would be a revolution in most social studies teaching in elementary and junior high schools throughout our land.

6. *"The Knowledge Explosion."*   The senior citizens in our society were born in the horse-and-buggy era and are now living out their later years in the airplane-jet-space-nuclear period. In that relatively short space of time man's knowledge has increased so greatly that we sometimes refer to it as "The Knowledge Explosion." The funds devoted to research have risen astronomically. Vast new areas of knowledge have been explored and many of them conquered. Thousands of new jobs have been created. Some authorities assert that our knowledge doubles every twenty years. They maintain that it will double in even less time in the immediate future.

With so much knowledge, the question of what to teach becomes involved and sometimes confusing. Information learned today is often out-of-date tomorrow. Theories held now are likely to be exploded in the near future. What, then, can we do in the social studies field?

Many social studies experts feel it is time we concentrated more upon teaching boys and girls how to think than we have in the past and upon the discovery by pupils of concepts and generalizations. Knowledge would not be downgraded, but it would be developed around "big ideas" and acquired more by problem-solving than heretofore. Along with this we certainly need to select fewer topics for study and do them in greater depth than we have previously done.

In these and other ways the Knowledge Explosion of our times is helping to bring about changes in the social studies.

7. *The Sums of Money Devoted to Educational Research and Experimentation.* Even more closely related to our task as social studies teachers and
specialists is the expansion in research and experimentation in our field. Some of
this is being undertaken by scholars in the social sciences. Some of it is being encouraged by foundations. Much of it is now being financed by the federal government through projects of the U.S. Office of Education. Many of those projects
are being conducted by teams of specialists in our colleges and universities.
Often they are working in conjunction with schools or school systems. Some of this
is at the elementary school level.

For example, a project has been carried on in developing a sequential curriculum
in anthropology for grades one to seven by specialists at the University of Georgia.
Another project with some funds from the federal government has been undertaken
by the schools in the Greater Cleveland area to frame a new course of study in the
social sciences from kindergarten through grade twelve. At Northwestern University, a program for the study of American Society has been underway for some time,
including grades five through twelve. At the University of California at Berkeley,
teachers guides and materials for the study of Asia from grades one through twelve
have been prepared. A model for the schools of the metropolitan area around St.
Louis has been the concern of specialists at Washington University. This has been
a K–12 program. Another program for social studies, K–14, has been carried on at
the University of Minnesota.

These and several similar projects have involved large numbers of teachers in
many school systems. They have not all been successful in their endeavors, but
they have helped to create ferment in the social studies in many schools in different
parts of the United States.

8. *The Increasing Numbers of Children in Kindergartens.* Most of our social
studies programs for elementary schools today were framed when very few children
attended kindergartens and fewer still, nursery schools. With the increasing numbers of children who attend these schools and therefore have experiences which
were once limited to the first grade, the question arises as to whether the social
studies programs of our schools ought not to be adjusted to this change.

For example, it has been generally accepted that children in the first grade
should study the home, the school, and the neighborhood in which they live. Many
experts feel that so much is now done on these themes before the first grade that
new emphases are needed in the first grade or at least in the second grade.

So this is another factor pointing toward curriculum change.

9. *Our Inability to Provide Meaningful Programs for Many Pupils.* Finally,
there is the feeling on the part of many teachers, parents and community leaders,
curriculum experts, and social studies specialists that we are not really providing
meaningful experiences for many of the pupils in our elementary schools.

We are not challenging the gifted. Many of them are bored by the limited content they are expected to learn and frustrated by the methods employed in the
social studies. Their abilities are not used, their skills not developed, their creativity not released.

Nor are we providing for many other pupils adequate, exciting programs in the
social studies. The same content and the same methods are all too often being used
with less able pupils as with more able ones. Materials suitable to reluctant readers are not being produced and substitute experiences not fully utilized.

For children in our inner cities, the social studies are often most unsuitable.

Pupils are exposed to Indian tepees and Dutch windmills — experiences which have no value in helping them to cope with their local environment. The policeman is extolled as a great community helper when they know better. And the visiting nurse and the relief worker are ignored in the study of community helpers. We are not helping children with the purchase or preparation of food for their younger brothers and sisters or assisting them in developing a feeling of pride in their own backgrounds.

The Initiation of More Radical Reforms

Like the 1930s, the 1970s seem a good time for radical reforms in elementary schools. In many places there is now a climate of opinion which is conducive to change. For example, there is a revival of interest in the methods of Montessori, and many Montessori schools are being established. Many schools are experimenting with ungraded or non-graded formats. And there is a wave of interest in the methods of the British primary schools, with their emphasis upon enriched environments for learning, more individualized instruction, centers of learning, and open classrooms. Many of these innovations, of course, resemble progressive education in the 30s.

Furthermore, there is movement now in the social studies field. Many school systems have begun to revise their offerings and examine their methods. Sometimes a single school has embarked on a program of curriculum change. Occasionally a large school system or a group of schools have undertaken curriculum revision. A few states have reexamined and revised their courses of study. Change is increasingly a characteristic of social studies programs in the U.S.A. today.

How is this usually done? In large systems it is often carried on by a group of specialists, working with selected teachers in the system, and with experts brought in from outside the school system. In state departments of education the process is similar. The members of the department are responsible for the final product, but they usually consult with scores of teachers and administrators before issuing their findings.

Here is a general pattern for curriculum change in the social studies, based upon the writer's experiences in several communities. No community has ever followed this exact pattern, but many have approximated it.

Over a period of months, teachers are alerted to the need for changes. This can be done in several ways. They include bringing in an outside expert to review current proposals, reading by individuals and groups of such volumes as Edwin Fenton's *The New Social Studies* (Holt, Rinehart and Winston) and the National Council for the Social Studies' *Social Studies in Transition: Guidelines for Change*, and attendance at workshops or conventions where changes are being highlighted. Usually several approaches are taken.

As a culmination of such a program, a systemwide meeting on the need for changes locally is arranged, featuring an outside expert and/or a panel of respected local leaders in the social studies. At the close of such a meeting, the teachers are invited to submit anonymously their comments on what they like in the current program of the school system, and what changes they would like to see instituted.

If there is enough support for systemwide changes, a steering committee is then appointed. On the committee it is wise to have a top-ranking administrator who has good rapport with the teachers, representatives of all the grades, and audio-visual and library personnel. After considering the comments of their colleagues, innovative plans in other school systems, and general trends in the social studies field, this steering committee presents a skeleton to meet local needs.

When the general proposal is completed, it should then be submitted to all persons and groups concerned with it, and their criticisms of it welcomed. This is probably done best in several small interest groups, followed often by a schoolwide meeting.

After such a survey of opinion, a ten-day or two-week workshop can then develop a general curriculum bulletin of forty to fifty pages in length in a workshop of fifteen to twenty persons, including again librarians and audio-visual personnel.

Following a year of experimentation in selected groups, more detailed syllabi or courses of study can be developed in another workshop of three or four weeks' duration. Or such syllabi can be developed over a period of time, with changes instituted in only a few selected grades.

Such grade guides should be full enough to give beginning teachers and insecure persons considerable help. But they should not try to spell out day-to-day programs. They certainly should include annotated bibliographies of books, films, and other resources which teachers can use.

After two or three years of work on a new curriculum, teachers should again be asked to react to the new courses of study. A small steering committee could then evaluate the changes which have been made and suggest modifications to strengthen the new program — or they should recommend its abandonment.

Meanwhile, it is important to maintain close contact with the Board of Education, in order to obtain ideas from its members and support for changes. Such support would include financial arrangements for the use of a consultant or consultants, remuneration for the members of the workshops, and the purchase of materials to implement the program, distributed over a period of years but with large expenditures in the first year.

Presentations of the new developments should also be arranged for Parent-Teacher meetings in order to win parent support. The newspapers need to be kept informed of developments, too.

Finally, in most systems some in-service courses in various social science disciplines, newer methods, and the learning process need to be offered locally for credit and/or for salary increments.

The Different Directions in Which We Are Now Moving

On the pages that follow are the major topics of four new types of courses of study. Perhaps you will want to collect and examine others.

The first is the program of Educational Services, Incorporated. That is a group of scholars based in Cambridge, Massachusetts. Some of them worked on the new programs in science or in math and are now working in the social studies. They are anxious to provide depth of experience by selecting a few topics for boys and girls to study. It is their feeling that children need to recapitulate the history of the human race, hence the topics which they have proposed. However, early experiences with the elementary school program were not altogether successful and this group is now concentrating upon the junior high school sequence — "From Subject to Citizen." One of the best aspects of their program is the use of the discovery method. Perhaps you will see by examination of their outline why they had considerable difficulty in developing the program for the early grades.

The second is an attempt to develop a program around three major ideas — the variety of human societies, cultural change and diffusion, and complexity. The primary grade sequence bears some resemblance to the E. S. I. program. The middle grade program is very strong in history. The seventh grade uses the local metropolitan community as a laboratory for studying contemporary problems.

The third program, from Contra Costa County, California, also has a strong anthropological base in the third grade and again in the fifth grade. Its seventh grade program is largely historical.

The fourth program is one developed by the author of this book. The rationale for it is written in brief form on page 73. In brief, it is an attempt to develop a program which shuttles between the local community and the United States and the world, thus introducing children much earlier than in the past to the peoples of our planet. It concentrates upon one segment of society for two grades, with an in-depth study of that segment in our own nation and then in other nations. It draws upon all the social science disciplines, with particular attention in the primary grades to anthropology—sociology. This program was developed in Ridgewood and Fair Lawn, New Jersey; and in Great Neck, Manhasset, and Locust Valley, New York. Then a series of textbooks was developed over a period of years, culminating in their publication by Ginn and Company early in 1972 as the Ginn social science series.

EXAMPLES OF NEW COURSES OF STUDY

A Program in the Social Studies from the Education Development Center, Inc.

Several years ago Educational Services, Inc. started to develop an experimental program in the social studies for elementary and junior high schools. Their approach was strongly influenced by the thinking of Jerome S. Bruner. The original course of study for the various grades was as follows:

Grade One	The Netsilik Eskimos of Pelly Bay, Canada (a nomadic tribe)
Grade Two	The African Bushman and the Australian Aborigines
Grade Three	The Primates and Archaeological Recapitulations of Emerging Humanity
Grade Four	Husbandry and Its Beginnings: Tehuacan Valley in Mexico
Grade Five	Urban Life and Its Beginnings: Ancient Mesopotamia Sumerian city of Nippur in Iraq
Grade Six	The Source of the "Western" Tradition: The Aegean area during the Bronze Age

*The junior high
school sequence*   *Subject to citizen*

Grade Seven	Urban Community Case Studies: Classical, medieval, and Renaissance periods: Athens in the fifth century B.C. Republican Rome Baghdad in the ninth century A.D. Paris or London
Grade Eight	Subject to Citizen: Seventeenth century England and eighteenth century America with the use of source materials

Grade Nine   Large Centers of Political Organization:
              (primarily urban), such as the nineteenth century industrial city
              in the Western World or a megalopolitan complex of the twentieth
              century.

This program was a most radical departure from existing courses of study. However, it was abandoned after considerable experimentation. You may want to hypothesize as to why it was abandoned. Some of the material prepared for this original plan has been utilized in a program on *Man: A Course of Study*, which is being used in many schools today in various grade levels, usually grades 5–8. Materials from that course of study are sold by Curriculum Development Associates, Inc. Suite 414, 1211 Connecticut Avenue, N.W., Washington, D.C. Much use is made of films in this program, with special emphasis upon the behavior of animals.

Other units developed by this organization are intended for use at various grade levels. One for the junior high school years is entitled *From Subject to Citizen* and includes units on Queen Elizabeth: Conflict and Compromise; The King Versus the Commons; The Emergence of the American; The Making of the American Revolution; and We the People. The focus is on the uses of power in human relationships. Material on this course of study is sold by the Denoyer–Geppert Company, 5235 Ravenswood Avenue, Chicago, Illinois 60640.

Other units include Black in White America: Historical Perspectives; Black in White America: The Struggle for Identity and Power; Racial Conflict; and Armed Intervention: Under What Circumstances?

Further information on these programs may be obtained by writing the Education Development Center, 15 Mifflin Place, Cambridge, Massachusetts 02138.

Course of Study in the Social Studies at New Trier Township, Illinois

For further information, write the Superintendent of Schools, New Trier Township, Illinois.

Kindergarten to grade three	*The varieties of human societies*
Kindergarten	Exploration of the Child's Own Society
Grade One	Food-Gathering Societies: Societies as different as possible
Grade Two	Food-Producing Societies: Not influenced by urbanization and industrialization
Grade Three	Metropolitan Society: Stressing Chicago

Grades four to six	*Cultured change and diffusion*
Grade Four	Movement of People: Technology: Indo-European migrations, Indo-Aryans in India, Achaeans in Greece, European Explorations and Discoveries, Competition in the Western Hemisphere, Immigration in the New World, and Westward Expansion

Grade Five   Political and Social Institution
              Warrior Cultures in Eurasia, Origins of the Caste System, Myce-
              nae — an early city, Colonial Organization of the French, Span-
              ish and French Colonies, Development of the Nation-State, Devel-
              opment of Political Parties, and Urban Democracy and Frontier
              Democracy
Grade Six   Myths, Religions, Beliefs, Ethics:
              Myths of the Farmers and Hunters, Indian and Greek Gods and
              Heroes, The Renaissance Ideal, The Protestant Reformation,
              American Folklore and Legend, The Image of America

Grades seven and eight	*Complexity*

Grade Seven   History of Chicago and the United States: Contemporary Problems:
              Chicago as a mirror of themes in U.S. history such as (1) Move-
              ments of People, (2) Urbanism-Industrialism, (3) Cultural
              Fusion, and (4) Growth of Democracy.
Grade Eight   Problems and Potentialities of Urban U.S.A.:
              An introductory unit on Constitutional Problems of Our National
              History and four or five Significant Problems of Contemporary
              U.S.A.

Course of Study in the Social Studies in Contra Costa County, California

Grade One   Our School
              Family Living
Grade Two   Life in American Communities
Grade Three   Comparative Communities of the World:
              Primitives of Africa
              People of the Hot, Dry Lands
              The Boat People of Hong Kong
              The People of Switzerland
Grade Four   California: Yesterday and Today
Grade Five   The People of America
              Anglo-America
Grade Six   The Cultural Heritage of the U.S.A.
Grade Seven   Man's Achievements:
              The more man learns the greater his achievements; some stages
                  of his learning have been more significant than others; pre-ice
                  age through early civilizations.
              Man spreads his cultural achievements through migrations,
                  trade, and invasions. Each culture rejects, adopts, or modifies
                  these transmitted ideas: Egypt through Rome.
              Ideas develop faster in a favorable environment and more slowly
                  in an unfavorable one: medieval, Renaissance, Industrial Revo-
                  lution, Age of Discovery.
              The ways man learns to use his resources influence his standard
                  of living.
              With specialization men become interdependent.
              A nation's economic growth is influenced by the value systems
                  within the nation.

Grade Eight   United States History:

Diverse elements in a society limit the areas of social agreement.

Transplanted institutions change in new surroundings.

The idea of representative government has moved continuously toward broader representation.

The establishment of a federal system of government reflects a fear of highly centralized government.

Sectional specialization contributes to divergent economic interests.

Advance in technology causes a shift in the role of national institutions.

Expanding interests lead to increasing interaction with other nations.

PROPOSAL FOR A TWIN SPIRALS SOCIAL STUDIES CURRICULUM K–12

Grade	Basic theme	Application locally and in the U.S.A.	Application to selected parts of the world
K–2	*Individuals (or Children) and Families*		
K&1	Individuals and Families Locally and in the U.S.A.	X	
2	Individuals and Families in Other Parts of the World		X
3–4	*Communities*		
3	Local Community and Selected Communities in the U.S.A.	X	
4	Selected Communities in Other Parts of the World		X
5–6	*Countries*		
5	The United States Today and Yesterday ("Postholing" certain periods)	X	
6	Selected Countries in Other Parts of the World A few nations studied in depth		X
7–8	*Basic Problems and Decisions in the U.S.A.*		
7	In the U.S.A. Today	X	
8	In the U.S.A. Yesterday	X	
9–10	*Cultures*		
9	Studies in depth of the 8 major cultural		X
10	areas of the world: today and yesterday, Western and non-Western		X
11–12	*The United States and the Emerging International Community*		
11	United States History	X	
12	Contemporary Problems: The U.S.A. and Other Parts of the World	X and	X

The proposal outlined on page 73 is an attempt by the author of this book to develop a course of study in the social studies from kindergarten through grade twelve which combines as many of the strengths as possible in our emerging thinking about this broad field. Here are some of the factors it tries to emphasize:

1. It rejects the old concentric-circles theory of curriculum and replaces it with a twin spirals approach which emphasizes the United States and the world in alternate years.
2. It encourages teachers and pupils to examine some segment of society in the United States first and then to apply the skills learned in such studies to similar segments of society in other parts of the world.
3. Therefore it introduces children early to the people of the world rather than postponing this until the sixth grade as in most current programs.
4. It introduces boys and girls to contemporary problems earlier than existing courses of study, especially in the seventh grade program.
5. This suggested curriculum strives for depth by limiting the number of families, communities, nations, cultures, or problems studied in any one year.
6. It encourages further depth by two-year sequences at several points, as in the study of families, communities, nations, decisions, and cultures.
7. It fosters in-depth studies of the United States at three different levels by using three different approaches: (a) post-holing in grade five, (b) 25 major decisions in our history, at grade eight, and (c) the study of our history in a world setting, at grades eleven and twelve.
8. Although not stated on the chart, the representative families, communities, and nations are selected from the eight major cultural areas of the world: (a) Anglo-Saxon, (b) Latin, (c) Germanic–Scandinavian, (d) Slavic, (e) Moslem, (f) Indic, (g) Sinitic (Chinese), and (h) African: South of the Sahara.
9. It emphasizes the industrialization and urbanization of the United States and other parts of the world through the study of several cities and metropolitan areas, in grades three and four and in grade seven.
10. The theme of interdependence is stressed at various points. For example, in grade four a small community and a nearby large city are studied, except in the section on London.
11. Major concepts and generalizations are drawn for every grade level from the various social sciences, thus insuring an interdisciplinary approach, rather than the multidisciplinary approach championed by some curriculum makers today.
12. Many different methods are suggested but the major emphasis is upon inquiry, discovery, or problem-solving.
13. Skills are developed in a sequential order, starting with the early years and continuing through high school.
14. Attitudes and values are stressed in each level of the sequence proposed here.

How do you react to this new proposal for a K–12 social science program? In your opinion, what are its strengths? What are its weaknesses?

13. Encouraging creative
    self-expression

14. Dramatizing situations

15. Utilizing television

Some Effective Methods We Can Use

16. Fostering individual and group work

17. Having pupils prepare pictorial time-lines

18. Planning panels

19. Taking trips

What Are Some of the Strategies, Methods, and Activities for Teaching the Social Studies? 6

The varieties of methods or activities upon which a social studies teacher can draw to promote learning in the social studies is tremendous. A list of more than eighty such methods or activities follows in this chapter.

There are several ways in which such a list might be organized. For example, methods or activities might be grouped as those which foster impression and those which stimulate expression. They could be placed in such categories as audio-visual methods, dramatics, oral activities, written activities, and the like. They might be listed in the order of importance, if one could determine their relative importance.

Because the list defies any easy categorization, methods and activities are presented in alphabetical order.

An opportunity is provided for you to utilize this list as a learning experience. You can review the methods and activities you are now using or have seen used. Then you can check those methods or activities which you would like to see used or would like to try yourself.

Activity or method	Am using this now	Should try or use more
1. Bulletin boards		
2. Buzz groups		
3. Cartoons		
4. Chalkboard		
5. Charts and graphs		
6. Choral speaking		
7. Collecting		
8. Committee or group work		
9. Cooking food of places studied		
10. Crossword puzzles		
11. Current events		

Activity or method	Am using this now	Should try or use more
12. Dances of places or periods studied		
13. Debates (for older pupils only)		
14. Diaries		
15. Dioramas		
16. Discussions		
17. Dramatics		
18. Exhibits		
19. Films		
20. Filmstrips		
21. Flags		
22. Flannelboards		
23. Flashcards		
24. Flow charts (especially of economic processes)		
25. Games		
26. Halls of Fame		
27. Interviews (sometimes tape recorded)		
28. Jigsaw puzzle maps		
29. Listening activities of many kinds		
30. Maps of many kinds, and globe		
31. Mobiles		
32. Mock ups (enlargements to more than life size)		
33. Mock panels, broadcasts, and telecasts		
34. Modeling in various media		
35. Making models		
36. Montages and/or collages		
37. Murals		
38. Music		
39. Newspapers and magazines		
40. Notebooks		
41. Open textbook study		
42. Panels and roundtables		
43. Pen pals		
44. People as resources		
45. Photographs taken by pupils		
46. Pictures		

Activity or method	Am using this now	Should try or use more
47. Plays		
48. Poetry		
49. Posters		
50. Problem solving or inquiry		
51. Puppets		
52. Questioning of many kinds		
53. Radio programs		
54. Reading of many kinds		
55. Recordings		
56. Reporting		
57. Reproductions or facsimiles		
58. Role playing or sociodrama		
59. Sandtables		
60. School affiliations		
61. Scrapbooks (individual and/or class)		
62. Service projects (like Trick or Treat)		
63. Sewing		
64. Slides		
65. Sociometric devices for grouping		
66. Source materials		
67. Stamps, coins, and other hobbies		
68. Story-telling		
69. Surveys		
70. Talks by teachers, pupils, visitors		
71. Tape recordings		
72. Television		
73. Tests of various kinds		
74. Textbooks		
75. Time-lines		
76. Transparencies		
77. Trips		
78. Word association device		
79. Workbooks or response books		
80. Writing for materials		
81. Written work of various kinds		

CRITERIA FOR SELECTING METHODS OR ACTIVITIES

With such a wide variety of methods or activities to select from, how should a teacher decide which ones to use? Here are four criteria to help you:

1. *Depending Upon Your Aims.*   The main basis of selection should be determined by the aim which you wish to achieve with your pupils.

If you are trying to develop empathy with other people in a family, role-playing or sociodrama may be the best method. A speaker from another country may be used to help your pupils understand family life in his or her part of the world. A good film may be your best method for the same purpose. Or a good story, read aloud by you, may help you accomplish your purpose.

If you are attempting to develop the concept of the relationships between a large city and its hinterland, other methods are called for. Perhaps your children should establish a small farm and then role-play the thoughts of the farmer as he decides what to do with his surplus crop and where he will buy fertilizer, seed, or machinery. Your pupils will then need to enlarge their map to include the nearest large city. Perhaps a film or a filmstrip will help at this point to see what they have already "discovered" themselves.

If you are working with older boys and girls on a world problem, still other methods may come into play. Newspapers and magazines can be used. Pupils can write away for materials, use encyclopedias, and do simple research in textbooks and trade books. Summaries can be made and note-taking used. Class discussion of the findings may then ensue.

2. *Depending Upon Needs and Interests of the Pupils.*   Of course individuals and groups vary in their needs. Some pupils need certain experiences; others need other activities. You as a teacher need to know these needs and plan activities accordingly.

In the primary grades you will need to rely largely upon non-reading activities. Trips should be high on the list. Simple dramatizations should prove useful very often in providing children with opportunities to act out what they are learning. Games and dances can help. Reading aloud should be a major social studies activity. Experience charts constructed by the class and the teacher should be used often.

In a class of so-called slower pupils in an intermediate or upper grade, you may want to have a frequent change of methods, with a good many in which the pupils are active. You may encourage them to look at pictures and to read the captions on filmstrips aloud and talk about them. You may want to do a great deal of problem-solving and considerable role-playing. Certainly you will want to do a great deal of open-textbook teaching, helping them to learn the skills of reading.

The interests of children will also determine, in part, the activities or methods used. A boy who is interested in stamp collecting should be encouraged to study the stamps of the United States or some other country as a way of developing his interest in United States history.

A girl who likes clothes may be encouraged to find out about the clothes of a particular part of the world or the people of some period in history and to report to the class her findings, perhaps with illustrations she has made.

As teachers we sometimes forget that everyone does not enjoy reading. We do and we assume that everyone else does, too. Some pupils learn best through listening — and there need to be many opportunities to listen. Some learn best through construction — and there need to be many chances to utilize this mode of learning. Some learn best through acting — and this needs to be given free rein, often. Others are shy and yet they can express themselves through puppets, in choral speaking, or in small committees.

In innumerable ways teachers can find "handles" by which they can develop interest in various aspects of social studies.

3. *Depending Upon the "Style" of a Teacher.*   Most teachers develop a personal style of teaching. Some do one type of teaching well; others become adept in other methods. That is as it should be.

For example, one teacher may be able to read aloud extremely well. She should do so — often. Another teacher may be able to utilize role-playing to great advantage. He should feel free to use this method frequently. A third teacher may be gifted in making simple sketches on the chalkboard to illustrate points. Fine. That person should utilize this special skill.

Teachers, however, should also try new methods and activities to extend their own range of skills. To do so, they may want to watch their colleagues. They may gain ideas from books, pamphlets, and articles. They may see a film or filmstrip or hear a tape recording which will help them to try out a new idea. They may even learn some new techniques from a college instructor who is able to practice what he preaches about a variety of methods and uses many methods in his own college classes.

4. *Depending Upon the Factor of Novelty.*   Pupils need to feel secure in what they do in the social studies. They need to understand very well a few methods which they can practice with competence and ease. This is especially true of "slow learners."

But greater learning probably will take place if there is also some novelty from time to time in the social studies. Pupils will be intrigued by the introduction of a new method or activity and their appetite for learning will be whetted.

If several units of study have been introduced by a film or filmstrip, why not introduce the next unit with a trip or an exhibit?

If you have been using one textbook as the basis of your work, why not use three or four and let the pupils compare what the different books say or stress?

If you have been working in committees for several weeks, why not vary the approach by having the class work most of the time as a committee-of-the-whole, with individual reports and an occasional panel?

Culminating activities can be varied, too. If you have been preparing a skit, a mural or an exhibit, why not arrange a "Meet the Press" program of experts, prepare a travel booklet on a community or country, or prepare a crossword puzzle summarizing the unit?

In providing variety, you are likely to increase the interest of the class, reaching pupils you have not really reached before. You are likely to teach new skills in this way, too. And you are likely to inject fresh enthusiasm into your social studies teaching. Yes, there is merit in a variety of methods just for the sake of novelty, as well as for other reasons.

SOME EXAMPLES OF A VARIETY OF METHODS AND ACTIVITIES IN ELEMENTARY AND MIDDLE SCHOOL SITUATIONS

Hundreds of examples could be assembled of ways in which social studies teachers utilize a variety of methods which are appropriate to the topic and to the needs of pupils. Here are just a few random examples:

1. *Walking Totem Poles.*   A third grade class studying the Indians of the northwestern part of what is now the United States, drew, on oaktag, large totem poles, similar to those they had found in various books on Indians. They hung these "sandwich ads" around their necks with heavy twine and walked into a nearby classroom studying the same topic. There they invited the pupils to ask questions about their totem poles. The result? The children who were asked questions had to return to their class to do further research on the questions they could not answer. Nevertheless they were proud of their accomplishments and probably will never forget what they learned.

2. *What Daddies Do All Day.*   A first grade class became very excited about Ruth Shaw Radlauer's *Fathers At Work.* Over a period of four days each pupil in the class told what his or her father, or some other adult male in his or her life, did to earn a living. Each pupil tried to bring in some illustration of what the adult did. When this was impossible, a drawing was made to serve as an illustration. Then three fathers were invited to speak to the class about their jobs.

3. *Collecting Pictures of U.S. Cities.*   A third grade class studying cities in the United States decided that they needed more pictures of the places they were study-ing. So they launched a door-to-door campaign in their community for old magazines, from which they cut out pictures of American communities. What they learned about the U.S.A. was tremendous, but the incidental learning in going through the magazines was equally profitable.

4. *Studying a Village in India.*   A fourth grade class studying villages in India borrowed a sandtable for two weeks from the kindergarten and in it reproduced an Indian village, using the maps in the books of Sonia and Tim Gidal and G. Warren Schloat, Jr. to aid them. The sandtable was then exhibited to other classes, with appropriate explanations of their layout.

5. *Carbon 14.*   A bright boy in a sixth grade class, interested in science and well-informed for a boy of that age, reported to the class on his research on Carbon 14. He told the class how it was used in determining the age of artifacts found in East Africa by the Leakeys, the English archaeologists and anthropologists who discovered the earliest known men in the world, in Tanzania and in Kenya.

6. *Local History in an Eighth Grade Class.*   A committee of pupils in an eighth grade prevailed upon the editor of a local paper to photostat the front pages of the local paper for eight major events in that community which had occurred in a period of about 100 years. These reproductions were placed on cardboard for safekeeping after they were used as realistic source materials in the study of local history.

7. *Tourist Booths on Cities Around the World.*   A fourth grade class was asked to present a program to the P.T.A. They decided to present to the parents and teachers what they had learned about eight communities in various parts of the world by having eight booths with travel agents in each of them to answer questions. Posters and other material were prepared for each booth. Every pupil in the class was involved in this summary (and evaluation) activity.

8. *An Actor's Box.*   An old trunk was brought into a first grade class by the teacher at the beginning of the year and the idea of "An Actor's Box" was explained to the mothers' group. With their help, many "props" were obtained, such as hats, shawls and pieces of ribbon and leather, for use in spontaneous dramatics by the pupils as they discussed and acted out various aspects of family life

in selected families in their own community and in families in other parts of the U.S.A.

9. *Use of Guests from Abroad in a Second Grade.*   When the pupils in a second grade in a midwestern community were studying "Families Around the World," two students in a nearby state university were contacted through the Office of Students from Abroad and were invited to the school for two days during the vacation period of the university. They sat in the class one morning and worked with the children on their current projects. Then, in the afternoon, they told about their lives and the lives of their families "back home." Part of what they said was tape-recorded for future use.

10. *Stamps of the United States in the Revolutionary Period.*   Three pupils in a fifth grade class studying United States history were encouraged to bring to class all the stamps in their collection which showed some aspect of the U.S. Revolution. They were able to find stamps of Franklin, Washington, Jefferson, and Hamilton. They discovered stamps of the Liberty Bell, the Minute Men, Yorktown, and Valley Forge. They brought in stamps of Pulaski and von Steuben and one of the Northwest Ordinance of 1787. Those stamps were used to discuss the importance of those men and events. Later the stamps were arranged as an exhibit for other classes to see.

11. *Cartoon Quiz on Current Events.*   Over a period of several weeks a seventh grade teacher collected cartoons from two local newspapers, from the "News of the Week in Review" section of the Sunday *New York Times* and from current events papers. As one question on a current events test, each pupil was asked to explain the meaning of the cartoon which had been given him or her. The teacher tried to determine in advance the degree of difficulty of each cartoon and to give the cartoons to pupils according to their abilities. After the test, each of the cartoons was placed in the opaque projector to stimulate a general review by the entire class of the major current events in recent weeks. This was a relatively sophisticated group of pupils in the seventh grade.

12. *A First Grade Class Visits Several Types of Homes.*   Because several pupils in a first grade class had never been in a variety of homes, the teacher arranged with four mothers in the class for visits by the group to their houses and apartments. Light refreshments were served in each home to make it a gala time as well as a good learning experience.

13. *Sharing Our Families with Children in France.*   Pupils in an independent school in Germantown–Philadelphia were involved in an ongoing affiliation with a school in France which touched all grade levels and the teachers and parents as well as the pupils. Since the theme in their second grade class was "Families of the World," the teacher developed a project in which three fathers of pupils in the class took pictures of some family scene in the home of each pupil. Children in the seventh grade wrote out a brief explanation of the scene, with the help of the child from that home. All this material was pasted on large cardboard sheets and mailed to France. The students in the high school French classes wrote French translations of all the material. Months later a somewhat similar project was sent from France by the children in the affiliated school there. This time it was done by drawings. In this way a great many people were involved in a significant project in social studies and international understanding.

14. *Sounds of the Local Community.*   A third grade class "listened" for the sounds of their community as they took several trips to nearby places. Then a

mother of one of the pupils took tapes of those sounds. The pupils used them for a local P.T.A. program and placed the tapes in the school library for use by other classes.

15. *A Hall of Heroes.*   As a review of certain aspects of United States history in a fifth grade class, the teacher asked each pupil to select three persons who ought to be placed in the classroom Hall of Fame of U.S. History, representing the Jacksonian Period. The next day nominations were made and the class voted on the persons to be placed in their Hall of Fame. Illustrations were then made and short statements on each person elected were written and posted with the illustrations. The discussion about the persons in that period was far more important to the learners than the final selection.

16. *An Illustrated Time-Line of State History.*   A rather slow group of seventh graders was studying the history of New York State. After they had completed a brief review of the history of that state, the teacher decided to review the topic by asking each pupil to make a drawing of two major events in the history of New York. A heavy cord was strung across the front of the room and divided into twenty-five-year periods. Various pupils were then asked to show their drawings and some of the slower pupils were asked to place them on the "time-line" in the front of the room. The illustration which caught the fancy of the class was a rubbing of a license plate of an automobile which had been done to represent the completion of the Thruway. Although this method was used on the history of New York State, the same thing could be done equally well on the history of any state. Because of the many learnings involved in this review, two periods were devoted to the explanations of the drawings and the placement of them on the time-line. Interest ran high and real learning took place in this project.

17. *Role-Playing Jefferson's Decision on the Louisiana Purchase.*   Two pupils prepared maps of the U.S.A. in 1803 and of the Louisiana Territory. An able, verbal pupil then used these maps as he played the role of Jefferson deciding whether to purchase this land or not. He thought aloud about his dilemma. Then he called upon the class as his advisers to help him decide. It was a dramatic episode.

18. *"Dropping" Children Into Lapland from an Airplane.*   A second grade class was ready to start a study of family living in Lapland as a part of their study of Families Around the World. The teacher took them on an imaginary airplane trip across the Arctic region on a Scandinavian Airlines plane. She then landed the group in the northern part of Scandinavia. Before they were dropped, each of them was supposedly given a good-sized knife. Then they were encouraged to start living there. With the teacher's encouragement and the help of pictures when they asked about the land, they began to decide how they might live, what they might eat, what their fathers might do to earn a living, and what fun they might have. The teacher provided as little information as possible, except for replies to questions the pupils asked. When this exploration was completed, she began to read to them about life in Lapland. Such a method sounds difficult, but it is often surprising what children can discover when the teacher permits, even encourages them to think and ask questions and pose possibilities. This method can be used with almost any age group and on almost any part of the world.

19. *Communities of the World.*   Toward the end of the fourth grade year, during which the pupils had studied a variety of communities in various parts of the world, the teacher brought in a stack of pictures mounted on cardboard. She dis-

tributed them to members of the class. None of them had captions. She then asked the class to discover where they were. After considerable thinking, they decided they were in some part of Southeast Asia. Actually it was Indonesia and the class proceeded to study two communities in that country, one a village and one a large city. The method can be used, of course, on any part of the world.

20. *The U.N. Faces Some Big Problems.*   The final unit in a sixth grade class was to be on the United Nations and its specialized agencies, as a culmination of the year's work on countries. In order to tie the U.N. in with the previous study of selected nations, the teacher led a discussion on some of the big problems of various nations in today's world. The list of problems was placed on the chalkboard. It was left there for several days as committee groups worked on some of the world's biggest problems and what the U.N. is trying to do about them. Among the topics listed were disarmament, water, land reform, poor health, low standards of living, lack of schools, and space. With the help of the librarian, the children did basic research. They placed their findings on 3" by 5" cards with the data on one side and the source on the other. As they reported their findings to the entire class, the various parts of the U.N. and its agencies were shown on a flannelboard. This unit lasted a little over three weeks. In it the children learned primarily three things. One was about the common problems of the world. Another was about the work of the U.N. and its agencies. The third was about the skills of simple research. This unit would probably not have been successful except at the end of a year of study of a selected group of countries, and of current events study.

21. *Studying Pollution Locally.*   A seventh grade class studying the problem of pollution decided to explore that problem locally. With the aid of a local ecology group, they made an extensive study of the pollution of the river in their locality, reporting their findings to the city council and the two local newspapers. Some changes were made as a result of their efforts.

22. *Interviewing an Old Resident in the Community.*   A third grade class studying the local community discovered that there were two persons in their locality who were over ninety years of age. With the help of the teacher, a list of questions was drawn up to ask those two persons. Various children then interviewed the teacher, who played the role of a ninety-year-old, so far as she could. Three persons were selected to go with the teacher to interview the woman and three others to interview the ninety-year-old man. Their visits were tape-recorded, with the permission in advance of the persons interviewed. The tapes were then played back to the entire class, who felt they had participated in the interviews by framing the questions. The questions concerned food, clothes, schools, means of travel, and similar topics.

23. *Deciding Which Side to Take in the Revolutionary War.*   At the beginning of a class period in social studies, the teacher drew on the chalkboard a stick figure of a man with a very large head and a very small body. The fifth grade children were convulsed with laughter over this odd figure. The teacher then explained that they were going to concentrate on what was going on in the head of a Philadelphia merchant at the outbreak of the Revolutionary War, as he pondered which side to take in that conflict. As points were suggested, they were placed inside the large head of the man pictured on the chalkboard. When this was completed, a similar figure was drawn of a farmer in New England and the same procedure used with him. Finally, a figure was made of a plantation owner in the South. By this simple

device, the pupils thought seriously about the difficulties which confronted people of that period in making up their minds as to whether to join the revolutionary cause or side with the Loyalists.

24. A Food Basket and Tool Chest. Members of another fifth grade class, studying the United States Today, were asked to find one product at home which was produced in the Midwestern part of our country and, if possible, to bring that product to class. Illustrations from newspapers or magazines or drawings were deemed satisfactory as substitutes for the "real thing." The teacher brought to class an old-fashioned market basket and a small box which she called a tool chest. Into these two receptacles the products were dropped, with an explanation of each product. The two receptacles were then placed on a table under a large placard, marked "The Midwest as the Food Basket and Tool Chest of the U.S.A." Later these items were used as part of a written test on that region.

25. Rewriting the Declaration of Independence. The teacher was not certain that the pupils in his class really understood the meaning of the opening lines of the Declaration of Independence. His homework assignment was to rewrite it in the words of the pupils. When the pupils read their version of the Declaration the next day, the teacher was able to evaluate how much they knew and to proceed from there in the study of the document.

26. Comparing Nigeria and Kenya. Anxious to have his pupils learn that all African nations are not alike, a sixth grade teacher launched a study of Kenya after the class had studied Nigeria. The center of that study was a large Comparisons and Contrasts chart which was placed in a prominent place in the classroom. On it the pupils recorded data on the two nations of East and West Africa.

PROBLEM-SOLVING AS THE MAJOR METHOD

All of the methods and activities mentioned on the foregoing pages have a place in social studies instruction. But they are not of equal value. The method which is most basic is problem-solving. It should be central in all social studies teaching — in every phase of the social studies and at all grade levels.

This method is not new. It is as old as man. Socrates was certainly one of its proponents. Comenius, Pestalozzi, and other great educators exposed their pupils to situations or experiences in which they had to discover for themselves, to probe, to inquire, to search, and to think.

Even the term problem-solving is a misnomer. We do not always or even often solve problems. We work on them. Perhaps it would be better to designate this method as working on problems. Some call it the discovery method; others refer to it as the method of inquiry. Whatever term is used, the process matters most.

There are two major approaches in this method. One is to create a stimulating environment in which children will be motivated to ask questions and then to search for answers under the competent guidance of teachers. The other is to confront children with problems and have them search for answers. The former is based on real situations. The latter is based on contrived situations. Both have merit.

For many years teachers and curriculum experts have listed critical thinking as one of their major aims. Yet, as Dr. Hilda Taba of San Francisco State College has pointed out, ". . . this objective has remained a pious hope instead of becoming a tangible reality."

Why, then, the renewed interest in this general approach at the present time? It has been prompted in part by recent studies in cognitive thinking and by the current emphasis upon the structure of the disciplines. It has also been given increased attention because of the knowledge explosion and the realization that there are just too many facts for children to learn. Therefore many people have turned to a consideration of what is really important and they have decided that learning *how* to approach problems is more important than most other kinds of teaching.

Problem-solving is based on a recognition of the fact that teaching is not telling. It is not merely the communication of knowledge. It is not the learning or memorization of facts, no matter how important they are. Teaching is probing, discovering, thinking, analyzing, inquiring, searching.

Problem-solving presupposes inductive rather than deductive thinking. Pupils are confronted with problems or placed in situations which arouse their curiosity. They begin to think. They obtain data and arrange the data in meaningful clusters. From this, concepts are formed. Further thinking should then take place until generalizations can be arrived at.

One of the people who has worked a great deal on the method of discovery is Dr. Charlotte Crabtree, Professor of Education at the University of California, Los Angeles. She has worked with very young children as well as older boys and girls. From her experience she has developed a simple chart which shows in a graphic way how discovery learning or inquiry takes place best.

Experience	T		T	
	H		H	
Experience	I		I	
	N	Discovery of	N	Formulation of
Experience	K	concepts	K	a generalization
	I		I	
Experience	N		N	
	G		G	

Problem-solving and critical thinking are not synonymous, but they are closely related. The term problem-solving relates more to the situations in which critical thinking takes place. Critical thinking is a broader term.

In his excellent volume on *Children's Thinking*, David Russell states that "problem-solving is likely to be more complex than . . . associative thinking . . . and is more often dominated by the objective, external obstacles or situation than by the autistic factors which affect children's fantasy."

Research indicates that children can and do carry on such problem-solving as early as three years old. The natural curiosity of children is the key factor in learning then. Often it is thwarted rather than utilized.

In the field of the social sciences recent experimentation with boys and girls in stimulating environments in the primary grades, under good teachers, and with concrete materials, has revealed that even young children can gather data, develop concepts, and discover and evaluate generalizations in such fields as geography and economics.

However, there is some evidence to show that children do not always follow the neat patterns outlined in textbooks on problem-solving. They do not always state the problem, collect data, formulate tentative conclusions, collect more data, test their hypotheses, and reach more or less firm conclusions, in such an orderly fashion.

With younger children it is almost impossible to overemphasize the importance of concrete materials and firsthand experiences. With older pupils, field experiences and a wide range of resource materials are important.

In problem-solving the teacher has three major roles. He or she must (1) provide a stimulating environment, (2) encourage and even provoke questioning, and (3) help children to formulate generalizations on the basis of their experiences rather than giving them the generalizations.

The long-range goal of all problem-solving education is to encourage boys and girls to explore and reach their own conclusions, under the guidance of competent adults.

Here are a few examples of problem-solving situations in elementary and junior high schools:

1. A simple diagram of a house or an apartment is placed on the chalkboard or on a piece of oaktag. The children are asked to find out what kind of a family might live in this place. They can also discover why the apartment or house is arranged as it is.

2. Children can see a number of pictures of a family. They can be encouraged to ask questions about that family until they discover many things about it, including its size, the composition of the family, what the adults do to earn a living, what each member does for the welfare of the group, and how they have fun as a family.

3. Children who are studying communities might well pretend that they have moved to a new community. What would they need to know as soon as they arrived? How would they obtain their information? What would they need to know within a relatively short time? How would they obtain that information?

4. Boys and girls can be given a piece of land from which they are expected to earn a living. They can ask questions about their land until they decide how to use it. Where it is, what kind of soil it contains, what the climate is like, where it is located, and similar questions may be asked.

5. Pupils studying the history of our nation can be encouraged to role-play their trip to America in 1607 or some other date. They can be asked to pack for this long and hazardous trip. The key question is, "What will you need to take for living in the New World?" Research will be needed to help them to decide what they will take with them.

6. Pictures of a nation can be passed out to a class or used in an opaque projector, with the base question, "What country is this?" With each suggestion that pupils make, the teacher should ask them to state why they made that proposal. Occasionally the teacher can also ask, "Wouldn't that be true of such and such a place, too?" — naming other parts of the world than those proposed by the pupils.

7. Boys and girls in the primary grades can discover the importance of specialization by timing their efforts in baking cookies individually and on an assembly line basis.

8. Older boys and girls who are trying to understand the poverty in large parts of the world, including parts of the United States, can be given $600 or so and can be asked to make a budget for a family of seven for an entire year.

9. Current newspapers may be used profitably to develop problem-solving, by

having pupils compare the accounts of the same event in different papers and explaining the differences.

What examples of problem-solving can you give?

10.

11.

12.

BIBLIOGRAPHY ON PROBLEM-SOLVING

1. ALLEN, DONEY F., JOHN V. FLECKENSTEIN, and PETER M. LYON. *Inquiry in the Social Studies: Theory and Examples for Classroom Teachers*. Washington: National Council for the Social Studies (1968), 114 pp.
2. BROGAN, PEGGY and ALICE MIEL. *More Than Social Studies*. Englewood Cliffs, New Jersey: Prentice-Hall (1957), 452 pp. Chapter 10, "Helping Children Increase Competence in Democratic Problem-Solving."
3. CRABTREE, CHARLOTTE. "Challenging Children to Think" in *Primary Education — Changing Dimensions*. Washington: Association for Childhood Education (1965), 76 pp.
4. CRABTREE, CHARLOTTE. "Inquiry Approaches to Learning Concepts and Generalizations in Social Studies" in *Social Education* (October, 1966), pp. 407–414.
5. CRABTREE, CHARLOTTE. "Inquiry Approaches: How New and Valuable?" in *Social Education* (November, 1966), pp. 523–531.
6. CRABTREE, CHARLOTTE and FANNIE SHAFTEL. "Fostering Thinking." Chapter 8 in *Curriculum for Today's Boys and Girls*. Robert Fleming, Editor. Columbus, Ohio: Merrill (1963), 662 pp.
7. DEWEY, JOHN. *How We Think*. Boston: Heath (1933), 301 pp. Still pertinent.
8. DUNFEE, MAXINE and HELEN SAGL. *Social Studies Through Problem Solving*. New York: Holt, Rinehart and Winston (1966), 385 pp.
9. ELLSWORTH, RUTH. "Critical Thinking" in *National Elementary Principal*. Washington: National Education Association (May, 1963), pp. 24–29.
10. GROSS, RICHARD E. and FREDERICK J. McDONALD. "The Problem-Solving Approach" in *Readings for Social Studies in Elementary Education*, John Jarolimek and Hubert M. Walsh. New York: Macmillan (1965), pp. 322–333.
11. GROSS, RICHARD E. and RAYMOND H. MUESSIG. *Problem-Centered Social Studies Instruction: Approaches to Reflective Thinking*. Washington: National Council for the Social Studies (1971), 96 pp.

12. HUDGINS, BRYCE B. *Problem Solving in the Classroom.* New York: Macmillan (1966), 74 pp.
13. HULLFISH, M. GORDAN and PHILIP G. SMITH. *Reflective Thinking: The Method of Education.* New York: Dodd, Mead (1961), 273 pp.
14. KRAVITZ, BERNARD and DIANE J. SOROKA. "Inquiry in the Middle Grades" in *Social Education* (May, 1969), pp. 540–542.
15. RATHS, LOUIS E. and others. *Teaching for Thinking: Theory and Application.* Columbus, Ohio: Merrill (1967), 348 pp.
16. ROGERS, OSCAR A., JR., and SYLVIA L. GENEVESE. "Inquiry in the Primary Grades: A Means to a Beginning" in *Social Education* (May, 1969), pp. 538–540.
17. RUSSELL, DAVID M. *Children's Thinking.* Boston: Ginn (1956), 449 pp. A classic in this field.
18. SCHMUCK, RICHARD, MARK CHESSLER, and RONALD LIPPITT. *Problem Solving to Improve Classroom Learning.* Chicago: Science Research Associates (1966), 88 pp.
19. SENASH, LAWRENCE. "The Problem Approach . . . An Effective Tool for Economic Education" in *Grade Teacher* (April, 1964), pp. 322–333.
20. SHAFTEL, FANNIE, CHARLOTTE CRABTREE, and V. RUSHWORTH. "Problem Solving in Elementary School." Chapter 3 in *Problems Approach and the Social Studies.* Richard Gross and others, Editors. Washington: National Council for the Social Studies (1960), 121 pp.

THE CHALKBOARD AS AN AID IN LEARNING

The chalkboard is an ally that is always near you to help in teaching. It is visible at almost any distance if you write or print large enough. It can be used for many purposes by the teacher, by pupils acting as secretaries for the class, and by boys and girls presenting individual or committee material. Here are a few suggestions on its use:

1. Be sure that the lighting is good so that pupils can see what is written on the chalkboard.
2. Write or print large enough for everyone in the room to see. Don't mix writing and printing, as it is confusing.
3. If the chalk sends shivers down your back, break it.
4. Start writing in the upper left hand corner if you are going to place considerable material on the board. Do not write too far down on the board where the pupils cannot see the writing.
5. If you have trouble with spelling, jot down your key words on a card which you have at hand.
6. Use colored chalk for some work.
7. Keep the chalkboard as clean as possible.
8. Mark the material which should not be erased.
9. Be sure to alert your pupils to items which ought to be copied.
10. If you use a class secretary, stand close to that person to help him or her.
11. Learn to underline titles.

What you place on the chalkboard depends upon you and the class. Here are some of the materials teachers of the social studies often include on the chalkboard.

1. Assignments.
2. Questions from the class to be organized for units — and outlines for units.
3. Statements made by pupils which you want them or the class to examine carefully.

4. Key ideas in a discussion, often placed there as you go along, but sometimes placed on the chalkboard as a summary of a discussion.
5. Maps made by you and by pupils.
6. Diagrams to help promote learning.
7. Time-lines.
8. Difficult words and terms. Sometimes these can be left on the chalkboard for several days.
9. Work which individuals or committees have prepared for the class.
10. Brief summaries of situations which you plan to use for problem-solving.
11. Guide questions for a film or filmstrip.
12. Ideas which do not fit into the current discussion, as a reminder that you plan to take them up later.
13. Charts, graphs, and diagrams.
14. A key date (in large numerals) to remind pupils of the period they are studying.
15. Problem situations on which the pupils will work.

DRAMATIC PLAY AND DRAMATIZATIONS

Dramatics in one form or another are important in social studies learning at every age and grade level. They run the gamut from free, natural, spontaneous dramatic play to more formal dramatizations.

For young children dramatic play is central. Much of their time should be spent in what adults often call "play." But "play" is very real to children. When they are pretending, making-believe, trying out, exploring, and imitating, they are learning. They are internalizing what they have seen and heard and "trying it on for size."

Here are some questions often asked by young teachers-to-be and some partial answers, to which you may want to add ideas of your own.

A. *What Values Can Come from Dramatic Play and Dramatizations Under Proper Guidance?*
  1. Realistic reproductions of the world of older children and of adults.
  2. Motivation and interest in new ideas and experiences.
  3. Heightened sensitivity to human beings and their relationships.
  4. A variety of skills, from printing signs to speaking.
  5. Applications of information gained previously.
  6. An air of reality to reading.
  7. Identification with persons here and now — or in the past.
  8. The raising of questions and problems for further exploration.
  9. The release and cultivation of feelings.
  10. Initiative and creativity.
  11. Group morale and cooperation.
  12. Revelation to themselves, to their classmates, and to teachers of how children really feel.
  13. Development of a self-image.
  14. Therapeutic values.
  15. Opportunities for teachers to evaluate attitudes, skills, and knowledge in behavioral situations.

B. *What Different Types of Dramatic Play and Dramatizations Can Teachers Include in Social Studies Programs?*

   1. Free, spontaneous play with situations developed on their own by children.

   2. Some guided or directed play with teacher suggestions or encouragement.

   3. Informal skits produced by older pupils with little preparation.

   4. More formal plays, written by children.

   5. Plays written by adults and acted out by children.

   6. Pantomime.

   7. Puppetry, using stockings, papier-mâché, felt, or other materials.

   8. Role-playing or sociodrama (see pages 105–107).

C. *When Should Such Methods Be Used in Social Studies?*

   1. Much of the time in the pre-school years and considerable time in the primary grades.

   2. In independent play periods in the primary grades.

   3. Frequently at the end of a unit, to summarize what has been learned.

   4. During a lesson in the middle and upper grades to clinch a point which has been difficult to understand. Role-playing is especially good for this.

   5. As a method of reporting by a committee to the class as a whole.

   6. Occasionally as motivation for a unit or topic.

D. *What Equipment Is Essential or Helpful for Such Dramatics?*

   1. In the early years in school as much space as possible is needed for play and construction activities.

   2. In the primary grades boxes, cartons, blocks, material for fences, bridges, tiny autos, trucks and other means of transportation, telephones, dolls (Negro and Asian faces as well as white faces).

   3. Don't forget special items for boys, such as gloves, tools, a doctor's kit, a chef's hat, desks and chairs for an office, men's hats and boots.

   4. For older children a small "Costume Chest" is a wonderful item to have in a corner of the room or in a closet, with a few different types of hats, shawls, scarfs, jewelry, etc., to lend an air of authenticity to skits or plays. Have the pupils collect these items. Some materials from abroad would be very helpful, too, such as a sari, hats, etc.

   5. Keep your materials up-to-date, including make-believe meters for autos, space helmets, etc.

E. *What Types of Dramatic Skits Can Be Used Especially Well in the Social Studies?*

Two examples are given on each of several types of themes. Space is provided for you to fill in other examples of special relevance to your teaching:

   1. *Family situations*

      (a) Answering the telephone.

      (b) Paying the "tax" at the grocery, supermarket, or drugstore.

      (c)

      (d)

2. *School situations*

    (a) Settling a fight on the playground.
    (b) A meeting of the School Council.
    (c)
    (d)

3. *Community situations*

    (a) The work of a fireman.
    (b) An interview for a job.
    (c)
    (d)

4. *Life in the U.S.A. and/or some other country today*

    (a) How a newcomer to the United States might feel.
    (b) Crossing a picket line.
    (c)
    (d)

5. *Scenes from history*

    (a) A discussion in a family on whether to join in the fight for independence in 1775.
    (b) Robert E. Lee argues with himself as to whether to join the Union or Confederacy forces.
    (c)
    (d)

6. *World affairs*

    (a) Men on a mission to Southeast Asia discuss the Mekong River project.
    (b) A mock assembly of the United Nations on a current problem.
    (c)
    (d)

7. *Current events*

    (a) Discussing the town's budget for next year.
    (b) Two Congressmen debate "deficit spending."
    (c)
    (d)

Some references on dramatic play and dramatics in the social studies

1. FRANK, LAWRENCE K. "Play Is Valid" in *Childhood Education* (March, 1968), pp. 433–440.
2. HANNA, LAVONE A., GLADYS POTTER, and NEVA HAGAMAN. *Unit Teaching in the Elementary School.* New York: Holt, Rinehart and Winston (1963), 595 pp. Chapter 11.
3. LANGDON, GRACE and IRVING W. STOUT. *Teaching in the Primary Grades.* New York: Macmillan (1964), 359 pp. See pp. 189–198.
4. RASMUSSEN, MARGARET (Editor). "Play — Children's Business." Washington: Association for Childhood Education (1963), 40 pp. Selection of toys and games.
5. SAGL, HELEN. "Dramatic Play" in *Social Studies in Elementary Schools.* Washington, National Council for the Social Studies (1962), pp. 205–212.
6. WARD, WINIFRED. *Drama With and For Children.* Washington: U.S. Government Printing Office (1960), 68 pp.

FILMS TO FURTHER LEARNING

Films are another of the wonderful resources for social studies learnings. Their advantages are many. They can take children on airplane trips to faraway places and give them front seats at historical events. Through sight and sound and color they can motivate learning, enrich reading and experiences, compress many years or processes into a brief span of time, and move children emotionally.

Recent developments in technology make film projectors easier to handle and more efficient than in the past. Films can now be stopped for discussion and reversed to rerun a scene. The eight-millimeter "loop" of three, four, eight, or ten minutes makes them more usable, especially in forty-five or thirty-minute classes.

Here are a few points which you may want to keep in mind when using films:

1. Be sure that the room is dark.
2. Try to arrange a preview if at all possible.
3. Try to have a competent operator.
4. Limit any introductory remarks to a few words. The children are set to see the film, not to hear you talk. Don't try to compete with the film!
5. Suggest procedures to the pupils, such as jotting down questions or listing a few of the main points — or merely watching the film.
6. Do not try to talk during the showing of a film. You will detract from the film and usually you will not be heard by the pupils. Stop it if necessary.
7. Use films for a variety of purposes — for motivation, for review, and for emotional as well as factual material. You may want to use a film more than once, especially at the beginning and end of a unit.
8. Be sure to find out what free films are available in order to stretch your audio-visual budget as far as possible.
9. For protection of the equipment — and of the pupils — station a student near the machine to keep the pupils from tripping over the cord. Do not plug in the cord until you are ready to show the film.
10. Films may be obtained from a number of sources, such as the following:
    (a) Your school library or audio-visual room.
    (b) Your city or county system.
    (c) Your state university or state library.
    (d) Government agencies.
    (e) Information centers and embassies of many governments.
    (f) Organizations in the community or state, such as World Affairs Centers, U.N. Associations, etc.
    (g) Companies.

The best single reference to films is in the *Educational Media Index*, published by the McGraw-Hill Company. There are separate volumes on the following topics: *Pre-School and Primary, K–3, Intermediate, Grades 4–6*, and *Geography and History*. You can use the catalogues of various film companies, too, for references.

FILMSTRIPS TO ENRICH LEARNING

Filmstrips can be extremely useful aids in social studies teaching. They are inexpensive. The machines are easily run by teachers or pupils. Hundreds of filmstrips in black-and-white or color are available. Moreover, filmstrips can be run at any speed, the machine stopped easily for discussion, and the strip turned back to refer to an earlier frame. They can be used by individuals, groups, and entire classes. If possible, a filmstrip machine should be in your classroom at all times. Here are some suggestions on the effective use of filmstrips·

1. Preview filmstrips if at all possible.
2. Show in a dark room. This is especially important with color filmstrips.
3. Station a pupil by the cord and machine to prevent accidents to the pupils or the machine.
4. Avoid long introductions; get to the filmstrip quickly.
5. Ask for volunteers to read, and call on others, but in no obvious order, to keep pupils alert.
6. If some captions are long or difficult, read them yourself or paraphrase them.
7. Most filmstrips are too long if you use every frame. Show them in installments or use only a few frames. But tell the pupils what you are going to do. Otherwise they seem to feel "cheated."
8. Pick out a few frames and discuss them at length. Ask the pupils what they see — and why that is important. Then, "What else do you see?"
9. Cover up the caption on some filmstrips and let the pupils see only the pictures.
10. If you stop for discussion, turn off the machine so that the machine doesn't get too warm and so that pupils can be heard.
11. Speak a little louder than usual because of the noise of the machine. Encourage pupils to speak up, too.
12. If you need a sharper image, focus or move the machine nearer to the screen. For a larger image, move it away from the screen.
13. Most filmstrip machines get out of focus easily. Focus them frequently.
14. Remember that you may want to use a filmstrip more than once.
15. If a filmstrip is out-of-date or the captions are poor, have the pupils "rewrite" the filmstrip.
16. Filmstrips filled with propaganda can be used to teach pupils propaganda devices.
17. If you use a pupil as the operator, stand near him or her to help direct the showing of the filmstrip.
18. Plan an alternative activity if the filmstrip machine doesn't arrive or doesn't work.

For hundreds of filmstrips, see the books — *Educational Media Index* — mentioned in the preceding section.

FLANNELBOARDS AND MAGNETIC BOARDS TO IMPROVE LEARNING

Many teachers use flannelboards or magnetic boards in mathematics, and the language arts, but too few teachers use them in the social studies. They can be used to great advantage in a number of ways, enhancing learning by movement, color, and novelty. Pupils should use them even more often than teachers. Here are some suggestions on their uses:

A. *How May Flannelboards (or Magnetic Boards) Be Used in the Social Studies?*
   1. To represent members of families in the U.S.A. or in other parts of the world.
   2. To help pupils to develop relationships in conjunction with studies of communities here or abroad.
   3. To highlight a daily current event of importance.

4. To show, by a series of pictures or drawings, any process such as the production of oil, a newspaper, or the story of milk.
5. To show and discuss charts and graphs.
6. To trace step by step the territorial growth of the United States.
7. To show the various components of the United Nations.
8. To develop time-lines.
9. To make vivid committee reports.
10. To illustrate different sizes — of states or countries, for example.
11. To help children to develop a social studies vocabulary by having words and pictures on flannel which can be matched.

B. *How Can Flannelboards Be Made?*

Teachers and/or pupils can make them with Masonite, plywood, heavy cardboard or Celotex and long-fibered flannel, felt, or suede. Place the material on the board and attach it with glue, staples, tacks, or some other material. Place flannel or some similar material on the back of the pictures, charts, or other material to be shown. When placing the material on the flannelboard, be sure the nap is loose enough to catch. Apply the material with pressure, rubbing the material so that it will stick to the flannelboard.

Use contrasting colors as background for your materials. Make the flannelboard black or gray. Be sure your figures are large and that printing is large too.

C. *Where May I Purchase Flannelboards (and Magnetic Boards)?*

1. Milton Bradley Company, Springfield, Massachusetts 01101.
2. Displays, 331 Madison Avenue, New York, New York 10017.
3. Instructo Products Company, 1635 North 55th Street, Philadelphia, Pennsylvania 19131.
4. School Service Company, 5701 West Vernon Street, Detroit, Michigan 48209.

GROUP WORK AND COMMITTEE WORK

Everyone needs to learn to work in small groups or committees as well as to work individually. Pupils will be working in such groups all their lives on their jobs or in organizations — or in both. Breaking large classes into small groups can also enhance learning if the committee work is well done. This method can also promote good social learning.

Committee or group work can be done at all grade levels from the nursery school on. Teams of two are usually used in the pre-school years. Gradually their size is increased, but they should never be over six or seven. Four or five is a better number.

Teachers should use committees, but everything should not be done in committees. Most topics or units should be introduced to the entire class and a few days spent on common background. Then committees may be formed.

There are many skills involved in committee work. Pupils should be introduced to them gradually. Panels can be used, with the teacher chairing them. Then panels can be used with pupils chairing them. This is preparation for committee work. Committees can be given very short assignments and strict time limits at first. Then they can be given larger tasks and longer time periods. Groups can be stopped and common problems discussed. Full-fledged committee work cannot be done without such introductions to group work.

Here are some of the questions young teachers often ask about committee-ing:

A. *When Should We Use Committees?*
1. When the topic lends itself to group work.
2. After a general introduction of a few days during which time a common foundation has been laid.
3. When there is adequate time for research by a group. Until the group knows how to utilize the time well, the periods should be relatively short and the topics fairly specific. Later, on broader topics, more time can be given.
4. When there are adequate materials for group work. It is wise to send for materials two or three weeks in advance of the time committees will be working on a topic. Otherwise they will find themselves without adequate materials.
5. When there is adequate space. If you are crowded, one or two groups may go to the library to work. Sometimes pupils are permitted to work outside the classroom in a corridor. Thin boards can be placed across fixed desks to give table space and drop-leaf tables can be constructed by the chalkboards in some schools.
6. When the teacher feels relatively self-assured about this type of work, being confident to carry on committee work successfully.
7. When the pupils need to learn the variety of skills involved in group work of committees. If they are relatively proficient in these skills, then you may want to emphasize some other skills, possibly individual research and reporting.

B. *How Is the Work of a Unit Best Divided into Committee Jobs?*

This will depend in large part upon the type of topic being studied.

If a world area or region is being studied, the group as a whole should do an overview. Then committees may be formed on individual countries in that region and a common outline used for research. As a culminating activity each group might fill in a large chart in the classroom. For example, Africa as a whole might be studied and then committees work on five or six individual countries. The same could be true of other world areas.

If the topic is "The People of the U.S.A. Today and Yesterday," the class might well study that theme for a few days and then form groups to study people according to their occupations or by ethnic groups.

If cities are being studied, two or three cities might be studied by the class as a whole and certain skills developed. Then committees on other cities could be formed to see if the generalizations worked out by the class apply to those locales.

C. *How Should Committee Members Be Chosen?*
1. Very brief jobs can be assigned by rows or by groupings according to alphabetical order, but this should not be done on longer-term assignments.
2. Neither should pupils be homogeneously grouped by reading ability or by intellectual ability, as is done in the language arts.
3. Committees can be organized on the basis of free choice by pupils, with the teacher ascertaining that the same pupils do not always work together.
4. The teacher may group the class by sociometric tests of choice.
5. Slips of paper with an indication of first, second, and third choice may be used, thus permitting the teacher to make some changes when they are necessary.

D. *How Are Chairmen of Groups Selected?*
   1. At first the teacher may want to select the chairmen. They need not neces-
      sarily be the brightest pupils. More often they are pupils with whom other
      children can work well.
   2. Eventually committees should select their own chairmen. The teacher should
      try to see that the job of chairman is rotated, to give many children this
      experience.

E. *What Is the Relationship of the Teacher to Committees Once They Are
   Formed?*
   1. To work with all committees a little.
   2. To concentrate at first with the committee having the most difficulty and/or
      the committee which is to report first to the class.
   3. To stop the groups and work on common problems with the class as a commit-
      tee-of-the-whole.
   4. To help see that adequate materials are on hand.
   5. To help each committee organize its findings adequately.
   6. To help each committee prepare a report to the class as a whole in a manner
      that will promote learning; reporting only on the major points.

F. *How Can Committee Findings Be Reported to the Rest of the Class?*
   Many a project falls apart when committees report back to the class. Some of the
pitfalls to avoid are: (a) reports by each member of the committee. This takes too
much time. (b) All reports in one form — usually an oral report. They become too
detailed and boring. (c) Reports only by chairmen.
   Here are some suggestions on ways to avoid these pitfalls:
   1. Have some committees report their findings in brief outline form, with an
      oral explanation. Have the class study the written report.
   2. Prepare visual summaries as at least a part of a report — a mural, a time-
      line, an interview, a series of pictures, glass slides, or charts.
   3. On controversial issues, have a majority and a minority report.

G. *What Do Other Members of the Class Do When Reports Are Made?*
   1. They can take notes.
   2. They can write down questions to ask the panel or committee.
   3. If the topic or report is presented in an interesting manner, they can be
      asked merely to listen and look!

H. *How Can the Work of Committees Be Evaluated?*
   1. Committees can be asked to evaluate their own work, sometimes with a
      checklist provided by the teacher or worked out by the class.
   2. The class as a whole may be included in the evaluation of a committee's
      work, using a general checklist. Start with "Things They Did Well" and
      then go on to "Suggestions for Improvement."
   3. The librarian may be helpful in evaluating certain parts of the work of a
      committee and/or individuals.
   4. Pupils may be asked to list the tasks they performed on a committee.
   5. The teacher can observe and evaluate individuals and/or the committee as a
      whole.

6. The comments of parents will prove useful sometimes, as they indicate interest or lack of interest of their children on the committee work they have done, discussions at home about their work, etc.

Some references on group or committee work

1. HOCK, LOUISE E. *Using Committees in the Classroom.* New York: Rinehart (1958), 55 pp.
2. SIMON, SIDNEY B. and PHYLLIS LIEBERMAN. "Committees ARE a Way to Learn" in *The Instructor* (September, 1965), pp. 122, 124, 126, and 128.
3. "Ways of Grouping" in *Childhood Education* (December, 1968).

OPAQUE AND OVERHEAD PROJECTORS

The opaque projector is a long-time favorite of social studies teachers. It has one distinct advantage. It can reproduce on a screen almost any clipping, picture, chart or graph, or page of a book, without preparing any special materials. Most projectors now are equipped with arrows which the teacher can use to indicate special points. But the opaque projector has one distinct disadvantage: it requires a very dark room for adequate projection.

The overhead projector is relatively new and is not used nearly enough as yet by teachers. It has many advantages. Here are a few of them:

1. It can be used in a fairly light room, even without curtains.
2. The teacher can face the class while using the projector.
3. The projector is very easily run. It merely requires plugging in and turning on the switch.
4. The images produced are large enough for everyone to see in any part of the room. For example, a ten-inch transparency at a distance of seven feet will become a sixty-inch screen image.
5. A variety of types of material can be used effectively.
   (a) Outline maps of any area can be projected.
   (b) Overlays can be used for comparisons and contrasts.
   (c) Different colors can be used with special pencils or with crayons.
   (d) Materials from newspapers, magazines, pamphlets and books can be reproduced for use in the projector.
   (e) Pupils as individuals and as committees can prepare many of the materials used and gain invaluable learning in the process.
   (f) Some types of projectors permit the use of slides.
   (g) Materials once produced can be stored for future use.
   (h) Rolls of plastic material can be used to show processes.

In the use of overhead projectors a few hints may prove helpful, including the following:

1. If possible, have an overhead projector set up in the room to use immediately.
2. Purchase some map transparencies from publishers to save time and effort and to ensure accuracy.
3. Use a large screen and see that it is tilted forward so that the picture or map is not wider at the top than at the bottom when projected.
4. Use magic markers, crayons, grease pencils, or India ink to prepare your materials.

In selecting overhead projectors for purchase, bear in mind the following factors: safety, the size of the working surface, the size of the projected image, the image brightness and sharpness, color fidelity, steadiness, noise level and heat problems.

The most comprehensive account of the overhead projector is Morton H. Schultz' *The Teacher and Overhead Projection*. Englewood Cliffs, New Jersey: Prentice-Hall (1965), 240 pp.

PICTURES TO ENHANCE LEARNING

Pictures ought to be high on the list of priorities for social studies teaching in elementary and in junior high schools. In the primary grades they should be used a great deal, inasmuch as children cannot read well enough to gather much of their information from the printed page. But pictures ought to be used also in the middle and upper grades. A great deal of learning can be promoted by the wise use of pictures. That means that they should be studied rather than merely looked at. Here are some questions which concern the use of pictures in social studies teaching:

A. *What Are Some of the Best Uses of Pictures?*
 1. To arouse interest in a new place, idea, word, or concept.
 2. To recall something previously learned.
 3. To see detail possible in no other way.
 4. To serve as a substitute for firsthand experience.
 5. To see quickly a process which might take months to see without pictures.
 6. To solve a problem. For example, pictures of water and lakes in Finland can be shown. Children can be asked how they think they might earn a living in such a place, what type of houses they might build, and how they could ship their lumber to other places.
 7. To correct a wrong impression.
 8. To enrich reading.
 9. To review a topic.
 10. To develop skill in viewing or looking or observing.
 11. To teach comparisons, such as size of men compared with the Pyramids.
 12. To increase retention of learning.

B. *What Are Some Characteristics of Good Pictures to Use?*
 1. Size. As large as possible while retaining clarity of the photograph.
 2. Clarity.
 3. Simple composition.
 4. Up-to-date (in most instances).
 5. Color (in most instances, but not all).

C. *When Should Pictures Be Used?*

Pictures are appropriate at almost any time. They can be used especially in such situations as the following:
 1. In the early stages of a unit, to arouse interest.
 2. At any point, to clarify an idea, concept, or point.
 3. To review an idea.
 4. To evaluate learnings.
 5. To settle an argument over a fact or idea.
 6. As a summary of a unit.

D. *How May Pictures Be Used Best?*

1. Through an opaque projector (be sure to have a dark room).
2. By having someone walk slowly around the class with the pictures.
3. With younger children, by having them sit on the floor around the teacher to view the pictures.
4. Mounting them for bulletin board use.
5. Using them with individuals or small groups.

E. *How May the Teacher Encourage the "Study" of Pictures?*

1. By using often the questions "What do you see?", "Why is that important?", "What else do you see?", and "Why is that important?"
2. By asking children to look for specific details.
3. By asking children to arrange pictures in chronological order, in the order of a certain process being discussed, or to illustrate a concept.

F. *Where May Pictures Be Obtained?*

1. Have a "Treasure Hunt" to collect magazines and use pupils and/or parents to cut them out and mount them for use.
2. Use any picture collection in the school and /or local library.
3. Cut them out of current events magazines and newspapers.
4. Cut them out of current affairs magazines.
5. Obtain travel magazines such as *Travel, Holiday,* and the *National Geographic.*
6. Use magazines with pictures of blacks, such as *Ebony, Jet, Tan,* etc.
7. Old textbooks are a source of pictures.
8. Books of pictures such as the *Family of Man, Three Billion Neighbors,* etc.
9. Professional magazines such as *The Grade Teacher* and *The Instructor.*
10. Large pictures produced by such companies as the Chandler Publishing Company, the John Day Company, Ginn, Scott Foresman, and Silver Burdett. Write them for catalogues of their picture portfolios.
11. The information bureaus and embassies of various countries.
12. Travel agencies.
13. Pictures from old calendars, especially of airlines.

G. *How Can Pictures Be Preserved?*

1. Mount them for better preservation.
2. Keep them in a safe place in the classroom or school library, or at home.

Some references on the use of pictures

1. BIERBAUM, MARGARET. "How to Make a Picture Really Worth a Thousand Words" in *Grade Teacher* (April, 1966), pp. 70–74.
2. NATIONAL COUNCIL FOR GEOGRAPHIC EDUCATION. "Geography Via Pictures," Normal, Illinois. Illinois State University (Undated), 24 pp.
3. WILLIAMS, CATHERINE M. *Learning from Pictures.* Washington: National Education Association (1963), 163 pp.

QUESTIONING TO STIMULATE LEARNING

Questioning is at the heart of good teaching and good learning in the social studies as in all other subject fields. Yet this is a much neglected aspect of teacher education. Very little has been written about it in the professional literature. Here are some comments — and some questions — about this important aspect of the teaching-learning process:

A. *Who Should Be Asking Questions in Social Studies Classes?*

    1. Probably the best test of real learning is whether pupils are asking questions — and good questions. Are they so stimulated that THEY raise questions? Do you provide opportunities for them to raise queries? For example, do you ever start a lesson with the question, "What bothered you about last night's homework?" or "What was the most difficult part of your work for today?" Or — do you stop frequently and ask, "What parts of the lesson don't you understand at this point?" Be sure to wait — and encourage answers.

    2. Textbooks and other materials sometimes ask questions for which pupils should seek answers.

    3. Of course, teachers should be asking questions, too. But all questions should not come from them.

B. *Can I Plan Questions Ahead of Time? Should I Do This?*

    1. Definitely, yes. Your key questions should be planned in advance. Take your aims of a lesson and write out the key questions from them. Turn the aims into key questions. In many cases one key question is enough for a lesson. For example, "On April 19, 1775, war broke out between the American colonists and the British. What were the steps that led to that event?" Then you can trace the "steps," making them as steps on the chalkboard. Or, "For the last few days we have been studying India. Why should Americans be interested in that country?" This type of question should not be asked until after the country has been studied for some time. It should not be asked the first day as is so often done.

    2. Subsidiary questions will have to be asked on the spur of the moment. They cannot be planned, usually, in advance.

C. *What Type of Questions Should I Avoid?*

    1. Yes–no questions. They encourage guessing.

    2. Double or multiple questions. They are confusing. Ask one question at a time.

    3. Broad or scoop questions, such as "What about. . . . ?"

    4. Questions with the answer incorporated in the question.

D. *What Can I Do if I Do Not Know the Answer to a Question?*

    1. Ask the class if they know the answer.

    2. Say that you don't know but will find out.

    3. Ask that pupil or another pupil to find the answer and report it to the class.

E. *What Are Some Types of Provocative Questions to Ask?*

    1. Situation questions. "Suppose you were dropped into such and such a type of land. Describe it. What would you need to know about that land in order to live?"

    2. Thought questions. "We have been studying . . . . . . for the last few days. Who is the most important person we have met?" This needs to be followed by the question, "Why do you think this is the most important person?"

    3. Questions to encourage observation. For example, you are looking at a picture

and you ask, "What do you see in this picture which seems strange to you?" Follow this with the question, "Why?"

4. Questions about sources of information. A pupil makes a broad or sweeping statement and you can ask, "What is the source of information?"
5. A pupil asks a question and you toss it back to the class, saying, "That's an important observation. Who can answer that question?"

F. *What Should I Do When Questions Are Asked by Pupils Which Are Not Germane to the Discussion?*

1. Answer the question quickly and return to the topic.
2. Postpone the question until later.
3. Say that it is a good question but not on the topic and that you will answer it privately.
4. If it really is important, you may want to change your plan for the period and take it up immediately.

References on questioning

1. ASSOCIATION OF SOCIAL STUDIES TEACHERS OF NEW YORK CITY. *A Handbook for Social Studies Teaching.* New York: Holt (1967), pp. 38–64.
2. GODBOLD, JOHN VANCE. "Oral Questioning Practices of Teachers in Social Studies Classes" in *Educational Leadership* (October, 1970), pp. 61–67.
3. GRIFFIN, RONALD D. "Questions That Teach: How To Frame Them; How To Ask Them" in *Grade Teacher.* (January, 1970), pp. 58–61.
4. GROISSER, PHILIP. *How to Use the Fine Art of Questioning.* Englewood Cliffs, New Jersey: Prentice-Hall (1964), 63 pp.
5. KLEBANER, RUTH P. "Questions That Teach" in *Grade Teacher* (March, 1964), pp. 10, 76–77.
6. PAYNE, S. B. *The Art of Asking Questions.* Princeton, New Jersey: Princeton University Press (1951), 249 pp.
7. SANDERS, NORRIS M. *Classroom Questions: What Kinds?* New York: Harper and Row (1966), 176 pp.
8. WELLINGTON, J. "What Is a Question?" in *Clearing House* (April, 1962), pp. 471–472.

RESOURCE PERSONS TO PROMOTE LEARNING

No matter where you live, there are persons in the school, in the community, or in the larger area who can be helpful to you and your pupils as resource persons to promote learning in the social studies. Their presence in the classroom will lend authenticity to a study and also will promote learning through the novelty of their appearance.

Here are a few of the people you might think about. Undoubtedly you can add others.

A. *Studying the School*

Principal and/or assistant principal, doctor, nurse, guidance personnel, older pupils (to tell about organizations in the school and work in upper grades), members of the student council, librarian, custodian, bus driver, school cafeteria personnel, members of the Board of Education, and others.

B. *Studying the Local Community*

Representatives of the Chamber of Commerce or similar organizations, town officials, older residents (on the community in the past), newspaper people, recreation workers, transportation and communication personnel, architects and builders, factory owners and store owners, park officials, doctors and health officials, bankers, lawyers, town planners, photographers, storekeepers, policemen, firemen, post office employees, welfare workers, a state employee or a legislator.

C. *Studying the United States*

Teachers, parents, and other adults or older boys and girls who have lived or traveled in other parts of the United States, federal employees in the community, in a few very special cases a Representative of the district in Congress and others.

D. *Studying the World*

Children in the school (not just in your classroom) who were born abroad and lived there for several years, parents and other adults who have lived or traveled abroad or were born abroad, students in nearby colleges or universities from abroad, representatives of international organizations.

Some Suggestions on the Best Use of Such Persons:
1. Select your visitors carefully, keeping in mind their ability to communicate.
2. Share with other classes in the case of persons who are in much demand.
3. Develop considerable background on the part of the pupils so that they can understand and question the visitor, but do not preplan and destroy the spontaneity of the occasion.
4. Encourage your visitors to bring visual aids if possible.
5. Keep a record of the presentation for future reference.
6. Be sure to express your gratitude to the resource persons in several ways. Encourage your pupils to do the same.

ROLE-PLAYING AND SOCIODRAMA IN THE SOCIAL STUDIES

One of the most important tasks of teachers of social studies is to help boys and girls get into the shoes of other people. Plays and stories and biographies and novels can help children do this. Talking about the feelings of others is sometimes successful. But teachers who have successfully used sociodrama feel that is perhaps better than any other method.

Sociodrama is the acting out by children or older people of situations. It is spontaneous and unrehearsed. It is done without costumes or scripts. The background of a situation is discussed and then parts are selected. Usually they are chosen rather than assigned. A short scene is then acted out involving a problem situation. After the sociodrama, the individuals in it discuss how they felt.

The parts played by the actors are called role-playing; the episode acted out is called sociodrama.

Here are some suggestions on the use of sociodrama in social studies classes:

A. *What Advantages Are There in the Use of Sociodrama?*
1. It can help children get into the shoes or skins of other people and begin to think and feel as they do.
2. It can make human relations situations real.
3. It can make history live for children.
4. It can release the imagination of pupils.

5. It can add novelty to classroom teaching and learning.
6. It encourages the feeling or emotional level of learning.
7. It encourages children to handle controversial issues in a realistic manner.
8. It is conducive to attitude change.

B. *What Types of Situations Lend Themselves to Sociodrama?*

Almost any situation involving human beings lends itself to sociodrama. Here are a few typical ones for class use:

1. The Feeling of a New Child in a School.
2. The Day I Got Lost.
3. Visiting the Doctor — or dentist — or any other new experience.
4. Our Family Moves.
5. We Were Bused to a New School. (An interracial situation.)
6. Making Out Our Family Budget.
7. My First Day in the Country of . . . .
8. Shall We Support the American Revolution?
9. Shall We Move with Our Neighbors Beyond the Appalachians?
10. How a "guest worker" in Switzerland feels.
11. A Slave Decides Whether to Join (or Start) a Revolt.
12. Shall We Join a Labor Union? (In the year . . . in the U.S.A.)
13. The Night We Came Home Later Than We Were Supposed To.
14. What We Told President Truman About Dropping the A-Bomb.
15. The Debate in the United Nations on . . . .

Situations I have used or could use in my class for role-playing:

16.
17.
18.
19.
20.

C. *What Suggestions Can Be Made on the Best Way to Conduct Sociodrama?*

1. Encourage pupils to pick their own roles. They will do better if they have selected their own parts. The ones they select may have meaning.
2. Take part in some of the first sociodrama situations yourself if you can become a "ham." It will help get the pupils into the mood of role-playing.
3. Discuss the situation in advance but do not tell pupils what to say or how they should feel. Let THEM decide.
4. Keep the situations brief.
5. Discuss the sociodrama after it is completed. Start with the actors and actresses themselves and discuss how they felt.
6. Replay the situation, with the same cast or with a different cast.
7. Organize some situations in which the entire class role-plays. A class may be divided according to percentage of persons in various types of jobs in the U.S.A. today and each group plays its roles accordingly.
8. In the initial role-playing situations, be firm about laughter on the part of the "audience." Otherwise they will break the realistic mood of the players.

Some selected references on role-playing and sociodrama

1. BOYD, GERTRUDE A. "Role Playing" in *Social Education* (October, 1957), pp. 267–269.

2. CHESLER, MARK and ROBERT FOX. *Role-Playing Methods in the Classroom.* Chicago: Science Research Associates (1966), 86 pp.

3. HERMAN, WAYNE L., JR. "Sociodrama: How It Works; How You Can Use It" in *The Grade Teacher* (September, 1964), pp. 84, 86, 153.

4. *How to Use Role-Playing.* Chicago: Adult Education Association of the U.S.A. (1960), 48 pp.

5. JENNINGS, HELEN HALL. "Sociodrama as an Educative Process" in *Fostering Mental Health in Schools*, 1950 Yearbook of the Association for Supervision and Curriculum Development.

6. JENNINGS, HELEN HALL. *Sociometry in Group Relations.* Washington: American Council on Education (1959), 105 pp.

7. NICHOLS, HILDRED and LOI WILLIAMS. "Learning About Role-Playing for Children and Teachers." Washington: Association for Childhood Education (1960), 40 pp.

8. SHAFTEL, FANNIE R. and GEORGE. *Building Intelligent Concern for Others Through Role-Playing.* New York: National Council of Christians and Jews (1967), 74 pp.

9. SHAFTEL, GEORGE and FANNIE R. *Role-Playing the Problem Story.* New York: National Conference of Christians and Jews (1952), 78 pp.

10. SHAFTEL, FANNIE R. and GEORGE. *Role Playing for Social Values: Decision-Making in the Social Studies.* Englewood Cliffs, New Jersey: Prentice-Hall (1966), 431 pp.

TEACHING MACHINES AND PROGRAMMED INSTRUCTION

Programmed instruction, used with or without teaching machines, is an extension into the school of the automation revolution of our times. Such instruction consists of breaking down subjects or topics into small segments, arranged in sequential order, to promote learning. On paper or on machines, pupils respond to the statements made and are given immediate responses to their answers before moving on to the next item. The critical factor in such teaching is the "programming" of content into individual frames.

A battle royal is now being waged between the proponents and opponents of programmed instruction. Here are the arguments of both sides in this current controversy:

A. *Advantages of Programmed Instruction as Seen by Its Proponents*

1. Pupils can study on their own and proceed at their own pace.
2. Programmed instruction provides for the constant participation of the learner.
3. There is immediate "feedback" or response, for the learner. This is particularly helpful to the slow learner, who usually needs immediate rewards.
4. The sequence of the frames follows the structure of the disciplines and therefore probably provides better and faster learning.
5. Machines do not punish children as teachers sometimes do.
6. Meaningful repetition is included.
7. The personality of the teacher is not injected into the learning situation.
8. The novelty of the machine arouses interest and promotes learning.
9. Programs can be written by experts for a variety of pupils.
10. Programmed material can be written so that it encourages pupils to engage in problem-solving, discovery, or inquiry.
11. The use of programmed materials by some pupils can free teachers to work individually and in small groups with other pupils.

B. *Disadvantages of Programmed Instruction as Seen by Its Opponents*

1. Programmed instruction is based on the conditioned response theory of learning, a faulty and outmoded theory.

2. Most questions in the social studies do not have clear, simple answers that can be programmed.
3. Programmed instruction promotes indoctrination and automated learners rather than thinking pupils.
4. The important teacher-pupil relationship is missing and socialization is neglected.
5. Programmed learning lacks the change of pace essential to learning.
6. When the novelty wears off, pupils become bored with programmed learning.
7. Machines cannot teach appreciations and values and cannot develop creativity.
8. Machines are costly and the money spent on them could be used more profitably for other worthwhile teaching equipment.

TELEVISION

The impact of television on children cannot be measured accurately and fully. But anyone who has worked closely with children knows that it is a powerful influence upon them, for good and for evil. At home many children spend several hours a week viewing the television screen. They pick up an astounding amount of misinformation. In schools television has already proved of tremendous value in many fields, including the social studies. Its potentialities, however, have scarcely been touched in most schools and classrooms. With worldwide programs already possible by way of satellites, the possibilities are greatly increased for its usefulness in social studies teaching. It may well become the electronic chalkboard of our times.

Here are a few questions teachers and teachers-to-be sometimes ask about this medium of mass communication:

A. *What Are Some of the Advantages of Television for Social Studies Teaching?*
1. It can provide a sense of immediacy, so that children feel they are taking part in current events and in the making of history.
2. It can arouse interest in social studies topics.
3. It can present programs which take weeks of preparation, which the classroom teacher cannot prepare.
4. It can show events as they are happening.
5. It can assemble persons from differing points of view to discuss controversial issues.
6. It can take children into all parts of the world.
7. It can use a variety of "props" which classroom teachers could not possibly assemble.
8. It can provide variety in teaching.
9. It can provide background papers for teachers, written by experts in the field.
10. It can serve as a catalyst, encouraging reading on subjects viewed.
11. It can utilize master teachers for large groups of pupils.
12. It can reach pupils in isolated areas.
13. It combines action, sound, and color.

B. *What Are Some of the Disadvantages of Television in Social Studies Teaching?*
1. The cost and the time involved in producing educational television programs.
2. Difficulties in scheduling programs for many schools and classes.

3. Technical difficulties in viewing programs.
4. The fact that television seems to arouse interest better than it motivates action.
5. The fact that television programs cannot be viewed in advance by teachers.
6. The lack of interaction between pupils and the program. Discussion is not possible with the person or persons presenting the program.
7. The fact that television cannot provide well for individual differences.
8. Programs are prepared for large groups of children without regard to the specific needs of smaller groups.

C. *In What Ways Can Teachers Plan for the Effective Use of Television in Classrooms?*

1. By providing the best possible viewing arrangements: turning on and adjusting the set before the program starts, avoiding light reflections on the screen, and seating pupils not more than six feet from the screen.
2. By studying program guides in advance if they are available.
3. By providing background material for programs to be viewed.
4. By providing ample time for followup discussions of programs, inasmuch as this cannot be done during a program.
5. By helping pupils in the development of listening and looking skills, such as looking for the main points, jotting down questions to ask later, taking notes, etc.
6. By avoiding a long "buildup" for a program, when the pupils are anxious to see the program itself.

D. *How Can Teachers Assist Children in the Better Use of Television at Home?*

1. By alerting children to programs which they might view, occasionally using these as homework assignments.
2. By discussing with parents some of the better programs on the air and discussing with them the amount of time that children view television during the week.
3. By discussing in school some of the programs children have viewed at home.

E. *Are There Distinct Advantages to Teachers, Especially Teachers of the Social Studies, in Television?*

1. Teachers can observe various programs to see the use of different teaching techniques and then utilize them in their own teaching.
2. Teachers sometimes can view master teachers over television.
3. Teachers can acquire a wide range of information through television programs.
4. Teachers sometimes can take courses over television.
5. Teachers often can learn about their pupils by watching their reactions to parts of television programs and thereby gain insights for more effective teaching.

Some selected references on television in education

1. Costello, Lawrence and George N. Gordon. *Teach With Television*. New York: Hastings House (1965), 192 pp. Second edition.
2. Dale, Edgar. *Audiovisual Methods in Teaching*. New York: Dryden Press (1969), 719 pp. Third edition.

3. GATTEGNO, CALEB. *Towards a Visual Culture: Educating Through Television*. New York: Avon Books (1971), 192 pp. A paperback.

4. GORDON, GEORGE N. *Classroom Television: New Frontiers in ITV*. New York: Hastings House, 1970, 248 pp.

5. MUKERJI, ROSE. *Television Guidelines for Early Childhood Education*. Bloomington, Indiana: National Instructional Television (1969), 57 pp.

TEXTBOOKS AS TOOLS FOR SOCIAL STUDIES TEACHING

Textbooks are the major scapegoats in education today. They are constantly criticized because they are too nationalistic or too world-minded in tone, because they include too much content or not enough, or because they overplay or underplay some special interest or minority group. One could continue to enumerate current criticisms.

Despite such criticisms, most modern textbooks in the social studies are remarkable productions. Years of effort go into them. Teams of authors, editors, and social science consultants develop them. Then the artists and map-makers begin their work. Often the books are tried out in classrooms before they are finally printed for wide use in schools.

Supplementing the basic texts today are "packages" of learning materials, including pictures, filmstrips, films, workbooks or response books, games, and/or other materials. Almost always there is a teachers' edition of the text, plus an annotated volume with suggestions for teaching in the margins.

Textbooks, however, should be considered merely as launching pads or springboards for learning. They should be thought of and used as appetizers rather than the entire meal. Teachers should not use them slavishly. They need not be the only basis for social studies learnings; they need not be followed chapter by chapter, paragraph by paragraph, and page by page. They should serve as the "common reading" for the class, supplemented and complemented by many other methods.

Here are some questions you may want to consider in regard to the use of textbooks in social studies teaching:

A. *What Aspects of Textbooks Need to Be Kept in Mind by Teachers in Working with Pupils?*

1. Learning to use the Table of Contents and the Index.
2. Studying the pictures rather than merely looking at them.
3. Using the Appendices as reference materials.
4. Learning to use the various headings as guides to what is coming.
5. Studying the maps, charts, graphs, and other graphic materials.
6. Stopping to answer the questions or to solve the problems posed in the text.
7. Using the study aids at the end of units or chapters for further learning or reinforcement of learning.
8. As teachers, obtaining and using the teachers' manuals provided with almost all textbooks today.
9. Using a variety of textbooks to provide for different reading levels in the class and to obtain different slants of various authors on a given topic.

B. *How Can a Teacher Include Such Suggestions as the Above in an Ongoing Program in the Social Studies?*

1. By spending some time in the early part of the year on learning how to use a new textbook.

2. By having specific lessons or parts of lesson on the skills of using textbooks and other lessons.
3. By determining early in the course the skills which pupils have learned and those they have not learned, and working assiduously on the latter.
4. By frequently stopping in a lesson to study a map, chart or picture, the material in the Appendix, or some other part of the textbook.
5. By asking pupils from time to time to:
   (a) Pick out the main idea of a paragraph or a page.
   (b) Ask two or three important questions based on what they have read.
   (c) Stop and discuss a difficult word or idea.
   (d) Sometimes let the pupils listen while you read, especially with poor readers.
   (e) Ask them to give an example of what an author has said.
   (f) Report on additional material they have read outside the textbook which extends the material in their book or books.
   (g) Make lists in their notebooks of the important people and places in a part of the textbook on pages devoted to such summaries.
   (h) Work in small groups with you, concentrating on the use of textbooks.
   (i) Compare accounts in various textbooks of the same topic.

C. *How Often and in What Ways Can "Open Textbook Lessons" Be Used?*
   1. The textbooks should be open much of the time early in a semester or year.
   2. With slower readers the textbooks should be open much of the time throughout the year as ready references.
   3. Frequently open textbook lessons should be used to concentrate on skills.
   4. On difficult passages or sections of a text, open textbook lessons are a "must." The teacher can say, "We're having a little trouble on . . . . Let's turn to our books and . . ." (look at a picture, look at a map, read a section).
   5. Open textbook lessons can be useful when the teacher reads aloud to a group or class as they follow her reading. Then you may ask various members of the class to read the same passages and then discuss them.
   6. Open textbook lessons are excellent learning devices when pupils are comparing what has been written in a variety of books. No two authors handle the same topic in exactly the same way or in the same depth.
   7. Open textbook lessons are suitable as pupils summarize the work of several days and pick out the important points they have discussed.
   8. If the books belong to the pupils, you may want to teach them how to underline certain sections or check a few sentences.

D. *What If the Textbook Is Too Difficult for Some Pupils?*
   1. Have the teacher or some pupils read aloud frequently from the text.
   2. Tape record chapters or sections and have pupils listen as they read.
   3. Study the pictures and other illustrative material before having the pupils read the text so that they will have background for their reading.
   4. Occasionally use other reading materials in place of the text.
   5. Display charts in the room with difficult words and terms explained.
   6. As the teacher, summarize some sections for the pupils.
   7. Mimeograph or xerograph a simplified version of the text for pupils.

TIME-LINES TO IMPROVE LEARNING

Every teacher realizes how difficult it is to develop time concepts with boys and girls. This is especially true with young children, but it also applies to boys and girls in the upper grades and in junior high schools.

One of the best devices for developing a sense of time is the time-line. Ordinarily it is considered a method for use with older pupils but it can also be used with young children if the categories are kept simple. For example, the categories might be Yesterday, Today, and Tomorrow. Later on they might be In the Past, Now, In the Future. Then they can be given dates as children gain a sense of time.

Here are some suggestions for the use of time-lines in social studies classes:

1. With young children use drawings and pictures underneath the categories in simple time-lines. This can also be done with slower learners in the middle and upper grades.
2. Develop time-lines with the class as you study a community or a country and leave the unfinished time-line in a prominent spot in the classroom.
3. Start a course in history with a simple time-line of the half-dozen periods to be studied. Then fill in details as you go along.
4. Study time-lines in textbooks and other reading materials.
5. Ask individuals or committees to prepare simple time-lines as a part of their reports to the entire class.
6. In the study of a period of history or a country, select a few important dates and place them on a time-line. This is a good homework assignment. In class the pupils can compare the dates they selected.
7. In a course in United States history have a time-line above the chalkboard to which reference can be made frequently to help develop a sense of time.

Here is a simple time-line on the history of the state of New York. Illustrations might well be added to give more meaning to this time-line.

Verrazano Discovers New York	English Take New Netherlands	The Erie Canal Opened	All Slaves in N.Y. Freed	St. Lawrence Seaway Opened
1524	1664	1825	1827	1959

TRIPS AS EDUCATIONAL ENTERPRISES

Trips or school journeys can serve very useful purposes if they are carefully planned and executed. Otherwise they can be merely jaunts and oftentimes fiascos.

They can be taken by pupils of any age, although younger children's trips should be limited to brief ones, usually in the local community.

They are especially important for children with limited experiences in the community or beyond the immediate community.

Here are some suggestions for your consideration as you think about taking a trip:

A. *Why Are Field Trips Important in the Social Studies?*
   1. To provide firsthand experiences for pupils with limited background.

2. To furnish opportunities for firsthand observation and new experiences for all children.
3. To correct false impressions.
4. To check with actual experience impressions gained by reading.
5. To provide experiences not possible in the classroom or school.
6. To stimulate interest in a topic, process, place, or period of history.
7. To provide common experiences for a group, and develop group morale.
8. To give opportunities for learning social skills.

B. *What Kind of Trips Are Possible?*
1. Trips in the neighborhood to parks, stores, government institutions, and other places of interest.
2. Longer trips to factories, banks, historic spots, radio and/or television stations, courts, election polls, etc.
3. Trips to the highest point in the area, from which an overview of geographical features and the layout of a city are possible.
4. Trips to nearby schools and colleges.
5. Trips to the state capital.
6. Trips to museums and art galleries in a nearby town.
7. Trips to villages that have been "restored."
8. Trips to visit people and interview them.

C. *What Are Some of the Criteria for Selecting Trips?*
1. Does this relate directly to the topic being studied or does it motivate for a new topic?
2. Is this a trip which could not be done by pupils and their families or by a group of pupils?
3. Is this trip within the right distance for the class, considering time, energy, and cost factors?
4. Is this a place where the pupils have not been recently, or where there is something new and important to see?
5. If there is no guide, can the teacher or some other person serve in that capacity?

D. *Should All Pupils Go on the Trip?*

Unless there is some overriding reason, all children should go. Children should not be deprived of such a valuable experience for disciplinary reasons except in very rare instances. Sometimes, but not always, it pays to use a disruptive pupil in a leadership role on a field trip. His feeling of being important often prevents him from becoming a discipline problem.

E. *What Plans Need to Be Made for the Trip?*
1. Approval of the proper officials in the school.
2. Approval, usually in written form, of the parents or guardians of the pupils.
3. Checking with other teachers who have made the same or a similar trip.
4. Planning, with the class, rules of conduct and dress.
5. Planning safety precautions, including a safety kit.
6. Inviting other adults to go with the class on this trip.
7. Clearance with other teachers, if necessary.
8. A visit to the place, if at all possible, in advance.

9. Planning of a time schedule, remembering that groups travel slower than individuals.
10. Planning the finances. Provide a "pool" of money which will include the expenses of those who cannot afford the trip, or earn money for the trip as a class enterprise.
11. Making arrangements for transportation.
12. Making provision for meals. Be sure to remember the special restrictions on any members of a religious group with strict dietary rules.
13. Make sure that no child will be excluded because of his or her race or religion.
14. Discuss what to do in case anyone gets lost!!
15. Discuss what to look for and whether to take notes or pictures.

F. *What Kinds of Follow-up Should There Be on a Trip?*
1. Provide for a variety of follow-ups. Do not always expect pupils to write essays or draw pictures on each trip. This may develop a mind-set against further trips.
2. Use an anonymous checklist on values of the trip.
3. Make sure that thank-you notes are written by you or by members of the class (not each individual).
4. Discuss with the class how to plan better for another trip.
5. Ask the class if they would recommend the trip to another class — and why or why not.
6. Discuss any questions that were unanswered on the trip.
7. Use any printed material, maps, pictures, etc., obtained on the trip.
8. Show pictures taken by pupils on the trip.
9. Make notes on the trip for the principal's office, for use by other teachers.

DEVELOPING
SOCIAL STUDIES
SKILLS

20. Acquiring reading skills with the help of a teacher

Developing Sc

21. Extending skills through preparing a report

22. Increasing map skills through the creation of a flannelboard map

23. Learning library skills

es Skills

24. Developing a sense of time

What Skills
Should Be Stressed
in the Social Studies?

7

The development of skills is an essential part of any effective social studies program. Such skills are basic tools for living today and they will be even more important tomorrow. Facts can be forgotten or become out-of-date, but skills, once learned, will usually be of service throughout life.

There are scores of skills in the social studies that need to be learned by boys and girls. For purposes of clarity, they are grouped in this chapter under seven headings: (1) Locating, Gathering, and Evaluating Information and Ideas, (2) Organizing Information and Ideas, (3) Communicating Ideas and Information (including Listening, Reading, and Presenting Ideas in Oral and in Written Form), (4) Interpreting Graphic Materials, (5) Critical Thinking, (6) Living With Others, and (7) Globe and Map Skills.

No skill is learned merely by some process of osmosis. Skills are learned best under the competent guidance of someone else, usually a teacher. Nor are skills learned quickly. They cannot be "covered" in a period or two. They must be taught and retaught or reinforced over a long period in many situations. Under the best conditions they can be perfected through practice.

Most of the skills mentioned in this chapter can be learned best inductively. After several experiences, children can begin to grasp what they are doing — and why. This is especially true if what they do is raised to the level of consciousness by questions from the teacher. Occasionally skills are learned deductively. For example, a teacher may point out that water is indicated on most maps by the use of blue coloring. Then the pupils can check to see if this is always true.

Some skills also are learned best kinesthetically. That is especially true of map skills. Children should construct many maps in clay, in papier-mâché, in salt and flour, and in other media, learning through the tips of their fingers as well as their heads.

In the acquisition of skills there is order, too. In the past we have too often guessed at the time when children learn certain skills, and in what order. We are learning that some skills can be acquired earlier than we thought formerly. That is true, for example, of many map skills. With the introduction of the new math, graphs, charts, and tables can also be used earlier than in former times.

Of course children learn skills at different ages, too. This is apparent, for example, in learning to read. It is just as true of other skills. Emotional factors are always involved in learning skills. Children have to be "ready" before they can learn.

Motivation therefore plays a significant role in the learning of skills.

All of these comments suggest the need for a comprehensive and cumulative program in social studies skills, from kindergarten through the high school — and beyond.

In order to teach any group of children effectively, teachers need to know how far advanced every pupil is in a variety of skills. That means that there should be an evaluation program in social studies skills very early in every school year. Then the teacher can act with intelligence in strengthening skills already learned, and in introducing new skills.

Since many of the skills in the social studies overlap with other fields, teachers would do well to work cooperatively with special teachers of other subjects. This is especially important in utilizing the skills of specialists in reading. But it also applies to other types of skills.

Now let us turn to some of the specific skills which need to be developed, with some examples of how to develop them.

LOCATING, GATHERING, AND EVALUATING INFORMATION AND IDEAS

All of us need to know about the many sources of information and ideas available to us. Furthermore we need to know how to gather information and ideas. Most difficult of all, we need to learn how to evaluate our sources of information.

These skills have begun to be developed before children come to school. Boys and girls are accustomed to using older boys and girls and adults as sources of information. They have used television and other sources, too.

In school they need to learn about other sources of information. They should be introduced to books as sources of information. They should have experiences with films and filmstrips, with trips, with slides and tapes, with recordings, and with visitors as other sources of ideas and information.

They will not use these sources of information and ideas, however, unless they see a use for them. That means that teachers must provide a stimulating and secure environment in which children ask questions without fear and because they learn that they can discover answers. Teachers need to give some answers to questions themselves and find other people who can answer some of the questions pupils ask. But they should not try to answer all questions, even if they are able to do so. Children need to learn to discover answers themselves.

Pupils should interview the school nurse about health problems. They should talk to the postman about his job. They should use pictures to find out about trains and airplanes and other means of transportation. They should observe films and filmstrips and listen to recordings from which they can obtain answers to the questions they have raised. Through such experiences boys and girls will discover some of the many sources which people use to obtain information and ideas — and their world will be expanded. What a wealth of sources there are to which children should be exposed in these early years.

Above all, children should be exposed to books. Books should be available in abundance in every primary grade classroom. Boys and girls should have many opportunities in a week to look at them. Teachers should often read to them for facts as well as for pleasure. And there should be many times when children go to the school and/or community library to browse and handle books. These should be pleasurable experiences, enhanced by story-telling periods or reading-aloud periods.

Parents should be encouraged to purchase and borrow books for their children.

There are many skills which need to be acquired before books can be read by the children themselves. One of the first is the skill in alphabetization. Another is word recognition. All too often such skills are looked upon as "language arts" activities. They are elementary social studies skills, too.

Very early in school, children can learn to read the titles of books, the names of authors, and the tables of contents. They can learn the sections of the library in which particular books are kept. They can scan picture dictionaries and picture encyclopedias during their early years in school.

In the middle grades children should learn about other useful sources of information. They should be able to use atlases prepared for boys and girls. They should begin to use the more simple encyclopedias. Some of them can use almanacs or books of facts to find information. They should learn about indexes in books and be able to use them from time to time. Some introduction can also be made to newspapers and periodicals. Here, again, skills are involved. Boys and girls need to learn about the different parts of a newspaper, about datelines on articles, about headlines, and about the importance of the "lead" sentences in news articles.

Children in the middle grades should use the library frequently. To do so competently means that they need to acquire skill in using the card catalogue.

Instruction can help them with new skills in using books and textbooks. They can learn about chapter headings, side headings, and aids at the end of chapters. They can learn to compare the information found in various books and learn about the dates of publication to see if information is up-to-date.

In the middle grades children can certainly learn from graphs, charts, and tables as important sources of information. They can also write to organizations, institutions, and companies to obtain data they need or want.

Each of these tasks requires the learning of skills. Children need help in acquiring this wide range of abilities in finding information and ideas.

In the upper grades or in junior high school there should be a great deal of emphasis upon skills. Then is the time to review what children know and to work diligently on the skills they will need in later years. It is not too much to expect some pupils to be able to prepare simple bibliographies at this stage in their schooling. They should be familiar with appendices, glossaries, and even prefaces to books. They should be able to use a wide variety of reference books.

By the time boys and girls have reached the middle grades or junior high school, they should be well advanced in being able to evaluate information. They should be able to find out about the authors whose books they are reading. They should be able to compare the authenticity of sources of information. Furthermore, they should be able to differentiate between fact and fiction and between fact and opinion. Hopefully they can examine the reasons for the inconsistencies they find.

More than before, pupils should be using the community as a source of information. By that time in their development, they should be able to explore on their own and in small groups outside of school hours, gathering data from people.

ORGANIZING INFORMATION

There is little value in gathering information and ideas unless they can be organized in meaningful ways. Otherwise they become collections of relevant and irrelevant data and useful and useless information. Pupils need to be assisted at every grade level in the organization of data.

This can begin in the kindergarten and be continued throughout the grades. The tasks selected for the organization of information should be simple in the early stages, of course, but they should be included in every good social studies program. With young children pictures can be utilized and be placed in groups by them. Experience charts can be developed and the title or the parts decided upon by the pupils. These are the types of experiences children should have in organizing information.

There are other ways, too, in which the skills of organization can be taught. Children can arrange pictures in sequence to tell a story they have heard. Captions can be decided upon for pictures and drawings. Picture dictionaries can be made. The organization of simple scrapbooks is another device by which children can learn to organize information in categories. Exhibits can be assembled of various people they know or of the many means of transportation they have discussed. Drawings of different types of materials used in building houses can be collected and placed on bulletin boards.

These are some of the ways in which the organization of information may be approached in the primary grades.

By the middle grades boys and girls should be ready to take simple notes. The teacher may give them the title and ask them to list several pertinent facts under it or he may give them several pertinent facts and ask them to write the title. Children can occasionally prepare reports to give to the class or take part as a committee in the organization of a report. Some pupils will be ready, also, to take some notes on materials in trade books and in encyclopedias.

Many pupils in the middle grades can also begin to sharpen their ability to differentiate between fact and fiction and to detect some of the more glaring examples of propaganda in articles they read, in films and filmstrips they see, or in statements made by their classmates.

At this point in their educational development pupils should also have experiences in organizing other types of materials, such as flow charts and time-lines.

In the upper grades of elementary school and in junior high schools pupils should have a great deal of practice in organizing materials. They may work with ideas in their textbooks or with data from trade books. They may bring together in good fashion data from interviews inside the school or in the community. They may take notes on television or radio programs.

At this point they ought to be able to compare and contrast differing accounts of events or people in the social studies field. This is not easy and they will need help in learning such skills, but this will help them throughout their lives in evaluating information and organizing it.

Teachers need to ask themselves from time to time whether they are helping children in this essential skill in the social studies and how they could do the job better than they are now doing it.

LISTENING

In the past little attention was given to the skills involved in listening. People assumed that you listened or you didn't listen. They did not recognize how many skills are involved in this complicated process.

It is estimated that two-thirds of the time in an elementary school is spent in oral activities which involve listening. Dr. Helen Carpenter has called listening "the most widely used language function in life today."

In his book on *Listening: Readings* (Scarecrow Press), Professor Sam Duker suggests four key principles. He believes that listening should be pleasurable rather than threatening. "Getting told," he says, is seldom synonymous with learning. In the second place, the amount of listening should not be overwhelming. Third, he warns that listening should not be confined to listening to the teacher. His fourth principle is that listening should be "for" something, not "at" something or somebody.

Duker goes on to suggest ten characteristics of listeners. According to him a listener is one who: (1) not only knows how to listen but who actually does listen, (2) is selective in his choice of what to listen to, (3) can identify main ideas in what he is listening to, (4) is a critical listener, (5) is considerate and courteous, (6) is an attentive listener, (7) is retentive, (8) is curious, (9) reacts to what he hears, and (10) reflects on what he hears.

So listening, as we ordinarily think of it, is not enough. It may not mean hearing. It may not indicate understanding. It may not involve reflection. It is not enough for teachers of the social studies, or any other subject, to tell pupils to "Pay attention" or "Listen carefully." Boys and girls need to be helped to learn to listen.

In the primary grades children need help in listening "for" something. It may be the sounds in the room, the building, or the community. It may be new names in the stories they are hearing. It may be the main ideas that a composer is trying to get across in a song. It may be something which an older pupil or an adult is saying when that person visits the class to speak to the pupils. Much of the time it will be learning to listen to what their classmates and the teacher are saying. Often it helps when boys and girls know that they are going to make a drawing or prepare a dramatic skit. Then their listening has purpose. Are you now helping children to listen? Could you do a better job of developing this skill or cluster of skills?

In the middle and upper grades the skills of listening should be improved and expanded. Boys and girls should be able to listen better to each other as well as to the teacher. "How do you feel about what Jim said?" is the type of question a teacher should often ask such pupils. Another good question is, "Could you put into your words what Mary has just said?" That, too, gives focus to listening. Pupils in these grades can utilize radio and television programs to help them learn to listen. They can talk with adults — and listen.

In social studies classes there should be many opportunities for pupils to learn to listen to each other, especially through such activities as small and large group discussions and panels or roundtables. As a teacher, you have opportunities to judge the ability of pupils to listen in such situations.

All their lives pupils will be hearing others — or not hearing them; listening to others — or not listening. Learning to listen is very important.

READING

Some social studies instruction can certainly be carried on without reading, but inability to read seriously limits the scope of learning. Much of the knowledge about people in the present and in the past is contained in printed materials. Reading is a major tool with which the rich vein of man's recorded experience is mined.

Therefore social studies and language arts programs overlap in many ways. Each complements the other. Each supplements the other. Each enhances the other. Improvement in the skills of reading can assist pupils in the social studies and improvement in the social studies can improve the ability of pupils in the language arts.

There are, however, some special skills in reading social studies materials. High on the list are the ability to see relationships, to develop a specialized vocabulary, to compare sources, and to summarize data.

Almost every lesson or activity in the social studies should include some attention to the skills of reading. In addition, some time should be set aside for more specific work in the development of skills. In planning lessons or activities, teachers need to stop and ask themselves what new ideas or concepts and what new words are being introduced. Then they need to plan how that is to be done.

There are several types of reading which pupils need to learn to carry on in the social studies. These include: (1) skimming, (2) cursory reading, (3) study reading, and (4) critical or reflective reading. Reading can also be considered in two other ways: (5) as a social activity for a class, and (6) as relaxation and recreation.

Textbooks are often difficult for pupils. In the short compass of one page many new words are introduced and several concepts may be developed. Pupils therefore need much help in reading textbooks. That is why open textbook lessons or study are recommended highly, especially with so-called slower students.

Often teachers need to arrange for experiences for children which will develop new words and concepts before they come across them in texts. Then they will bring some meaning to the words they encounter. Merely defining words which appear in books will not do much to promote learning. Such a practice encourages memorization rather than learning.

What a vast array of new words there are. Children are confronted with islands, imperialism, industries, interdependence, independence and inns — to select just a few words beginning with the letter "i." And they are confronted with words which have more than one meaning, such as the Indians of North America and the Indians of India. Then there are the words that have a different meaning in social studies than in more ordinary use, such as the "hold" of a ship, "gaps" in mountains, and "mouths" of rivers. Is it any wonder that many pupils are confused at times?

Furthermore there are new people, new places, and new events which children encounter in their reading. Meanings need to be developed for each of these and pronunciations learned.

But this is not all. Interpretations of these people, places, and events differ. Authors vary in their treatment of historical events and contemporary affairs. Children may want to know why. That is where critical thinking and critical reading enter the picture.

No program of social studies can be really effective if the pupils must rely solely on a single textbook, no matter how good it may be. They need to use a variety of textbooks. They need to read in books devoted to a single topic in depth. They need to read a variety of graphic materials, current events papers and newspapers, encyclopedias and almanacs, pamphlets and booklets, magazines and documents.

Yes, reading involves a wide range of materials and a complex cluster of skills. Time needs to be taken in the social studies to develop the skills of reading. And a wealth of resources needs to be on hand to provide for the varied reading abilities in any class.

How does this work in actual practice at various levels in schools? Let us turn to a more detailed description by grade levels.

In the primary grades pupils need to have a wealth of easy-to-read books available. At first they should be encouraged to peruse these volumes, look at the pictures, and read the simple captions or text. Because children's interests vary widely, the books should cover a wide range of topics — mothers and fathers, older and younger brothers and sisters, trains and airplanes, farms and stores, and a host of other topics. The main aim is to interest children in books.

They should be encouraged to tell about the books to their classmates and to answer questions about them which their peers and the teacher raise. This may mean that they need to return to the books for further information.

Soon they will begin to learn the skills of locating information. They need to know about the titles of books and the authors. They need to learn about the table of contents. They need to learn where material is located in a book — approximately one-quarter of the way through, halfway through, or three-fourths of the way through.

Then they should learn to look for specific information — the name of a person, the color of a house, the objects in a store — or other relevant information. This involves practice in skimming.

Pupils in these early grades should learn to spot new words and to ask about them. Teachers should keep lists of words which are new and provide experiences to help children learn the meaning of these words and terms.

Pupils can learn a great deal through reading aloud to their classmates. This involves a variety of reading skills, too.

Finding the main ideas in a paragraph or a page is a skill that needs to be fostered in the primary grades. Teachers can help pupils to acquire this skill by providing guide questions.

In the middle grades there will be much variation in the reading skills of pupils. Some will have grasped the initial skills easily and will be able to use a wide variety of fairly difficult sources. Others will still be wrestling with the elementary skills in reading. This means that a great range of reading materials needs to be available. Three or four different texts should be used in order to provide for individual differences. Two or more current events papers should be utilized to meet the differing reading abilities of the pupils. Trade books covering three or four different grade levels in readability are called for.

In these grades pupils should learn how to locate library books in different sections. They should begin to use the card catalogue frequently. Some pupils should begin to prepare simple bibliographies for reports.

Within books they should now be able to use the indexes, the lists of maps, and the appendices, as well as the tables of contents and the title pages.

Many pupils should be able to begin the use of regular newspapers during these years. Most children should be able to use a wide variety of reading materials, too, including films, filmstrips, graphs, charts, pictures, and posters, in order to acquire information.

Much more research in books, almanacs, and encyclopedias can be carried on in the middle grades, too. Pupils will need to be taught how to use these reference materials. And this should be when they really need them. That will make learning functional — and more lasting.

Pupils in this period of their schooling will need help, also, in grasping the main ideas in paragraphs and sections of books and in note-taking.

A much larger vocabulary can be expected in these grades in the social studies. But the acquisition of such a vocabulary will not come automatically. Teachers will need to help pupils in this respect.

Because boys and girls can read more widely, many trade books can be used. Some will be for pleasure; others will be used to gather information. Biographies, simple novels, and even elementary source materials can be usefully employed, too.

In the upper grades, teachers of the social studies need to reinforce all of the skills learned up to this point and to add some new ones. With some pupils it will be necessary to do a great deal of training in the skills of reading because many pupils need such basic education. With others, advanced materials can be used because they read widely and well.

The range of reading abilities in grades seven and eight will probably be even wider than in the middle grades. That necessitates the use of a wide variety of textbooks and trade books as well as magazines and newspapers.

In these grades most pupils should be using encyclopedias, yearbooks, and almanacs as reference materials. They should be able to handle card catalogues well. In these upper grades or junior high school years skill training should be given high priority. Without adequate ability to read, pupils will be handicapped for the remainder of their years in school.

Some selected references on reading in the social studies

1. FAY, LEO, THOMAS HORN, and CONSTANCE MCCULLOUGH. *Improving Reading in the Elementary Social Studies.* Washington: National Council for the Social Studies (1961). Bulletin 33.
2. HUUS, HELEN. "Reading." Chapter 6 in CARPENTER, HELEN M. *Skill Development in Social Studies.* Washington: National Council for the Social Studies (1963), 332 pp.
3. JAROLIMEK, JOHN. *Social Studies in Elementary Education.* New York: Macmillan (1971), 534 pp. Chapter 11 on "Reading Social Studies Materials."
4. MICHAELIS, JOHN U. *Social Studies for Children in a Democracy.* Englewood Cliffs, New Jersey (1963), 624 pp. Chapter 10 on reading.
5. PRESTON, RALPH C. (Editor). *A New Look at Reading in the Social Studies.* Newark, Delaware: International Reading Association (1970), 69 pp.

GRAPHIC MATERIALS

Charts, cartoons, diagrams, graphs, posters, tables, time-lines and other graphic material are being used more and more in our everyday life. They constitute a special kind of language of our times. But the symbols they use are worth learning and therefore worth teaching because they help us to understand many ideas and gain much information. They simplify complex ideas, show relationships, compare and contrast statistics, and indicate steps to take in many processes. And they do this in a very short space in newspapers and magazines, in posters, in ads, over television, and in textbooks and other books. They are truly remarkable teaching devices.

All of these symbols need to be taught in schools. Many of them are an essential part of any social studies program. Some of the most simple can be used in the primary grades. Most of them should be introduced in the middle grades and used increasingly in the upper grades.

Since they can be used for persuasion and propaganda as well as education, children need to learn how they can be misused as well as used educationally.

Elsewhere in this book we have discussed the uses of pictures (pages 101–102) and time-lines (page 112). Here we shall limit ourselves primarily to the use of charts and graphs, with a mere mention of cartoons and posters.

There are many uses of charts and diagrams in the primary grades. The experience chart is probably the most frequently used. Through it, children, with the aid of a teacher, can learn to summarize and to develop sequences. Charts can also be used effectively to summarize a trip made by the class or a film or filmstrip viewed by them.

In the study of families, effective charts and diagrams can be made of a variety of types of homes in the United States and in other parts of the world. Some of the most effective charts and diagrams can also be made to show buildings or mines. Such graphic materials look as if a slice had been made of the building or mine or an x-ray picture taken of them.

Children can be introduced very early to the use of simple bar graphs. This is more easily done if they include pictorial representations. For instance, attendance can be portrayed in bar graphs, with a figure to represent the boys and another to represent the girls. The birthdays of children in the class can be shown by months in bar graphs. The books read by children in the primary grades can be shown in a similar way, with a symbol of a book utilized.

In the middle grades, charts and graphs can be of even more use in the social the social studies, such as the population of various communities or the ethnic groups within a community, the expenses of local government for various services, or the height of different buildings.

In the upper grades they can be used even more extensively. They can show the growth of population in the United States or the growth in the number of states over a long period of time. They can illustrate the number of persons involved in various types of work here or in other parts of the world. They can be used to show exports and imports of various countries or the size of different nations.

In the primary grades simple charts can be made of new words used in social studies work. In studying about the building of a house in another part of the world, a simple chart can indicate the steps by which it is constructed. Children can learn through a simple pictorial chart how milk moves from the farm to their home. This is ordinarily called a process chart. If they are studying clothing, they can do the same for cotton or wool, tracing it from the original source to the final product which they are wearing. The same thing can be done with the travels of a letter from their home through various steps to its final destination. Young children should also have charts around the room which will help them to know about the division of labor in their homeroom jobs, ways in which to listen well, or possible steps in giving an oral report.

In the middle grades, charts and graphs can be of even more use in the social studies. Processes are shown well in charts, whether it is the steps in making steel or in electing a mayor. Another type of chart can show the many uses of trees, of coal, or of copper. Organizational charts can be introduced, showing the ways in which the local government works or how a business is organized.

At the upper grades level an even greater variety of charts and graphs can be employed usefully in teaching the social studies. Organization charts can be used to show how the federal government is divided into three main divisions or to show the President and his helpers in the many departments and special agencies of our national government. The United Nations can be charted and thus better understood. The exports and imports of various countries can be presented in chart form. The Civil War can be made more meaningful by a chart of the resources of the North and the South at the outbreak of hostilities. These are but a few samples of such activities at that point in school.

Posters can serve many purposes, too. One is to create a lively and interesting environment. This can be done at all levels. In the study of a city in the United States or in some other part of the world, the classroom can be filled with posters of that locality and a different atmosphere created. Some information can also be obtained from posters. More than anything else, posters can help pupils to visualize a place. The same can be done extremely well in the study of countries. Contact with travel agencies and tourist companies, airline companies and boat companies, will uncover much material of this kind.

Cartoons are much more difficult to use in elementary schools. With bright youngsters in the middle grades, a few cartoons can be used and the skill of interpreting them introduced. In the upper grades they can be used much more. Most cartoons are pretty sophisticated, however, and teachers need to use them with great care. Pupils need to bring to them considerable background on the topics presented, in order to understand them. The cartoons reproduced in current events papers are probably the best for introducing boys and girls to this graphic form inasmuch as they have been carefully screened for publication. Occasionally cartoons in daily newspapers can be used effectively with older pupils.

Children, however, will learn about all of these graphic forms better if they engage in preparing examples of them. There will be more motivation if the teacher suggests that factual information be presented in graphic style and the pupils themselves select the media to be used. This should be encouraged in the middle grades and used widely as a method in the upper grades. The use of these various graphic forms will also appeal to different children and thus help provide better for individual differences in social studies teaching.

PRESENTING INFORMATION AND IDEAS ORALLY

Throughout life all of us need to communicate with others. Most of this communicating is done orally. This is true of everyone, from the infant who makes his or her wants known in inimitable fashions, to elderly persons.

Schools bear a great share of the responsibility for the improvement of oral communication. And in that responsibility the social studies have a primary role to play. This is carried on through the development of skills in four main categories according to Dr. Helen Carpenter in the 33rd Yearbook of the National Council for the Social Studies, on *Skill Development in Social Studies*. They are: (1) discussion, (2) reporting, (3) interviewing, and (4) dramatizing. The skills involved in these four broad areas overlap, but each has its own distinctive aspects. Let us see how they may be developed with pupils of different ages and of different maturity levels.

In the primary grades children need to have a rich background of experiences individually, in small groups, and as a total group. Otherwise there is little to communicate. This is especially true of children whose home backgrounds do not provide many rich experiences. Visitors to the class, working with blocks or playhouses, trips, stories the teacher reads, books the children examine, filmstrips and other pictorial materials they see, should all provide the kinds of experiences about which even the most reluctant will want to talk. As Alvina Burrows has pointed out, "Reporting in the social studies is the natural outcome of the child's compelling urge to communicate both with other children and with adults."

Children in these early years in school will carry on much discussion in teams or small groups. Sometimes they will converse as a total class. Teachers can help them with many aspects of discussion from the use of their voices to clear and specific instructions on what others are to do. Praise needs to be lavished upon the quiet child to give him security to talk with others. Sometimes children can be asked to repeat their statements or give their brief reports again after suggestions have been made to make them more audible or more interesting. With more successful experiences, they will have "models" for later presentations.

The Show and Tell period may be helpful if it does not become a ritual whether there is anything to tell about or show or not. Often children need to be encouraged to have something visual to enhance their reports. Children can make simple announcements and give directions for small groups or an entire class. They can tell about a trip they took, report on questions asked of adults or older children, or show and tell about a book they have perused or read. They also can take part in brief discussions which the teacher leads. In them they can learn to speak quietly, to the point, and in interesting ways. Most reporting on books, however, should come after repeated experiences with other types of reports.

In the middle grades all of these activities should continue but should be improved. Children in those grades should make much more use of a wide variety of reading experiences for reports. Their reading should include current events papers, encyclopedias, newspapers, graphic materials and visual materials, and trade books. Interviews with older boys and girls and with adults can be increased. Some simple panels and roundtables can be introduced and elementary forms of debates started. Oral reports can often be combined with learning the skills of note-taking. But children should be kept from reading their reports. Often a teacher needs to say, "Put down your notes and tell us in a few sentences what you learned." More able pupils can be used to preside at some discussions for a brief period or to chair panels or roundtables. Dramatics can be used often as ways of reporting to a group. Committee reports can also be introduced at this point.

In the upper grades teachers need to remember that many pupils are extremely self-conscious. Therefore, many reports need to be given from the seats of the pupils rather than in front of the room. Boys and girls will feel more comfortable, too, in group reports, such as panels and roundtables, than in individual reports.

More skills can be developed in the selection of material for oral reports, coupled with better note-taking. Interviews can be used much more as pupils, at this stage in their development, move out more into the community and want to associate more with older people. Role-playing and dramatics are ways of overcoming some of the natural shyness of many pupils. With others it is necessary to temper outward self-confidence with a little humility. Standards of oral reporting can be kept high and demands made for excellent results on the part of such pupils.

Debates appeal to many boys and girls at the upper grade or junior high school level. There is more clash and direct confrontation in a debate than in a panel, and someone wins. But there are drawbacks in the social studies in debates. They tend to teach children that issues are right and wrong, black and white with no shades of gray. They teach pupils that one side must win — and that compromises are forbidden. Despite the fact that your author is a former intercollegiate debater, he feels strongly that the roundtable or panel is a better form of oral discussion than debating. Do you agree? Disagree? Why?

At this stage in life, some pupils are ready for more training in parliamentary procedure. Sometimes this comes in co-curricular activities. Often, however, it can be used in classes in the social studies.

Some able pupils at this point are ready to present their ideas to pupils in other classes and to large groups of boys and girls in assemblies.

PRESENTING INFORMATION AND IDEAS IN WRITTEN FORM

A perusal of many books on the social studies in elementary and junior high schools reveals that very little has been written in recent years about the presentation of ideas in written form. Is there a reason for this? Is it because people consider this the domain of the language arts? Or is it because we are not too concerned about developing this aspect of social studies learnings?

Despite the paucity of comments, communicating in written form is another important part of social studies teaching and learning. It, too, has at least four major divisions. We need to communicate with ourselves, taking down ideas and arranging them for use. We need to communicate with others. In the third place most of us will need these skills in later life. All pupils will need some skills in written communication in high school and many in college, as well as in later life. The fourth category is writing for fun as a means of self-expression. Only a few people will enjoy themselves in this way, but some will, and they should not be overlooked. They may be among our most creative people.

Where does written communication begin? Not in the junior high school or even in the middle grades. It begins, like almost everything else, in the primary grades or even in kindergarten.

The child draws a picture or paints something. This is one form of written communication. It actually becomes writing when he suggests a title and the teacher executes it for him or her. Later, pupils can begin to decide upon their own titles and present them, with help from the teacher.

Writing continues as the teacher writes for children while they decide upon lists of ideas or items, describe an experience they have had in a classroom, summarize the comments of a visitor they have met, or jot down the sequence of a dramatization in which they will take part.

Then children can learn to copy material the teacher has prepared, alone or with the pupils.

Labels can be placed on exhibits and models, on pictures clipped from magazines and newspapers.

All these are beginnings of written communication.

Children can then begin to look for facts in their reading and engage in "fill-in" work on the books they have read.

Eventually, with much help from the teacher, they can begin to express themselves in written form, learning about capital letters, about simple punctuation, complete sentences, and spelling.

This, too, is a gradual process which some learn fairly quickly and others learn very, very slowly. By the middle grades children should be able to do some brief, simple work in the social studies in writing. They will do better if they sense the purpose. Perhaps it is a letter to obtain materials, to invite someone to visit the class, or to thank them for their visit. Perhaps it is a brief statement to go with an exhibit for an assembly program or for a P.T.A. Perhaps it is a series of brief statements to show the processes in the development of a product which will be used with a report of a committee to the entire class. Possibly it is a brief account of a book read.

In the latter part of the middle grades and in grades seven and eight most tests should include at least one short essay question. Some homework and some class work should include note-taking and the skills of note-taking should be taught as specific and important skills in the social studies. Here is where the overhead projector can be used to advantage, with all eyes concentrated upon the electronic chalkboard. Materials which children have written can also be placed in the opaque projector for everyone to see. This can serve, too, as motivation for better work, even though it is not very high in the scale of internalized motivations.

Much attention should be given by the sixth grade to written work in the social studies and continued in depth in grades seven and eight. This is a skill—or perhaps one should say a whole cluster of skills—upon which social studies teachers need to concentrate. Each of these skills is important to children now and will be increasingly important to them as they continue their education and venture forth into life as adults.

THINKING

Almost without exception, lists of objectives for the social studies include some statement about helping children to learn to think. Many such statements are more specific, stating that children should be taught to think critically or creatively.

Too often, however, these statements appear in printed lists rather than in classroom practice. They are pious hopes rather than tangible realities. They are honored more in principle than in practice. Unfortunately educators do not agree upon how children are taught to think. They do not agree upon what children should think about. And they do not agree on whether children should be allowed or encouraged to act upon their thinking.

Thinking is an enormous topic and cannot possibly be treated in much depth in a couple of pages of this book, augmented by other sections, such as those on learning and on problem-solving. In this part of this book, therefore, some general comments will be made in the hope that they will stimulate thinking on the part of the readers and encourage them to turn to the writings of such persons as John Dewey, Boyd Bode, David Russell, Jean Piaget, Jerome Bruner, and Hilda Taba, to mention only a few persons who are experts in the field and have carried on years of research on the thinking processes.

There are children, of course, in every school who cannot be expected to do much thinking. They are already "crippled." Life has taught them before they entered school, or at some point thereafter, that they must not think. Someone will do their thinking for them. They are like turtles who have drawn into their shells. Some of them can be persuaded to stick their heads out and explore the environment if that environment is exciting and the people in it are protective, warm human beings. Others are damaged so much that they need psychological or psychiatric counseling.

Fortunately such persons are in the great minority in most schools. Most children are eager to explore, intent upon discovering, and eager to have new experiences. For them the environment needs to be filled with a variety of concrete objects and stimulating experiences.

Basically, thinking is problem-solving. A person feels a difficulty or problem. He needs then to locate or define that problem and then to draw up tentative hypotheses for solving it (with suspended judgment). Then he needs to test the implications of those hypotheses and by observation and experimentation arrive at a conclusion. If children are not confronted with problems which are meaningful to them, precious little critical thinking will take place.

We emphasize the need for concreteness largely as a result of Piaget's research in which he simplifies the stages of learning into three categories. The first he calls the sensory-verbal stage or the pre-intelligence level. The second he terms the stage of concrete operations, with objects and concrete events as the bases of learning and thinking. This is a stage in which most children from two to eleven find themselves. The third he characterizes as that of conceptual or formal thought, beginning usually at the age of eleven.

Children should not be overwhelmed with stimulating experiences, but they need to have a wide variety of contacts with people, places, and events which will cause them to ask questions and seek more experiences. From such activities they will have data with which to think. Nursery schools and kindergartens often provide such opportunities. Starting in the primary grades, children too often are deprived of the wide opportunities for probing and discovering and "trying out" that they need so much for growth.

If such an environment exists or you can create it, another problem arises. That is the teacher's commitment to thinking. Most teachers talk too much. They tell children what to think — and when. The effective social studies teacher creates a stimulating environment. Then such a teacher encourages children to talk about what they see, feel, and hear. Language and its use are inextricably bound up with thinking. Without words or other symbols to express ideas, thinking is not likely to result. That is why the emphasis in the "Head Start" programs has been upon experiences to evoke the use of language. Such teachers as we have been describing also ask questions which will start children on further explorations for facts and ideas.

Almost immediately, children begin to form hypotheses. They may not be correct ones, but the thinking process is underway. It is then that the effective teacher raises the questions that help pupils to revise their thinking. "All people live in apartment houses," the young child says. A trip around the block or further away may provide material for revising that conclusion. Or a picture or filmstrip or a book may help that child to revise his opinion. Soon the teacher may feel that that child or a group of children are ready to restate their conclusions. Then is the time to clinch a generalization or a concept and move on to another stage.

Writing about cognitive learning, Dr. Hilda Taba says that concept formation consists in its most simple form of three stages. One is the differentiation of the characteristics of objects and events. The second is grouping. The third is categorizing or labeling.

In the illustration given above regarding houses, the children can then move on to discover that people live in a variety of houses. Then they can discover why that is so. Thus they can move from "what" stage to the "why" stage in thinking. Too many teachers do not help children to move in that way. They keep pupils on the plateau of "whats" and "whens" instead of leading them on to the peaks of "whys" and "hows."

If a teacher moves a group of children too fast from one stage to another, some children may revert to a lower stage of thinking. With proper pacing and timing, boys and girls can move from one stage of thinking to another, reaching a real understanding of a concept and being able to state it in simple terms. That is thinking at its best. And it can be done with young children as well as with older ones.

Although some stress has been laid on concrete, firsthand experiences, it is evident from research that facts for thinking can also be gleaned from past learnings or experiences and recalled by children. Children do not have to have immediate experiences as background for all their thinking.

And how does one categorize the many experiences children have in watching television programs? Are these first-hand experiences? Not exactly, as boys and girls are not actually taking part in what they see, even with "audience participation" devices. Yet the people and events may be very real to the viewers, much like a first-hand experience. From such experiences they can gather a great deal of useful information to be applied later in developing concepts and generalizations.

In the middle grades the environment can be focused on the major ideas, generalizations, and concepts that the teacher feels are essential at that juncture. She can set the stage for specific learnings. For example, children may be asked to find the ten largest cities in the United States. With a flannelboard, they place symbols of those cities in their proper location. Then comes the question, "In what respect are all of the cities alike in regard to their location?" Boys and girls can then be given time to find their answers. This is one way to develop the concept of water in the past as a factor in the location and development of cities. Such an approach is conducive to thinking and learning rather than to memorization.

In the social studies such teaching is similar to the use of sets in modern mathematics. Children begin to collect information, to compare data, and to place the data in categories. Concepts are convenient categories into which children organize their experiences.

As children grow older and more proficient in the processes of thinking, they can deal with more abstract ideas and with events which are further removed in time and place. Often the people they study can provide contrasts with those they have already learned about. Thus a study of the people of some part of Africa or Asia can provide a broader framework in which thinking can be carried on. Previously accepted generalizations will then need to be tested and broader generalizations framed.

In the later years in elementary school and in junior high school, boys and girls can cope with historical data far better than in their earlier years in school. Thinking does not have to be restricted chiefly to the here and now. It can include other times, other people, other places, and other events. For example, a study can be made of the reasons people came to the New World. Several groups in a class may work on different groups of immigrants. Then they can report their findings as to why those particular people came here. As a result of "pooling," members of the total group can make a general statement about the various reasons for immigration to the United States. Some of the reasons would highlight conditions in the homelands of the emigrants which drove them out of their former localities. Other reasons would focus on the attractions of their new homeland. The teacher does not need to tell the pupils the reasons; they can discover the reasons themselves.

If thinking is basically problem-solving, as we have already stated, then the problems with which older boys and girls wrestle must be real, relevant, and meaningful to them. These might be classified in two categories. The first would be personal problems, such as the question of self-identity, relationships with peers and with adults, vocational explorations, and questions of values (or religion). The second category would be society's problems, particularly local ones. At this age boys and girls usually want to identify with adults (at least part of the time). One way to promote this is to study some of the problems of the adult world — such as drugs, alcohol, pollution, poverty, war and peace, and housing.

Long after specific information is forgotten, boys and girls still should retain certain skills in ferreting out reasons for the actions of people. Such skill training is an absolute essential of effective social studies instruction.

GLOBE AND MAP SKILLS

Globes and maps are important tools for almost everyone. Boys and girls begin to use them early in their lives and will continue to do so as long as they live. Therefore they should learn to use them while in elementary school. However, learning to use globes and maps is not an easy process. Such learning involves a wide variety of new skills. In many ways learning these new skills is like learning a new language.

But these skills are important and learning them can be fun. Here are some suggestions on the use of globes and maps in elementary schools and some suggestions for teachers for further reading in order to expand their knowledge of geography.

A. *Why Are Studies of Maps and Globes Important to Boys and Girls?*
1. They help pupils to observe more closely.
2. They help pupils to learn many relationships which they might not acquire otherwise.
3. They help pupils with day-to-day living.
4. They enhance their understanding of their homes, their neighborhoods, their country, and other parts of the world.
5. They help pupils to understand television programs and newspaper articles relating to current events.
6. They help pupils to develop skills which can be used throughout life.
7. They can lead later to vocations for a few pupils.
8. They provide enjoyment — just the fun of making and studying maps.

B. *How Can One Explain Maps and Globes in Simple Terms?*
1. Children can discover the meaning of maps and globes through concrete experiences, rather than being told what they are. In introducing globes and maps, it may be useful to start with other "models," showing how dolls represent real people, and miniature airplanes real airplanes. Pictures of children are also "models."
2. Globes are models of the earth, on a very small scale. Maps or models or drawings represent parts of the earth on a small scale.

C. *How Early Can Globes and Maps Be Introduced in Schools?*
1. Globes can be used even in preschool work if they are placed in the room and referred to from time to time as pupils talk about people and places. This is part of a "readiness" program.
2. Pupils can begin to learn the principles of map-making as they play with blocks and objects and learn about space. More formal education in the use of maps needs to be introduced at least by the first grade.

D. *How Can Teachers Arouse Interest in Maps and Globes?*

Much of the interest of boys and girls in maps depends upon their early successes in making them and in interpreting them. The first experiences should be enjoyable rather than tedious and boring. In the beginning, absolute accuracy should not be emphasized. Gradually, as more skills are learned, the standard of achievement can be raised.

Here are some suggestions on ways to arouse interest in the use of maps:

1. Get children to make maps of the places which are close to their daily experiences — their homes, their neighborhoods, their schools, and the places they visit in vacation times.
2. Encourage children to role-play situations involving maps — such as the newspaper carrier or the postman, learning how they know where to deliver the newspapers and the mail.
3. Maps can be made of the places to which people send mail.
4. Visits to supermarkets and a study of the products on the shelves can lead to map-making of the stores themselves and later, of the places from which goods come.
5. With older pupils, interesting maps can be collected and hung in appropriate places in the room and discussed by the teacher and the pupils.
6. Boys and girls can send away for maps and have the joy of receiving them. Then they can share them with their classmates and study them together.
7. Films and filmstrips on maps can be utilized to advantage.
8. Many boys and girls enjoy making maps of different kinds and then exhibiting them in class.
9. Some companies sell map puzzles and games. One of these is the McKinley Publishing Company.
10. Children can cut out maps from magazines and newspapers, mount them on cardboard or plywood, and make jigsaw puzzles of them.
11. When visitors come to a class, maps can be made of the places they come from or the places they have visited and told the class about.
12. Overhead projectors can be used to make maps which the entire class can see and study.
13. Boys and girls can interview shopkeepers in the neighborhood or community about the places where their products come from; they can interview manufacturers about the places where their products are sold. Maps can be made of such localities.
14. Maps can be made of the places described in stories read to pupils or read by them.
15. Maps can be cut out and used in opaque projectors.
16. With older pupils, persons who are map-makers can be used as resource persons in the classroom. Perhaps they can bring samples of the type of work they do.

E. *What Are Some of the Globe and Map Skills Which Need to Be Developed?*

No special reference is made here to grade levels since boys and girls learn to read maps at varying ages, just as they learn reading at different age levels. Some suggestions of a general nature on grade placement are made in section G. The items are listed in general, however, according to their degree of difficulty.

1. Land and bodies of water on a map or globe.
2. Directions. At first directions in relation to familiar places. Eventually more specific directions: northeast, southeast, etc.
3. Various bodies of water: oceans, lakes, rivers, etc.
4. Location of cities, starting with one's own city or the nearest large town.
5. Keys or legends.
6. Determining distances by using the scale of miles.
7. Resources — coal, iron, cotton, etc.

8. Climate, starting with climate in general and gradually developing the ability to interpret maps, indicating rainfall, wind currents, etc.
9. Transportation: rivers, canals, railroads, airlines, etc.
10. Political features: countries, regions, capitals, etc., starting with continents.
11. Comparing maps of different sizes of the same area.
12. Beginning to use the grid system for locating places on the earth.
13. Elevations.
14. Human factors: density of population, migrations, etc.
15. Cultural factors: cultural areas, location of religions of the world, ethnic groups, etc.

F. *What Are Some of the Common Mistakes in Introducing Children to Globes and Maps?*
   1. Introducing children too soon or too late to maps and globes.
   2. Limiting maps to Mercator projection maps; various types should be used, especially polar projection maps.
   3. Hanging maps on the walls of classrooms so that children learn to think and talk about "up" and "down" on maps. In early map study, especially, use maps on a table or over desks, on sandtables, or even on the floor.
   4. Using maps which are cluttered. They should be relatively simple, especially for beginning studies, stressing one or two major ideas.
   5. Teaching about specific colors for water, mountains, etc., rather than teaching children to read the legends which indicate the colors used.
   6. Too much emphasis upon memorization rather than relationships. Children need to interpret rather than merely memorize facts.
   7. Undue emphasis upon longitude and latitude. Pupils should know about the grid system for locating places on our planet, but they need not be subjected to hours of practice. Most of us do not use longitude and latitude that much in our daily lives.

G. *What Types of Map and Globe Work Can Be Done in the Preschool Years and in the Primary Grades?*
Current research indicates that children in these years can do a great deal more such work than we thought formerly. Here are some of the ideas that can be taught:
   1. Size of objects and general proportions compared.
   2. Relation of objects in a classroom.
   3. Relation of rooms in a school building and in a home.
   4. General layout of streets in a community.
   5. Some idea of the various parts of a community or city — business, residential, industrial, etc.
   6. General directions.
   7. Use of simple symbols and legends.
   8. General idea of the globe on which we live.
   9. Location of the hometown on the globe.
   10. Location of the United States and sometimes a few other nations.
   11. Some simple ideas of interdependence and therefore of transportation.
   12. Different types of soil.
   13. Climatic areas.
   14. Some idea of elevation.

H. *What Materials Can Be Used in Developing Map and Globe Concepts?*
   1. Blocks.
   2. Cardboard cartons.
   3. Sandtables.
   4. Chalk on floor of the classroom and/or playground.
   5. Simple house plans.
   6. Yarn or string for rivers.
   7. Simple maps of communities.
   8. Some overhead projector maps.
   9. Simplified air-age maps.

I. *What Are Some Typical Experiences for Beginners in Globe and Map Work?*
   1. Arrows for directions in the classroom and school.
   2. Use of compasses.
   3. Maps of the classroom and school.
   4. Maps of the home and community.
   5. Maps of transportation to and from school.
   6. Occasional use of the globe in connection with current events.
   7. Trips to the highest spot in the community to see the lay of the land.
   8. Trips to nearby places with different geographical features.
   9. Simple films and filmstrips.
  10. Making maps of visits made by persons in the class.

J. *What Different Types of Maps Are Available for Study?*
   There are many kinds of maps. Among the general categories are the following:
   1. Physical maps, showing landforms, climate, soil, resources, rainfall, winds, etc.
   2. Political maps, showing political divisions, such as townships, counties, states, countries, etc. Often maps show physical and political features.
   3. Economic maps, showing crops, land use, resources, trade, railroads, etc.
   4. Social maps, showing population distribution and density, races, languages, wealth, education, etc.
   5. Historical maps, showing empires, campaigns, treaties, alliances, etc.

K. *How Should Maps and Globes Be Selected for Use in Classes?*
   1. Every classroom in which social studies is taught should have as a minimum a world globe, a world map, and a map of the United States. Other maps should be obtained according to the topics discussed by the class.
   2. Many maps should be made by pupils, including papier-mâché and clay maps. Transparencies can be made and/or bought and overlay maps can be made easily by some pupils.

L. *What Criteria Should Be Used in Selecting Maps and Globes for Purchase?*
   1. Accuracy.
   2. Recency, for most maps.
   3. Large enough to be seen by the entire class, for most maps.
   4. Simplicity. National Geographic maps, for example, are much too detailed for use in most elementary school classes.
   5. Sturdiness.
   6. Unusual enough in some respect to arouse interest.

7. Almost always in color.
8. With large lettering if possible.
9. Some polar projection or aerospace age maps.
10. As inexpensive as possible, in most instances.
11. At least one large globe and, if possible, several small globes for use by individuals and committees.

M. *How Can Maps Be Preserved Best?*

1. Mount on linen, cardboard, or old window shades.
2. If you want to use maps in several rooms, purchase folding maps. They are easier to store, handle, and preserve.
3. Organize a central file for all maps in a closet, room, or the library.
4. Small maps from newspapers, magazines, books, and other sources should be mounted on cardboard for safekeeping and for use in the overhead projector.
5. Transparencies made by the teacher and/or pupils should be placed between sheets of paper and filed, preferably flat.

N. *What Types of Maps Can Boys and Girls Make?*

The possibilities of map-making by pupils are many. Here are a few suggestions:

1. Papier-mâché maps. Tear newspapers into fine strips. Soak for twenty-four hours. Squeeze the water out and mix with salt and flour. Use three parts paper, one part flour, and one-third part salt.
2. Freehand maps drawn with pencils or crayons.
3. Salt and cornstarch maps. Four cups of coarse salt, one cup of cornstarch. Heat salt until it is very hot. Mix the cornstarch with water to the consistency of thick cream. Pour into the hot salt. Then form the map.
4. Clay maps on glass or on cardboard or in trays, with aluminum foil on the bottom to get the clay out easily.
5. Sawdust and glue maps.
6. Maps enlarged from textbooks or other sources by projecting them in an opaque projector and tracing them on oaktag or other material.
7. Typed slides — made by typing on special paper and inserting between two pieces of glass. Materials may be purchased from stores or map companies.
8. Overlay maps, made on special paper or on the material in an overhead projector, using a special crayon. Pliofilm for clothes from the drycleaner can also be used, with Magic Markers as crayons.
9. Maps enlarged by squares.
10. Relief or topographic maps from sponge rubber.
11. Frosted glass slides, with special crayons. These can be erased and used again.
12. Jigsaw puzzle maps, made by pasting maps on cardboard or gluing them on plywood and then cutting or sawing them into pieces.
13. Electric maps, with batteries.

O. *Where Can Different Types of Maps Be Obtained?*

1. Desk outline maps.	Any map company. *The Instructor* Magazine in Dansville, N.Y. has stencils for sale.
2. Political maps.	Any map company.

3. Current events maps.	Current events papers and some newspapers.
4. Road and subway maps.	Most service stations give away road maps. Chambers of Commerce and transit authorities are also good sources.
5. Weather maps.	Daily newspapers and U.S. Weather Bureau.
6. Geological maps and soil conservation maps.	The U.S. Geological Survey or local offices of the U.S. Department of Agriculture.
7. City and state maps.	City and state Chambers of Commerce and governments.
8. National and state park maps.	State governments and the National Park Service, Washington, D.C. 20025.
9. Laminated surface maps and chalkboard maps.	Most of the leading map companies and especially the School Products Bureau, 517 South Jefferson Street, Chicago, Illinois.
10. Maps of other nations.	Any map company — or make your own. Use the small maps in current events papers, possibly projecting them through the opaque projector. Some embassies and information offices of governments give away maps.
11. Air age or polar projection maps.	Almost any of the map companies.
12. Raised surface maps.	Aero Service Corporation, now affiliated with A. J. Nystrom and Company, 3333 Elston Avenue, Chicago, Illinois 60618. Many such maps can be made by pupils.
13. Symbol maps of world areas.	Friendship Press, 475 Riverside Drive, New York, New York 10027. Older pupils can make some maps of this type.
14. Intercultural map of the United States.	Friendship Press (see address above).

If you have difficulty as a teacher drawing maps on the chalkboard, make cardboard outlines or plywood outlines and run the chalk around them. Or make outlines with perforated edges and dust them with the chalk from erasers, to make bare outlines which can then be filled in with more details.

BIBLIOGRAPHY OF MATERIALS FOR BOYS AND GIRLS ON MAPS

Books

1. BACON, PHILIP. *The Golden Picture Atlas of the World.* New York: Golden (1960). Six volumes, each on a major region of the world. Grades 5–8.
2. BROWN, LLOYD A. *Map Making: The Art That Became a Science.* Boston: Little, Brown (1960), 217 pp. Grades 7–10.
3. CARLISLE, NORMAN and MADELYN. *True Book of Maps.* Chicago: Children's Press (1969), 47 pp. Grades 2–4.
4. COLBY, C. B. *Mapping the World: A Global Project of the Corps of Engineers, U.S. Army,* New York: Coward–McCann (1959), 48 pp. Grades 7–9.
5. EDITORS and CARTOGRAPHERS of C. S. HAMMOND and COMPANY. *The First Book Atlas.* New York: Watts (1968), 96 pp. A revised edition. Grades 4–6.
6. EPSTEIN, SAM and BERYL. *The First Book of Maps and Globes.* New York: Watts (1959), 63 pp. Grades 4–7.

7. FRAZEE, STEVE. *Where Are You?* New York: Meredith (1968), 96 pp. Grades 5–8. Surveying and its application to map-making.

8. HACKLER, DAVID. *How Maps and Globes Help Us.* Chicago: Benefic Press (1963), 72 pp. Grades 4–6.

9. HAMMOND, C. S. and COMPANY. *Illustrated Atlas for Young Americans.* Maplewood, New Jersey: Hammond (1956), 16 pp.

10. HARRISON, RICHARD E. and others. *The Ginn World Atlas.* Lexington, Massachusetts: Ginn (1966), 62 pp. Updated from time to time.

11. HATHWAY, JAMES A. *The Story of Maps and Map-Making: How Man Has Charted His Changing World from Ancient Times to the Space Age.* New York: Golden (1960), 54 pp. Grades 5–8.

12. HIRSCH, S. CARL. *The Globe for the Space Age.* New York: Viking (1963), 88 pp. Grades 7–9. A fascinating history of the man-made globe.

13. HIRSCH, S. CARL. *Mapmakers of America: From the Age of Discovery to the Space Era.* New York: Viking (1970), 176 pp. Grades 6–8.

14. MCFALL, CHRISTIE. *Maps Mean Adventure.* New York: Dodd, Mead (1961), 128 pp. Grades 6–9.

15. MARSH, SUSAN. *All About Maps and Mapmaking.* New York: Random House (1963), 143 pp. Grades 6–9.

16. NEAL, HARRY E. *Of Maps and Men.* New York: Funk and Wagnalls (1970), 179 pp. Grades 7–10.

17. OLIVER, JOHN E. *What We Find When We Look at Maps.* New York: McGraw-Hill (1970), 39 pp. Grades 1–3.

18. RHODES, DOROTHY. *How to Read a City Map.* Los Angeles, California: Elk Grove (1967), 46 pp. Grades 4–7.

19. RHODES, DOROTHY. *How to Read a Highway Map.* Los Angeles, California: Elk Grove (1970), 53 pp. Grades 4–7.

20. RINKOFF, BARBARA. *A Map Is A Picture.* New York: Crowell (1965), 40 pp. Grades 1–3.

21. ROSS, GEORGE E. *The World Today.* New York: Platt and Munk (1969), 64 pp. Grades 5–7. Largely an atlas of nations.

22. SCHERE, MONROE. *The Story of Maps.* Englewood Cliffs, New Jersey: Prentice-Hall (1969), 66 pp. Grades 5–8.

23. TANNENBAUM, BEULAH and MYRA STILLMAN. *Understanding Maps: Charting the Land, Sea and Sky.* New York: McGraw-Hill (1969), 159 pp. Grades 6–9.

24. WARNTZ, WILLIAM. *Geographers and What They Do.* New York: Watts (1964), 149 pp. Grades 7–10.

25. WERNER, ELSA JANE. *The Golden Geography.* New York: Golden (1964), 63 pp. Grades 3–5. Large, colored, simple maps and some textual material.

Films and filmstrips

1. "Latitude and Longitude." Coronet, black and white. A film.
2. "Maps: How We Read Them." Coronet, color. A film.
3. "Maps and Their Uses." Coronet, color or black and white, 11 minutes. A film.
4. "Maps Are Fun." Coronet, color or black and white, 11 minutes. A film.
5. "Our Big, Round World." Coronet, color or black and white, 11 minutes. A film.

Jam Handy has five filmstrips on "An Introduction to the Globe," consisting of these titles: "Continents and Oceans"; "Up and Down"; "North, South, East and West"; "Night and Day"; and "Hot and Cold Places."

Jam Handy has another set of filmstrips on an "Introduction to Maps" with the following titles. "What Is a Map?"; "Coast Lines and Their Symbols"; "Land Forms and Their Symbols"; "Lakes, Rivers and Their Symbols"; and "Towns, Cities and Their Symbols."

Other titles may be found in the *Educational Media Index* or other sources.

For books, pamphlets, and articles on maps and geography for teachers, see the bibliography on page 27.

8

How Can We Help
Pupils Develop Values?

Much attention is being given these days to processes in social studies teaching. They are highly important and considerable space is devoted in this book to various processes, with emphasis upon inquiry or the discovery method. Important as processes are, however, values are even more important. Helping boys and girls develop a core of values is the most important task a social studies teacher can perform. Such values are a little like yardsticks which boys and girls use now to make their decisions and to measure their actions. They will carry them through life, with few changes. What is more important than helping children construct yardsticks which are worthwhile?

A few years ago the Educational Policies Commission of the National Education Association turned its attention to this dimension of education, asserting that:

> Whether we consider the social effects of recent wars, the remoteness of workers from the satisfactions of personal achievement, the mounting complexity of government, the increasing amount of leisure hours, the changing patterns of home and family life, or current international tensions, the necessity for attention to moral and spiritual values emerges again and again. Moral decisions of unprecedented variety and complexity must be made by the American people. . . . The public schools must increase their efforts to equip each child and youth with a sense of values which will lend dignity and direction to whatever else he may learn.

Are we doing this now? Some teachers are. Many are not. According to James Shearer, writing in "Meaningful Teaching of Contemporary Affairs," the "examination of conflicts between values and data, and values and other values, is the area most neglected in the social studies courses." Would you agree with this indictment of social studies teaching or any teaching in elementary and junior high schools? Were you assisted in shaping your own values? Did your social studies learnings equip you or help to equip you with a set of values with which to measure your actions and make decisions? Or were you subjected to facts and facts and facts without any frame of reference in which to place them?

Actually processes and values should not be mutually exclusive. The best way to learn values is to live them. They are not taught. They are not even caught. They are learned in situation after situation, day in and day out. They are learned at home. They are learned in churches and synagogues and in other community groups. They are learned from the mass media. They are learned from peers and older people. But the schools have a very special obligation to help develop values. Many hours of each week for many weeks in the year, the school, and especially the individual classroom, is the laboratory in which values are being learned. How important it is that teachers consider carefully what values should be acquired and how they may best be developed.

One of the educators who has thought most carefully about education for values is Professor Louis Raths of New York University. According to him an operational definition of a value is any belief, attitude, purpose, feeling, or goal that: (1) is prized by an individual, (2) is chosen after careful consideration of alternatives, (3) is affirmed when challenged, (4) is recurring, and (5) penetrates into life. This is a demanding definition. It excludes many superficial beliefs that pass for values.

What, then, are some of the beliefs or goals or values for which we should be striving? A. H. Maslow suggests eleven which lead to what he terms the "self actualizing" person. He names them as: (1) openness to experience, (2) flexibility, (3) objectivity, (4) recognition of the complexity of existence, (5) appreciation of the perfection of form in many aspects of life, (6) spontaneity, (7) rationality, (8) integrity, (9) autonomy, (10) responsibility, and (11) charity.

The Educational Policies Commission of the N. E. A., in the statement referred to earlier, suggested ten values which we would strive as teachers to develop with children. They were the basic values of: (1) human personality, (2) moral responsibility, (3) consideration of institutions as the servants of man, (4) the importance of common consent, (5) devotion to truth, (6) respect for excellence, (7) moral equality, (8) brotherhood, (9) the pursuit of happiness, and (10) spiritual enrichment.

Are they too broad? Are they too general? Then try your own list. That would be an important exercise for you as a teacher-in-service or as a teacher-to-be.

High on my own list would be respect for the worth of every individual. With young children that would mean valuing their own families and the other young people and adults they know as well as the children in their own class and their friends. Later this would fan out to include scores and then possibly hundreds of people. Hopefully it would include people of other religions, of other races, or other nationalities, and of other socio-economic backgrounds. That is broad, too, but it can be used in almost every classroom situation as a guide as to whether children are learning this important value in their day-to-day relationships, with the inevitable setbacks which always occur as we learn. Another way of stating this is the development of sensitivity to other human beings, leading to concern for them, and eventually to commitment for a better life for all people, everywhere.

Valuing curiosity would have a high priority for me, too. If people can be curious and want to find answers, much else will follow. With such curiosity would go, I hope, a willingness to follow truth wherever it takes you, and to make decisions that are as rational as possible. This value would also lead to the recognition of the need for change.

Another value I would place high would be cooperative work. Recognizing the worth of others would be the basis for this value and would lead to a realization that shared responsibility can lead to better results for all. This is true in every aspect of life — governmentally, economically, educationally, and religiously. Boys and girls do not learn to work cooperatively by preachments; they learn to work with others by practice in group work skills. Are these being learned in your classroom? Could you improve on this score?

My own list of values would include a recognition of purpose in the universe. This would lead to a reasonable optimism about humanity and the future of mankind. I suspect that purpose was high on the list of men like Gandhi, Schweitzer, and other great leaders of recent times. William James thought of it as the basis of "lives worth living."

There are many others. What would *you* suggest as some values you prize? Perhaps one of the best ways to test such values would be to ascertain whether they could be universalized — for everybody in the world — or whether they are just for a select few in one nation or culture. Of course some values are central in one country or culture and not in another. That is what makes for cultural differences.

And how are values acquired in the classroom and in the school? Certainly not by preaching or telling. They are acquired when children are in an atmosphere where they are practiced and therefore learned. It does little good to talk about cooperation unless children have opportunities to practice it. It does little or no good to preach about regard for persons of other races or faiths or socio-economic levels if boys and girls have no contact with such persons or no successful experiences in such interpersonal and interfaith and interracial relations.

Models, too, are needed by children. They should include teachers and persons brought to the school. And they should include models in literature. The child who has a handicap needs to meet and/or read about persons who have or have had a similar handicap and have overcome it or adjusted to it. Boys and girls of a minority group need to know about the achievements of their race or religion or ethnic group and about individuals in it who have contributed to the welfare of the United States or some other part of the world.

In studies of other people, whether in learning about families or communities or countries, children need to discover that other people have values, too. How do people spend their spare time? Why? What do people purchase with their spare money, if they have any? Why? What ideas do people stress? Why? Such questions, varying in degree of complexity, can help children to develop a philosophy of life, a set of values, a frame of reference for lives worth living.

In an excellent pamphlet on "What Values Are We Giving Our Children?" Roy Menninger emphasizes the importance of models in value formation, varied opportunities to test values, and yet some controls by adults.

Perhaps the best statement of ways in which values can be developed is contained in a chapter of the book by James A. Smith on *Creative Teaching of the Social Studies in the Elementary School,* in which he lays great stress upon several methods. Among them are role-playing, dramatizations, role-reversals, puppets, open-ended stories, problem pictures, and the use of books (which he calls bibliotherapy) in which value situations are developed.

In case you are interested in this high priority aspect of teaching the social studies and want to think about it further, a list of readings is provided on the next page. Why not select some readings from this list and react to them in terms of your teaching?

BIBLIOGRAPHY ON VALUES

1. ALLPORT, GORDON W., PHILIP E. VERNON, and GARDNER LINDZEY. *Study of Values.* Boston: Houghton Mifflin (1960), 12 pp.
2. ASSOCIATION FOR CHILDHOOD EDUCATION. *Basic Human Values for Childhood Education.* Washington: A. C. E. I. (1963), 76 pp. A colloquy of men and women from various professions.

3. ASSOCIATION FOR CHILDHOOD EDUCATION. *Implications of Basic Human Values for Education.* Washington: A. C. E. I. (1964), 64 pp. Applications of the ideas found in No. 2, above, to classroom learnings.

4. BARR, ROBERT D. (Editor). *Values and Youth.* Washington, D.C.: National Council for the Social Studies (1971), 112 pp.

5. BAUER, NANCY. "Guaranteeing the Values Component in Elementary School Social Studies" in JOYCE, WILLIAM W. and others. *Elementary Education in the Seventies.* New York: Holt, Rinehart and Winston (1970), 579 pp.

6. BERKOWITZ, LEONARD. *The Development of Motives and Values in the Child.* New York: Basic Books (1964), 114 pp.

7. CLEGG, AMBROSE A. and JAMES L. HILLS. "A Strategy for Exploring Values and Valuing in the Social Studies" in JOYCE, WILLIAM W. and others. *Elementary Education in the Seventies.* New York: Holt, Rinehart and Winston (1970), 579 pp.

8. DEPARTMENT OF ELEMENTARY SCHOOL PRINCIPALS. "Values and American Education" in *The National Elementary School Principal* (November, 1962). A special issue.

9. ERICKSON, ERIK H. *Insight and Responsibility.* New York: Norton (1964), 256 pp.

10. FOSHAY, ARTHUR W. and others. *Children's Social Values: An Action Research Study.* New York: Teachers College Press (1954), 323 pp.

11. GOODMAN, MARY E. *A Primer for Parents: Educating Children for Good Human Relations.* New York: Anti-Defamation League (1959), 31 pp.

12. JONES, JESSIE ORTON. *The Spiritual Education of Our Children.* New York: Viking (1960), 124 pp.

13. LANG, MELVIN. "Value Development in the Classroom" in *Childhood Education* (November, 1964), pp. 123–126.

14. METCALF, LAWRENCE (Editor). *Values Education: Rationale, Strategies, and Procedures.* Washington: National Council for the Social Studies (1971), 208 pp. 41st Yearbook.

15. NATIONAL COUNCIL FOR THE SOCIAL STUDIES. "Focus on Values" in *Social Education* (January, 1967). A special issue.

16. PECK, ROBERT F. and others. *The Psychology of Character Development.* New York: Wiley (1960), 267 pp.

17. PHENIX, PHILIP H. *Education and the Worship of God.* Philadelphia: Westminster (1966), 192 pp.

18. PHENIX, PHILIP H. *Realms of Meaning.* New York: McGraw-Hill (1964), 391 pp.

19. RATHS, LOUIS E. "Values Are Fundamental" in ·*Childhood Education* (February, 1957), pp. 246–247.

20. RATHS, LOUIS E., MERRILL HARMIN and SIDNEY B. SIMON. *Values and Teaching.* Columbus, Ohio: Merrill (1966), 275 pp.

21. SMITH, JAMES A. *Creative Teaching of the Social Studies in the Elementary School.* Boston: Allyn and Bacon (1967), 281 pp. Chapter 9 on "Values." Especially good on methodology.

22. WHYTE, DOROTHY K. *Teaching Your Child Right from Wrong.* Indianapolis: Bobbs-Merrill (1961), 192 pp.

25.   Some resources for teachers and pupils

What General Resources
Are Available for
Social Studies Teaching? **9**

There are many references to social studies resources throughout this book under such headings as Studying Individuals and Families, Studying Communities, and Studying Countries. Those cited here are more general in nature and are primarily for teachers rather than pupils.

LIBRARY RESOURCES

You should have access to some, at least, of the following libraries. In some cases they will be Instructional Materials Centers rather than merely libraries, in the older sense of that word. There may be other resource centers available to you.

1. Your school library or resource center.
2. The library or curriculum center of the Board of Education.
3. The local library or libraries.
4. The State Library (through its Loan Division).
5. The library of the nearest college or university.

Other libraries or resources centers to which I have access:

6.
7.

Be sure that your pupils have cards to all possible libraries, and that they use them.

INDEXES

Four of the major indexes which should prove useful to you are:

1. *The Reader's Guide to Periodical Literature*. All subjects indexed, including social studies topics.
2. *Education Index*. Articles pertaining only to education.
3. *The National Geographic Index*.
4. *The New York Times Index*.

SOCIAL STUDIES AND RELATED ORGANIZATIONS

Some materials may be obtained free and others purchased from the following organizations:

1. Association for Childhood Education International, 3615 Wisconsin Avenue, N.W., Washington, D.C. 20016.
2. Joint Council on Economic Education, 2 West 46th Street, New York, New York 10036.
3. National Council for the Social Studies, 1201 Sixteenth Street, N.W., Washington, D.C. 20036.

PROFESSIONAL MAGAZINES HIGHLIGHTING THE SOCIAL STUDIES FIELD

1. *Childhood Education.* 3615 Wisconsin Avenue, N.W., Washington, D.C. 20016.
2. *Early Years.* Box 1223, Darien, Connecticut 06820.
3. *The Grade Teacher.* Darien, Connecticut 06820.
4. *The Instructor.* Dansville, New York 14437.
5. *The Journal of Geography.* Room 1226, 111 West Washington Street, Chicago, Illinois 60602.
6. *Social Education.* 1201 16th Street, N.W., Washington, D.C. 20016.
7. *Social Studies.* 112 South New Broadway, Brooklawn, New Jersey 08030.

READING LISTS OF SOCIAL STUDIES BOOKS FOR PUPILS

1. Association for Childhood Education International. *Bibliography of Books for Children.* Washington: A.C.E.I. (1971), 130 pp. Many on social studies.
2. Baker, Augusta. *The Black Experience in Children's Books.* New York: New York Public Library (1971), 109 pp. An outstanding list.
3. Bank Street College. *Books for Children.* New York: Bank Street College. A short, annotated list, issued annually.
4. *Books for Children: Preschool Through Junior High School.* Chicago: American Library Association. Issued annually. Approximately 150 pages.
5. Eakin, Mary K. *Subject Index to Books for Primary Grades.* Chicago: American Library Association (1967), 113 pp.
6. *The Elementary School Library Collection: A Guide to Books and Other Media.* Newark, New Jersey: Bro-Dart Foundation. Issued annually. Over 700 pages.
7. Griffin, Louise (Compiler). *Multi-Ethnic Books for Young Children: An Annotated Bibliography for Parents and Teachers.* Washington: National Association for the Education of Young Children (1970), 74 pp.
8. Information Center on Children's Cultures, U.S. Committee for UNICEF. *Africa: A List of Printed Materials for Children.* Also lists on South America and the Near East. New York: U.S. Committee for UNICEF.
9. Keating, Charlotte M. *Building Bridges of Understanding Between Cultures.* Tucson, Arizona: Palo Verde Publishing Company (1971), 233 pp.
10. Kenworthy, Leonard S. *Studying Africa in Elementary and Secondary Schools.* New York: Teachers' College Press (1970), 74 pp. Also similar pamphlets on *Studying the U.S.S.R. in Elementary and Secondary Schools, Studying South America in Elementary and Secondary Schools,* and *Studying the Middle East in Elementary and Secondary Schools.*

11. *Recommended Reading About Children and Family Life.* New York: Child Study Association (1969), 74 pp.
12. *Subject Index to Children's Books in Print.* New York: Bowker (1972). Published annually. See also *Children's Books in Print.*
13. U.S. Office of Education. *Books Related to the Social Studies in Elementary and Secondary Schools.* Washington: Government Printing Office. Issued annually. An inexpensive publication.
14. Wolfe, Ann G. *About 100 Books — A Gateway to Better Intergroup Understanding.* New York: American Jewish Committee (1969), 48 pp. Issued every two or three years.

CURRICULUM GUIDES IN THE SOCIAL STUDIES

There are hundreds of curriculum guides in the social studies published by local and state boards of education and curriculum bureaus. These can be helpful to teachers as well as to persons developing curricula. Here are some ways in which you can obtain such guides or examine them:

1. Obtain such guides from your city, county, or state education office.
2. Examine such guides in the library of your Board of Education or in the library of your nearest college or university.
3. Look for such guides at local, state, regional, or national conventions. For example, the National Council for the Social Studies and the Association for Supervision and Curriculum Development display curriculum guides at their national conventions and publish a yearly list of such materials.
4. Write to various cities and states and inquire about the availability of curriculum guides in the social studies — and their costs.
5. Watch for announcements of such guides in professional magazines.

SOME PUBLICATIONS OF THE NATIONAL COUNCIL FOR THE SOCIAL STUDIES

Curriculum Guides

1. Social Education of Young Children: Kindergarten–Primary Grades.
2. Social Studies for the Middle Grades: Answering Teachers' Questions.
3. Social Studies for Young Adolescents: Programs for Grades Seven, Eight and Nine.

How to Do It Series

These are leaflets, 4 to 14 pages in length, on the following topics, prepared and sold by the National Council for the Social Studies:

How to Use a Motion Picture
How to Use a Textbook
How to Use Local History
How to Use a Bulletin Board
How to Use Daily Newspapers
How to Use Group Discussion
How to Use Recordings
How to Use Oral Reports

How to Handle Controversial Issues
How to Introduce Maps and Globes
How to Use Multiple Books
How to Plan for Student Teaching
How to Study a Class
How to Use Sociodrama
How to Develop Time and Chronological Concepts

How to Locate Government Publications	How to Teach Library Research
How to Conduct a Field Trip	How to Ask Questions
How to Utilize Community Resources	How to Use Folksongs

The official publication of the National Council for the Social Studies, *Social Education*, has a special section for elementary schools several times a year, as well as articles in the general issues.

TEXTBOOKS IN THE SOCIAL STUDIES

The best single reference to textbooks in the social studies is the volume issued annually by the R. R. Bowker Company (1180 Avenue of the Americas, New York, New York 10036) entitled *Textbooks in Print*.

Cards or letters to the following publishers should give you a wide range of information about textbooks. Do not request complimentary copies unless you are actually considering the adoption of a text or several texts for use in your class or school. Textbook companies cannot afford to send out samples to teachers merely to augment their libraries and you should not request such materials.

It is important to bear in mind that almost all textbooks have teachers' manuals to accompany them. There are very useful guides. Often they may be purchased from textbook companies even if you are not using the text that company publishes.

Some of the major companies issuing social studies textbooks for elementary schools are as follows:

1. Addison–Wesley Publishing Company, Reading, Massachusetts 01867.
2. Allyn and Bacon, 470 Atlantic Avenue, Boston, Massachusetts 02210.
3. American Book Company, 450 W. 33rd Street, New York, New York 10001.
4. Benefic Press, 10300 W. Roosevelt Road, Westchester, Illinois 60153.
5. Bobbs-Merrill Company, 4300 W. 62nd Street, Indianapolis, Indiana 46206.
6. Field Educational Publications, Inc., 609 Mission Street, San Francisco, California 94105.
7. Follett Publishing Company, 1010 West Washington Boulevard, Chicago, Illinois 60607.
8. Ginn and Company, 191 Spring Street, Lexington, Mass. 02173.
9. Globe Book Company, 175 Fifth Avenue, New York, New York 10010.
10. Harcourt Brace Jovanovich, 757 Third Avenue, New York, New York 10017.
11. Harper & Row (including books of Row Peterson), 10 East 53rd Street, New York, New York 10022.
12. D. C. Heath, 125 Spring Street, Lexington, Massachusetts 02173.
13. Holt, Rinehart and Winston, 83 Madison Avenue, New York, New York 10017.
14. Houghton Mifflin, 2 Park Street, Boston, Massachusetts 02107.
15. Laidlaw Brothers, Thatcher and Madison Avenues, River Forest, Illinois 60305.
16. The Macmillan Company, 866 Third Avenue, New York, New York 10022.
17. Charles E. Merrill Books (including the Iroquois Publishing Company), 1300 Alum Creek Drive, Columbus, Ohio 43216.
18. Prentice-Hall, Englewood Cliffs, New Jersey 07632.
19. Random House (including the Singer Publishing Company), 201 E. 50th Street, New York, New York 10022.
20. W. H. Sadlier, Inc., 11 Park Place, New York, New York 10007.
21. Scott, Foresman, 1900 E. Lake Avenue, Glenview, Illinois 60025.

22. Silver Burdett, 150 James Street, Morristown, New Jersey 07960.
23. Steck–Vaughn Co., Box 2028, Austin, Texas 78767.

For further suggestions on the use of textbooks see pages 110–111. Teachers may want to consider seriously the use of several textbooks in their classes. Remember that the textbook is only a springboard. It cannot do the job of teaching for everyone.

ENCYCLOPEDIAS

Encyclopedias can be extremely useful reference materials for enriching social studies learnings if properly used. Children can be introduced to them at a very early age, even before they can find their way through them by themselves. They can examine the pictures and browse in them. Some encyclopedias are intended primarily for younger children. They include *Childcraft, Britannica Junior, Our Wonderful World*, and *The New Book of Knowledge*.

Perhaps their most unique advantage as a teaching tool is in helping to answer the questions children raise. These may be questions on which teachers do not have ready answers and they may be questions on which teachers have good answers but want children to find answers themselves.

There are many skills involved in using encyclopedias and those skills must be taught by teachers, parents, and librarians. Time needs to be found to teach such skills in concrete situations — when children want to find answers to their questions.

Teachers should be aware that the authors of encyclopedia articles have already condensed much material into compact articles. It is asking too much of children to expect them to condense the material in encyclopedias still further. Children need to be taught to use relevant parts of articles rather than merely copying them for special reports.

Teachers, too, can find encyclopedias invaluable. They can provide needed information in the form of maps, charts, articles, and in other ways. They may also give teachers ideas as to how to visualize certain ideas.

Among the encyclopedias currently available are the following:

1. *American People's Encyclopedia.* Spencer International Press, 201 North Wells Street, Chicago, Illinois 60606. Junior high primarily.
2. *Britannica Junior.* Encyclopaedia Britannica, 425 North Michigan Avenue, Chicago, Illinois 60611. Middle grades and junior high.
3. *Childcraft: The How and Why Library.* Field Enterprises, Educational Division, Merchandise Mart, Chicago, Illinois 60654. Primary grades.
4. *Collier's Encyclopedia.* F. P. Collier and Company, 1000 North Dearborn Street, Chicago, Illinois 60610. Middle grades and junior high.
5. *Compton's Encyclopedia.* Reference Division, Encyclopedia Britannica Educational Corporation, 425 North Michigan Avenue, Chicago, Illinois 60611.
6. *Encyclopedia Americana.* 4606 East-West Highway, Washington, D.C. 20014. Junior high primarily.
7. *Grolier Universal Encyclopedia.* Grolier Educational Corporation, 845 Third Avenue, New York, New York 10022.
8. *The New Book of Knowledge.* Grolier Educational Corporation, 845 Third Avenue, New York, New York 10022.
9. *Our Wonderful World.* Spencer International Press, 201 North Wells Street, Chicago, Illinois 60606.

10. *The World Book Encyclopedia.* Field Enterprises, Educational Division, 510 Merchandise Mart Plaza, Chicago, Illinois 60654.

FREE AND INEXPENSIVE MATERIALS

There are literally hundreds of materials which you and your pupils may obtain for use in your social studies programs. Some of them can be collected by pupils and parents — such as pictures from old magazines and calendars, maps from local filling stations and other sources, and slides contributed by parents.

Other materials may be obtained free of charge by writing embassies and information centers, state and city Chambers of Commerce, and various organizations and business firms.

Some of this material needs to be carefully screened. But even the most blatant propaganda can be used to study distortions of the truth. This, too, is teaching for critical thinking.

You may want to mimeograph a card or letter to ask for some of this material. If possible, keep the letter brief, but include as much data about your groups as possible — the age level, the reading level, and the number of children in the class. Use school stationery if possible.

Here are some sources of information about free and inexpensive materials:

A. *Lists in Current Magazines*

See such magazines as *The Grade Teacher, The Instructor,* and *Social Education.*

B. *Booklets on Free and Inexpensive Materials*

Most of these booklets are issued annually; others every two or three years. The booklets themselves are not free.

1. Aubrey, Ruth H. *Selected Free Materials for Classroom Teachers.* Palo Alto, California: Fearon (1965), 104 pp.
2. Kenworthy, Leonard S. and Richard A. Birdie. *Free and Inexpensive Materials on World Affairs.* New York: Teachers College Press (1968), 64 pp.
3. Miller, Bruce. *Free and Inexpensive Teaching Aids.* Box 369, Riverside, California. Also booklets on pictures, travel posters, etc.
4. Peabody College. *Free and Inexpensive Learning Materials.* Nashville, Tennessee: Peabody College. Issued annually. The most comprehensive of all the booklets listed here.
5. Salisbury, Gordon and Robert Sheridan. *Catalog of Free Teaching Aids.* Box 943, Riverside, California.
6. Scholastic Magazines. *Where To Find It Guide.* Appears annually in the fall in *Scholastic* magazines.
7. Suttles, Patricia H. *Educator's Guide to Free Social Studies Materials.* Randolph, Wisconsin: Educators Progress Service. This is a very useful publication, but it is also an expensive one. The school library might like to purchase it.

You can have your name placed on the mailing list of the U.S. Government and thereby receive every two weeks an extensive list of their publications. Write the Superintendent of Documents, Government Printing Office, Washington, D.C. 20402. Many of their publications are inexpensive and will pertain to your work in the social studies.

C. *Some Suggestions for Sending for Free and Inexpensive Materials*
   1. Use school stationery.
   2. Do not ask for more than one copy of materials in most instances unless you know they are eager to send you more.
   3. Be as specific as possible in stating your needs.
   4. Be sure to include a return address. This is guaranteed if you use school stationery.
   5. If children write, do not mail more than one letter. Write your initials and "Approved" at the bottom corner of the letter.
   6. Be sure to pay your bills. It is amazing how many teachers fail to do so on such items.

D. *Some Ways of Evaluating Free and Inexpensive Materials*
   1. In public school, avoid the use of materials from religious groups.
   2. Beware of companies that bombard you with free materials. They are likely to have an "axe to grind."
   3. Have a committee in the school or school system screen materials.

   Some materials which may be of value to you in this regard are as follows:
   (a) "Choosing Free Materials for Use in the Schools." Washington, Association for Childhood Education International.
   (b) "Sponsored Resources for the Social Studies." Washington, National Council for the Social Studies.
   (c) "Using Materials in the Classroom." Washington, Association for Supervision and Curriculum Development.

E. *Some Suggestions on Ways of Keeping Free and Inexpensive Materials*
   1. Sort your materials at least once a year or every other year. Throw away material you really do not need. Otherwise you will have the problem of storage space.
   2. Start folders for this material so that you and/or your pupils can find it easily.
   3. Mount some materials on cardboard before they become crumpled.
   4. If necessary in your situation, mark some materials as "slanted." Otherwise let the children discover this themselves.
   5. Try to get the school to purchase a steel file for you to place in your classroom so that you and your pupils can file this material. Or you may want to keep this material in the school library, if you have one. If you cannot obtain even a second-hand file, use an old orange crate or a cardboard box in which to keep your files.
   6. Enlist the help of your pupils, especially boys and girls in upper grades, in developing such a file of free and inexpensive materials. If they are involved, they will be more likely to collect materials, use them, and keep them in some semblance of order.

GAMES AND SIMULATIONS

Games

Pupils can often learn many things through games. In some cases they can develop their own games. In other situations you may want to have games on hand, including flag games, jigsaw puzzles, and other materials. Among the companies which sell such games are the following:

1. Milton Bradley Company, 74 Park Street, Springfield, Massachusetts 01101.
2. Cadaco-Ellis, 1446 Merchandise Mart, Chicago, Illinois 60654.
3. The Grade Teacher, Darien, Connecticut 06820.
4. McKinley Publishing Company, 112 South New Broadway, Brooklawn, New Jersey 08030.
5. Parker Brothers, Salem, Massachusetts.
6. World Wide Games, Box 450, Delaware, Ohio 43015.

Simulations

In the last few years widespread interest has developed in games or simulations in social studies education. Pupils are placed in locales or situations which simulate or parallel real-life conditions. They are confronted with problems and challenged to solve them, cooperatively or in competition. Evaluations of the learnings which accrue from such simulations are not definitive as yet. Many of us feel that their greatest value is in arousing interest in a subject and in challenging pupils to react to life-like situations.

Most of the games on the market today are intended for secondary school pupils. But there are a few which can be used in grades five or six and several which can be used in middle school or junior high school classes. Those which are listed below fit into those categories. They are samples of the many simulations now available. Further data may be obtained on such games from these publishers and from the books and articles listed on the following page.

1. Abt. 55 Wheeler Street, Cambridge, Massachusetts 02138. *Adventuring* — the causes of war illustrated by the civil war in England. *Economy* — various facets of the free economy. *Neighborhood* — problems involved in developing an urban area. *Pollution* — problems in controlling pollution locally. *Potlatch* — customs of the Kwakiutl Indians of the Northwestern United States. *Reconstruction* — Southern society after the U.S. Civil War. *The Slave Trade* — functioning of the slave trade with Africa and colonial possessions.
2. Education Development Center, 15 Mifflin Place, Cambridge, Massachusetts 02128. *Bushmen Exploring and Gathering* — the organization of the bushmen of the Kalahari desert in Africa for survival. *Caribou Hunting* — difficulties of Eskimos in hunting. *Seal Hunting* — Eskimos versus seals.
3. Portala Institute. 1115 Merrill Street, Menlo Park, California 94025. *Atlantis* — pupils become archaeologists and go on "digs."
4. Similie II. Box 1023, La Jolla, California 92037. A division of the Western Behavioral Sciences Institute. Several games for grades 5–8, including: *Blue Wodjet* — a simulation on pollution; *City Council, Explorers I* — North America; *Explorers II* — South America; *Homesteaders* — in the U.S.A. in the 1870s and 1880s; *Import* — on six trading firms; *Panatina* — on politics in South America; *Powerhorn* — an elementary version of Star Power, on the uses and abuses of power in government; and *Roaring Camp* — mining in the U.S.A. in the 19th century. For grades 7–8 *Crisis* — international conflict involving six fictional nations; *Napoli* — on national politics; and *Star Power* — on the uses and abuses of power in government.
5. Western Publishing, School and Library Department, 850 Third Avenue, New York, New York 10022. *Community Response* — alternative actions possible when a local community is hit by a natural disaster. *Consumer* — the various processes in consumer buying. *Democracy* — law making processes. *Economic*

System — components in our free enterprise system. *Ghetto* — the many pressures on the poor in a ghetto. *Parent and Child* — five issues on which parents and children can take a variety of positions.

A Brief Bibliography on Simulation Games

This is a relatively new field and a growing one. New games are appearing rapidly. In order to keep up-to-date you will want to scan the pages of various professional magazines and to keep in touch with publishers like the ones mentioned on page 153. You may also want to get your name onto their mailing lists. The addresses of other companies interested in this broad field may be found in some of the following publications.

1. ARRIG, JOHN C. "The Use of Games as a Teaching Technique" in *Social Studies* (January, 1967), pp. 25–29.
2. BOOCOCK, SARANE and E. O. SCHILD (Editors). *Simulation: Games in Learning.* Beverly Hills, California: Sage Publishers (1968), 279 pp.
3. CARLSON, ELLIOT. *A New Approach to Problem Solving: Learning Through Games.* Washington: Public Affairs Press (1969), 183 pp.
4. CHRISTINE, CHARLES and DOROTHY. "Four Simulation Games That Teach" in *Grade Teacher* (October, 1967).
5. COLEMAN, JAMES S. "Games — New Tools for Learning" in *Scholastic Teacher* (November 9, 1957).
6. GUETZKOW, HAROLD S. (Editor). *Simulation in Social Science: Readings.* Englewood Cliffs, New Jersey: Prentice-Hall, 1962.
7. INGRAHAM, LEONARD W. "Teachers, Computers, and Games: Innovations in the Social Studies" in *Social Education* (January, 1967).
8. NESBITT, WILLIAM A. *Simulation Games for the Social Studies Classroom.* New York: Foreign Policy Association (1971), 144 pp. An excellent introduction to the entire field, with specific information about several games, as well as an overall approach to the field.
9. ROGERS, VIRGINIA M. and MARCELLA L. KYSILKA. "Simulation Games: What and Why" in *Instructor* (March, 1970).
10. SACHS, STEPHEN M. "The Uses and Limits of Simulation Models in Teaching Social Science and History" in *Social Studies* (April, 1970).
11. SHUBIK, MARTIN (Editor). *Game Theory and Related Approaches to Social Behavior.* New York, Wiley (1964).
12. YOUNT, DAVE and PAUL DEKOCK. "Simulations and the Social Studies: The Use of Game Theory in Teaching U.S. History" in Dale L. Burbaker's *Innovations in the Social Studies: Teachers Speak for Themselves.* New York: Thomas Y. Crowell (1968), 63 pp.
13. ZUCKERMAN, DAVID W. and ROBERT F. HORN. *The Guide to Simulation Games for Education and Training.* New York: Western Publishing Company (1971), 334 pp. Contains information on over 400 simulations, arranged under 18 broad topics.

RESOURCES OF OUR NEIGHBORHOOD, COMMUNITY, AND CITY FOR THE SOCIAL STUDIES

1. *Transportation and communication*   7. *Health*

2. *Government institutions*   8. *Human relations — ethnic groups*

3. *Food, clothing, and shelter*   9. *Citizenship*

4. *Recreation*   10. *World affairs*

5. *Cultural institutions*   11. *Others*

6. *Newspapers and mass media*

SOME SUGGESTIONS OF SPECIAL RESOURCES AND ACTIVITIES FOR ELEMENTARY SCHOOL SOCIAL STUDIES CLASSES

Perhaps you will want to use these suggestions as a kind of check-list of things you might do now or in the foreseeable future:

	Have already done	*Should do at once*	*Can do eventually*
1. Compile a list of older pupils in the school who could be helpful in aspects of our social studies.			
2. Invite students from abroad in a nearby college to meet with our class.			
3. Make a drive for old magazines to obtain pictures and articles.			
4. File old current events magazines which can be used on special topics.			
5. Mount maps on cardboard or plywood and make jigsaw puzzles.			
6. Collect wallpaper scraps to use for bulletin board displays and as covers of scrapbooks.			
7. Make raised surface maps for blind children.			
8. Persuade an airplane pilot, soldier, or world traveler to send items or postcards to our class from different places from time to time.			
9. Use cardboard or wooden boxes for dioramas, with a sheet of cellophane for the front.			
10. Arrange a Hall of Fame or Heroes for a country or a period of history, with drawings or pictures of the people selected.			
11. Tape record interviews with older people in the community about life when they were children.			
12. Use the space over the chalkboard for a frieze, mural, or time-line.			
13. Encourage the pupils to prepare typed and/or frosted glass slides on a community, country, or period of history.			
14. Have the children visit a store and purchase the items they think a child in another country might like to have.			

	Have already done	Should do at once	Can do eventually
15. Collect old holiday cards and give them to the local hospital, possibly the children's ward.			
16. Donate class and/or individual scrapbooks to the local hospital, possibly the children's ward.			
17. Make a map of the United States or the world on the school playground, especially if it is made of concrete, and have the children invent games to use on it.			
18. Arrange an affiliation with a school in another community in the U.S.A. or with a school in another country.			
19. Start a Hall of Flags of the world.			
20. Visit the tallest building in the community to see the layout of the land from that point.			
21. Have various groups of pupils keep the bulletin board, placing the names of those responsible near their display. Rotate these groups every 2–3 weeks.			
22. Develop a system of cards of materials used by the class, with their annotations on the backs of the cards, for future use by other classes.			

Further ideas you have along these lines:

23.

24.

25.

26.

27.

26. Younger pupils work with the teacher on current events, using current events papers

Studying Current Events

27. Older pupils report
to the class
on their research topics

10

What Role Should Current Affairs and Special Occasions Play in Social Studies Programs?

Current affairs should occupy a central spot in the social studies programs of every classroom and of every school. They are not something extra to be added when there is a little time left over at the end of the day or when a film has not arrived on time. Current events should be an integral part of every effective social studies program.

They should be carefully selected with the age, ability, interests and needs of particular groups of pupils in mind. They should be incorporated into the daily and weekly schedules of all classes. Methods of handling them should be wisely chosen and adequate resources should be made available. Learnings in current affairs also need to be effectively evaluated.

Current events or current affairs should be considered a part of the social studies learnings of every grade level. For young children current affairs will focus upon events in the home, the neighborhood, and the community. Those are the current affairs of prime importance to those boys and girls. But even with them there may be some look at events on further horizons. With older boys and girls the scope of current affairs will naturally be much broader.

The Values in Current Affairs Instruction.   If education is interpreted as helping children to live effectively now and in the future, as we have maintained throughout this volume, then there is nothing more important than helping children to understand the world in which they live right now. Current events and current affairs should help pupils to understand what is happening around them and assist them in coping with situations which confront them.

Through current affairs, children can also feel a part of the world of older children and of adults. This is especially important for boys and girls in the upper grades and junior high schools. They want very much to be considered mature at this stage in their development. Most of them long for identification with older people — at least upon some occasions. One way to help them in this respect is to assist them in understanding the current happenings so that they can share in the life of older people.

Through current affairs many important attitudes can be learned, too. These range from loyalty to one's community and country to respect for divergent opinions of members of the groups in which they function.

A wide range of skills also can be developed through current affairs programs in schools. These include the collection of information, the organization of data, the interpretation of maps, charts, and similar materials, and, above all, critical thinking. These are only a few of the skills which can be taught through current affairs; others will be mentioned later in this chapter.

Another important value of current affairs instruction is the widening of horizons for children. Through television programs, current events magazines, newspapers, and other resources, they can learn about events in the United States and in other parts of the world and begin to identify with larger groups of people in all parts of the globe.

Since many children today are exposed to a wide range of media portraying events in all parts of our globe, current affairs teaching can help them digest what they have seen and gain background for these quick presentations which they do not always understand. Current affairs teaching can therefore supplement and complement learnings from other sources.

Current events instruction can also update the materials presented in textbooks. Such materials cannot possibly be wholly current. Current affairs teaching therefore can serve as revised editions of printed materials.

Finally, current affairs instruction can serve as a motivation for many other types of social studies learnings. Boys and girls are much more likely to want to study themes from the past if they see the relevance of such studies in understanding the present.

Delineating Current Events, Current Affairs, and Current Issues.   In the lines you have already read in this chapter, the term *current affairs* has been used more than any other term. Perhaps you have wondered why it has been used more often than the term *current events*.

Usually the term *current events* connotes something of passing importance. Anything can be a current event or happening. There is likely to be little differentiation between the important and the unimportant or much selectivity among the millions of daily occurrences unless one has a frame of references in which to place such events.

The term *current affairs* is a much broader term and connotes significant happenings. They are likely to be events of more than passing interest.

Current issues is a term that is usually used to indicate some controversy. However, it is less used than the phrases *current events* or *current affairs*.

For young children the emphasis in social studies will usually be upon current events. But such children should be introduced early to current affairs. In the middle and upper grades there should be much more emphasis upon current affairs and some attention to current issues. Many pupils in the fifth or sixth grades can wrestle in an elementary way with some current problems or issues.

The Sources of News for Children.   To teach effectively, instructors need to know the news sources of their pupils. This should be done very early in the school year.

One way to ascertain the news sources of pupils is to have them talk about the programs they see on television, the broadcasts they listen to on radio, and the newspapers and magazines they see at home. This might well include the number of hours they view television, the materials that come into their homes, whether they discuss current events with older brothers and sisters and/or their parents.

You will need to determine very early in the year whether your pupils should subscribe to current events magazines, and if so, how they will be paid for.

With such data teachers should be able to plan more effectively for current affairs teaching. They will know whether children can view television programs or listen to radio programs as homework assignments. They will know whether there are newspapers to be read at home or magazines available. Teachers can also determine in this way what resources are needed in the classroom to supplement the materials in the homes of their pupils.

Teachers should likewise be conversant with the materials available in the school library and in the local community library for current affairs instruction.

Criteria for Selecting Current Events and Current Affairs for Discussion. In a chapter in the 1962 yearbook of the National Council for the Social Studies on *Social Studies in Elementary Schools*, Dorothy Fraser suggests three criteria for selecting current affairs for discussion in elementary school classes. The first of these is the maturity level, the ability, and the experience backgrounds of children. This will vary, of course, from grade to grade and even within grades. Many topics in the news are beyond the comprehension of pupils. For young children this would be true of such important topics as the gold supply of a nation, the intricacies of foreign aid, and the factors involved in the selection of a Secretary-General of the United Nations.

However, teachers often underestimate the background which children have gained today through the mass media and in other ways. Some teachers fail to grasp the fact that simple answers are all that young children require. They can be told in simple terms about foreign aid, for example. Even young boys and girls can understand that the people in another part of the globe do not have enough to eat or schools to attend. That can be "foreign aid" to them. Almost any topic can be simplified in that way. Children, then, do not need to feel isolated from what is going on around them or baffled by what they hear about or view.

With older children or with children with wider experimental background, much more can be done.

A second criterion which Dr. Fraser suggests is whether the topic is really significant or can lead into the study of a significant topic. She suggests that many insignificant incidents can be utilized to introduce children to topics of wider import. For example, a local robbery can be used to explain how the local government copes with such a problem or how people have different value systems or different incomes or are out of work. With older children an interest in such an event can be used to open up the topic of crime.

A third criterion which Dr. Fraser suggests is whether there are adequate materials available for a study of a given topic. If a teacher decides to concentrate upon a current event, it should be one on which there are available resources — including newspaper and magazine articles, films or filmstrips, or people in the community.

The author of this book would add a fourth important consideration. That is the ability of the teacher to handle the topic adequately. There are topics on which the teacher feels that he or she does not have enough background. If the teacher cannot obtain background quickly, then that topic may well be sidetracked, at least for the time being. There are also topics upon which teachers have such strong feelings that they cannot handle them adequately or objectively. If so, they should avoid such topics until they are able to deal with them in an adequate manner. Sometimes there are issues upon which a community is divided and about which there is strong feeling. This is the type of topic which should be handled in the classroom. That is social studies teaching at its best, for good teaching should cope with such problems. It may be a local strike or an election in which feeling runs high. It may be an issue involving religion or race. Boys and girls should be able to blow off steam and then discuss in a calmer manner the basic issues involved. But if a teacher does not feel he or she can do this, then the issue should be avoided. Better to have no teaching about such a problem than to have it handled poorly.

Given such situations, teachers might well look at themselves and examine their own prejudices. Then they can begin to work on themselves with the hope and expectation that eventually they can handle even the most delicate and most emotionally charged problems with equanimity and objectivity.

Handling Current Affairs in the Elementary School Classroom.   Granting the importance of teaching about current affairs, how should they be handled in elementary schools?

Possibly the first answer to that question is that, important as current events are, the daily headlines should not be used solely to determine the curriculum in social studies. If they are used as the sole or even the chief basis for determining the social studies curriculum, poor results will ensue. Teachers and children will be exposed to the passing and the contemporary only. They will have little planned learning. Superficiality will possibly prevail. Every few days a new topic will have to be explored and depth will be lacking. The front page of the newspaper is important but it should not be used as the format of any ongoing social studies program.

Some teachers like to have a few minutes each day on current events. That plan has merit. It provides for a continuing program. But unless a teacher is careful, topics will be treated quickly and superficially. If this plan is adopted, some days should be devoted only to one topic. Occasionally a topic should be pursued for several days to develop it in depth.

Other teachers depend upon current events magazines for most of their work in current affairs. This approach has many merits. One is the possibility of regularity, inasmuch as the papers arrive weekly and the children will look forward to them. Another is the fact that they are written for children and are therefore likely to be readable. In addition, such newspapers cover only a few topics. Therefore boys and girls are not likely to be overwhelmed by the extent or diversity of the news in them. Even if you use these classroom current events papers, you will often want to concentrate upon one article in them and do that topic in considerable depth, rather than trying to cover all the topics in the papers. As a teacher, you will probably want to save your copy (and possibly other copies), filing them for future use. Often the articles in them are worth saving. Instead of saving the entire magazine, you may want to clip certain articles and file them.

Other teachers try to incorporate current affairs in their regular, ongoing lessons. This is especially commendable as it relates the topic under discussion to the current scene.

For example, if a third grade class is studying transportation, current happenings regarding that topic are mentioned and related to the unit. A new bridge may be under consideration locally, a road nearby may be under construction, or a survey of transportation in the area may be approved by a planning commission. Such teaching is excellent. However, it is not enough. Pupils will not be exposed to other current events which do not relate to the unit.

Therefore a combination of these methods should be used in classes in social studies. Only in this way will full coverage be possible, regularity of treatment planned, and some depth in a few topics fostered, and relationships with ongoing units guaranteed.

Specific Methods in Developing Current Affairs Programs. There are a number of ways in which teachers may handle current affairs in their social studies classes. Among the methods teachers use are the following (more details on most of these approaches will be found in this volume in the section on Methods):

1. Individual reports.
2. Experts on a topic or area. Pupils may be asked to report for a period of three or four weeks on one topic or world area. In this way they gain some depth in that theme.
3. Panels or roundtables on a theme.
4. Preparation of bulletin boards by pupils.
5. Assignment of television programs in current events and class discussion of the programs.
6. Mock radio and/or television programs in class.
7. A class scrapbook of current events, with sections on different themes.
8. Individual or committee notebooks or scrapbooks, with comments on the clippings by the pupils.
9. Interpretation of cartoons.
10. Use of filmstrips on current events (see the section on resources in current affairs).
11. Comparative studies of newspapers and how they report events.
12. Use of world events news maps (see the section on resources).
13. Talks by a pupil, the teacher, or invited speakers.
14. Preparation of a current events program for a group of classes by a pupil, a committee of pupils, or a teacher. (This is an especially good use of more able pupils.)
15. Time-lines related to one current affairs topic.
16. Interpretation of news pictures.
17. Use of bulletin boards in the classroom and/or in the school.
18. Listening to tapes of current events broadcasts.
19. Mock interviews regarding current affairs.
20. Studies of newspapers and how they are prepared.
21. Preparing newscasts for delivery over the school broadcasting system.
22. Developing a Speakers' Bureau for younger classes.
23. Keeping newspaper and magazine clipping files on a topic.

Resources for Teaching Current Affairs. Among the many resources available to teachers for current affairs programs are the following:

1. Events in the lives of children, especially in the primary grades.
2. Local newspapers and national newspapers, such as the *Christian Science Monitor* and *The New York Times*, copies of which should be in every classroom of older pupils.
3. Current events papers for children.
4. Current newsmagazines suitable for older pupils.
5. Radio newscasts.
6. Television programs on current affairs.
7. Collections of cartoons.
8. Collection of news clippings.
9. Collections of pictures, especially of people in the news.
10. Maps which can be used in conjunction with current affairs.

11. Special news maps or charts, such as the series published by the World News of the Week, Skokie, Ill. 60067.
12. Filmstrips on current affairs for older pupils, such as those sold by the Office of Educational Activities of *The New York Times*.
13. Special newsmaps, such as those sold by Scholastic Magazines, 50 West 44th Street, New York, N.Y. 10036.
14. Personnel in the school and/or in the community who have some special background on a theme of current importance. Pupils may want to interview such persons, taking notes or even taping the interview. Sometimes such persons can be invited to a class or to a school assembly.
15. Booklets for teachers on current affairs are published from time to time by the various publishers of current events papers for children. Write to them to inquire if such materials are currently available.
16. The National Council for the Social Studies has printed two booklets for teachers on current affairs. They are "How to Use Daily Newspapers" and "How to Take a Public Opinion Survey." Each of them is about eight pages in length and inexpensive.
17. The Office of Educational Activities of *The New York Times* publishes several booklets on teaching current events. Some local newspapers may also have such materials.

On the next page is a listing of the current events newspapers for children and young people. Teachers may need to obtain editions which are easier or more difficult than the grade level of their children, depending upon the reading abilities of their pupils.

You will need to decide, too, whether to use local or national newspapers. Some use can be made of them in most third or fourth grade classes. More use can be made of them in older classes. In some groups you may want to rely entirely upon regular newspapers rather than current events papers. This is especially true in the upper grades. Pupils will be (or should be) reading such newspapers the rest of their lives and they should learn to read them intelligently. In fact that is one of the many aims of social studies teaching.

CURRENT EVENTS PAPERS: KINDERGARTEN THROUGH GRADE EIGHT

The Civic Education Service of Washington, D.C., which formerly published current events papers, has been amalgamated with Scholastic Magazines. The George A. Pflaum Company no longer publishes current events papers for Catholic schools. Those which are now being published include the following:

1. American Education Publications, Education Center, Columbus, Ohio 43216

Kindergarten	*My Weekly Reader* — *Surprise*
Grade 1	*My Weekly Reader* — *Grade 1*
Grade 2	*My Weekly Reader* — *Grade 2*
Grade 3	*My Weekly Reader* — *Grade 3*
Grade 4	*My Weekly Reader* — *Grade 4*
Grade 5	*My Weekly Reader* — *Grade 5*
Grade 6	*Senior Weekly Reader*
Problem readers, ages 11–15	*Know Your World*
Grades 7–8	*Current Events*

2. Scholastic Magazines, 50 West 44th Street, New York, N.Y. 10036
       Grade 1   *News Pilot*
       Grade 2   *News Ranger*
       Grade 3   *News Trails*
       Grade 4   *News Explorer*
       Grade 5   *Young Citizen* (for slower readers)
       Grades 5–6   *Newstime*
       Grades 6–7–8   *Junior Scholastic*

There are summer editions of some of the publications listed above. With each of these publications for pupils, there are teachers' editions. Of course you may want to use a paper which is listed for a grade other than the one you teach, if your pupils are faster or slower than the average.

Synopsis is the title of a relatively new publication, printed by Curriculum Innovations, 1611 Chicago Avenue, Evanston, Illinois 60201. Each issue is devoted to a single topic or contemporary social issue. Each issue is 12 pages in length. Many junior high school or middle school pupils can use them to advantage. They represent a stage between textbooks and current events papers.

The News Map of the Week is published weekly from August through May by a company of the same name, at 7300 North Linder Avenue, Skokie, Illinois 60076.

In many communities the local newspaper or newspapers are glad to provide copies for school use free or at special rates for pupils. Be careful, however, not to play favorites in the use of such papers.

A few of the larger newspapers in the United States have special departments for schools. The best known of these is the *New York Times*. It has kits for teachers on various aspects of newspaper production and on methods for using newspapers in classes. They also print a special *New York Times School Weekly*, which can be used well with older pupils in middle or junior high schools. For more information write to the New York Times, College and School Service, 229 West 43rd Street, New York, N.Y. 10036.

DESIRABLE OUTCOMES OF CURRENT AFFAIRS PROGRAMS

Any program in the social studies should focus on changed behavior. That means that attitudes should be improved, skills learned or perfected, and important knowledge gained. This is as true of programs in current affairs as it is in other aspects of social studies programs. What, then, are some of the outcomes with children to be sought by teachers?

One would be an increased interest in current events, in current affairs, and in current issues, depending upon the children involved. With young children the hope is that they will become concerned with other children and with adults and with new aspects of the world around them rather than being merely interested in themselves. With older children the hope is that any current affairs program will broaden their horizons and arouse new interests. With upper grade and junior high school pupils it is hoped that they will become interested in a few of the major issues of their community, nation, and the world.

Attitudes need to be changed, too, through current affairs programs. One attitude, of course, is an interest in current events and affairs. Another is in seeking all the facts available and learning to organize and present them in a variety of ways. This aim overlaps the objective of skills. A third is the ability to listen to other people and to try to understand what they are saying — and why they hold a certain belief, without necessarily agreeing with that point of view. A fourth attitude is one of suspended judgment. This is difficult to acquire, but important. It will not come easily with anyone, but teachers should attempt to educate their pupils in this respect. Concern for others is another attitude which should be cultivated.

Many skills also can be developed through current events and current affairs programs. One is the ability to locate and to gather information from a variety of sources. Pupils need to know where they can find out about current events and current affairs. Then they need help in learning to organize their materials in oral and/or in written form. They need to learn to find the main points and to get examples of these ideas. Next they need to know how to present their ideas and their information. This may be in oral or in written form. It may be in individual reports, in panels or roundtables, or in class discussions.

Through the use of current events and current affairs, boys and girls can be assisted in the development of a wide range of concepts taken from the various social sciences. Sometimes they will be introduced to such concepts through current events or current affairs; at other times their work in contemporary affairs can reinforce previous learnings. They can also develop their own generalizations, or reject or revise generalizations they have reached by using other experiences.

Children can be helped, too, to recognize some of the techniques of propaganda. These include such devices as name-calling, glittering generalities, transfer, testimonials, the "plain-folks" approach, card-stacking, and the bandwagon approach.

Another skill they need to acquire is the use of the vocabulary of current affairs. Other skills include the use of maps, charts, graphs, and other visual materials. Knowledge is not unimportant but attitudes and skills are even more basic in any desirable current affairs programs.

Handling Controversial Issues

Some teachers are frightened about teaching some current affairs or current issues because of the controversies involved in them. This is not unusual, especially for the young teacher or for the older teacher in areas where some topics arouse strong feelings or are almost taboo.

Teachers often wonder if they can teach such controversial topics. The answer depends upon many factors. One is your own personal security as we have already indicated in this chapter. Perhaps there are topics you are not yet ready to teach. You may want to postpone them, meanwhile trying to develop enough background or security to handle them later.

There may be issues which are too hot for you to handle in a given community or in a special school situation.

However, almost any topic can be discussed in social studies programs if you know how to approach it. Here are a few suggestions you may want to consider:

1. Establish some ground rules for discussions. Work these out in conjunction with your pupils. Write them on the chalkboard or on a large piece of oak tag. Refer to them when discussion becomes too heated. They might include such items as the following:
   (a) No monopoly of discussion by any pupil.
   (b) No "name calling."
   (c) Citing sources of information.

(d) The right of anyone to talk if he or she observes the basic rules of discussions.

(e) The development of such phrases as "It seems to me that . . ." and "I think that such and such is true."

2. Be sure that all sides of a question are presented. Ask questions about points that are not raised, or comment on points which do not come up in the discussion. If you want to, preface your remarks by stating that "There is another point of view which some people hold."

3. Try to get pupils to separate facts and opinions. Upon occasion, place two columns on the chalkboard as follows:

        Facts   Opinions

4. Try to keep in mind the importance of "suspended judgment" on many issues.

5. Make use of the continuum to show how people tend to divide on almost any issue.

6. Permit pupils to air their opinions at first. Then begin to examine what they have said.

7. Make sure that there are resources which represent various points of view on a controversial issue.

Always remember that controversy is a part of democracy and a valuable part. Pupils need to learn that adults (as well as children and young people) disagree on their interpretations of democracy, the need for changes, and ways to effect changes. But they also need to learn that debates can be conducted in a way which is fair and decent to all parties involved.

OBSERVING SPECIAL DAYS AND WEEKS

The celebration of special days or weeks should be an important part of the social studies programs of elementary schools. Such celebrations should include some, if not all, of the special days and weeks listed on the next page.

A few such occasions should be of local importance. Several of them should be days which have meaning for all Americans. A few might also be special holidays of ethnic groups in the United States, such as Columbus Day and St. Patrick's Day. This is especially important for the children of these minorities in our land. A few occasions should also relate to the international community, such as World Health Day and United Nations Day and/or Week.

These special holidays should be highlighted in the primary grades. Through them children are introduced to their own community and to their country. Through them they begin to identify with other people. Through them they gain their first introduction to history.

Such occasions should be happy ones for children. More will be learned through their feelings than through the facts which they acquire, even though both are essential to the success of these special days.

Stories, songs, art work, and filmstrips can be utilized for these observances. Food, candles, and other appropriate accessories will enhance the celebrations.

But teachers should not overlook the social studies content in these celebrations. Celebrations should be more than party days. They should be times in which children learn about the greatness of human beings and a little of the history of the United States. This emphasis should be increasingly evident as children grow older.

The careful planning of celebration of Christmas and Hanukkah can help children to develop feelings of respect for the traditions of the two major religious groups in our country. Halloween can be more than food and other gifts; it can mean helping the children in other parts of the world. There are important values to be inculcated through the celebration of such meaningful occasions.

In order to avoid meaningless repetition, teachers need to plan yearly programs for a school or a school system. Columbus Day can become boring and meaningless if it is observed in the same way year after year. The same thing can be said of any holiday.

One of the great dangers in the celebration of holidays is the fact that social studies in the primary grades is too often limited to such occasions. These special days and weeks are important but they are only a small part of any good social studies program even in the primary grades. Often short units of a week or two weeks' duration can be developed around the celebration of special events.

You should also be aware, if you are not already, that the celebration of some special days may be a cause for controversy in your school and/or community. Examples of this would be the question of whether to observe Martin Luther King, Jr. Day or Malcolm X Day. You should inquire about the rulings of the local school board on such issues and the feelings of your principal — as well as your own conscience on such matters.

Some Special Days and Weeks to Observe

In addition to the days and weeks listed below, you may want to add special days celebrated locally and/or in your state. By an act of Congress in 1971 the official holidays for Washington's Birthday, Memorial Day, Columbus Day, and Veterans' Day were shifted to Mondays as indicated below.

JANUARY

1 . . . . . New Year's Day
20 . . . . Inauguration Day every four years

FEBRUARY

Negro History Week
Boy Scout Week
Brotherhood Week — the week which includes the 22nd
12 . . . . Lincoln's Birthday
14 . . . . Valentine's Day
Third Monday — Washington's Birthday

MARCH

Girl Scout Week
Red Cross Month
Conservation Week, the week which includes March 7
7 . . . . . Arbor Day
17 . . . . St. Patrick's Day

Easter and Passover occur in March or early April

APRIL

Library Week
National Youth Week
Kindness to Animals Week
14 . . . . Pan-American Day

MAY

National Music Week
1 . . . . May Day
1 . . . . . Child Health Day
18 . . . . International Goodwill Day
Third Sunday — Mother's Day
Last Monday — Memorial Day

JUNE

14 . . . . Flag Day
Third Sunday — Father's Day

JULY

4 . . . . . Independence Day

AUGUST

19 . . . . National Aviation Day

SEPTEMBER

Constitution Week includes the 17th
First Monday . . . . Labor Day
17 . . . . Citizenship Day

OCTOBER

9 . . . . . Fire Prevention Day
24 . . . . United Nations Day
31 . . . . Halloween
Second Monday — Columbus Day

NOVEMBER

Fourth Monday — Veterans' Day
Book Week and American Education
Week
First Tuesday after the first Monday,
Election Day
Third Thursday . . . . Thanksgiving

DECEMBER

10 . . . . Human Rights Day
15 . . . . Bill of Rights Day
25 . . . . Christmas
Hanukkah occurs in the latter part of
December

BIBLIOGRAPHY ON TEACHING CURRENT AFFAIRS AND OBSERVING SPECIAL DAYS

1. ARONSON, JACOB I. "Sixth Graders Analyze Foreign Policy, in *Social Education* (October, 1959), pp. 287–289.
2. BURNS, P. C. "Check List for Teaching About Holidays," in *The Instructor* (February, 1960), pp. 89, 99.
3. COLE, WILLIAM (Editor). *Poems for Seasons and Celebrations.* Cleveland: World (1961), 191 pp. Grade 5 and up.
4. CROWDER, WILLIAM W. "A Good Nose for News" in *Grade Teacher* (September, 1965), pp. 158–160.
5. FRASER, DOROTHY M. "Current Affairs, Special Events, and Civic Participation" in JOHN U. MICHAELIS (Editor), *Social Studies in Elementary Schools.* Washington: National Council for the Social Studies (1962), pp. 131–149.
6. GARNETT, EVE. *A Book of the Seasons: An Anthology.* Cambridge, Massachusetts: Bentley (1961), 80 pp. All ages.
7. HOWITT, LILLIAN C. *Enriching the Curriculum with Current Events.* Englewood Cliffs, New Jersey: Prentice-Hall (1964), 64 pp. Especially good for the junior high school grades.
8. JAROLIMEK, JOHN. *Social Studies in Elementary Education.* New York: Macmillan (1971), 534 pp. Chapter 9 "Current Affairs in the Social Studies."
9. KRAVITZ, BERNARD. "Factors Related to Knowledge of Current Affairs in Grades 7 and 8" in *Social Education* (March, 1962), pp. 143–145.
10. LEE, JOHN R. in "How You Can Keep Current" in *The Instructor* (March, 1963), p. 5.
11. MCAULAY, JOHN D. "Current Affairs and the Social Studies" in *Social Education* (January, 1959), pp. 21–22.
12. MCLENDON, JONATHAN C. "Using Daily Newspapers More Effectively" in *Social Education* (October, 1959), pp. 263–265.
13. MICHAELIS, JOHN U. *Social Studies for Children in a Democracy.* Englewood Cliffs, New Jersey. Prentice-Hall (1963), 624 pp. Chapter 6 "Current Affairs and Special Events."
14. MUSTARD, SHIRLEY C. and EDWARD F. DE ROCHE. "The Newspaper as a Daily Textbook" in *The Instructor* (May, 1965), pp. 33, 99.
15. SMITH, LLOYD L. "Current Events for the Elementary Schools" in *Social Education* (February, 1961), pp. 75–78. The work of a sixth grade.
16. WASS, PHILMORE B. "Improving Current Events Instruction" in *Social Education* (February, 1961), pp. 79–81.

17. WILSON, RICHARD C. "Using News to Teach Geography" in *Social Education* (February, 1960), p. 56–57.

See also several publications for teachers, available from the College and School Service, New York Times (229 West 43rd Street, New York, N.Y. 10036) on the printing and use of newspapers.

There are many books on holidays. See for example the series on holidays by the Crowell Publishing Company and by the Garrard Publishing Company.

Silver Burdett has sets of colored illustrations on various holidays, printed on laminated surface cardboard for display and study purposes.

How, Then, Should a Teacher Plan Learning Activities in the Social Studies?

11

By now you have certainly been wondering how you can plan specifically for the group or groups of pupils you will teach. You have raised a good many questions about planning for a year, for a semester, for a week, or even a day. In this section we will try to anticipate some of those questions and try to answer them, at least in part.

A whole year stretches before you in the social studies. Does it seem like a long time? Have you wondered if you know enough to "last" for an entire year? Most young teachers do. But you need not be fearful. If you utilize what you already know, continue to learn, and avail yourself of all the aids that exist, your problem will not be running out of material. Your problem is more likely to be one of utilizing the time available to do all the things you want to do. It takes much more time to develop learning with boys and girls than most young teachers realize. And you probably know much more than you realize, especially in comparison with your pupils.

Then there are many people from whom you can obtain help. There is a principal and possibly an assistant principal. There may be a supervisor or the teacher on a "team" of teachers. There is the librarian of the school and the librarian of the local library. There are scores of parents and other adults in the community. Above all, there are several older teachers. They will be busy, too, but many of them will stand ready to help you if you show a readiness to listen and a willingness to express your appreciation for their assistance.

All of these points apply, too, if you are a student teacher and are working in the social studies field.

Right away you will want to obtain the course of study in social studies of your particular school or school system. Then you will want to obtain several other courses of study which are similar. If you have collected some of them before you start your student teaching or regular teaching, you will be fortunate. All of these will prove to be life-savers and time-savers.

If your school does not have a course of study, you will want to rely on those from similar school systems at first and then begin to develop your own plans for your class or classes.

Before you do a great deal of planning, you should find out about the particular class or classes you will be teaching. Then you can plan with those groups in mind.

Now you are ready to think of long-range goals and plans. What do you hope to accomplish in a year's time? The tendency at first may be to think largely or even solely in terms of knowledge. Important as that is, you should think and plan in terms of much wider goals. You should jot down the major attitudes you want to develop, change, or reinforce. You should plan for the major skills which you will need to stress with your particular group or groups. All of these goals should be focused upon the changes in behavior which you will want to develop.

Once this long-range planning has been completed, you probably will want to think in terms of dividing the year's work into blocks of time which last from three or four weeks to six or seven weeks. These often are called "units." Most social studies teachers find the unit a very helpful method of teaching. For example, a year spent on various communities in the United States might well be divided into studies of eight or ten communities. An average of three or four weeks would be spent on each community.

Ordinarily the first unit will take the most time. Pupils are being introduced to new content and new methods and they need considerable time to develop new skills. Thus the first unit might be five or six weeks in length. Some of the later units could be shorter. Later in this chapter we will consider the "unit" in more detail.

In your preplanning, you will probably not have time to develop in detail more than the first unit. This is especially true if you are the teacher of a self-contained classroom and have to think in terms of other subject fields.

As you develop specific plans for the first five to six weeks, you will need to know what type of teaching your pupils have had before they reached you. If they have been accustomed to a wide variety of methods and considerable pupil-teacher planning, you will be fortunate. You can move on from there. However, if they have been accustomed to a single-textbook approach, you had better not plan on too much change in the first few weeks. This is especially true of younger and of slower pupils. They will need considerable security and some of that security comes from doing things in the ways they have done them before.

Gradually you can introduce new methods and new materials. On the "ladder" on the following page, we suggest how a teacher might move cautiously from a more formal approach with limited experiences and materials to a much broader approach. You would do well to move up the ladder step by step rather than trying to jump your pupils from a lower rung to the highest rung quickly.

As a teacher you will undoubtedly want to think in terms of three types of experiences for your pupils. They should work alone some of the time. Eventually they need to work in small groups much of the time. And they certainly need to work, too, in the total class group. Some of them, especially in the upper grades, need to have a fourth type of experience. That is the experience of working with pupils in even larger groups than the single classroom.

Today there is a tendency for groups of teachers to work together in "teams." Sometimes this is done only in planning for several sections of a grade. Often this plan of teaching includes large group instruction of pupils as well as smaller group work. You may find yourself on a team. If this is true, you may become the expert on two or three units in the year's work of a given grade. You may teach that unit to two or three different groups or you may be responsible for the major work on those units for several sections at one time. In either case you will undoubtedly want to follow the general suggestions in this chapter for the preparation of social studies units, remembering that you are preparing a type of teaching unit, with only your school in mind.

MOVING PUPILS UP THE LADDER OF
LEARNING EXPERIENCES

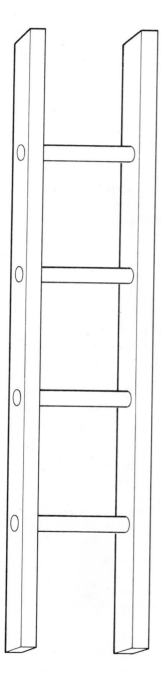

Problem-centered learning, much pupil-teacher planning, variety of reading materials, trips, and audio-visual materials. Much group work. Some individual work, too. Deemphasis upon grades. Consultations between teachers, pupils, and parents on progress. Rigorous thinking promoted.

Some problem-centered teaching with some participation of pupils in planning. Several textbooks plus enrichment reading materials. Much audio-visual work. Some trips. Some small group work and individual research. Considerable critical thinking promoted.

A basic text plus other texts and supplementary reading. Some audio-visual resources. A few panels. Many individual reports. Some individualized assignments. Some emphasis upon grades. Some thinking encouraged.

A basic textbook plus some supplementary reading materials. An occasional film or filmstrip. A few individual reports. Considerable emphasis upon grades. All work still planned by the teacher. A little thinking promoted.

Slavish use of one textbook for all pupils. Very little enrichment reading. Paucity of trips, audio-visual materials and individualized reports or small group work. Much emphasis upon grades, largely determined by tests. Very little thinking promoted.

PLANNING SOCIAL STUDIES UNITS

Let us assume that you have decided to divide your social studies work for the year into eight units. With the exception of the first unit, each of them will last from three to five weeks. The first unit will probably take a little longer than that, so you allow for five to six weeks for it.

Now you are ready to preplan for this particular unit of work. You want to leave much room for teacher-pupil planning, but you do not want to approach the unit "cold," especially this first unit. Hence your preplanning.

The Title and the Scope of the Unit

You decide that you will set the stage in your class for a unit on "Houses in the United States and Around the World." The particular title for the study will be decided upon by the class. Perhaps it will only deal with houses in the United States for the first three weeks or so. Then you may decide to enlarge it to include houses in other parts of the world. Since this is a third grade group of children, you decide that you will touch upon houses in olden times but concentrate upon houses today, since the concept of time is so difficult for children that age. That is all you are willing to do at this point in preplanning and narrowing the scope of this particular unit.

Determining the Aims of the Unit

It takes much longer to decide upon the aims of the unit. You are aware that many units have very broad aims, so broad, in fact, that they become almost worthless. So you decide that you will try to state your aims as concisely as possible. The order in which they are stated is not important at this point. You want to include in these aims some statements which concern attitudes, some on skills, and some on knowledge (organized around basic concepts to be discovered by the pupils). You are also aware of the different social science disciplines involved in this study, so you try to see that all or most of them are included. When you have finished your thinking on aims, you have a list like this:

1. To understand the variety of types of houses in the U.S.A. today.
2. To discover how many different building materials are used in making houses and the reasons for the use of these materials.
3. To learn how climate helps to determine the building plans for houses.
4. To learn how income or standards of living help to determine the homes in which we live and the houses we build.
5. To see that higher standards of living and transportation bring changes in the types of materials used.
6. To learn that many people are involved in the construction of houses.
7. To learn about the use of certain building materials and the processes by which they are made.
8. To discover how people beautify their homes in different ways.
9. To see how governments sometimes make regulations about houses.
10. To learn that governments sometimes build homes for people or help them build homes.

You may want to make a list of the major concepts on which you will want to work in conjunction with this unit. That is another way of stating aims. For example, here are a few such concepts. Undoubtedly you will think of others to add to the list:

apartments	houses
architects	inspectors
blueprints	layout
building codes	planning
building materials: brick, cement, glass,	plans
steel, stucco, wood, etc.	protection
construction workers: bricklayers,	residential
electricians, plumbers, etc.	site
costs	taxes
government housing	transportation
homes	wiring

You may also want to develop a list of generalizations for this unit which the pupils will discover themselves. For example, one such generalization would be:

> "The least expensive materials for homes are usually found in the community where the house is being built"

These first ten aims include material from geography, economics, sociology, and government. Some attention may be given also to history by a visit to a home or homes built many years ago, by pictures which show such homes, or by other visualizations of homes in our country in the past.

So far the aims are all you want to develop. Now you think about some of the skills and attitudes you want to develop. You add a few more aims to cover these approaches. Here are some you might add:

11. To learn to observe on trips to houses.
12. To learn to "read" pictures about houses.
13. To learn to study and interpret maps and make maps of the location of houses and the layouts of houses.
14. To gain further skill in using books about houses (indexes of books, tables of contents, etc.)
15. To develop critical thinking in connection with the study of houses.
16. To learn to read better through the use of various materials.

You ask yourself, what attitude or attitudes are central in this unit? And you decide, above all, you want children to develop respect for people in different types of homes. You also want to develop in them an interest in the safety of their own home and its beautification. So you add those as two more aims, as follows:

17. To develop respect for people, no matter what kinds of homes they live in.
18. To develop an interest in their own homes, especially in regard to safety and beautification.

You may want to include some science work in this unit, thus relating the unit to science as well as social studies and the language arts. You may also consider how arts and music play a part in the enjoyment of the home. If you want to keep the unit within manageable proportions, you may decide to drop these last few aims.

You may now go over your list of aims, checking the most important ones.

Gathering Ideas and Resources on the Unit

Your list may not be as extensive the first time around as the list of aims on the preceding pages would indicate. Perhaps you could make a tentative list of aims and then start searching for materials for the unit.

There are many places to go for help. Among them are the following:

1. Curriculum laboratory for resource or teaching units of other school systems.
2. Libraries — for pictures and books for the unit, both for adults and for pupils. Make 3″ by 5″ cards on each item, with notes on the book as to readability, if possible.
3. Looking in booklets on "free and inexpensive materials" for charts, posters, pictures, and booklets.
4. Looking up in film and filmstrip catalogues suitable visual materials for your class.
5. Call to see if there is a government agency which could assist you.
6. See if there are parents or other adults in the building business who could be resource persons for the class.
7. Begin to spot pictures and other materials in newspapers and magazines.
8. Talk with school librarians (or the local librarian) about other sources of materials.
9. Walk or drive around the community, taking notes on houses to visit.
10. Examine the *Education Index* to see if there are articles on houses which might be helpful.
11. Start a collection of picture magazines, to which the pupils will add others later, from which pictures can be obtained.

You may now want to work out the experiences and possibly even some of the resources for your unit. You might develop them in relation to your aims, taking each aim and showing what experiences could be used to help children to discover a basic idea. One such aim is used here as an example of what you can or should do:

Aim	Possible experiences
The variety of houses in the U.S.A.	(a) Discussion in class of different types of houses.
	(b) A walking trip through the community to see different kinds of houses.
	(c) Assembling of pictures of different kinds of houses.
	(d) Examine books quickly for pictures of different kinds of houses.

Probably you will want to list some of the kinds of houses, including ones that the children do not know about already. If you have a looseleaf notebook, you can place your notes on a topic like this on one page and keep adding to your notes, thus saving much time and energy.

If you do not feel too secure about teaching this unit, you may want to go further in your pre-planning, listing the topics you want to study in the order in which you think you will study them. You may even want to develop a tentative schedule, indicating alongside each topic the number of days you expect to spend on that part of the unit. As you develop more security you should be able to teach from the suggestions of the class and shift your plans according to pupils' needs and interests.

Launching the Unit

With this preparation, you should now be ready to think about ways of initiating or launching the unit. You think about several alternatives. They might include the reading of a book about houses and a discussion based upon that reading, the showing of a film or filmstrip on houses in the U.S.A., an example of pictures around the room on houses in our country, or a trip to see various houses in your community. Let us assume that you decide upon the trip.

On the trip you will want to see a variety of houses. When the pupils return, you can move in one of several directions. You may want to talk about the materials used in making the different houses. You may want to talk about the size of the houses. You may want to discuss with them the effects of climate on housing.

Probably you will want to assemble a list of questions which have been raised by the trip and your class discussion. To the questions they suggest, you may want to add others. You may find quite early that a unit is a little like one of those nests of wooden dolls which are made in Eastern Europe or in Japan. As soon as you start with the large outside doll, you find there is a second doll inside it — and then a third and a fourth and a fifth. When children start on a topic, they are likely to discover there are more things to learn than they had at first imagined.

Developing the Unit

Many teachers immediately divide the class into committees, once they have launched the unit. This is a questionable practice. There are usually many broad topics which need to be explored by the entire class before they are broken into subgroups. A class also needs some common background. Otherwise the unit and the learnings in the unit become fragmented.

You may want to keep the group together for a week or so before dividing it into committees or subgroups. During that time you may well have some of the children write away for materials. If they wait till later, the replies may arrive after the unit is completed!

After a few days together, you may want to divide the group into smaller units. Committees are more likely to work if they are kept to five or six pupils, and if they have a specific topic to explore and a specific time to report back to the class.

Each committee should collect its own material, organize it, and report back to the group in an original way. Committee work often breaks down at the point of reporting back to the class. Each child reads a report and that takes much time. The oral presentation is not always the best way to learn and the class often does not gain much by this procedure.

Committees should be helped with the methods they will use in reporting their findings. Different committees should use different methods according to their topics. In that way oral reports will be minimized and a variety of interesting methods will promote better learning. Interviews, pictures drawn by the pupils, panels, plays, rexographed materials, and pictures shown through the opaque projector are some of the ways in which committees reports can be given. For other suggestions, turn to pages 97–100 of this book.

In a unit on houses, you may want to keep the entire group together for the study of houses in the United States. Then you can divide them into committees on houses in other parts of the world, selecting five or six countries to explore. You may want to have committees, however, on houses in the United States, dividing them according to their interest in the materials used in houses, the layout of homes, trailer camp houses, apartment houses, and safety in homes.

Concluding the Unit

Most advocates of unit teaching suggest that there be some big summarizing or culminating activity. Sometimes that is valuable. It may be an assembly program or a program for another class. It may be an exhibit. It may be a play. But all units need not have a culminating activity. Teachers may not even feel that a test is necessary; they may have plenty of evidence of whether children have learned or what they have learned. The culminating activity may be just a closing summary of what they have done and how well they have done it, with suggestions for the next unit.

Correlation with Other Subjects in Units

Most units lend themselves to considerable correlation among various subject fields. Units in the social studies always include some work in the language arts. Often they lend themselves to correlation with music and art. Sometimes science is included in a social studies unit, as in this one on houses. Mathematics seldom lends itself to real correlation with social studies.

Evaluation of the Unit

Evaluation by the teacher and in some instances by the pupils should be going on throughout the unit. There need not be any written evaluation at the end of the unit. As the teacher, you will want to refer back to your list of aims and see how well they are being accomplished. Are children learning map skills in this study of houses? How well? Who is making a great deal of progress? Who is making only a little? Are some pupils making progress in reading as a result of their interest in this topic? Which ones? Why? So you can go through your list of aims and evaluate the learnings of individuals, small groups, and the class as a whole. Comments of the librarian and of parents may also help in your evaluation. You may want to have an anonymous evaluation of the unit, based on a checklist or list of open-ended statements on the parts that were most interesting and least interesting. There are numerous methods you can use to evaluate progress on a unit. For suggestions along this line see pages 214–223.

Different Types of Units

The type of unit which we have just described in some detail is a teaching unit. It is a plan for a specific group on a given topic over a period of several weeks.

There are several other types of units. One of the most helpful for you as a teacher of the social studies is the resource unit. It is a general unit developed for teachers in many schools, with a wider variety of aims, methods or experiences, and resources from which various teachers can obtain ideas in preparing their own teaching unit.

Dr. Wilhelmina Hill has written widely over a period of several years on unit teaching. In a recent booklet available from the U. S. Government Printing Office, entitled *Unit Planning and Teaching in Elementary Social Studies*, she describes several types of units. In addition to resource and teaching units, she mentions:

Descriptive units. Written by a teacher or a group of teachers upon the completion of a study by a class to share with others what transpired.

Integrated units. A term used to describe units which emphasize the integration of several subject fields in the study of a topic.

Experience units. Units based primarily upon the direct experiences of children.

Activity units. A term usually used in contrast to the experience units to indicate topics studied at a distance, such as a unit on pioneers in the American colonies or a study of a country other than the United States.

Contemporary events units. This type of unit concentrates upon current affairs such as an election or a meeting of the U. N.

Subject matter units. Many textbooks are divided into what are called units, even though they are planned in advance and have very few of the characteristics of real units.

Unit Teaching and Textbook Teaching

Unit teaching and textbook teaching need not be mutually exclusive. In a unit, several textbooks may be used as resources or a single text used as a common reading. However, the textbook is not likely to be central. On the other hand a teacher using a textbook may start with that and then develop a unit from the common background provided by the text.

A SUGGESTED OUTLINE FOR RESOURCE UNITS

1. *Title of the unit*
2. *Statement of scope*
   Age or grade range, reading range, maturity of group for which the unit is intended.
   Any limits placed on the topic.
   Alternative proposals for organization of the unit, with a statement of the approach you have used in organizing the materials.
3. *Aims of the unit*
   Include changes in behavior, attitudes, skills, and knowledge.
   Be as specific in stating aims as possible, with aims for this particular unit rather than for any unit.
4. *Possible introductory experiences*
   List a good many so that there is a wide choice available for teachers and ones for their classes.
5. *Possible experiences*
   List a good many so that there is a wide choice available for teachers and pupils using the resource unit.
   Possible experiences may be listed by aims, by topics, or by subject fields.
6. *Resources*
   List all possible resources, inasmuch as many of them will not be available to some teachers and pupils.
   Annotate wherever possible. Use the annotations others have made, if necessary, but give them credit.
7. *Possible culminating experiences*
   List several.

8. *Evaluation*

Refer to your original aims for the unit and be sure there are means of evaluating all of these aims.

9. *Addresses of organization and publishers cited*

In order to facilitate the assembling of materials, list the addresses of the firms cited in the unit.

Alternate Form for Resource Units — Plan A

| *Aim* | *Possible experiences* | *Resources*
(These may be numbered, referring to bibliography) |

Alternate Form for Resource Units — Plan B

Aim (number 3. above)
Possible experiences to carry out this aim
Resources on this aim

BIBLIOGRAPHY ON UNIT TEACHING

1. ALDRICH, JULIAN C. "How to Construct and Use a Resource Unit." New York: Joint Council on Economic Education (undated), 44 pp.
2. GILBAUGH, JOHN W. *How to Organize and Teach Units of Work in Elementary and Secondary Schools.* San Jose, California: Modern Education Publishers (1958), 246 pp. There are also several books of units for each grade level.
3. HANNA, LAVONE A., GLADYS L. POTTER and NEVA HAGAMAN. *Unit Teaching in the Elementary School: Social Studies and Related Sciences.* New York: Holt, Rinehart and Winston (1963), 595 pp.
4. HILL, WILHELMINA. *Selected Resource Units — Elementary Social Studies: Kindergarten–Grade Six.* Washington: National Council for the Social Studies (1961), 91 pp.
5. HILL, WILHELMINA. *Unit Planning and Teaching in Elementary Social Studies.* Washington: Government Printing Office (1963), 79 pp.
6. JAROLIMEK, JOHN. *Social Studies in Elementary Education.* New York: Macmillan (1971), pp. 68 ff.
7. MICHAELIS, JOHN U. *Social Studies for Children in a Democracy.* Englewood Cliffs, New Jersey: Prentice-Hall (1962), 624 pp. Chapter 7 "Planning Units of Instruction."
8. MICHAELIS, JOHN U. *Social Studies in Elementary Schools.* Washington: National Council for the Social Studies (1962), 334 pp. Chapter 8. Sections on "Designs for Social Studies Units," "Incorporating New Content in Units of Study," and "Teacher Planning for a Specific Class."
9. MICHAELIS, JOHN U. *Teaching Units in the Social Sciences.* Chicago: Rand McNally (1966). Separate books on "The Early Grades," "Grades III–IV," and "Grades V–VI."
10. NERBOVIG, MARCELLA. *Unit Planning: A Model for Curriculum Development.* Worthington, Ohio: Charles A. Jones Company (1970), 96 pp.
11. NOAR, GERTRUDE. *Teaching and Learning the Democratic Way.* Englewood Cliffs, New Jersey: Prentice-Hall (1963), 244 pp. Includes materials on units.
12. PRESTON, RALPH C. *Teaching Social Studies in the Elementary School.* New York: Holt, Rinehart and Winston (1968), 370 pp. Chapter 4 "The Unit Method."
13. RAGAN, WILLIAM B. and JOHN D. MCAULAY. *Social Studies for Today's Children.*

New York: Appleton (1964), 409 pp. Chapter 9 "Unit Teaching in the Social Studies."

14. WESLEY, EDGAR B. and WILLIAM H. CARTWRIGHT. *Teaching Social Studies in Elementary Schools*. Boston: Heath (1968), 354 pp. Chapter 10 "The Unit Approach in Organization."

Printed units may be obtained from the following places. Some are free; others have a small charge.

Compton's Pictured Encyclopedia, 100 North Dearborn Street, Chicago, Illinois 60610.

Educational Publishing Corporation, Darien, Connecticut 06820 (*The Grade Teacher*).

The Instructor magazine, Dansville, New York 14437.

Modern Education Publishers, Box 651, San Jose, California 95100.

World Book Encyclopedia, 510 Merchandise Mart, Chicago, Illinois 60654.

PLANNING DAILY ACTIVITIES WITH SPECIAL REFERENCE TO THE SOCIAL STUDIES

Now that we have looked at the overall objectives of the social studies and at units as they attempt to carry out these aims, it is time to consider how daily plans can fit into these larger plans.

In most elementary schools you will probably want to think in terms of the entire day's program, fitting the social studies activities into the overall plan. In the upper grades you may have to think in terms of a limited amount of time and little relation with the other subject fields.

In any case, you will need to do some planning. That aspect of teaching is highly important to all of us, experienced or inexperienced. The experienced teacher may carry her plan in her head or merely jot down an outline of the day and the topics she wants to cover. The inexperienced teacher needs to spend considerable time in planning the activities of a day and in making plans for individual parts of it. This is not easy, because time and energy are limited. But it is highly important.

If you are working with a class all day, probably you will want to sketch out the entire program, realizing that it is subject to change. In the lower grades, you will want to provide for alternating quiet times and action and seeing to it that there are some play periods during the day. You also will want to think in terms of large blocks of time for related activities and for two or three reading and language arts periods.

Here is a possible plan for one day in the first grade, in this case centered on the social studies as a followup of a trip the previous afternoon to a new house under construction in the neighborhood:

9:00– 9:15 — Getting the day started. Attendance. Health checkup. Planning with the class.

9:15–10:15 — Social studies and related activities block. Discussion of the trip and development of experience chart on the trip. Reading aloud about the building of a house.

10:15–10:45 — Midmorning snack and plan; bathroom.

10:45–11:30 — Language arts and reading block. Reading individually and in groups. Teacher works with one group.

11:30–11:45 — Music.
11:45–12:00 — Getting ready for lunch.
12:00– 1:00 — Lunch period.
 1:00– 1:10 — Group planning and discussion.
 1:10– 1:30 — Mathematics (or science).
 1:30– 2:00 — Language arts and reading.
 2:00– 2:15 — Bathroom.
 2:15– 2:45 — Individual and small group activities in clay, wood, drawings, etc.
                       Some work may be done in conjunction with the trip yesterday.
 2:45– 3:00 — Cleaning up and getting ready for dismissal.
 3:00–   — End of the day.

Part of the day outlined on the previous page is devoted to learning activities growing out of a meaningful trip. But the entire day is not devoted to this social studies experience. The children may use their time for individual and small group activities to illustrate the trip, but everyone is not forced to do this.

In a later grade only the social studies and language arts work may be related, or the social studies and the music may be related. There should be some correlation of subject fields but they should not be forced. For example, a plan for the fifth grade might look like this for one day in which the social studies and language arts are tied together in a natural way:

 9:00– 9:15 — Opening activities of the class. Planning.
 9:15–10:00 — Reading by individual pupils on the books they have obtained in
                       the library on Life in Colonial Times.
10:00–10:45 — Class discussion of Life in 1700 in the Middle Colonies, drawing
                       upon the reading done in the previous period and on previous days.

The rest of the day would be devoted to other subjects and in no special way correlated with the social studies and language arts fields.

Particularly as a young teacher you may want some help in planning a social studies activity or a social studies lesson. Here are some of the main points you will want to bear in mind as you do such planning:

Group with which I am working:

Time probably available:

The one or two main aims of this activity or lesson; stated as specifically as possible:

1.
2.

The way in which I intend to introduce this topic (sometimes called motivation):

The methods I intend to use:

The materials or resources available for this activity or lesson:

Ways in which I can involve the bright pupils:

Ways in which I can involve the slower pupils:

How I can relate this work to the daily lives of these pupils:

Summary of the activity or lesson (not always necessary):

Where this activity or lesson leads; future planning:

Perhaps at this point you will want to take a specific activity or lesson for your class and see how you could use this outline for it.

SOME EXAMPLES OF LESSON PLAN OUTLINES

Explanation of the Lessons Which Follow

No outsider can write a lesson plan or tell you how an activity is to be conducted.

Therefore the plans which follow are merely suggestions to get you started. You will have to adapt them to your own group and change them to meet your needs and your personality and style of teaching.

The outlines which follow are for several grade levels of schools. One is for the Kindergarten and/or First Grade. Another is for the Second or Third Grade. A third is intended for children in Grades Three and/or Four. The next is for Grades Five and/or Six. The last one is for Grades Seven and/or Eight.

They represent concentration on various aspects of the social sciences. The one on airplanes is devoted largely to economics and geography. The plan on the Masai boy is focused on anthropology and sociology with some economics and geography included. The lesson on government services and taxes highlights government and economics. The map lesson is primarily geography-centered. The lesson for the Seventh and/or Eighth Grade is history and geography.

The methods these lessons and activities represent are varied. They represent problem-solving, role-playing, interpretation of maps, and pictures.

The lessons outlined here are done on small cards. This is to achieve the maximum flexibility, with teachers shuffling the cards if the lesson develops in a different way than planned. It also encourages the teacher to move around the room during an activity or lesson.

Introducing the Study of Air Transportation (Kindergarten—Grade One)

I had expected to introduce the study of airplanes and air transportation sometime during the year, probably in the spring. But today I changed my mind when Juan told me that he was going to Puerto Rico during the holidays. His uncle is paying for his trip.

This provides us with a wonderful opportunity to do two things: (1) introduce the study of airplanes and air transportation, and (2) give Juan a central spot in the activities of the class — for a change. He can be the "hero" for this part of our program.

That means I need to do some thinking about this short study of airplanes. Overall aims come first, so here is a list of the major ideas I want the children to discover:

1. People travel in many ways. Air transportation is one important means of transportation. Many people travel to distant places by airplane.
2. People also send goods by airplanes. This includes letters, flowers, vegetables and fruits, packages, and even animals.
3. Airplane transportation is very fast.
4. Airplane transportation costs money.
5. Many people work in connection with airplanes. There are ticket agents or clerks, pilots and co-pilots, porters, engineers, mechanics, weathermen, and stewardesses.
6. All kinds of people work in air transportation (different races).
7. These people go to school to learn to do their jobs well.
8. San Juan is a large city in Puerto Rico. Many people come to our city from Puerto Rico and help our city in many ways.

Most of this material is taken from economics. Some is from geography. All of it is related to the big topic of transportation. Some of these concepts and generalizations will not be easy. I will need to move slowly and let the children see a great many pictures. They will need to act out, with blocks and other materials, the ideas we are developing. We will need to role-play considerably, acting as ticket agents, porters, pilots, etc.

I will want to go to the school library and the local library to find several books on airplanes. The pictures in them can be used, and I can read some of the simple materials to the class. In a few cases I will have to do the reading myself and rephrase it to tell the children.

Eventually we will want to go to the airport. This can come after we have done some initial work on this unit and when they are ready to see a few places and people there. It can also motivate them for further study when the idea of airplanes begins to be boring. Then they can "play out" what they have seen.

This should be a fascinating study for my pupils inasmuch as many of them are interested in airplanes. This seems to be especially true of the boys. Considerable learning can take place as we develop the generalizations listed above.

With that background or planning in mind, I am ready to think about the first day when Juan returns. I must play most of this work by ear, but I want to have some ideas in mind as to the questions we will ask Juan after he tells about his trip. Here are the notes I have jotted down. I don't need a "lesson plan" as such, but I will need some background notes as follows:

Teacher Background Notes on
An Introduction to Airplanes and Air Transportation

Get Juan to tell about his trip over the holidays to San Juan.
   May mention: eating on the plane, the people he met on the plane, his relatives, how he "felt."
Children to ask questions — teacher to ask questions too, probably on:
      How fast the airplane went?
      Who flew the airplane?
      Why Juan went by airplane?
      What did he see?

Followup on one or two topics which they raised:
      1. What the airplane looked like — what kind?
      2. What it was like inside?
Look at pictures of airplanes, especially the inside of planes.
   Note the people in it — pilot and co-pilot, flight engineer, and stewardesses.
Later in the day read from *I Want to Be an Airplane Pilot.*
Watch for:
      Areas of interest by children.
      Vocabulary.
      Children who act out airplanes — free play.

Of course I thought about extending this brief unit on air transportation (a part of a longer unit on transportation in general) to include a study of San Juan. I even talked about it with the assistant principal who is in charge of my grade. But she wondered if I could do justice to such a study on the spur of the moment and whether that study would not be better for the third or fourth grade, where they are studying communities in the United States and in other parts of the world. So I decided against that idea.

An Introductory Lesson in a Second Grade Unit on
"Children and Their Families in Kenya"

During the next two or three weeks I plan to concentrate with my second grade class on "Children and Their Families in Kenya." Since this is the first time that any of my pupils have studied any part of Africa, I realize how important this short unit will be. I want to shatter some stereotypes which children already have about Africa and Africans. I want the children to learn that people everywhere have the same needs, although they meet them in many different ways.

Here are some of the generalizations I want my pupils to discover in the next few days:

1. There are many kinds of people in Kenya (or this part of Africa).
2. The children live in families. (Some live in small families, "the nuclear family"; some live in large families, or "extended families."
3. The children wear a variety of kinds of clothes. The clothes are suited to their climate (and their economic level). Some wear clothes bought in stores; some wear clothes made at home.
4. Many children live on farms; some live in small towns; a few in large cities.
5. The children and their families have work to do. Many children help their parents, especially on the farms (shambas).
6. Many children go to school (but not all of them do).
7. Children in Kenya have fun in many ways.
8. Their families have problems, just as our families have problems.
9. Some children and adults in Kenya live quite different lives from the ones we lead. They have learned to live well in their environment. (For example, the Masai.)

Of course there will be other generalizations which we discover as the unit develops.

For my own background I read Edna Kaula's *The Land and People of Kenya* (Lippincott, 1968) and Richard Cox's *Kenyatta's Country* (Praeger, 1965). Then I glanced at three books on childhood in East Africa: E. B. Castle's *Growing Up in East Africa* (Oxford University Press, 1966), Lorene K. Fox's *East African Childhood: Three Versions* (Oxford University Press, 1967), and R. Mugo Gatheru's *Child of Two Worlds: A Kikuyu's Story* (Praeger, 1964).

Now I think I am ready to begin the unit. I thought of using as an introduction the large, laminated surface picture of a Masai boy in the Silver Burdett set of photos on "Living in Kenya." But I decided that was too different for a start. So I plan to use the picture in that same set which shows children outside a school at recess time or at lunch time.

I want to use this to help me evaluate the present knowledge and attitudes of my pupils and to help them grow in the skills connected with studying pictures. I want to encourage them to give their frank responses. Then I want to direct them in studying carefully a few aspects of that color photograph. I doubt if I need any cards, but here are my notes:

What do you see in this picture?

Free-flowing comments from pupils with minimum leads
   by me. Hold answers or "structuring" till later unless
   one aspect "catches fire."

Clothes: School uniforms and reasons for them.
   No competition for "best dressed." Some schools in the
      U.S.A. have uniforms for pupils.
   Neatness of their appearance.
   Different uniforms for boys and girls.
   No shoes: climate, custom, and lack of money.

Having fun

London Bridge Is Falling Down.
Ring Around the Rosy or Mulberry Bush.
Boys jumping.
Soccer ball.
Talking with friends.

Where do you think it is?
   Evidence: Palm trees.
            Palm tree leaves for roof.
            Lack of glass in windows.
   Tropical setting, probably along coast of the Indian Ocean.
      Show on globe.

I am not concerned about the order in which these main points are taken up. I
intend to play this by ear. Nor do I want to contradict what they say. For the time
being I want to raise questions in their minds and let them discover whether they
were right or not — later.

Later we can use other pictures in this portfolio, including the one on the Masai
boy. In addition I want to use a couple of films and filmstrips, skipping the captions
in order to encourage the pupils to study the pictures, and turning off the sound
track on any films.

Among the books I can use with the class are the following: Anna Riwkin-Brick
and Astrid Lindgren's *Sia Lives on Kilimanjaro* (Macmillan, 1966 reprint), Natalie Donna's *Boy of the Masai* (Dodd, Mead, 1964), Muriel Feelings *Zumani Goes to
Market* (Seabury, 1970), Emily Hallin's *Moya and the Flamingoes* (McKay,
1969), Frederick Moffitt's *A Busy Day for Okoth* (Silver Burdett, 1967), and Hilda
Van Stockum's *Mogo's Flute* (Viking, 1966).

The Services of Government Cost Money
(An introduction to the study of taxes)

For several days our class has been studying the services of government. We have concentrated upon the services of the local government but have mentioned some services of other governments. In order not to confuse pupils, I have avoided differentiating among the various units — local, state, and federal — lumping them together and calling them "the government." Differentiation can come later in their school years.

Our social studies today came late in the afternoon as we concentrated on reading in the morning. As a group we made a list of the various services we had studied. That list looked something like this:

1. Parks and playgrounds
2. Streets and highways
3. Schools
4. Post Offices
5. Fire departments
6. Police departments
7. Housing
8. Health protection, including the inspection of water, milk, candy, etc.
9. Water supply
10. Lights
11. Hospital and clinic
12. Relief workers and health workers

After a quick nap and dinner, I am now ready to think about tomorrow. I think the social studies will come early in the day as a followup of our discussion of all these services and the pictures of one service which each child drew at home as homework. Here is what I jotted down for tomorrow in social studies:

Review of the "Services of Government"

Have each child show his or her picture of a service and
   explain what it is.
Post all pictures on chalkboard with Scotch tape.

What title shall we give to our exhibit?

Something about "What The Government Does to Help Us"
   or "Services of the Government."

How many of these services cost money?
Examine several:
   Fire department:
      Helmets and uniforms cost money.
      Firehouse costs money.
      Men must be paid money.
      Fire engines cost money.
   Schools:
      Buildings cost money.
      Teachers are paid money.
      Textbooks and other materials cost money.
   Same with two or three other items.

Problem-solving and role-playing
   You are members of the City Council we talked about. You
      have to find a great deal of money to pay for all our
      "services." Where can you get the money?
   Go to filling station to get gas — some of the money goes to
      the government.
   Go to store — some of the money goes to the government.
   Go to the movies — some of the money goes to the govern-
      ment.
   People who own property pay money to the government;
      houses and factories.

A special name we give to all this money the government
   collects: *TAXES.*
Write the word "Taxes" on the chalkboard.
Discuss the word.

Homework:
   Talk to parents about whether they pay taxes. What kind of
      taxes?

An Overview of the Land and People of South America

Usually the best introduction to a geographical region is through the use of a film or a filmstrip. But since we have used that approach several times this year, I have decided to start the study of South America with a study of maps.

One reason for this decision is merely to have a new approach, since I know that variety does often make for better learning. The other reason is that I want to see if generalizations and concepts learned in other units can be applied to the study of this area.

We plan to limit our study of South America to Argentina, Brazil, and Peru, but I think that an overview of the entire continent will be a good introduction. Then we can concentrate or specialize on those three nations.

Here are the cards or lesson plan I drew up for my sixth grade class:

Lesson on:
　　Some of the major geographical features of South America
　　　　as they affect location of people.

Aims:
　　The major rivers and river valleys of South America.
　　Problem-solving of where people would probably live,
　　　　checking this with population distribution map.

Materials and methods:
　　Large map of South America in front of room.
　　Overhead projector with map of South America — physical
　　　　map and overlay of population distribution.

As promised you on Friday, we will start our study of South America today. In our evaluation of the unit on Mexico, you said you learned a lot about maps. Want to see how well we can apply that now to a new region — South America.
As you look at the map of South America, where would you expect to find people living? WAIT for them to think.

General areas: Probable answers:
　　River valleys of Amazon, Rio de la Plata, Magdalena and
　　　　Orinoco.
　　Along coasts of Atlantic, Pacific, and Caribbean Sea.
　　Where are there mineral resources?

Place overlay map of population distribution in the overhead
    projector.
Let's see if your answers were correct. Were they?
Wait — and let them discover if they were right.
Yes on most of the features.
No on the Amazon River. Discuss why not.
    Read short excerpts from:
        Armstrong Sperry, *Amazon: River Sea of Brazil* (Garrard
            Press)
No for areas with metals. Mountains. Few people.

Homework:
    In South America there are a variety of people. Jot down
        this list of people:
    Africans or Negroes
    Indians
    Asians
    Spanish and Portuguese
    Italians and other Europeans (besides Spanish and
        Portuguese)
Take two of these groups and see what you can discover
    about where they live in South America. Find your own
    sources this time. (Discuss tomorrow — Where they live
    and why.)

This lesson as outlined should take all the time we have available. But just in
case I "run out" of material, we can start on the homework in a general way or we
can locate some of the very large cities of South America.

A Lesson on Pioneers "Moving Upstream"

During the past few days we have been studying about life in the colonies in what is now the U.S.A. in our seventh (or eighth) grade. We have discussed life in New England, the Middle Colonies, and the South as fairly distinct ways of life. Now we are ready to see where the newcomers settled and where some of the old-timers moved. Usually this was "upstream," along the various rivers. In thinking about tomorrow's lesson, I jot down first the general outline, including the aims and methods I want to use. Then I take the other cards for more specific plans. Here are the cards I wrote out for this lesson:

Lesson on:
　　People in Colonial Times "Moving Upstream"

Aims:
　　To review rivers studied in lessons on life in the colonies.
　　To figure out where new settlers might live.
　　To decide why some old settlers might move.

Materials:
　　Textbooks and large map of colonies

Methods:
　　Map work
　　Discussion
　　Role-playing

Here are the more specific notes on the lesson as I see it developing:

What were the major rivers we have learned about in colonial times?

Connecticut	Delaware	Slower pupils to locate
Hudson	Potomac	these on map in front
Susquehanna	James	of room.

Why did people settle at the mouths of these rivers?
　　Fertile land in coastal plains.
　　Access to ocean and trade with Europe or other colonies.

Divide class for role-playing. Two-thirds as newcomers. One-third as older settlers.

As newcomers, why might you settle along the coast?
If enough money, purchase land from older settlers.
Could work on the farm of a former settler.
If skilled in a trade, might work in a town.

As newcomers, why might you move "upstream?"
Could get land cheaper in these areas.
If no skills except as farmer, and had no money, would you have to move "upstream?"

As settlers who had been in the colonies many years, why might you stay near the coast?
Nearer other people — friends and relatives.
Had good land and home. Why move?
Little sense of adventure.

As settlers who had been in the colonies many years, why might you move "upstream?"
Family farm not large enough for all the children as they grow up and marry.
Wanted more land than was available.
Sense of adventure.

Summary Question:
What do we mean by "moving upstream?" (Test with slower pupils to see if the idea has been taught. Get others to restate or amplify.)

Homework:
Years later people began moving west.
1. Why might you stay in the East?
2. Why might you move west?
3. If you decided to move west, what five routes might you take around and through the Appalachian mountains to go west?

Evaluating an Activity or a Lesson in the Social Studies

All of us need to evaluate what we have done. The following checklist is included here to serve as one means of helping you to evaluate an activity or a lesson you have planned and carried out. You may need to modify it to suit yourself and the type of activity you have arranged.

	Not too good	Fair	Very effective
1. Activity or lesson seemed related to their day-to-day lives?			
2. Introduction of activity or lesson?			
3. Interest throughout the lesson?			
4. Geared in part to slower pupils?			
5. Geared in part to faster pupils?			
6. Visualizations were included?			
7. New words were "discovered" by pupils?			
8. Major idea or concept discovered by them?			
9. Questions from the pupils?			
10. Several illustrations or examples of major point used?			
11. Activity or lesson today led to further work, easily and naturally?			
12. Previous knowledge of children taken into consideration and lesson based on that evaluation?			
13. Examples used in this lesson were taken from present or past experience of pupils so far as possible.			
14. Pace or "timing" of the lesson.			
Other aspects of the lesson I want to evaluate:			
15.			
16.			
17.			

How Can a Teacher
Provide for
12 Individual Differences?

Student teachers and beginning teachers are often surprised by the wide variety of backgrounds and abilities in a given grade. Sometimes they are stunned by the range of reading or speaking skills of the pupils in their class. Frequently they are appalled by the range of interest in the social studies (or in any other field), or by the differences in experiential backgrounds.

Have you already had such an experience? Perhaps you were assigned to a third grade or a sixth grade. You realized, intellectually at least, that you would find some differences among your pupils. But you were shocked to see how great those differences really were.

Why is this true? Perhaps it is because people tend to think in terms of "third grades" and "sixth grades" and expect all the pupils in those grades to be similar. Actually there is no such category as "third grades" and "sixth grades." There are groups of children of approximately the same age, placed in a room under the care of a teacher.

Possibly it is because it is a long time since you were a pupil in an elementary school and you have forgotten how "slow" — or how "fast" some children are in learning.

Perhaps you were always in classes for the gifted when you were in elementary school and therefore you had little or no contact with persons who were slower intellectually than you.

Be that as it may, you are now faced with thirty or so pupils. They differ in physical development, in intellectual abilities, in social background, and in emotional maturity. They vary in interests and in skills. They represent different socio-economic classes, different religious background, and different ethnic and national heritages.

Yet you are expected to teach them all. This is the expectation of the parents of these children, of the taxpayers, and of the administration. Furthermore, and even more important, it is your goal, your ideal, your desire.

"But how can I possibly attain this goal?" you ask. The task is great. There is no denying that. But it is not an impossible job. Others have done it and you can, too. In the next few pages we will offer some suggestions as to ways in which you may provide for individual differences. You will undoubtedly want to supplement these ideas with others gained from talking to teachers and reading books and articles.

Possibly it will help you to think of yourself as a forester or nurseryman. Or, you may want to think of yourself as an orchestra conductor. Possibly it will help you to think of your role as the coach of a group of amateur actors. How can you develop each person and the group as a whole? Here are some suggestions which are offered in the hope of helping you:

1. Individual Differences Are Most Likely to Be Met When Teachers Gather and Utilize All Pertinent Information About Their Pupils.

Can you imagine a nurseryman planting hundreds of seeds or transplanting scores of tiny trees without knowing all he possibly can about the conditions under which they will grow best? Wouldn't he be foolish to proceed without knowing, for example, which trees need a great deal of sunlight and which need a great deal of shade, or which need sandy soil and which need rich, loamy earth?

Can you imagine a salesman who intends to make a large sale and does not find out all about the person or firm to whom he expects to make such a sale?

How much more important it is in working with human beings that we learn all we can about them. Several ways in which teachers can gather data about their pupils are suggested in the section of this book on "Helping Boys and Girls to Understand Themselves and Others." Perhaps it will suffice here merely to mention a few of the facts that teachers need to know about the pupils in their classes.

Certainly teachers need to know about the physical conditions of each pupil. Furthermore, teachers need to know all they can about the family life of each pupil, including the socio-economic background of the family, the authority pattern of that group, the attitude of the family toward education, and its value system.

Teachers need also to know about the general intelligence of their learners, preferably on individualized tests.

Even more important are data on the achievement of pupils in the social studies.

Information about the special interests of pupils can be utilized in social studies classes by differentiated reading and activities which are geared to those interests.

The breadth and depth of experiences in the home and community — and beyond — can prove helpful to social studies teachers, too.

Along with all this information, teachers need to know as much as possible about the expectations of their pupils and about the expectations of parents for their children.

2. Individual Differences Are Most Likely to Be Met When Teachers Can Develop an Atmosphere in the Classroom and School of Acceptance of Each Pupil.

Every one of us has his or her prejudices in regard to pupils. Some of us prefer bright pupils; others prefer average pupils. Too few really enjoy teaching "slower" pupils. Some prefer girls; others boys. Many teachers really like upper middle-class children. Too few prefer lower-class pupils.

Do you know which types of pupils you prefer — and why? Your pupils probably do, as they sense very quickly the attitudes of their teachers toward them. This is a real factor in their ability to learn. After you have analyzed your likes and dislikes, then comes the big job of working on the development of empathy for those whom you now like least.

Have you ever stopped to analyze whether your pupils consider their classroom an interesting place? Have they been involved in making it "their classroom"? How?

Have you ever made a list of your pupils and by their names written what each of them does best? Have you ever asked them to list on a piece of paper "What I Like to Do Most"?

Has the guidance counselor helped you to understand your pupils?

Even more important to children is their acceptance by their peers. This is a crucial factor in learning. Do you know from observation and by listening, how your pupils view each other? Have you used tests of sociometric groupings to discover what they rate each other?

Have you provided enough variety in reading and in related activities so that each child can contribute to the work of the class and find satisfaction in his or her contributions?

Is praise frequent on your part and on the part of pupils? There is nothing equivalent to praise to oil the wheels of learning.

3. Individual Differences Are Most Likely to Be Met When Teachers Provide for a Wide Variety of Reading Abilities.

As we have said before, growth in social studies learning is not absolutely dependent upon the ability to read. Nevertheless, since so much of the social studies has to be focused upon reading, teachers must provide a wide variety of reading materials.

This means that there should be more than one textbook. Preferably there should be textbooks with different reading ranges in each classroom. This is especially important in the upper grades where the range within a classroom is likely to be great. Often teachers need to help children discover which texts they can read best — and to encourage their use of such books.

Since textbooks are often not available for different ranges of ability on the same topic, other reading materials need to be used widely and often. Materials from social studies papers, pamphlets, and supplementary books should be in the classroom at all times. The teacher should have files of such materials and should know as much as possible about the reading level of those materials.

With slow readers, the textual material can sometimes be recorded and children can listen to the material as they follow it in their texts. This can be done while the teacher is working with other groups of pupils.

In some groups better readers can read to the class. Or the teacher can read aloud in the social studies. This can be done by the teacher, working with the class as a whole or with small groups. Or it can be carried on by a reading specialist in the school if your school has one.

Good readers often can be excused from reading text materials and other more difficult reading be substituted. Or they can quickly read the material assigned for the entire class and move on to enrichment reading.

4. Individual Differences Are Most Likely to Be Met When Teachers Provide a Wide Variety of Learning Activities.

If individual differences are to be met in the social studies, there *must* be a wide variety of learning activities. That word "must" should be underlined, put in italics, and coupled with an exclamation point.

Too often in social studies classes, reading is the exclusive method employed.

Teachers have excelled in this approach and so they rely too heavily upon it in their teaching. It is important, but it should be only one method of several which are used.

All of us learn through a variety of experiences. We learn by listening. We learn by observing. We learn by acting out situations. We learn by drawing and painting. So the list might be continued.

Too often these types of activities in the social studies are considered supplementary to reading. They are not. They are basic, fundamental learnings. They are not or should not be activities just for the slower children. They are for all children.

One way to test yourself on this approach to meeting individual differences is to ask how many different types of activities you used the past week in the social studies. Or you can ask yourself how many different types of activities you are planning to use next week.

Often there can be alternative approaches, with different pupils or groups of pupils working through different media.

This can be true of homework assignments, too, where they are given. You can vary the types of homework assignments and sometimes give two or three, from which pupils can select one on which they work.

5. Individual Differences Are Most Likely to Be Met When Teachers Utilize a Variety of Classroom Procedures.

Various types of classroom organization can also be used to provide for individual differences.

For example, individual reports can be extremely helpful in promoting individualization. This can be just as true of pupils with limited abilities as with boys and girls of special talent. To provide for such individualization, teachers need to know the special interests of pupils and their special needs. A slow pupil who collects stamps may report on the stamps of a given country or of a period of American history and may contribute to the knowledge of the group as well as to his or her self-image. A girl interested in costumes can collect swatches of different materials and have the class identify them. A child who does not read too well can go through a stack of magazines and find illustrations of people working in different types of jobs or living in different kinds of homes. Individual reports can be given by the fast pupils, average pupils, and slow pupils.

Panels can be used to advantage, especially in the upper grades. The chairman can be an especially gifted pupil or a pupil with average ability who gets along well with his peers. Shy children often will do better in a panel than in general classwork. They may also be much more at home in a panel than in giving an individual report.

Pupils can be encouraged to work in teams on a project and select their partners. Under such conditions they usually will do better than when they are assigned their teammates. Occasionally, however, the teacher will want to assign such teammates for special purposes.

Of course committee work can provide for individual differences extremely well. Such committees may be organized by the types of jobs that need to be carried on, or each committee may divide its work into different categories to provide for individual differences.

6. Individual Differences Are Most Likely to Be Met When Teachers Use Carefully Planned Questions.

One of the best devices for meeting individual differences is through the art of good questioning. Yet this approach is too often overlooked in discussions of individualization.

For example, at the beginning of a period a teacher may want to review what the class has done on the previous day or in another part of the same day. If she wants to get a clear, concise summary, she may call on one of the more able pupils. If she wants to see whether an idea has really been clinched, she may call on one of the pupils who learns more slowly.

Slower pupils are more likely to be able to answer questions of what and when than questions emphasizing why or when. But even why or when questions can be pitched so that slower pupils can answer them.

Students with more ability to generalize can be asked to summarize a discussion or to clinch a generalization or concept.

When hands are raised, it is usually best to call upon the pupils who volunteer least often or the slower pupils, leaving the more difficult questions for the brighter pupils. Occasionally a teacher can put off an eager pupil with the statement, "Just wait a minute, I have a very special question saved for you."

In writing lesson plans or daily plans, a teacher may want to see if there are questions for pupils of various abilities in her plan. Beginners may want to mark an "S" or an "F" beside key questions to indicate that they are intended for slower or faster pupils.

7. Individual Differences Are Most Likely to Be Met When Evaluation Is Geared to Individual Abilities.

So far as possible, teachers need to evaluate pupils on the basis of their ability rather than on some general standard for a group.

Many teachers place far too much emphasis upon grades and thus stunt the growth of slower pupils, especially in heterogeneous classes. In elementary schools there is really no reason for such an emphasis. Papers can be returned with comments on them which are brief and to the point, without any grade attached. A poor paper, for example, may bear the statement, "This is better than your paper last week" or "You can do better next time, I'm sure." Papers of brighter pupils can be marked accordingly, too, encouraging them to make more accurate statements, or praising them for the work they have done.

Reports to parents can also be made in statement form rather than in numerical or letter grades, providing the school permits such practice.

Conclusion

Everything that has been said in the last few paragraphs calls for a variety of learning activities, based on knowledge of every pupil. This makes teaching more difficult but learning more possible. By working on such approaches as are suggested here, teachers will find that individual differences are often assets rather than liabilities.

WORKING WITH "SLOW" PUPILS

A. *In Heterogeneous Classes*

1. Obtain as much information about these pupils as possible.
2. Use a variety of textbooks and reading materials.
3. Arrange as many varied activities as possible so pupils may be motivated and may contribute in special ways.
4. Try to arrange for remedial help for these pupils.
5. Include as many visualizations as possible.
6. Use simple vocabulary and plenty of "for examples."
7. Give them a chance to answer the questions for which they volunteer; save other questions for more able pupils.
8. Provide differentiated homework.
9. Use praise generously, but judiciously.
10. Do some work in pairs and let pupils select their partners at times.
11. Spend time in individual and group work with these pupils.
12. Provide for these pupils through committee work.

B. *In Homogeneous Classes*

1. Obtain as much information as possible about these pupils as the basis of working with them.
2. Beware of "labeling" this group. Occasionally do something exciting and different which they can talk about to pupils in other classes.
3. Have some well-established routines which seldom change. This gives many slower pupils a feeling of security.
4. Do a great deal of open-textbook and reading work.
5. Read aloud often to these pupils.
6. Use all kinds of firsthand experiences, including many trips.
7. Visualize as much as possible.
8. Use television programs in class and as homework from time to time.
9. Vary your programs to meet the short attention span of such pupils.
10. Find simple reading materials to use, or, if necessary, prepare some simple reading materials yourself.
11. Plan as many immediate rewards as possible.
12. Relate new ideas to previous experiences wherever possible.
13. Make your assignments concise. Copy them on the chalkboard. Check to see that they have copied this material.
14. Use a great many hand activities.
15. Assign special duties to various pupils and rotate these duties.
16. Do some drill work.
17. Use many short tests, some of them ungraded, rather than big tests.
18. Find every opportunity to praise them for work done.

WORKING WITH "DISADVANTAGED" PUPILS

There is no term which is really satisfactory to use with pupils designated as "disadvantaged," "deprived," or "culturally different." Actually, many children from wealthy homes may be "disadvantaged," too. But the term usually refers to children from homes in lower socio-economic groups and/or minority group pupils. They are not necessarily "slow." This is important to keep in mind in working with them. There is not enough research yet to determine a great deal about such pupils but here are some suggestions based upon current "findings":

1. Often they lack successful "models," so try to recruit successful models as teachers and as visitors. Use biography. Use pictures of people who can serve as "models."
2. Usually they lack a breadth of experiences, so make much use of trips, resource people, and audio-visual materials.
3. Their self-image is often poor, so provide as many successful experiences as possible and use praise judiciously but often.
4. They feel unwanted and therefore are often antagonistic to teachers, so bend every effort to be sympathetic in a realistic way. You will have to prove yourself to many such pupils.
5. School experiences have taught them they are failures and schools are dull and impractical, so provide a rich variety of experiences related to their lives with occasional outstanding and different experiences.
6. "Time" and "space" concepts are extremely difficult for many such pupils.
7. Their expectations are narrow. For some this is realistic but for others their sights should be raised considerably. Show examples of successful persons within their school, minority, ethnic groups. Help them to become aware of their potentialities properly developed.
8. Lack of experiences with language or acceptable language is often lacking, so plan as many and as varied experiences as possible in the social studies which concentrate on oral expression. Reports, tape-recordings of talks, and panels should be widely used. Skits, role-playing, and other forms of drama should be used frequently.
9. Such pupils cannot yet postpone rewards, so provide for constant and immediate feedback to them on their work.
10. They usually lack school "know-how," so work a great deal on skills.
11. Their attention span may be short; provide short periods of activity.
12. They may be restless and may find the confinement of the classroom intolerable. Provide for frequent periods of physical activity and for places to go when they cannot stand the confinement of the classroom any longer.
13. Reading may be "rough" for these pupils. Work on their reading skills but do some reading aloud for them. Plan much use of open-textbook lessons, too. With younger children you may want to record some parts of textbooks and/or trade books and have them listen as they follow the books.
14. Taking tests may be threatening to them. Work with them on test skills. Plan short and often easy tests to help them overcome their "blocks" on this aspect of their work.

WORKING WITH "GIFTED" PUPILS

A. *In Heterogeneous Classes*
1. Hold your most provocative questions for them, stressing "why" and "how" questions. Encourage *them* to raise questions.
2. Plan for differentiated assignments which will release their creativity and will challenge their abilities.
3. Place them in charge of committees, panels, and other groups.
4. Make them specialists on a topic or on current events, and encourage them to learn how to present their findings simply and cogently to their classmates.

5. Excuse them from class to work on special assignments.
6. With older pupils, let them take charge of the class occasionally.
7. Encourage other teachers to use them from time to time on their "specialties."
8. Encourage them to prepare special visualizations for use by the class.

B. *In Homogeneous Classes*

1. Involve them in student-teacher planning as much as possible.
2. Use more difficult materials and a wider range of materials, including source materials for older pupils.
3. Encourage them to make and compare accounts in different sources.
4. Ask them to find out about the authors of materials you use.
5. Arrange for them to present their findings on special topics to other classes, including older groups of pupils.
6. Use them in assembly programs and in interschool affairs.
7. Encourage them to develop skits, plays, and audio-visual materials. Persuade them to give their best materials to the school for wider use with more classes.
8. Develop much independent reading.
9. Include them in the evaluation of their own work and the work of others.
10. Vary the routines to avoid monotony.
11. Encourage depth studies.
12. Require often that they refer to the sources of their information.
13. Emphasize the skills of investigation or research.
14. Work on their understanding and appreciation of other pupils.
15. Provide for some long-term assignments, with occasional checks on their progress.
16. Relate their hobbies or special interests to the social studies.
17. With older pupils, encourage independent trips to places in the community.
18. Have them evaluate the materials they have used and write out comments on films and filmstrips seen and books read.

Of course you may have in your class bright children who are underachievers, based upon available data on their intelligence. Sometimes they have physical "blocks" to learning. Often the blocks are psychological. You would do well to examine their physical records and talk with the school guidance people about them.

A SELECTED BIBLIOGRAPHY ON MEETING INDIVIDUAL DIFFERENCES

1. BLOOM, BENJAMIN S., ALLISON DAVIS, and ROBERT HESS. *Compensatory Education for Cultural Deprivation.* New York: Holt (1965), 179 pp.
2. BRICKLIN, BARRY and PATRICIA. *Bright Child — Poor Grades: The Psychology of Underachievement.* New York: Dell (1967), 164 pp. A paperback.
3. CHASE, W. LINWOOD. *A Guide for the Elementary Social Studies Teacher.* Boston: Allyn and Bacon (1966), 234 pp. Chapter Two. Very specific.
4. CUTTS, NORMA E. and N. MOSELEY. *Providing for Individual Differences in Elementary School.* Englewood Cliffs, New Jersey: Prentice-Hall (1960), 273 pp.
5. DUFFEY, ROBERT V. "Helping the Less-Able Reader" in *Social Education* (April, 1961), pp. 182–187. (Reprinted in Jarolimek and Walsh, *Readings for Social Studies in Elementary Education,* pp. 299–304.)
6. FROST, JOE L. and GLENN R. HAWKES. *The Disadvantaged Child: Issues and Innovations.* Boston: Houghton Mifflin (1966), 450 pp.

7. Fuchs, Estelle. *Teachers Talk: Views from Inside City Schools.* New York: Doubleday (1967), 224 pp. Recordings of talks with beginning teachers.

8. Gallagher, James J. *Teaching the Gifted Child.* Boston: Allyn and Bacon (1964), 330 pp.

9. Getzels, Jacob W. and Philip W. Jackson. *Creativity and Intelligence.* New York: Wiley (1962), 293 pp.

10. Grossman, Herbert. *Teaching the Emotionally Disturbed: A Casebook.* New York: Holt, Rinehart and Winston (1965), 184 pp.

11. Hildreth, Gertrude. *Introduction to the Gifted.* New York: McGraw-Hill (1966), 572 pp.

12. Johnson, G. Orville. *Education for the Slow Learners.* Englewood Cliffs, New Jersey: Prentice-Hall (1963), 330 pp.

13. Kephart, Newell C. *The Slow Learner in the Classroom.* Columbus, Ohio: Merrill (1960), 292 pp.

14. Lewis, Gertrude M. *Educating the More Able Children in Grades Four, Five, and Six.* Washington: Government Printing Office (1961), 84 pp.

15. Long, Nicholas J. *Conflict in the Classroom.* Belmont, California: Wadsworth (1965), 512 pp.

16. McGeoch, Dorothy M. and others. *Learning to Teach in Urban Schools.* New York: Teachers College Press (1965), 139 pp.

17. Passow, A. Harry (Editor). *Education in Depressed Areas.* New York: Teachers College Press (1963), 359 pp.

18. Ragan, William B. and John D. McAulay. *Social Studies for Today's Children.* New York: Appleton (1964), 409 pp. Chapter 8 on "Providing for Individual Differences."

19. Riesman, Frank. *The Culturally Deprived Child.* New York: Harper (1962), 140 pp.

20. Sand, Ole and Bruce Joyce. "Planning for Children of Varying Ability." Chapter 9 in Michaelis, John U., *Social Studies in Elementary Schools.* Washington: National Council for the Social Studies (1962), pp. 293–312.

21. Torrance, Ellis Paul. *Gifted Children in the Classroom.* New York: Macmillan (1965), 102 pp.

22. Torrance, Ellis Paul. *Guiding Creative Talent.* Englewood Cliffs, New Jersey: Prentice-Hall (1962), 278 pp.

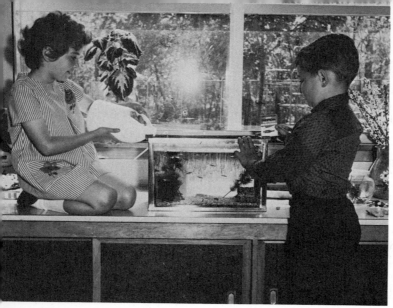

28. Science and social studies complement each other

29. Pupils enter into the lives of other people through their dances

Enrichment of the Social Studies by Other Subjects

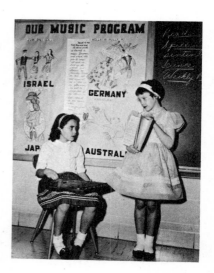

30. Music enhances understanding

How Can a Teacher Utilize Other Subject Fields in Teaching the Social Studies?

13

Obviously there is much overlapping among the various subject fields taught in elementary and junior high schools. Examples of this abound. Is the study of health in the domain of the social studies, is it a part of science, or is it a separate subject? Is the study of space a part of the science or the social studies programs? Are the folk tales of the United States and of other countries a part of the language arts or do they belong in the social studies? To what extent is music separate and to what extent is it part of the social studies?

Every teacher is confronted at some point with such questions. This problem arises as the teacher of the self-contained classroom plans the school day. Should there be separate periods for all the subject fields? Should they be integrated in large blocks of time? Or is some combination of these two plans possible and desirable? The specialized teacher of social studies is confronted with a similar problem. What are the boundaries of the social studies? Should art, music, the language arts, and science be utilized to teach the social studies or not? To what extent should the teacher work with teachers of these other subject fields?

Four General Approaches to the Organization of the Social Studies

There are four ways in which this important question is answered today in elementary and junior high schools:

1. Separate Subjects. Some schools and some teachers still organize their daily schedules in tiny boxes. Each subject field is given a small block of time and taught as something separate and distinct. For example, the language arts may be taught early in the morning for an hour and the social studies in the afternoon for a half hour. There is little or no relationship between the two. The only merit in this approach seems to this writer to be that each field receives a portion of the time for teaching and is not likely to be neglected.

2. Correlation of Subjects. A modified type of arrangement of separate subject fields is known as correlation. Under such a scheme some music or reading may be related to an ongoing unit in the social studies. Periods may be arranged so that a reading period follows a social studies period and the reading may be related to the social studies topic under discussion. Or the art period may be used to produce a mural on a social studies theme. Such a plan preserves the separate subject fields but permits some reciprocal relationships.

3. *Integration or Fusion.*   Other teachers find that the lines between subject fields are blurred. Breaking everything up into little boxes defeats the purpose of education and prevents mutual enrichment. Such teachers generally organize their work into themes or units.

Thus a study on Colonial America will focus upon the social studies, but the language arts, music, and art will be used to enrich the unit. Or a unit on Space may combine science and social studies, without worrying about the labels.

The great value in such an approach is the use of all subject fields to promote learning. The danger is the neglect of certain subjects and lack of attention to skill teaching. Thus the language arts can become a handmaiden to the social studies and little time may be devoted to the skills or to creative writing.

4. *A Combination of Approaches.*   A well-planned, integrated program actually combines the best features of all these approaches. It might be thought of as a modified program of integration. Thus a large block of time is devoted to the ongoing unit, but some time is left in the school day for separate subject fields. Some music is included in the unit, but some time is devoted to music as a separate field. Some time is devoted to art in conjunction with the ongoing general unit, but all the art work is not integrated with the social studies unit. The same thing is done with the language arts. Some of the reading and writing is related to the major unit but time is also provided for separate and distinct work in the language arts during the day.

The Social Studies as the Integrating Area

By and large, however, it is the social studies field that lends itself best to integration.

John Jarolimek has said that "the social studies provide a natural setting for the application and use of basic skills in solving problems and, in this way, may legitimately be used as a means of integrating the activities of the school day."

Helen Heffernan has pointed out in the yearbook of the National Society for the Study of Education on *Social Studies in the Elementary School* that "Since the school has a social purpose, the content of the curriculum must be chiefly concerned with man and society." Therefore, she maintains that the social studies is the integrating discipline. She is very clear, however, in pointing out that "No worthwhile learning need be sacrificed in a program which emphasizes helping children see relationships."

In the primary grades it is fairly easy to organize daily schedules so that most of the time is devoted to social studies themes in conjunction with other subject fields.

In the middle grades large blocks of time need to be reserved for the social studies and related fields. In self-contained classrooms this is relatively easy.

In the upper grades most social studies work is carried on by specialists. They will therefore need to work closely with teachers of other subjects so that art, music, the language arts, health, and science are often used to enrich the social studies.

Through team teaching, the correlation or integration of the social studies with other subjects can be carried on creatively. The social studies specialist brings a special expertise to the team, but so do the other specialists. Together they can supplement and complement each other.

No matter what approach is used, the social studies field needs to be enriched by other areas in the curriculum.

Science and the Social Studies

Inasmuch as science affects the lives of all of us today and will affect us increasingly tomorrow, teachers need to examine carefully the relationships between science and the social studies. Far too little attention has been paid to this topic, and teachers who are interested in the reciprocal learnings of these two fields should be encouraged to pursue this topic and share their findings with others.

Writing in a chapter on "Science and Social Studies in Today's Elementary School" in the National Council for the Social Studies yearbook on *Science and the Social Studies,* Glenn Blough has suggested four contributions of science to the social studies. He believes that science education should make children more proficient in problem-solving, more scientific in their attitudes, more appreciative of their physical and natural environment and more interested in it, and give them useful, organized knowledge to interpret their world of forces, changes, energies, living things, and phenomena. He makes a plea in that chapter for fusion at times between science and social studies, times when they are temporarily related, and times when they are completely independent.

Certainly there are themes on which science and social studies obviously overlap. One is in the study of the earth and its effects upon man as well as man's effects on the earth. Another is in the study of the solar system. A third is in the study of space. A fourth is in the study of inventions which have changed the lives of man. Science should also be included in units on such topics as clothing, food, transportation and communication, houses, and the weather.

At many other points, science needs to be utilized to make the social studies meaningful to children. Primary grade children studying airplanes will need to examine some scientific principles to understand these modern means of transportation. Older children studying tunnels and transportation will need to explore the scientific aspects of tunnels. Older boys and girls will need to learn about the discovery of atoms as they explore the uses of atomic energy for military and for peaceful purposes. The study at any point of the resources of a community or country will necessitate considerable attention to the field of science in the discovery of resources and in their use.

Certainly there is a great deal of mutual enrichment between these two major subjects in the curricula of our schools.

Health and the Social Studies

Many topics studied in the elementary and junior high school years provide opportunities to correlate or integrate health and social studies. The study of water is certainly one of those topics. The control of diseases and the functions of government in promoting better health is a second. The study of any community or country should include some attention to the ways in which that group of human beings provides for the health of its people. Certainly pollution should be highlighted in the study of communities at the third and/or fourth grade levels and considerable attention paid to it in a problems course at the seventh grade level.

In the upper grades the study of health and social studies can be combined as boys and girls learn about health problems on a world scale. This would include some attention to the World Health Organization of the United Nations.

Art and the Social Studies

One aspect of art is for art's sake. It is self-expression. It is creativity. It is fun. Another aspect of art relates more closely to the social studies. That part of art might be considered art for understanding.

From the earliest years in school, children should participate in this second aspect of art as well as the first. They should be encouraged to draw and paint pictures of themselves, their families and friends, and other people about whom they are studying. They should draw and paint their homes and various spots of interest in their home communities. All of these aspects of art are based upon direct experience.

Then there is the re-creation of stories, folk tales, and history through the various media of art — through paintings, drawings, work in clay and papier-mâché, wood, and metals. Some of this can be imaginary, permitting the free flow of ideas. Much of it ought to be based on pertinent research. Children should learn the details of a Conestoga Wagon before they draw one, or the details of women's dress in the Middle Colonies of the Americas. In that way, social studies and art are closely related.

Children should also view the paintings and drawings of other children and should be encouraged to ask questions and seek answers as to why they drew the type of houses they drew or the means of transportation they reproduced. Collections of such paintings can be found in a few books and in collections obtained from Art for World Friendship (Media, Pennsylvania) or the Junior Red Cross (Washington, D.C.). They may, in some instances, want to provide pictures to be sent abroad.

Some of this work will be highly individual; other productions will be group undertakings, such as murals, friezes, and scenery for plays.

Yes art, too, can enrich the social studies in many ways.

Music and the Social Studies

Music can enrich the social studies at least as much as art, possibly more. Often it can be used as motivation for the study of a period of history or of a country.

At other times it can be used to help provide the emotional tone which is so necessary for learning. The landscape of Finland can take on added meaning with the music of Sibelius's *Finlandia* or the Rocky Mountains can assume new dimensions with Grofé's *Grand Canyon Suite*. This will not necessarily be true for all pupils, but it certainly will be for many of them.

There can even be problem-solving through music. Play and sing some of the songs of the Civil War — and then ask the children what they have learned from those songs. The results may be amazing to you if you have never done this.

But some music will be difficult for children to understand. Often people say that music is international. But that is certainly not so. For instance, the music of Africa and of the Middle East sounds very strange to most of us in the so-called Western World. Our ears are not attuned to it. The desire and the ability to express oneself in music are international but music itself is cultural.

Social studies teachers need to work with music teachers or educate themselves on the music of other parts of the world which sound strange. Children need to be exposed early and often to such music so that their lives can be enriched by a variety of music and their understanding of other people through music can be enhanced.

Sometimes music can be used to review a period of history or a country of the world. The music can be played and children can discover for themselves how it reflects the period or the place.

None of these suggestions is intended to take away from the pleasure of music as such. Actually, music will be better appreciated if people understand what it is all about and how it reflects time and place.

Music to enrich the social studies is not difficult to find. Most of the music books for elementary and junior high school pupils contain music of our own country and of other parts of the world. Pamphlets like the *Fun and Festival* series of the Friendship Press and the *Hi Neighbor* booklets of the United States Committee for UNICEF, contain music of other parts of the world. Little songbooks of music from various countries can be purchased from the Cooperative Recreation Service in Delaware, Ohio. And many records can be bought from Folkways Records and other companies. The two best sources for lists of songs are the Nye, Nye, and Nye volume on *Toward World Understanding With Song* (Wadsworth Publishing Company, Belmont, California, 1967) and the older Tooze and Krone volume on *Literature and Music as Resources for Social Studies* (Prentice-Hall). The first half of the Tooze and Krone volume is on the United States and the second half on the world. There are several volumes of songs of the U.S.A., collected by Paul Glass and published by Grosset and Dunlap in 1966 and 1967 which are very suitable for use by elementary school pupils.*

The Language Arts and the Social Studies

Probably there are no two subjects in the curricula of our schools that are more closely dependent on each other than the social studies and the language arts. Together they form or should form a large part of the day of every pupil in our schools. They are or should be like twins.

One of the major roles of the language arts is to help children express themselves clearly and concisely. This work begins with preschool children and continues throughout the years in school. The activities in the social studies program should provide many of the experiences on which children draw in learning to express themselves. This can be true of children describing a trip they took, or of pupils discussing a current national or world problem. It can occur in a dramatization of a story the teacher has read about children in some other part of the U.S.A. or in some other place in the world, and it can take place as boys and girls in the upper grades present a mock session of the Constitutional Convention of 1787 or a meeting of the United Nations today.

But it is not only in oral expression that the language arts can draw upon the social studies, and the social studies can be improved by language arts instruction. It is also true in written expression.

From the experience charts which young children help the teacher write to the term papers of junior high school pupils, the social studies must draw upon the language arts as a tool for expression. Where does one draw the line between the language arts and social studies? Is learning to speak over a phone purely language arts or is it also a part of the social studies? Is chairing a meeting or panel a language arts activity or a social studies experience — or both? The lines often are blurred between these two curriculum areas. And that is as it should be.

---

*Hy Zaret has pioneered in producing some intriguing short pieces of music for the social studies in trying to motivate interest in the social studies through music. His records are sold through the Argosy Music Corporation in Mamaroneck, N.Y. 10543.

Reading is not the sole activity of the social studies, but it should be a very important part of that field. In books man has stored much of his knowledge. To gain such knowledge, one must learn to read and read well. Fortunately there are hundreds and even thousands of books for boys and girls which contain useful as well as exciting stories of people and places now and in the past. Teachers need to know about many of them. Furthermore they need to know where to turn to find the titles of books which may prove helpful. Such lists do exist for social studies teachers. This volume contains the titles of several hundred books. The volume by Tooze and Krone on *Literature and Music for the Social Studies* contains even more. The National Council for the Social Studies has a remarkably fine publication by Helen Huus, titled *Children's Books to Enrich the Social Studies*. The State Education Department of New York, located in Albany, has a fine booklet on social studies in the elementary school, titled *Social Studies Bibliography: Elementary School*. The American Council of Education frequently publishes its *Reading Ladders for Human Relations*, which is an annotated list of books appropriate for social studies teaching. Librarians will be able to supply you with other such lists which are current. On this score you should have no worries. And with increased appropriations by state government and the federal government for libraries and books, the chances are much better than they were a few years ago that you will have many books to use.

But there is reading, too, in other places. Films and filmstrips dealing with social studies require skill in reading. So do current events papers and daily newspapers. All of these experiences draw upon the language arts.

Reference has been made, in the section of this book on skills, to the importance of skills in conjunction with the use of libraries. Teachers in the social studies need to work on such skills with their pupils in concrete library situations when pupils need to use those skills and therefore are likely to learn them better.

In these and other ways the language arts and the social studies should be linked together as companion aspects of the curriculum. They should complement and supplement each other. They should enhance and enrich each other. The opportunities for integration and correlation between the two are innumerable.

How Can
Social Studies Learnings
14 Be Evaluated?

"What did I teach them today?" "What did we really accomplish in the unit we have just completed?" "Why didn't they do better on that test I gave?" "Why does Jim despise so many people?" "What makes Jane so receptive to other people?" "Why did the class perk up today when I used that flannelboard?" "Is this book difficult enough for Beth?" "Where can I get something for the slow readers?"

Have you ever asked yourself such questions? If you are a veteran teacher, you have done so hundreds of times. If you are a beginner, these are the types of questions you will ask yourself frequently in the months and years ahead. Every teacher who is worth his or her salt (or salary) asks many such questions every day, every month, and every school year.

This is good because it indicates a concern with the evaluation of teaching and learning. As teachers we naturally want to see results. We want to observe gains in behavior. We want to see improvement in attitudes. We want to note progress in the acquisition of knowledge. We want to observe gains in skills. We long for evidence of new concepts and generalizations discovered by pupils and old ones reinforced by them. We are on the lookout for new and better methods and materials. Our job is to bring about changes and we should be concerned with evidence that we have accomplished this task.

Actually there is nothing more important in teaching than the processes of evaluation. As J. D. McAulay has said, "Evaluation is the key to successful social studies teaching. Without it the social studies is without a rudder, without a compass."

Since there are many facets to social studies teaching, there must be many means of evaluating such learnings. Many teachers tend to think in terms of tests and testing. Bruce Joyce has written of the evaluation process in these terms: "The purpose of evaluation is to determine what the child can do, what he knows, what skills he can practice, how well he can think, and what he feels and values." He might well have included the ways in which the child learns and the usefulness of the resources he has used in learning.

Robert L. Ebel has expressed some doubts about the possibility of evaluating every aspect of social studies teaching. On this point he has recently written: "The plain fact is that we do not have many evaluation instruments which will do the job you want done. What is even worse, our disappointing experience in trying to measure some of these outcomes is beginning to convince us that the job cannot be done. I even suspect that part of the job should not be done."

Ebel wrote those words out of his background at the Educational Testing Service in Princeton, New Jersey and knows whereof he speaks. Evaluating behavior, attitudes, and understandings is not easy and may not be possible with any large degree of accuracy.

Nevertheless we have far more instruments and methods of evaluation than any teachers utilize. We can evaluate today so many more aspects of social studies teaching than we could in the past. We no longer have to limit ourselves to tests of knowledge. We now have projective techniques and sociometric methods and a whole host of new devices for helping us to know where children stand now and where they can be expected to progress.

With this brief introduction, let us turn to some of the more specific questions about evaluation that you probably have in mind.

1. What Should Be Evaluated?

Every possible aspect of social studies teaching needs to be examined to see how it can be evaluated. That means teachers need to be clear on their objectives. Only in that way can they develop a realistic and effective program of evaluation. Here are the main aspects of social studies teaching and learning which need to be considered:

(a) The behavior of pupils — and teachers.
(b) Concepts and generalizations.
(c) Understandings.
(d) Appreciations.
(e) A wide variety of skills.
(f) Personal interests.
(g) Attitudes, values, beliefs, goals.
(h) Personality.
(i) Knowledge of many things and processes.
(j) Methods used in learning and teaching.
(k) Resources used in teaching and learning.
(l) Widened horizons.

2. To Whom Is Evaluation Important?

Evaluation is important to a wide range of persons. These include the following:

Pupils

(a) in determining their strengths and their weaknesses.
(b) in showing progress — and retrogression.
(c) in developing feelings of satisfaction and accomplishment.
(d) in spurring them on to further improvements.

Teachers

(a) in determining the present status of pupils and classes.
(b) in fixing goals for individuals and groups.
(c) in deciding upon methods to bring about improvements in learning.
(d) in guidance and in locating resources for more effective learning.
(e) in comparing the results of learning with comparable groups of pupils in other parts of the United States.

Parents

(a) in revealing the strengths of their children.
(b) in revealing the weaknesses of their children.
(c) in discovering the potentialities of their children.
(d) in showing ways in which they can help their children.

School officials

(a) in ascertaining the present status of individuals and groups.
(b) in indicating the strengths and weaknesses in current social studies programs.
(c) in pinpointing areas where experimentation needs to be carried on.
(d) in comparing the teaching in one school with the teaching in other comparable schools in the U.S.A. by results of standardized tests.
(e) in revealing the strengths and weaknesses of teachers so they may be helped to improve their teaching.

3. What Are Some of the Characteristics of an Effective Evaluation Program in the Social Studies?

It should be obvious at this point that an effective program of evaluation is many-faceted. It cannot be carried on by teachers alone. It must involve the pupils, the parents, the teachers, and the administration. A truly effective program of evaluation might be characterized as follows:

(a) It is based on clear and specific objectives, stated hopefully in terms of behavior.
(b) It is continuous and cumulative, with records kept by teachers and passed on to other teachers.
(c) It includes as much subjective material as possible, provided by teachers and others.
(d) It includes as many objective methods of evaluation as possible.
(e) It includes pupils, parents, and the administration.
(f) It is carried on daily and also at frequent intervals with more formal instruments.
(g) It includes good organization so that the evaluation data can be utilized fully and effectively.
(h) It includes the evaluation of means as well as ends.
(i) It is wisely interpreted to pupils, parents, and others.
(j) It does not stress the I.Q. of children unduly but considers the individual abilities of various pupils.
(k) It does not stress unduly any one socio-economic class.
(l) It includes experimentation with new methods and devices.
(m) It is carried on in part by persons specially equipped in evaluation.

4. Who Should Carry on Evaluation in the Social Studies?

Many persons are involved, or should be involved, in the many dimensions of evaluation. Among them are the following: (Possibly you can think of others.)

(a) the pupils — individually, in groups, and as a class.
(b) teachers of the pupils in social studies.
(c) other teachers.

(d) bus drivers, librarians, cafeteria personnel, school nurses, and other personnel.
(e) personnel in the administration of the school.
(f) parents.
(g) administrative officers in schools beyond elementary and junior high schools.
(h) testing agencies.
(i) city, state, and national offices of education.

5. What Are Some of the Difficulties in Evaluation?

Evaluation is not an easy process. The subjectivity of the teacher often enters into evaluation. Instruments are often faulty, as witnessed by the tests of I.Q. which long crippled children of certain racial and socio-economic groups. Some aspects of social studies teaching are very difficult to measure. Here is a partial list of difficulties in evaluation. (What can you add to the list from your experience as a pupil or as a teacher?) It is difficult:

(a) to obtain data that are impartial.
(b) to obtain evidence that is valid.
(c) to prevent teachers from teaching for "tests."
(d) to provide adequate instruments or methods for evaluating all aspects of social studies teaching.
(e) to evaluate without spending undue time on the process.
(f) to differentiate between data which are confidential and material that is available to teachers and parents.
(g) to find people who are really expert in developing evaluation methods and instruments.

6. What Are Some of the Methods of Evaluation Social Studies Teachers Can Use Profitably?

There are scores of ways in which teachers can evaluate the processes of learning in social studies classes. They range from simple observation to tests. They depend upon the age and maturity of children, upon the capabilities of teachers, and the needs of groups. In this section we will outline briefly some of them. You are the one who must determine when to use them or whether you want to use them. Space will be left for additional methods which you want to note.

(a) *Observations by Teachers.* Alert teachers are constantly evaluating the learning process in their classes. They note which questions provoke the best response from different children. They observe which pupils are taking part in a discussion or are playing together today. If the children are working on a mural, a montage, or a map, they observe and raise questions with the group about items which are incorrect or could be added.

Such alert teachers also make note of words with which children have difficulty in the social studies and the books and other materials which seem to fit the needs of individuals and the class as a whole.

Obviously a teacher cannot keep a record of everything which transpires in a period or in a day. But many teachers find it helpful to have a pack of 3" by 5" cards or a "penny pad" at hand on which they can take down important observations about a child or a group. These can be placed in a drawer and eventually find their way into a cumulative record.

Other teachers find it easier to follow one child for a week or more, noting the particular behavior of that child. It may be a behavior problem. It may also be a child who is overly active in group discussions. By concentrating on this one child for a period of days, the teacher can gather more pertinent data and handle that pupil in better fashion.

(b) *Tape Recordings.*   Some teachers use tape recordings of groups as a means of evaluation; more teachers should utilize this method. It is particularly helpful if done with the same group at intervals of three or four weeks to see what changes have taken place in the group's ability to handle discussion.

The teacher may want to use the tape recordings for himself or herself for evaluation purposes. But the teacher may also want the pupils to hear themselves in a playback of the discussion or play.

Such tape recordings can also be used with parents. They are likely to be more accurate than any verbal description of a pupil or even more accurate than a visit by a parent when the pupil is aware of the presence of his or her parent.

Often the teacher can be relieved of the work of recording classwork by a pupil who likes to handle mechanical equipment. This can free the teacher for better teaching.

(c) *Group Discussion.*   One of the best evaluation techniques is group discussion, usually led by the teacher but sometimes led by pupils in older classes. In the self-contained classroom this can go on any time during the day. In schools where the social studies work is directed by one teacher, time needs to be provided for such group discussions.

One of the best times for such discussions is at the beginning of the day. Another time is at the end of the day. When work is developed in units, the end of a unit is a splendid time for evaluating the work accomplished, comparing it with work on previous units. Then goals can be set for individuals and groups in upcoming units of work.

When a class is not responding as it should, a teacher may also want to stop and try to probe why the group is not functioning as it should. To obtain good results, however, there must be an atmosphere of confidence and freedom to say what is on one's mind. The work may be too difficult, the topic uninteresting, the work too easy — or any number of other reasons may explain why the group is not working up to par. Children will often spot the difficulties better than the teacher.

Group discussions should also highlight achievements. They should be used to compliment pupils and groups of pupils and to spur them on to even better accomplishments.

If the group has set up rules, group discussions can evaluate how those rules are working out in actual practice. If need be, they can be revised in the light of experience.

(d) *Individual and Group Conferences.*   Individual and group conferences take time but they are often extremely productive in their results. A teacher can use them with any age, even if they are very informal conferences with primary grade children.

Time must be allotted in the day's or week's program for such meetings. Eventually everyone should be included, even if children with special problems are given more time or priority in planning such meetings.

When a committee has a group report to make, such a conference is helpful.

After they have given a report, another conference is in order. Praise can be given and suggestions for improvement can be made better than in the larger group.

In schools where formal report cards are written, the reports will mean much more to children if they are talked over before they are sent home. This means individual and/or group conferences on the reports.

(e) *Work Summaries.* As we have pointed out several times in this book, children learn best when they are involved in planning and evaluating. One of the best ways to get pupils involved in their own evaluation is to have "work sheets" for units. On them pupils can keep track of what they have done and evaluate their own work as well as the materials and methods used in the unit. An outline for such a worksheet follows. It can, of course, be modified by a teacher for his or her purpose.

(f) *Samples of Work.* Teachers may want to keep occasional samples of the work of their pupils as another means of evaluation. It is particularly important to keep some samples at the beginning of the year as a kind of "benchmark" for future reference.

The children should know that this is being done but this fact should not be used as a threat to them. Samples of their best work should be kept, as well as their poorest work. These can be used to show them how well they can do.

Such samples of work are also useful in cases where you want to talk with school officials and with parents.

A Sample Work Sheet

My name _____

Unit on _____

I was a member of the committee on _____

My special jobs on the committee were:

These are the materials I used:

The most helpful of the materials were:

The least helpful materials were:

Other sources I used (films, filmstrips, interviews, etc.) were:

The most helpful of these other sources were:

(g) *Diaries and Logs.*   With older pupils, diaries or logs of what they are doing are often useful evaluation devices. Goals may be set up by a child and schedules outlined. They may want to gain a feeling of progress by checking the task when it is accomplished. Two columns are sometimes kept in such diaries or logs, indicating the date when a job is to be done, and the date when it was actually finished.

The more personal these diaries or logs are, the more helpful they will be to the teacher. Children often enter thoughts and ideas in them that they do not want to share with the class or a committee group.

In most instances, these diaries or logs should not be graded. Their purpose is to encourage a child to keep a record of his thoughts and actions for himself and the teacher and not for grading purposes. These can often be the basis of individual interviews or conferences.

If teachers can find time to keep such diaries themselves, they can be most helpful. Some college classes use this method to encourage young teachers to evaluate their teaching.

(h) *Sociograms.*   One of the best ways to evaluate the roles of pupils in a class is through the use of sociograms. They can be developed in several ways. One way is to suggest to children that they have a part in the selection of their fellow committee members. The question is then posed, "With whom would you like to work?" They are then asked to write their name at the top of a sheet of paper and underline it. Then, under their name, they place the names of the three children with whom they would like to work.

Teachers can chart the results and find out who the isolates in the class really are. They can discover, too, with whom pupils think they can work best. Committees can then be determined through the use of these data.

Sometimes startling results are obtained in this way. A young teacher in one of the author's classes maintained that a gifted child in her class had no friends. "Nobody likes him" was her assertion. But on the results of the sociogram, it was shown that seven children picked him as a partner for the next unit!

More detailed information on the use of such devices can be found in Helen Hall Jennings' *Sociometry in Group Relations* (American Council on Education, 1959 edition) or in other books on sociometrics.

(i) *The Use of Pictures, Drawings, Murals, and Montages.*   The drawings and paintings of children also can be used as evaluation devices. Children often reveal in this way what they are thinking about or what their attitudes really are.

Sometimes children can be asked to draw pictures or paint pictures at the beginning of a study of a topic and again at the end. These drawings can be compared and used as useful evaluation devices.

This can be done with almost any level with children. Kindergarten children can do it with pictures of their families and their homes and older pupils can do it with any number of topics. For example, the author has a large collection of drawings of children from grades 4 through 9 on "Germans" and "Russians." From these he learned that a large percentage of children at various grade levels do not separate these two groups. They are "the enemy" and are identified in many cases by the swastika, whether they are Russians or Germans.

If a group of children do a montage of Africa, what do they place on it? Are there skyscrapers as well as simple houses, dams as well as streams, modern office buildings as well as scenes of open, outdoor markets? The use of such devices is highly recommended as a means of evaluation that is not used often enough in elementary and junior high school.

(j) Word Associations.   Somewhat similar is the use of word associations. This also can be used at the beginning and end of a unit of work. Place the name of a place or topic on the board and have the children mention quickly whatever comes into their minds. Do this again at the end of a unit and see if the results are different — and in what ways they are different.

For example, almost any group when asked to associate with Switzerland will immediately mention tangible objects, such as snow, chalets, cheese, watches, tunnels, and mountains. Seldom do they mention words like the Red Cross, the League of Nations and the United Nations, democracy, and neutrality.

Why not try this on a unit on Indians or on Puritans, on any city in the United States or on any country in the world? You may learn a great deal about what your pupils know and what stereotypes they hold. Then, try it again at the end of a unit and see what they mention at that time.

(k) Direct Questioning on Attitudes.   Often we try to arrive at our knowledge of attitudes indirectly. Usually that is a more successful way to arrive at the true feelings of people. But with children there is likely to be less camouflage and sham. They are more likely to say what they really think without disguising their beliefs and prejudices. Dr. Taba has suggested in her volume on *Diagnosing Human Relations Needs* (American Council on Education, 1955) that we ought to use this direct approach more often. Her advice has been used in many situations by teachers. In the human relations study in Wilmington, Delaware, for example, children were asked to name some of the things they would like to change in their neighborhood and some of the things they liked (and disliked) about school. The results were extremely helpful to their teachers, to school officials, and to community workers. This is worth doing, too.

(l) Checklists, Inventories, and Questionnaires.   Pupils and teachers can often learn a great deal from the use of these and related devices. For instance, a checklist of interests, if honestly answered, may indicate much about the pupils in any class. The teacher may obtain "leads" for special reports and for the use of hobbies as motivating factors for social studies work.

Self-evaluation checklists can be used helpfully with many pupils to foster improved work habits. They can be arranged with columns of Yes and No or with more gradations, including such items as Always, Usually, Sometimes, Never.

Teachers may set up checklists on their pupils on many items. One might be on the use of pupils in special assignments or tasks in the classroom. Another might well be on the areas in which pupils are outstanding.

Pupils can be asked to fill out questionnaires or checklists on the work of a week or of a unit, indicating their reactions to the work undertaken.

These are also useful devices and should be developed by teachers according to their needs and the needs of their pupils.

(m) Teacher-Made Tests and Standardized Tests.   Every teacher should become adept in making up tests himself or herself. No outsider can develop for you the tests which will fit your particular needs for a group. You have to do this work and you can do it better than anyone else.

With young children the tests will have to be given orally. Perhaps you will place them in a problem situation and have them react to it. Possibly you will have flash cards of words they should have learned in the social studies. You may want to have them act out a summary of a story they have heard. These are all in a sense "tests" because they give you some "feedback" on what you have been trying to do with them.

By the second or third grade children should be able to take some simple written tests. They may consist of completion items or simple recall items. A few persons feel that children that young can write very brief essay questions, although most of us would postpone such test items until later in school.

Certainly by the fifth grade pupils should be able to cope with multiple choice and matching questions on tests.

Since testing is a social studies skill or cluster of skills, pupils should be assisted in learning how to take them. This is especially true of disadvantaged children.

Another type of test which children can take is the open-textbook test, encouraging them to find the answers quickly in the textbook.

In the fifth, sixth, seventh, and eighth grades, essay questions can be added. Teachers should be warned that most pupils will not be able to write more than three or four sentences on most such questions. Again, writing essay questions is a skill which has to be taught. Some essay question responses can be written on the board and examined by pupils. Answers can be projected through the opaque projector, too, so that children learn to evaluate answers written by their classmates.

Some written work should be given at least once a week in the social studies. It need not always be on Friday, as children will begin to associate that day with testing and learn to abhor it. Nor should every unit end with a written test. Often the teacher can evaluate what children learn without using a written test on the unit.

Following are some samples of various types of items mentioned in the foregoing paragraphs:

Completion

The President of the United States during the Civil War was _____.

Simple recall

Who helps us to find groceries at the grocery store? _____.

Multiple choice

One may travel from England to France by: (a) crossing mountains, (b) going through a tunnel, (c) crossing water, (d) crossing Germany by automobile _____.

Matching

In the space before each of the items on persons, place the letter of the branch of government:

Persons	*Branch of Government*
_____Budget director	A. Executive
_____President	B. Legislative
_____Senator	C. Judicial
_____Representative	
_____Judge	

Standardized tests

Standardized tests are likely to be more objective than those prepared and administered by classroom teachers. They have been written by experts and tried out with pilot groups of pupils. With them, teachers can compare the results of their students with students in other parts of the United States.

However, teachers need to bear in mind the fact that their pupils may not have studied the material which is tested in the standardized tests. That means that comparisons cannot always be correct.

In recent years some standardized tests have been constructed so that they emphasize skills more than content. That is a gain and makes such tests more applicable to all social studies pupils. They also provide evaluation of skills as a very important aspect of social studies teaching. And incidentally, they may have an important byproduct — the encouragement of teachers to stress skills merely because pupils will be tested on them.

There are several such standardized tests available for use in elementary and junior high schools. Samples of some of the test items appear below and on the next page, reprinted by permission of the copyright owners.

Teachers interested in sample test items in United States history would do well to purchase the booklet on "Selected Test Items in United States History" from the National Council for the Social Studies.

Teachers would also do well to purchase the very extremely helpful booklet on "Diagnosing Classroom Learning Environments" (Science Research Associates) by Robert Fox, Margaret Luski, and Richard Schmuck, which is full of helpful ideas.

DIRECTIONS: Instructions for answering each part of the test are preceded by a ✓. Read these instructions carefully and then do exactly as they say.

✓ For each of the following questions, four answers are given. Choose the ONE answer you think is correct or best, then mark its number.

2. Which one of the following men lived first?

   1 John Smith, 2 Columbus, 3 Daniel Boone, 4 De Soto.   _____ 2

3. Which one of the following nations tried to obtain a part of the New World?

   1 Greece, 2 Italy, 3 Switzerland, 4 France.   _____ 3

4. Which one of the following territories was explored chiefly by the English?

   1 Mississippi River Valley, 2 Florida, 3 Texas, Arizona, and New
   Mexico, 4 Massachusetts, Connecticut, and Rhode Island.   _____ 4

6. Which one of the following is connected with the Revolutionary War?

   1 Pony Express, 2 Secession, 3 Boston Tea Party, 4 Mayflower Compact.   _____ 6

9. The chief way in which the French colonists made their living was by

   1 fur trading, 2 farming, 3 fishing, 4 prospecting for gold.   _____ 9

From *Survey Test in Introductory American History* by Georgia Sachs Adams and John A. Sexson. Copyright © 1959 by McGraw-Hill, Inc. Used by permission of the publisher, California Test Bureau, a division of McGraw-Hill Book Company, Monterey, California.

✔ Below is given a map of a make-believe continent. There are five countries in this continent, numbered 1, 2, 3, 4, and 5. Read each question below, and then use the map to answer it.

76. Which one of the following cities is the capital of Country 4?

    D   E   F   _____ 76

77. Which one of the following cities is the largest?

    A   B   G   _____ 77

78. Which city is slightly northeast of City H?

    K   B   D   _____ 78

80. The distance from City C to City H is about

    a 600 miles.   b 450 miles.
    c 300 miles.   _____ 80

81. Between which two countries does a river form part of the boundary?

    a 1 and 2.   b 1 and 3.
    c 2 and 5.   _____ 81

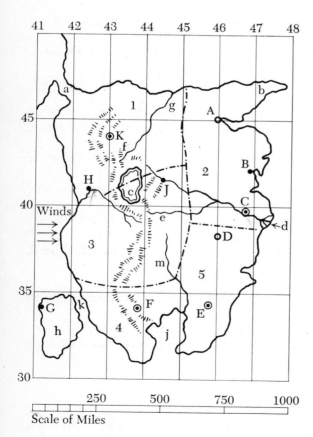

⊙ Capital Cities

O Cities over 10,000 Population

• Cities with 2,500-10,000 Population

〰〰〰 Mountains

–·–·– National Boundaries

Scale of Miles

250   500   750   1000

In interpreting the test results of standardized tests it is well to remember that the achievement scores of poor readers will not necessarily reflect their ability in social studies. Since they cannot cope with the reading, their results will be inaccurate. It is well to bear in mind, too, that a child's score should be considered in relation to his ability rather than his grade or his chronological age.

The best account of evaluation in the social studies is the 35th yearbook of the National Council for the Social Studies, edited by Harry D. Berg and titled *Evaluation in Social Studies* (Washington, 1965).

7. What Standardized Tests Are Available in the Social Studies?

There are a good many tests currently available in the social studies for elementary and middle grade pupils. Sometimes they are a part of overall tests covering several subjects. Occasionally they are devoted exclusively to the social studies, especially for the upper grades. Here is a list of some of the publishers of social studies tests, together with their addresses. Write to them for lists of their tests.

American Guidance Service, 720 Washington Avenue, S.E., Minneapolis, Minnesota 55414.

Bobbs-Merrill Company, Inc., 4300 West 62nd Street, Indianapolis, Indiana 46268.

Bureau of Educational Measurements, Kansas State Teachers College, Emporia, Kansas 66801.

Bureau of Educational Research and Service, State University of Iowa, Iowa City, Iowa 52240.

California Test Bureau, Del Monte Research Park, Monterey, California 93940.

Cooperative Test Division, Educational Testing Service, Princeton, New Jersey 08540.

Educational Test Bureau, 720 Washington Avenue, S.E., Minneapolis, Minnesota 55414.

Harcourt Brace Jovanovich, Inc., Test Division, 757 Third Avenue, New York, New York 10017.

Houghton Mifflin Company, Educational Division, 110 Tremont Street, Boston, Massachusetts 02107.

Ohio Scholarship Tests, State Department of Education, 751 Northwest Boulevard, Columbus, Ohio 43215.

Psychometric Affiliates, Box 1625, Chicago, Illinois 60690.

Public School Publishing Company, 1720 East 38th Street, Indianapolis, Indiana 46206.

Science Research Associates, 259 East Erie Street, Chicago, Illinois 60611.

Western Psychological Services, 12031 Wilshire Boulevard, Los Angeles, California 90025.

8. What Types of Evaluation Should Be Made to Parents?

Sometimes teachers forget that parents have a right to know what progress their children are making in school. That is as true in the social studies field as it is in other parts of the curriculum. Teachers need to look upon parents as friends and allies rather than as enemies. Often parents can be of help to teachers in promoting learning.

Sometimes in elementary schools informal conferences can be held as mothers (and occasionally fathers) bring their offspring to school or pick them up in the afternoon. Even though hurried, such conferences can suffice on some topics.

In many schools mothers (and sometimes fathers) can be encouraged to visit classes and sit in on them. This is often revealing to a parent about what the teacher is trying to do and what a child is doing in a particular grade or class, even though the situation may be a bit artificial at the time of such a visit.

Almost every school has some kind of written record, evaluating a pupil in a range of subjects and sometimes in skills. When such reports are made, it is highly desirable to have them presented to parents in a personal conference. Some schools even provide for such conferences on "school time." If this can be preceded by a conference with the pupil, so much the better.

Below is a list of the existing types of reports to parents. They are arranged from the best, in this writer's judgement, to the worst.

1. Conference, with the pupil involved, with a parent or both parents, and some written comments. No grades.
2. Conference of the teacher and the parents or a parent, plus some written record. No grades.
3. Written comments but no conferences. No grades.
4. Written comments and a checklist. No grades.
5. Written comments and a grade.
6. Only a grade.

In written comments and in conferences with parents and pupils, teachers need to remember how important praise is. They need to make suggestions to parents as to what they can do to help their children succeed in school. Such proposals should be made as suggestions rather than commands, as nothing is resented more by parents than being "told" by the teacher what they should do. Written comments will probably have to be brief because of the number of such notes that have to be written. Here is a typical comment on a report card on a boy's work in the social studies:

> John seemed enthusiastic about the current unit on transportation. His visit to the airport with his father was reported in class in excellent fashion and with considerable detail. His reading has been more extensive during this period than before and his comprehension good. He still needs help with maps and map skills. Would a map of the United States in his room be helpful? Could you find opportunities occasionally to use maps as a family, giving him a chance to use them? He's coming along well in the social studies.

Teachers should find two references especially useful:

1. HEFFERNAN, HELEN and VIVIAN E. TODD. *Elementary Teacher's Guide to Working With Parents.* West Nyack, New York: Parker Publishing Company (1969), 210 pp.
2. "Learning from Parents." Special issue of *Childhood Education* for December, 1971.

BIBLIOGRAPHY ON EVALUATION

1. ALLISON, ROSALIE. "Making the Grade With Parents" in *The Grade Teacher* (November, 1971), pp. 52–53, 76.
2. BEATTY, WALCOTT H. and others. *Improving Educational Assessment and An Inven-*

tory of Measures of Affective Behavior. Washington: Association for Supervision and Curriculum Development (1969), 164 pp.

3. BERG, HARRY D. (Editor). *Evaluation in Social Studies.* Washington: National Council for the Social Studies (1965), 251 pp. 35th Yearbook.

4. BLOOM, BENJAMIN S. (Editor). *A Taxonomy of Educational Objectives.* New York: Longmans, Green (1955), 2 vols.

5. BUROS, OSCAR K. *Fifth Mental Measurement Yearbook.* Highland Park, New Jersey: Gryphon (1959), 1292 pp.

6. BUROS, OSCAR K. (Editor). *Tests in Print.* Highland Park, New Jersey: Gryphon (1961), 479 pp.

7. CARPENTER, HELEN M. *Skill Development in Social Studies.* Washington: National Council for the Social Studies (1963), 332 pp. 33rd Yearbook.

8. FOX, ROBERT and others. *Diagnosing Classroom Learning Environments.* Chicago: Science Research Associates (1966), 131 pp. An excellent booklet.

9. GREEN, JOHN A. *Teacher-Made Tests.* New York: Harper (1963), 141 pp.

10. HUNT, J. McV. "What You Should Know About Educational Testing." New York: Public Affairs Committee (1965), 28 pp. An easy, popular pamphlet.

11. JAROLIMEK, JOHN. "Evaluation in the Social Studies" in Jarolimek and Walsh, *Readings for Social Studies in Elementary Education.* New York: Macmillan (1965), chap. 16.

12. JAROLIMEK, JOHN. *Social Studies in Elementary Education.* New York: Macmillan (1971), 534 pp. Chapter 17 "Evaluating Pupil Achievement and Teacher Effectiveness."

13. JOYCE, BRUCE R. *Strategies for Social Science Education.* Chicago: Science Research Associates (1965), 302 pp. Chapter 14.

14. KURFMAN, DANA G. and ROBERT J. SOLOMON. "Measurement of Growth in Skills" in Carpenter, Helen, *Skill Development in the Social Studies.* Washington: National Council for the Social Studies (1963), 332 pp. 33rd Yearbook.

15. McAULAY, J. D. "Evaluation in the Social Studies of Elementary School" in *Social Studies* (November, 1961), pp. 203–205.

16. MICHAELIS, JOHN U. *Social Studies for Children in a Democracy: Recent Trends and Developments.* Englewood Cliffs, New Jersey: Prentice-Hall (1963), 624 pp. Chapter 18.

17. SHAFTEL, FANNIE R. "Evaluation — for Today and the Future" in *Educational Leadership* (February, 1957), pp. 292–298.

18. SHAVER, JAMES P. and DONALD W. OLIVER. "Teaching Pupils to Analyze Public Controversy" in *Social Education* (April, 1964), pp. 191–194.

19. SUTTON, RACHEL S. "An Appraisal of Certain Aspects of Children's Social Behavior" in Jarolimek and Walsh, *Readings for Social Studies in Elementary Education* (New York): Macmillan (1965), pp. 466–471.

20. WILHELMS, FRED T. *Evaluation as Feedback and Guide.* Washington: Association for Supervision and Curriculum Development Yearbook (1967), 283 pp.

21. WRIGHTSTONE, J. WAYNE. "Evaluation of Learning in Social Studies" in Michaelis, *Social Studies in Elementary Schools.* Washington: National Council for the Social Studies (1962), pp. 313–327. 32nd Yearbook.

What Kinds of
Teachers Are Needed
for the Social Studies? **15**

In the beginning of this book we raised some questions about the teachers you have had, and we described briefly two types of teachers—Mrs. Appleby and Mrs. Zelch. We also suggested a few activities on which you might like to concentrate in your preparation for social studies teaching. Now we return to this topic because of its centrality in learning.

Many persons have written and spoken on the important role of teachers in our society. Ashley Montagu is one of them. On this topic he has said, "The teacher is the most important of all the public servants of the community; for what service can be more important to the community than the kind of molding of the mind and channeling of the social behavior of the future citizen which the teacher is able to direct?" Harold Taylor has written in a similar vein, going a little further than Montagu in raising a fundamental question about the role of the teacher in transforming as well as transmitting the culture. Here are his words: "The role of the teacher in any society lies at the heart of its intellectual and social life, and it is through the teacher that each generation comes to terms with its heritage, produces new knowledge, and learns to deal with change. Provided, that is, that the teacher has been well enough educated to act as the transforming element."

Certainly we are all agreed upon the importance of teachers in general and concerned about the role of teachers of the social studies, whether they are teaching the social studies as a specialty or as a part of a cluster of subjects.

SOME CHARACTERISTICS OF EFFECTIVE
SOCIAL STUDIES TEACHERS

It is almost impossible to separate the characteristics of good teaching in general from those of good social studies teaching. A few years ago two persons studied several groups of children in Ohio to ascertain what pupils thought were the characteristics of good teachers. In a single sentence they summarized their findings. They wrote, "Knowledge, a sense of humor, control of the class — mix these all together kindly, warm with understanding, and behold 'the really good teacher.' "

In a somewhat humorous approach, Harper and Row recently ran an ad which asked what a teacher did all day. Their reply was, "She's mother, father, warden, clergyman, traffic controller, philosopher, friend, psychologist, hygienist every day, maybe zookeeper some days."

Professor Louis Raths of New York University has spent a lifetime studying the processes of teaching and teachers in general. In an article in *Childhood Education* in 1964 he listed twelve points as a broad framework for teachers to discover more about themselves in relation to the functions of teaching. Those twelve were:

1. Explaining, informing, showing how.
2. Initiating, directing, administering.
3. Unifying the group.
4. Giving security.
5. Clarifying attitudes, beliefs, problems.
6. Diagnosing learning problems.
7. Making curriculum materials.
8. Evaluating, recording, reporting.
9. Enriching community activities.
10. Organizing and arranging classroom.
11. Participating in school activities.
12. Participating in professional and civic life.

That list may prove frightening at first. But none of us attains a high degree of perfection in all these tasks. We do the best we can at a given moment and then stop from time to time to ask ourselves how we can become more competent. If we waited until we were really ready to teach, we would never start.

In his *Guide to Social Studies Teaching in Secondary Schools* the writer listed eight characteristics of effective social studies teachers, abbreviated here as follows:

1. Content.   The best social studies teachers know a great deal. They have read much. They have traveled widely. They are "saturated" in subject matter. But they are also able to organize this subject matter around important concepts, generalizations, or "big ideas."

2. Confidence.   Such teachers have confidence in themselves and in their pupils. They know their shortcomings, but also their strengths. Because they are relatively secure themselves, they are free to help others.

3. Caring.   The best teachers care a great deal about boys and girls and about society. And they know how to communicate their caring.

4. Communication.   One can know a great deal and not be a master teacher. The master teacher has studied carefully the various means of communication and knows well how ideas are communicated to learners.

5. Creativity.   The effective teacher has imagination. To him or her history is drama — the pageant of mankind. People — all people — are important. Such a teacher has a wide variety of methods or strategies to make the present and the past live for children.

6. Curiosity.   Effective teachers are never satisfied with what they know. They are eager to explore, to inquire, to discover. They are continuously learning.

7. Commitment.   Effective social studies teachers are not dogmatic. But they have developed a set of values to which they are committed, including commitment to a better world for children as well as adults, and a world of diversity.

8. Catalytic Power.   Above all, the effective social studies teacher is a catalyst.

He or she knows how to excite children, to arouse them, to stimulate them, and to spur them on to more and better learning.

SOME SUGGESTIONS FOR STUDENT TEACHERS

If you are a student teacher, here are some suggestions on ways you can contribute in a classroom, as outlined by Dr. Sam Duker of Brooklyn College: (This list should not be construed, however, as a complete catalogue of everything you can do to be helpful and to learn about teaching in the process.)

1. *Working With Individual Children.* This may be in the nature of remedial teaching of slow learners or of providing interesting and challenging supplementary work for the rapid learner.

2. *Working With Small Groups.* In these days when accommodations for individual differences are found in so many classrooms, an extra hand, even when it is inexperienced, makes it possible for the teacher to complete longer projects while working with a single group.

3. *Preparing and Making Teacher-Made Tests and Scoring Standardized Tests.* This is a task for which the student teacher is eminently qualified. From it she gains a valuable opportunity to apply principles she has been studying in this field. Guided practice in the administration of such tests is ample reward for the time spent in scoring. There is, however, a complete lack of mutuality when the student is given no part in the administration of the test and is then asked to do the marking.

4. *Taking Responsibility for Routine Tasks.* Such time-consuming tasks as collecting and keeping account of money for milk, hot lunches, newspapers, etc., can all be safely entrusted to the student teacher, who thus gains valuable experience in efficiently handling such items. In the meantime the classroom teacher is enabled to devote herself to other matters.

5. *Taking Responsibility for Physical Aspects of the Classroom.* The student teacher can take responsibility for supervision of pupil activities in connection with room decorations, bulletin-board displays, setting up or maintaining various displays such as science corners, plant or animal exhibits, etc.

6. *Gathering Instructional Materials.* While excellent experience is given the student teacher by the responsibility of locating and bringing in instructional materials, equal benefit to the class results in the often ingenious resourcefulness of the enthusiastic student teacher in carrying out such tasks.

7. *Planning Out-of-Class Activities.* The student teacher can play a valuable role in helping to plan and execute assembly activities, playground activities, and so forth.

8. *Teaching.* A major contribution is made when the student teacher actually takes over a substantial share of the day's teaching, thus giving the teacher more time to work on some of the innumerable other tasks always confronting her. Of course most student teachers want to be in charge of the class as much of the time as possible. But that isn't all there is to teaching. Perhaps numbers one and two are the best kind of teaching. No? Yes? In one of the references cited on page 231 Carl Rogers says, "Forget You Are A Teacher." What do you suppose he means by that?

A SELF-ANALYSIS SHEET

Perhaps you are now ready to rate yourself, as a future teacher of the social studies, giving yourself credit where credit is due and facing up to areas on which you need to work in the months and years ahead. Here is a self-analysis sheet which you may want to fill in — or at least think about. It summarizes many of the points made in this chapter:

	COMMENTS ON MY PREPARATION AT THIS POINT		
	Fair	*Good*	*Very good*
1. Understanding of myself (present strengths, potentialities, weaknesses).			
2. Understanding of society (local, U.S., and world).			
3. Understanding of children.			
4. Understanding of the learning process.			
5. Understanding of subject fields in the social studies.			
a. anthropology			
b. economics			
c. geography			
d. history			
e. political science			
f. sociology			
6. Development of special interests within the social studies field.			
7. Development of my own philosophy of life.			
8. Development of my own philosophy of teaching.			
9. Knowledge of methods or strategies for teaching.			
10. Knowledge of resources for teaching the social studies.			
11. Ability to get along with other teachers.			
12. Understanding of administrative procedures.			
13. Ability to communicate.			
14. Personal health.			

KEEPING ALIVE PROFESSIONALLY — AND KEEPING ALIVE

Teaching is often tough. Especially in your first year or two there will be times when you go home dead tired. Sometimes you will wonder why you ever decided to enter the teaching profession.

Fortunately there will also be other days when you go home exhilarated. A pupil has volunteered for the first time. A committee or panel has done a superb job of reporting. The attitudes of the class seem to be shifting — finally. The supervisor or principal complimented you on your teaching. Then you know why people "gladly teach."

Because teaching is such a strenuous job, it will pay you to ask yourself from time to time whether you are keeping in shape physically and mentally and whether you are growing professionally.

Some suggestions along these lines follow. They are hints for good health and professional growth.

1. Have a happy home life. And if you are not married, be sure to develop a close friendship circle. Do not limit it entirely to teachers, and do not confine your conversation to "shop talk," even though that is important from time to time.
2. Develop a hobby or hobbies, preferably unrelated to your teaching.
3. Don't try to complete your work for a degree in a mad dash. You will get more out of your work and live longer if you don't fill every free hour and every summer with work for an advanced degree in those first two or three years.
4. Join at least one professional organization and attend at least one professional meeting a year. Be sure to spend time at the exhibitors' booths and with colleagues, as well as attending the more formal meetings.
5. Try to read at least one book a month, related directly to your teaching — or just for fun.
6. Set high standards for yourself, but standards you can reach in a reasonable amount of time. Don't be a perfectionist.
7. Remember as often as you can that you are the conductor, not the soloist, or the coach, not the star of the show. Get the pupils involved.
8. If you have a real problem, don't try to solve it alone. Talk it over with a trusted colleague who will view it more objectively. Even if he or she doesn't help you with specific suggestions, he or she will help you merely by listening.
9. Find at least one area in which you can become a specialist. Do some special work in this area. It may be in methods and it may be on content.
10. Try a new method of presentation every few days so as not to become stale in your methods.
11. Pay a great deal of attention to your relations with your colleagues. They can "make" or "break" you.
12. Have a medical checkup from time to time.

And good luck!

TEACHERS: A BIBLIOGRAPHY

1. ABRAHAM, WILLARD. *A Handbook for the New Teacher*. New York: Holt, Rinehart and Winston (1961), 59 pp.
2. ALEXANDER, WILLIAM M. *Are You a GOOD Teacher?* New York: Holt, Rinehart and Winston (1960), 57 pp.
3. ASHTON-WARNER, SYLVIA. *Spinster*. New York: Bantam (1961), 191 pp. Very readable story of a creative teacher in New Zealand, working with primary grade children from the Maoris. A paperback.
4. ASHTON-WARNER, SYLVIA. *Teacher*. New York: Bantam (1963), 191 pp. More on the philosophy of Sylvia Ashton-Warner.
5. ASSOCIATION FOR CHILDHOOD EDUCATION INTERNATIONAL. "The Education of a Teacher" in *Childhood Education* (May, 1964), pp. 447–500. A special issue.
6. ASSOCIATION FOR SUPERVISION AND CURRICULUM DEVELOPMENT. *Continuing Growth for the Teacher*. Washington: A. S. C. D. (November, 1962), 64 pp. A special issue.

7. BRAITHWAITE, E. R. *To Sir With Love.* New York: Pyramid (1966), 189 pp. A successful West Indian teacher and a group of adolescents in the East Side of London. A paperback.

8. COVELLO, LEONARD. *The Heart Is the Teacher.* New York: McGraw-Hill (1958), 275 pp. The autobiography of a teacher and later principal of a school in East Harlem.

9. FLEMING, ALICE. *Great Women Teachers.* Philadelphia: Lippincott (1965), 157 pp.

10. JERSILD, ARTHUR. *When Teachers Face Themselves.* New York: Teachers College Press (1955), 169 pp. Problems of teachers analyzed.

11. KAUFMAN, BEL. *Up the Down Staircase.* Englewood Cliffs, New Jersey: Prentice-Hall (1965), 340 pp. Cleverly written, popular story of a teacher in an "inner city" school in New York City. Also in a paperback edition by Avon Books.

12. KELIHER, ALICE. *Talks With Teachers.* Darien, Connecticut: Educational (1958), 148 pp.

13. MACMILLAN, C. J. B. and THOMAS W. NELSON (Editors). *Concepts of Teaching: Philosophical Essays.* Chicago: Rand McNally (1968), 154 pp.

14. MELBY, ERNEST O. *The Teacher and Learning.* Washington: Center for Applied Research in Education (1963), 118 pp.

15. RASEY, MARIE I. *It Takes Time: An Autobiography of the Teaching Profession.* New York: Harper (1953), 204 pp.

16. ROGERS, CARL. "Forget You Are A Teacher" in *The Instructor* (August–September, 1971), pp. 65–66.

17. VANDER WERF, LESTER S. *How to Evaluate Teachers and Teaching.* New York: Holt, Rinehart and Winston (1960), 58 pp.

18. WEBER, JULIA. *My Country School Diary.* New York: Harper (1946), 270 pp. Julia Weber Gordon's story of her first teaching experiences in a one-room rural school. Still very good.

19. WILSON, CHARLES H. *A Teacher Is a Person.* New York: Holt, Rinehart and Winston (1956), 285 pp.

20. ZIRBES, LAURA. *Spurs to Creative Teaching.* New York: Putnam (1959), 354 pp.

Helping Boys and Girls
to Understand
Themselves and Others   16

If the major task of the schools is to help boys and girls to become increasingly mature, sensitive, competent individuals, capable of living effectively in and contributing to a democratic society, then the most important lesson they can learn is self-respect. Without confidence in themselves, little else can be accomplished. They may go through the motions of learning, but what they seem to learn will be lost because it is not internalized.

Without a healthy self-image, children cannot reach out to others. Unless they value themselves, they cannot value others. Unless they respect themselves, they will not respect others.

Children who reject themselves are likely to become either apathetic or hostile. Even in the primary grades one finds such children. They are the timid, the shy, and the withdrawn, or they are aggressive, the hostile, and the belligerent. Both types are indications that children have rejected themselves and therefore have become stunted in their social development.

The apathetic ones have already given up in the race of life. They have been defeated, beaten, or frustrated in the first lap. They have turned in on themselves. They are a little like turtles which have withdrawn into their shells and only occasionally thrust their necks out to see the world around them. Sooner or later they will become the neurotic personalities of our times who fill our clinics and hospitals and cause untold suffering to themselves and to others.

The belligerent ones are already engaged in a lifelong battle against themselves and others. They are disruptive in class now. Later in life they will be disruptive in every human group in which they are members. Many of them will join hate groups. Some of them will become the social outcasts of society. A few of them will end up in mental institutions as victims of their own self-destruction or in prisons as a penalty for aggressions against society. They are a little like porcupines. Whenever they sense danger of any kind, they begin to bristle and their quills jut out into the air.

Conversely, when children develop a healthy self-image, they begin to see other human beings as aides rather than enemies, as helpers rather than opponents. They are open to new ideas. They consume so little energy in confronting themselves that they have plenty with which to cope with others. One finds them, too, in every classroom in every school. They are ready to learn, ready to reach out to others, ready to explore, ready to handle new situations and new ideas. They are the contributing children who are already on the road to maturity. They are contributing members of their class now. Eventually they will be contributing members of every group in which they are members. These are the healthy members of society. Our job as teachers is to help create as many such individuals as possible. We cannot do the job alone; others in society have an influence on children and youth. But we can do our part. No one can ask more.

The goal of all education and especially of the social studies is to help steer every child onto this road of self-acceptance and growth. You should label this goal by whatever term means the most to you. It is not the label but the fact that matters. Some speak of "mature individuals." Others prefer the term "effective persons." Some psychologists speak of "fully-functioning people" or "self-actualizing persons." "Integrated individuals" means more to some as a statement of the goal of life and therefore of education.

Writing in the Yearbook of the Association for Supervision and Curriculum Development on *Perceiving, Behaving, Becoming,* Earle Kelley refers to the ideal as the "fully-functioning personality." Such a person, he states:

> thinks well of himself.
> thinks well of others.
> therefore sees his stake in others.
> sees himself as a part of a world in movement —
>   in the process of becoming.
> sees the value in mistakes.
> develops and holds human values.
> knows no other way to live except in keeping with
>   his values.
> is therefore cast in a creative role.

Professor Alice Miel has developed a Direction Chart in which she indicates two levels of social learning. While teachers should aspire to the level indicated by column two, she indicates that they will often need to be satisfied with the level of column one. Here are the two columns, with the subtopics omitted for lack of space:

1.	2.
Maintaining self-control.	Bearing a friendly feeling.
Getting along with others.	Being a contributing member of a group.
Taking responsibility for doing one's own job.	Taking responsibility for a share of the labor involved in a common enterprise.
Showing obedience to authority.	Evaluating and cooperating with authority.
Being satisfied with majority rule.	Working for consensus.
Exhibiting tolerance of difference.	Valuing differences.
Having concern for others — family, friends and neighbors, those of the same nationality, race, religion.	Having concern for all mankind.
Conceiving of freedom as extending until it interferes with the freedom of another.	Seeing the necessity of a cooperative search for conditions guaranteeing maximum freedom for all.

Such statements as we have just quoted tend to be rather idealistic and at times far removed from the immature behavior which we see every day in the classroom. But they are statements of our ideals. They are the end result of years of growth and education. No one has made this more clear than Dr. Gordon Allport. The title of one of his excellent books sums up all this in one word — *Becoming*. As he says, "Personality is less a finished product than a transitive process." Later in this chapter we shall look at some of the ways in which teachers can help in this process.

The need for developing mature persons has been commented upon by scores of experts. For example, Clyde Kluckhohn, the eminent anthropologist, has stated that: "As long as the aggressions of children and of individual adults are met primarily by retaliation, this will remain the dominant pattern for dealing with interclass, interracial, and international aggressions."

Educators who have thought seriously about the goals of education in a democracy in our anxious age are agreed upon the centrality of this aim. In the opening section of its statement on "Educational Services for Young Children," the Educational Policies Commission listed as the first aim "The Development of the Individual," asserting that:

> Respect for others has self-respect as its cornerstone. It requires that we make a positive effort to recognize and appreciate the position and worth of others. Such recognition involves acceptance and appreciation that growth is individual. It cannot be forced into a groove. We assume our share of the responsibility for furthering the optimum development of the individual.

The teacher who is striving to help children "to become" needs to have a very clear idea of the type of persons society needs. He or she ought to have some "models" in mind from contemporary experience and from history.

Teachers need also to have a vast store of patience, for the process of developing self-identity in children is a long, arduous, and complicated one. It is a difficult business. It is an involved business. At times it is a dangerous business. But it is likewise a wonderfully rewarding task.

The development of children with strong self-images depends in large part upon the belief of adults that there are extraordinary possibilities in ordinary people. This is at the heart of the Judaic-Christian ethic. It is also at the heart of the democratic ideal.

The development of such healthy personalities also comes from expert advice on the part of teachers as well as other adults. Teachers cannot be psychiatrists or even psychologists, but they can be experts in guidance and growth. Classroom teachers can usually handle most of the problems of normal children. But they should not try to work on problems beyond their depth. They should refer them to specialists, if specialists are available.

The author of this volume has attempted to summarize the goals of education for social competence in the statement which follows. It is one man's statement of goals. Far more important than this summary would be one which you drafted for yourself in which you outline the type of persons you would like to help develop in your teaching.

THE ELEMENTARY SCHOOL CHILD WHO IS GROWING IN SOCIAL COMPETENCY

1. *Has begun to accept himself*

    (a) his assets
    (b) his potentialities
    (c) his limitations

2. *Has gained an elementary understanding of the need to live with a variety of human beings*

    (a) people in own family
    (b) people in the friendship circles of himself and members of his family
    (c) his neighbors
    (d) other people in the neighborhood and community
    (e) people in a larger geographical area whom he meets
    (f) people in larger geographical areas whom he does not meet but knows about

3. *Has developed a basic understanding of other human beings*

    (a) recognizes some of the similarities among people
    (b) realizes some of the differences among people
    (c) recognizes and understands in an elementary way some of the reasons for similarities and differences
    (d) has begun to accept some differences as desirable

4. *Has developed some basic skills for living cooperatively with others*

    (a) has begun to listen to others with appreciation
    (b) knows how to carry on a simple conversation
    (c) is beginning to work well in teams, committees, or groups
    (d) knows the simple rules of courtesy in his culture and acts upon them
    (e) knows how to show appreciation to others in simple ways
    (f) knows how to discuss a situation with someone he trusts
    (g) is outgoing, friendly, kind, with a sincere interest in and respect for others
    (h) has learned to respect and appreciate people with a true concern for their feelings, welfare, and happiness

5. *Has begun to understand the need for rules, regulations, and laws and some basic institutions in society*

    (a) understands some of the basic rights of persons, including himself
    (b) understands some of the basic responsibilities of persons, including himself
    (c) understands and acts upon certain accepted patterns in society, such as respect for persons or positions
    (d) has an elementary understanding of the need for officers of the law, law-makers, and governmental institutions.

SOCIAL LEARNINGS AND THE SOCIAL STUDIES

There are some people who feel that the building of self-understanding and understanding of others is important but that these tasks are jobs for the home. Obviously the home teaches the first lessons in human relations. It is the first laboratory in social living for many children.

However, it cannot do the job alone. When children enter school, they spend much of their time there. They come in contact with large numbers of children and adults and a wide variety of human beings. Boys and girls are confronted with new people and new situations. They need help in these new confrontations.

There are also people who think that the sole job of the schools, especially in the early years, is to teach the "three R's" of Reading, 'Riting, and 'Rithmetic. Of course these are basic skills which must receive high priority. No teacher would deny this. But it is not a question of teaching the "three R's" *or* social Relationships — the "fourth R." All of these deserve and really demand attention. But the "three R's" will not be acquired if a child is expending all his energy on himself. Furthermore, learning to live with oneself and with others is at least as essential as learning to read, write and count. In fact, it seems infinitely more important to this writer.

There was a period in the United States in the 1920s and 1930s when some teachers may have overemphasized the "fourth R" of Relationships to the detriment of the other three. They were so concerned with the emotional and social development of children that they minimized the intellectual growth of boys and girls. But the number of such teachers was small and they were certainly not among the best qualified of the profession.

Now the danger seems to be that teachers will become so concerned with intellectual development that they may minimize the emotional and social growth of children.

What we are calling for is a balance between the intellectual and the social development of children.

Even though social learnings should be an integral part of the social studies at all grade levels, they should be central in the early years in school. Personality patterns formed in those years tend to persist throughout life. In fact the basic personality patterns may have been formed before children ever enter school. They can be changed, but the later that is postponed, the more difficult is the process.

The heart of social studies teaching in the primary grades should therefore be the acceptance and growth of the self and the acceptance and understanding of others. Professor Helen L. Gillham has written about the nature of social learnings in her booklet on "Helping Children Accept Themselves and Others." She says:

> They (social learnings) may be defined as all the learnings which children have both in and out of school that help them to look at themselves and others with increasing respect, help them to gain ability to solve life's problems, and help them to build better understandings of living together.

No one has done more to promote the idea of social learnings as central in the elementary school program than Professor Alice Miel of Teachers College, Columbia University. Much of her thinking can be found in the book by Dr. Miel and Peggy Brogan, titled *More Than Social Studies: A View of Social Learning in the Elementary School* (Prentice-Hall, 1957). The title has always seemed to this writer unfortunate as it narrows the concept of social studies. But the title was obviously selected to show that social learnings take place in all the experiences of children in elementary schools, in the classroom and outside the classroom.

In that volume the two authors place social learnings in the setting of democracy, maintaining that there is a discipline in democracy that is difficult to learn but nevertheless central to our way of thinking and acting. It starts with "feeling good about oneself." With such security one can then begin to feel good about others. But one does not operate merely at the feeling level. There are a host of skills to be acquired for living democratically in groups, ranging from play groups with younger children to citizens' groups of adults. Miel and Brogan refer to the need of children to "extend their life-space."

Of course social learnings are intricately interwoven with value systems. Summarizing this aspect of social learning, these two authors say:

> The wholeness of our method of solving problems is at the heart of our value system. It is a precious part of our cultural heritage. Democratic problem solving constitutes a preferred way to carry good feelings into a selectively extended life-space — a life-space in which the present is extended to include the past and the future, the here to include the there, and me to include you and us and them.

The number of people whom children must learn to accept and understand or at least live with, is staggering. Adults too often forget that they have had to adjust to such a wide variety of people — and all within a few years. Here are the people with whom they are involved. Perhaps you can add to the list in the blank spaces provided for that purpose:

1. Themselves
2. Members of their family circle
3. Friends of the family
4. Neighbors
5. Children in the classroom
6. Children in the school
7. Other adults in the school — custodian, bus driver, cafeteria employees
8. Parents of their classmates, brothers and sisters of their classmates
9. Children and adults of the other sex
10. Children and adults of other socio-economic groups
11. Children and adults of other races and colors
12. Children and adults of other religions and value systems
13. Children and adults of different national backgrounds
14. Physically handicapped children and adults
15. Mentally handicapped children and adults
16. Children and adults in positions of authority
17.
18.

Some Problems for Children in Growing Up

No two pupils have the same problems in growing up. But there are enough similarities that one can foresee most of the problems of pupils. Here are some of them. You may want to add others at the end of this list. Or you may want to take one pupil and check the items which seem like problems to that particular child.

1. Being assured that they are loved.
2. Being the smallest in the family or class.
3. Being the youngest in the family or class — or being the oldest or middle one.
4. Being the only child in a family.
5. Being a brother or sister.
6. Being a child with a younger brother or sister added to the family.
7. Being a child with problems of physical appearance.

8. Being a child who is less sophisticated than his peers.
9. Being a child with problems of a physical handicap.
10. Being a child in a handicapped family — from a minority group or from a disadvantaged area.
11. Being a child who is less bright than his peers.
12. Being a person with a name you don't like.
13. Being a child with a different set of values than most members of a group. For example, different religion or liking to read when others don't.
14. Being an orphan.
15. Being an adopted child.
16. Being able to excel or live up to the high standards set by parents, teachers, or other adults.
17. Being able to cope with the insecurity of a family more interested in social activities or business than in children.
18. Being able to find the proper perspective in behavior when there are different sets of values around you.
19. Being able to assimilate the daily explosion of knowledge.
20. Being a child with sickness in the home.
21. Being a child with three parents, or one parent.
22. Being a child with poor health.
23. Being an aggressive child, hyperactive.
24. Being a child with prominent parents, brothers and sisters, or other well-known relatives.
25. Being a child who has grown more rapidly than others or who has matured physically more rapidly than his peers.

Fortunately no one has to cope with all of these problems, but it is amazing how many children have several of these difficulties. Perhaps you can spot the ones which are most common to your group and build your learnings program around these common obstacles.

Some Opportunities for Social Learnings in Schools

If teachers consider the classroom and the school as laboratories for social living and social learning, then almost every experience can be useful to children in the growing up process. Without spelling out in detail the ways in which these situations can be used, here are some of the opportunities for social learning in almost any school:

1. Equipping and decorating the classroom — or the school.
2. Organizing a class, with rotation of responsibilities, class officers, etc.
3. Planning classwork and activities together.
4. The types of learning activities.
5. Eating together in the classroom or the school lunchroom.
6. Planning parties.
7. Having visitors in the class — parents, other teachers, other children, other adults.
8. Assemblies — for small groups or larger ones.
9. Classroom "housekeeping."
10. Visits to the homes of children in the class or to other homes.
11. Trips of many kinds, in the school and in the local community, or even farther from home and school.

12. All-school projects, including Trick or Treat programs, United Fund programs, Red Cross drives, etc.
13. Experiences with school personnel — including the principal and assistant principal, nurse and doctor, custodian, bus driver, etc.
14. Experiences on the playground, in the gym, and in athletic contests of various kinds.
15. Running an activity for the school, ranging from "school post office" to "school store."
16. Gardening.
17. Events with children from other schools.
18. School camping.
19. Entertaining visitors from abroad in the classroom or in the school.
20. Clubs.
21. Student government.
22. Surveys conducted by a class or representatives from several classes.
23. Welcoming new pupils.
24. Safety patrol duty.
25. Interviewing people at home or in the community.
26. Use of the school library or the local library.
27.
28.
29.
30.

The Teacher's Role in Social Learnings

In helping children to understand themselves and others, teachers play many roles. They are often difficult and involved roles, demanding more knowledge and skill than any of us has. Teachers should attempt continuously to improve their knowledge of children and their skill in handling them. At the same time teachers must be satisfied to do the best they can and not expect perfection. They must know when they are beyond their depth and the children need more expert help. The classroom teacher should be a guidance counsellor but not attempt to be a psychologist or psychiatrist. High goals but reasonable goals are all that can be expected of classroom teachers.

The Teacher as a Model. Children need models or samples of the kind of people they can become. Parents serve as their first models. Older brothers and sisters, relatives, neighbors, and friends of the family serve in this role, too. The teacher also serves consciously or unconsciously as a model to children. This is especially true of teachers in the preschool years and in the primary grades but it can happen to any teacher at any level.

This places a heavy responsibility upon teachers to be the kind of mature persons they would like their children to become. To do this they need to examine carefully and deeply their own motivations and their own prejudices. If they can obtain professional help on this, so much the better. Perhaps some day teachers will be psychoanalyzed as part of their training to handle human beings at a formative period in their lives. But for the present that is an impossibility. But teachers can discover for themselves the types of children they favor and the types of children they resent. And they can delve into their own past and into their own value system to determine why that is so. Then they can work upon themselves so that they are accepting of as many children as possible — with all their weaknesses and all their strengths and all of their potentialities.

Here are a few questions you might ask yourself as a start on this self-analysis:

To what extent does the physical appearance of a child determine your attitude toward him or her? What physical characteristics appeal to you? What ones repulse you? Why?

Are you an aspiring middle-class or upper middle-class person? What effect does this have on your teaching and your relations with children?

Do you hold certain religious or moral values dogmatically? How does this affect your thinking about children who represent other value systems?

What is your attitude toward rewards and punishments? Are rewards many and varied and given to all or most children or limited to a few? Which ones? How many different kinds of talent have you recognized in children in the past week? What others might you value?

To what extent is your classroom (and your school) a laboratory in which children learn the ways of democracy? In what ways is it the same as a classroom in a totalitarian nation? In what ways is it different from the classroom in a dictatorship?

What skills in democratic living are children learning at the present time under your direction? What democratic attitudes are they learning? Can you cite evidence for what you have just concluded?

Such a self-analysis can be helpful but it also can be paralyzing. Don't become so introspective that you cannot function. That can happen, you know. If you were a saint, you probably would not be on earth right now. So do the best you can and don't worry yourself to death!

The Teacher as a Diagnostician. A second role that teachers play is that of diagnostician. Or they should play that role. If children are to change and grow and become, then someone must know about them as they are and as they can become.

Can you think of any manufacturer or artist or forester who would not learn as much as possible about the product or paints or trees with which he was dealing? In a sense, children are our products. We are the manufacturers or the artists or the foresters. We need to know all about the boys and girls we teach. How much more important they are than corn or clay or coal or shrubs or trees.

Teachers need to bear in mind that children are children. They are not miniature adults. That is an old idea that still persists and does much damage to child growth.

Teachers need to remember or realize that children are different and that they should be different. They are not cast in one mold. For that we should be thankful. They are not all beautiful or intelligent or creative. But they can be useful, interesting, important people.

Teachers also need to remember that children are the result of their heredity. The Russians don't believe this or at least they claim that everything is due to environment. But the rest of us are convinced that the biological component is one aspect of life and a very important one. All that you can learn about their biological or physical background is therefore important and should prove useful to know.

But teachers should know and believe that children are *primarily* the result of their environment. They learn their values. They learn their skills.

There are many ways of discovering a great deal about the children we teach. Here are a few of them:

1. Reading about child growth and development in general.
2. Reading and studying case histories of specific children.
3. Reading books about specific children, including novels dealing with children.
4. Observing one child over a period of several hours or days.
5. Investigating one child in depth.
6. Taking courses and attending lectures on children.
7. Seeing movies on children.
8. Utilizing to the full the cumulative records of children kept by the school.
9. Talking to parents about their children, approaching them with the comment, "Could you tell me about . . ." or "I need your help in understanding . . ."
10. Talking to other teachers about a child or a group of children.
11. Examining the art work of children or the construction they have done. Keeping copies of art work over a period of several weeks or months.
12. Asking children to write out the three things they would most like to do or have — the "Three Wishes Test."
13. Reading stories and discussing them with children, observing and listening to their reactions.
14. Looking at pictures and listening to the comments of the children on what they see.
15. Encouraging spontaneous dramatics and role-playing and seeing children's reactions.
16. Reading aloud stories without any ending and having the children complete the stories — orally or in written form.
17. Talking with groups of children or children individually.
18. Placing a large envelope or mailbox in the room, into which children can drop notes about problems which are troubling them — with the statements signed or unsigned — to produce frankness.
19. Encouraging pen pal correspondence and learning from the letters how children explain themselves to others.
20. Having children analyze discussions of themselves from tape recordings.
21. Doing time budgets.
22. Visiting homes of the children.
23. Observing behavior on trips.
24. Using the Children's Manifest Anxiety Scale.
25. Keeping diaries.
26. Asking children for statements on the "Three Things I Like Best" and "The Three Things I Dislike Most."
27. Drawing pictures of themselves.
28. Using sociometric tests.
29. Using self-rating and interest inventory tests.
30. Utilizing the standard achievement and intelligence test results.
31. Observing the play of children.

Lest this be just another list, we suggest that you place in the box the numbers of three or four items which you have not used or have not used successfully, which you want to try soon:

Ways of learning about children I want to try soon

1.

2.

3.

4.

Some Things Teachers Need to Know about Their Pupils

Here are a few of the facts which you should know about every pupil in your class. If that seems too overwhelming to you, start with one child and try to compile such information.

1. *Family and home*
   Size of the family and its composition.
   Place of the child in the family constellation.
   Data on parents, stepparents, or foster parents.
   Socio-economic conditions of the home, including employment.
   Housing conditions.
   Values of the home.
2. *Siblings*
   Who they are, age, sex, etc.
   Relation with pupil.
3. *Close relatives and friends*
   Who they are.
   Influence on the child.
4. *Atmosphere of the home*
   Relations among members of the home and the family circle.
   General attitude toward authority and specific attitude toward this pupil.
   Attitudes of family to education.
   Aspirations, expectations for child.
5. *Special conditions in the home*
   General "climate" and special conditions.
6. *Physical development of the child*
   General.
   Specific difficulties.
7. *Peers and friends*
   Number and who they are.
   Common interests.
   Difficulties.
8. *Out-of-school activities*
   Free play.
   Work.
   Organizations to which the child belongs.

9. *Intelligence and aptitudes*
   Scores.
   Specific comments on strengths and weaknesses.
10. *Special interests*
11. *Previous school record*
    Grades and comments of former teachers.
    General attitude toward school.
12. *Ambitions or plans (especially with older pupils)*
13. *Transportation to and from school (especially with younger pupils)*

BIBLIOGRAPHY ON STUDYING CHILDREN

If you are interested in reading further on this topic, almost any good current psychology book will be helpful. The references below are focused on ways of studying children:

1. COHEN, DOROTHY H. and VIRGINIA STERN. *Observing and Recording the Behavior of Young Children*. New York: Bureau of Publications — Teachers College, Columbia University (1965), 86 pp.
2. ELKIND, DAVID. *A Sympathetic Understanding of the Child Six to Sixteen*. Boston: Allyn and Bacon (1970), 154 pp. A paperback.
3. GILLHAM, HELEN L. *Helping Children Accept Themselves and Others*. New York: Bureau of Publications — Teachers College, Columbia University (1961), 56 pp.
4. GORDON, IRA J. *Studying the Child in School*. New York: Wiley (1966), 145 pp.
5. ISAACS, SUSAN. *The Children We Teach: Seven to Eleven Years*. New York: Schocken Books (1971), 160 pp. A paperback. See also her volume on *The Nursery Years: The Mind of the Child from Birth to Six Years*.
6. REDL, FRITZ. *Understanding Children's Behavior*. New York: Bureau of Publications — Teachers College, Columbia University (1961), 41 pp.
7. *Twelve to Sixteen: Early Adolescence*. Special issue of *Daedalus* for the Fall, 1971. 325 pp. A paperback.

The Teacher as a Goal-Setter. If children are to grow in desirable ways, the teacher must have appropriate goals in mind for each child and for the group as a whole. He or she cannot impose these goals on a pupil or on a group, but the setting the teacher furnishes and the activities provided will determine in large part how far and how fast they each develop. These goals must be consistent, of course, with the knowledge which the teacher has gained of each of the pupils by the methods suggested in the previous section.

Psychologists have made much in recent years of the role of expectations. When much is expected of children, they are likely to grow "taller" and faster than when little is expected. This is especially important to bear in mind with children from disadvantaged backgrounds. So little has been expected of them, so little confidence has been placed in them, so little esteem built in them, that they tend to have goals far below what they might be expected to achieve.

On the other hand, expectations can be too high. Aspiring parents of the middle class, especially in suburbs, tend to expect too much of their offspring. Consequently, the pressures that result are sometimes very great and the tensions that develop very often enormous.

Therefore teachers need to find the particular goals for children and groups which are suited to those children.

Even more important is the attitude of the child as to what he or she can accomplish. The psychologist Alfred Adler wrote of this years ago in his book on *What Life Should Mean to You*. First he pointed out that:

> The greatest difficulty in education is provided, not by the limitations of the child, but by what he thinks are his limitations.

Then he went on to state that:

> No teacher can succeed in removing the limits a child has set to his own development if he himself believes that there are fixed limits to that development.

Within the last few years several writers have spelled out in vivid and even lurid detail the effects of poor self images on children. Nat Hentoff did it in his volume on *Our Children Are Dying* (Viking, 1966). Jonathan Kozol told the same story of the destruction of the hearts and minds of black children in the Boston public schools in *Death At An Early Age* (Bantam, 1967). John Holt furthered this theme in his volume on *How Children Fail* (Dell, 1964). And Charles Silberman concentrated on this in *Crisis in the Classroom* (Random House, 1970). The autobiographies of almost any black testify to the stultifying effects of an education which tells children early in life that they will not amount to anything and should not make any effort to succeed. Therefore, all that teachers can do to help children see themselves as people who can "become" is of the utmost importance.

The Teacher as a Referral Agent. As we suggested earlier, there are sometimes problems in self-understanding and the understanding of others that are too difficult for most teachers or any teachers to handle. In such cases, teachers should not try to handle them by themselves. They can do more damage than good in such cases. But they should know to whom such children can be referred. Sometimes it is a school psychologist. Sometimes it is another teacher with whom this child has a special identification. Often it is professional help or even institutionalization that they need.

The Teacher as a Developer. When the teacher has discovered all that he or she can about a child and set the goals, hopefully with that child included, then the teacher assumes another related role, that of developer.

In a sense teachers are social engineers. They help to build the foundations on which long lives will be built. Most people think of this as their sole role, but it cannot be done without the previous steps of discovery and diagnosis as well as of goal-setting.

In the making of lesson plans, in the construction of teaching units, and in the selection of a wide variety of resources, the teacher is developing personality. Every lesson and every activity should be utilized to help develop attitudes, understandings, skills, and knowledge — with the particular children of the class in mind.

Much of the rest of this book is intended to develop this theme so we will merely mention it here. The brevity of this statement, however, should not mislead any reader into thinking that this is a minor role. It is one of the most important roles any teacher can play.

Since children differ radically from each other, there need to be many and varied activities. Since children learn by themselves, there need to be many activities in which children work alone. But children see themselves in relation to others and there should be many small group activities, too. And, of course, the entire class should often work as a unit.

One last word of warning seems in order here. That is that all records which teachers have of children should be kept confidentially and placed in a safe place where no one can find them and use them. Attention to this footnote in this chapter can save teachers many a bad moment, and in some cases their jobs.

BIBLIOGRAPHY: BOOKS FOR CHILDREN

On birth and physical growth

1. ANTONACCI, ROBERT J. and JUNE BARR. *Physical Fitness for Young Champions.* New York: Whittlesey (1962), 160 pp. Grades 4–7.
2. APPEL, CLARA and MOREY, and SUZANNE SZASZ. *We Are Six: The Story of a Family.* New York: Golden (1959), 60 pp. Grades 1–3. Preparation of a family for a new member of the group.
3. DE SCHWEINITZ, KARL. *Growing Up: How We Become Alive, Are Born, and Grow.* New York: Macmillan (1965), 54 pp.
4. ETS, MARIE HALL. *Just Me.* New York: Viking (1965), 32 pp. N–2.
5. GREEN, MARY M. *Is It Hard? Is It Easy?* New York: Scott (undated), 32 pp. K–2. Differences among children.
6. GRUENBERG, SIDONIE M. *The Wonderful Story of How You Were Born.* Garden City: New York: Doubleday (1965), 39 pp. Grades 3–5.
7. HEGELER, STAN. *Peter and Caroline: A Child Asks About Childbirth and Sex.* New York: Abelard-Schuman (1961), 35 pp. Grades 1–3. A fairly radical approach to sex education.
8. HOBSON, LAURA Z. *I'm Going to Have a Baby.* New York: John Day (1967), 43 pp. Grades 1–3. A young boy anticipates the arrival of a sibling.
9. JOHNSON, CROCKETT. *The Blue Ribbon Puppies.* New York: Harper (1959), 32 pp. N–2. Stresses individuality.
10. KRASILOVSKY, PHYLLIS. *The Very Little Boy.* Garden City, New York: Doubleday (1962), 32 pp. N–2.
11. KRASILOVSKY, PHYLLIS. *The Very Little Girl.* Garden City, New York: Doubleday (1961), 32 pp. N–2.
12. MENDOZA, GEORGE. *And Amadeo Asked How Does One Become a Man.* New York: Braziller (1965), 44 pp. Grades 4–6.
13. NOSHPITZ, JOSEPH D. *Understanding Ourselves: The Challenge of the Human Mind.* New York: Coward-McCann (1964), 114 pp. Grades 6–9.
14. SCHLEIN, MIRIAM. *Billy: The Littlest One.* Chicago: Whitman (1965), 40 pp. K–2.
15. STANLEY, JOHN. *It's Nice to Be Little.* Chicago: Rand McNally (1965), 32 pp. Grades 1–3.
16. WOODS, BETTY. *I Want to Be Different.* Chicago: Reilly and Lee (1964), 48 pp. Grades 2–4.

On Adjusting to Physical Handicaps

1. ARMER, ALBERT. *Screwball.* Cleveland: World (1962), 192 pp. Grades 5–8. About a victim of polio.
2. BAASTED, BABBIS R. *Kristy's Courage.* New York: Harcourt (1965), 159 pp. Grades 4–6. A child handicapped by scars from an automobile accident.
3. BUTLER, BEVERLY. *Light a Single Candle.* New York: Dodd, Mead (1962), 242 pp. Grades 5–8. A boy who suddenly goes blind.
4. CHRISTOPHER, MATT. *Sink It, Rusty.* Boston: Little, Brown (1963), 138 pp. Grades 2–4. About a polio victim.
5. CORBIN, WILLIAM. *Golden Mare.* New York: Coward-McCann (1955), 122 pp. Grades 5–7. A courageous boy with heart trouble.
6. DAHL, BORGHILD. *Finding My Way: An Autobiography.* New York: Dutton (1962), 128 pp. The story of a famous blind writer.

7. GELFAND, RAVINA and LETHA PATTERSON. *They Wouldn't Quit: Stories of Handicapped People.* Minneapolis: Lerner (1962), 55 pp. Grades 5–8.
8. PUTNAM, PETER. *The Triumph of the Seeing Eye.* New York: Harper (1963), 178 pp. Grades 6–9. About the Seeing Eye dogs and blind people.
9. ROBINSON, VERONICA. *David in Silence.* Philadelphia: Lippincott (1963), 64 pp. The story of a deaf boy.
10. VANCE, MARGUERITE. *Windows for Rosemary.* New York: Dutton (1956), 61 pp. Grades 4–7. See also *A Rainbow for Robin.* On blind children.

On making and keeping friends

1. ANGLUND, JOAN WALSH. *A Friend Is Someone Who Likes You.* New York: Harcourt (1958), unpaged. Grades 1–4.
2. ANGLUND, JOAN WALSH. *Love Is a Special Way of Feeling.* New York: Harcourt (1960), 30 pp. Grades 1–3.
3. BEIM, LORRAINE and JERROLD. *Two Is a Team.* New York: Harcourt (1945), 61 pp. K–3. A Negro and a white boy learn to work together. An old book but still relevant.
4. GRAMATRY, HARDIE. *Little Toot on the Thames.* New York: Putnam (1964), 96 pp. Grades 1–3. An allegory about a little boat and larger boats.
5. HOFF, SYD. *Who Will Be My Friends?* New York: Harper (1960), 32 pp. K–3. A boy's skill in baseball finally wins him friends.
6. SLOBODKIN, LOUIS. *One Is Good But Two Are Better.* New York: Vanguard (1956), 26 pp. K–3.
7. SMARIDGE, NORAH. *Impatient Jonathan.* New York: Abingdon (1964), 32 pp. K–3. Patience is the key idea in this story.
8. UDRY, JANICE MAY. *Let's Be Enemies.* New York: Harper (1961), 32 pp. K–3. Quarrels among boys and how they eventually make up and become good friends.

Several textbooks deal with this question of how children make friends. Two of them are C. W. Hunnicutt and Jean D. Gramb's *We Live with Others* and *We Have Friends.* Both of these are Singer publications. The first is for the first grade and the second for the second grade. The two volumes on *Willy, Andy and Ramon* and *Five Friends at School,* both Holt, Rinehart and Winston publications, stress intercultural friendships. The first is a beginning social studies reader and the second is a sequel. Included are Puerto Ricans, blacks, and whites.

Getting along with brothers and sisters

1. BELL, GINA. *Who Wants Willy Wells?* New York: Abingdon (1965), 40 pp. K–2.
2. BROWN, MYRA B. *Amy and the New Baby.* New York: Watts (1965), 56 pp. K–2.
3. BUCKLEY, HELEN E. *My Sister and I.* New York: Lothrop (1963), 32 pp. K–3. The fun of having an older sister.
4. CHRISTOPHER, MATT. *Long Stretch at First Base.* Boston: Little, Brown (1960), 149 pp. Grades 2–4.
5. CHRISTOPHER, MATT. *Two Strikes on Johnny.* Boston: Little, Brown (1958), 136 pp. Grades 4–7.
6. COOMBS, PATRICIA. *Waddy and His Brother.* New York: Lothrop, Lee and Shepard (1963), unpaged. K–2.
7. DARINGER, HELEN F. *Stepsister Sally.* New York: Harcourt (1962), 182 pp. Grades 4–6.
8. EMBRY, MARGARET. *Kid Sister.* New York: Holiday (1958), 165 pp. Grades 3–5.
9. HOYT, HELEN F. *Aloha! Susan!* Garden City, New York: Doubleday (1961), 167 pp. Grades 4–6.
10. KESSLER, ETHEL and LEONARD. *Kim and Me.* Garden City, New York: Doubleday (1960), 32 pp. N–2.

11. Matsuno, Masako. *Chie and the Sports Day*. Cleveland: World (1965), 32 pp. Grades 1–4. Americans of Japanese ancestry.
12. Orgel, Doris. *Sarah's Room*. New York: Harper (1963), 46 pp. K–2.
13. Parish, Peggy. *Willy Is My Brother*. New York: Scott (1964), 48 pp. K–3.
14. Zolotow, Charlotte. *Big Sister and Little Sister*. New York: Harper and Row (1966), unpaged. K–2.
15. Zolotow, Charlotte. *Big Brother*. New York: Harper (1960), 32 pp. K–3.

On adopted children

1. Flory, Jane. *Faraway Dream*. Boston: Houghton Mifflin (1968), 219 pp. Grades 4–6.
2. Guy, Anne. *A Baby for Betsy*. New York: Abingdon (1957), 31 pp. K–1.
3. Wier, Ester. *The Loner*. New York: McKay (1963), 153 pp. Grades 5–7. A nameless and friendless boy gets a new home and a name — and friendship.

On death

1. Brown, Margaret W. *The Dead Bird*. New York: Scott (1958), 48 pp. K–3.
2. Harris, Audrey. *Why Did He Die?* Minneapolis: Lerner (1967), 30 pp. Grades 2–4.
3. Zim, Herbert S. and Sonia Bleeker. *Life and Death*. New York: Morrow (1970), 64 pp. Grades 4–6. A more direct approach than the two previously listed volumes.

Teachers will find some help on explaining death to children in the following books:

1. Grollman, Earl A. *Explaining Death to Children*. Boston: Beacon (1967), 296 pp. An anthology of essays on death.
2. Jackson, Edgar N. *Telling a Child About Death*. New York: Channel (1965), 91 pp.
3. Wolf, Anna W. M. *Helping Your Child to Understand Death*. New York: The Child Study Association (1958), 45 pp.

Living with parents — or a parent

1. Behrens, June. *Soo Ling Finds a Way*. San Carlos, California: Golden Gate (1965), 32 pp. K–2. A Chinese-American girl helps her grandfather meet the competition of a new laundromat.
2. Beim, Jerrold. *Trouble After School*. New York: Harcourt Brace Jovanovich (1957), 128 pp. Grades 3–5. An 8th grader gets in trouble after school when he is left alone by his mother.
3. Bromhall, Winifred. *Peter's Three Friends*. New York: Knopf (1964), 32 pp. K–3.
4. Burch, Robert. *Skinny*. New York: Viking (1965), 128 pp. Grades 5–6.
5. Cote, Phyllis. *The People Upstairs*. Garden City, New York: Doubleday (1951), 214 pp. Grades 4–6. A mother works while grandfather cares for the children.
6. Fall, Thomas. *Eddie No-Name*. New York: Pantheon (1963), 48 pp. Grades 4–6. About an orphan.
7. Johnson, Johanna. *Edie Changes Her Mind*. New York: Putnam (1964), 48 pp. Grades 1–3. A little girl changes her mind about the fun of staying up all night when her parents let her do so.
8. Klein, Lenore. *Runaway John*. New York. Knopf (1963), 32 pp. K–3.
9. Lowe, Patricia. *The Different Ones*. Indianapolis: Bobbs-Merrill (1965), 153 pp. Grades 5–7. About an orphan.
10. McNulty, Faith. *When a Boy Wakes Up in the Morning*. New York: Knopf (1962), 32 pp. N–2.
11. Newberry, Clare T. *Ice Cream for Two*. New York: Harper (1953), 58 pp. Grades 4–6. A mother and her son adjust to life in New York City.

Other problems of children

1. BROWN, MYRA B. *First Night Away from Home.* New York: Watts (1960), 56 pp. K–2.
2. BURCH, ROBERT. *D. J.'s Worst Enemy.* New York: Viking (1965), 142 pp. And D. J. was his own worst enemy.
3. CHANDLER, EDNA W. *The Boy Who Made Faces.* Chicago: Whitman (1964), 32 pp. Grades 2–4. It was fun — until he got in trouble.
4. COLE, JOANNA. *The Secret Box.* New York: Morrow (1971), 40 pp. Grades 4–6. A young girl and the problem of stealing. Drawings are multi-racial.
5. ESTES, ELEANOR. *The Hundred Dresses.* New York: Harcourt Brace Jovanovich (1944), 80 pp. Grades 3–6. A classic about a girl who imagined she had one hundred dresses.
6. FEAGLES, DARMAR. *Casey: The Impossible Horse.* New York: Scott (1960), 96 pp. Grades 3–5.
7. HARRIS, AUDREY. *Why Did He Die?* New York: Lerner (1967), 29 pp. K–3.
8. KRASILOVSKY, PHYLLIS. *Susan Sometimes.* New York: Macmillan (1962), 31 pp. K–2. Susan invents seven sisters to help amuse her.
9. PARKER, RICHARD. *The Boy Who Wasn't Lonely.* Indianapolis: Bobbs-Merrill (1965), 141 pp. Grades 4–6.
10. RINKOFF, BARBARA. *The Remarkable Ramsey.* New York: Morrow (1965), 96 pp. Grades 3–5. About a lonely child.
11. YASHIMA, TARO. *Crow Boy.* New York: Viking (1955), 37 pp. Grades 1–4. A cross-eyed, shy Japanese boy who finally wins friends.

UNDERSTANDING PEOPLE FROM OTHER GROUPS

The American Dream has long been one of a land of equal opportunity for all — irrespective of socio-economic class, national or ethnic background, race, religion, or geographical location. This has been our ideal.

In practice, however, there have been two contradictory strands in the story of our country. One has been the welcoming hand to newcomers, religious and racial tolerance, and aid to minority groups. The other has been the practice of exclusiveness, intolerance, and even violence toward those who were different.

Even today, part of the past endures. There is still much intolerance, much exclusiveness, much prejudice, and even some violence. But there is also increasing concern for minority groups, more assistance to the disadvantaged in our nation, and more understanding of people who are different. The passage of civil rights legislation, the election of a Catholic to the Presidency, and the appointment of Jews and Negroes to high posts in our government are all indications of improved human relations in the United States in recent times. Many individuals and organizations have fostered such improvements. For example, the National Conference of Christians and Jews, the Anti-Defamation League of B'nai B'rith, and the National Association for the Advancement of Colored People come to mind as three organizations which have pioneered in better human relations. But there is a great deal yet to be done.

If we are to retain the gains we have made and build even stronger foundations for a multiracial, multiethnic, multireligious society, much of the education for human rights must be done in the schools.

And it must be done early. As Rodgers and Hammerstein reminded us in *South Pacific,* "You've got to be taught before it's too late, before you are six or seven or eight." Studies like the Trager-Radke study of young children indicate that many children bring their prejudices to school. Even before they are six, many boys and girls have developed prejudice against people who are different. Therefore teachers in all the grades, but especially in the primary grades, need to be constantly aware of the need to work on preconceived ideas that children have about other human beings.

The mental hygiene approach is basic in developing understanding and respect. As we have pointed out before, children are not likely to respect others if they do not respect themselves. But this is not enough. Children need to know why other people act as they do and to develop respect for differences. This is not an easy assignment for teachers, but it is an important one.

Hopefully children in elementary schools will come in contact with boys and girls from a wide variety of backgrounds. That is one of the purposes of the so-called "common school." Children need to learn that some children go to a place of worship on Friday night, some on Saturday, many on Sunday, and some not at all. They need to learn to respect people whether they are Jewish, Catholic, Protestant, adherents of other faiths, or unaffiliated with any religious group.

They need to learn very early that color is really unimportant and that people should be judged by what they do and not by their pigmentation. This is almost impossible to learn if there are not children of various colors in the classroom. This is the message Professor Miel accents in her pamphlet on *The Short-Changed Children of Suburbia*.

Such intercultural understanding cannot be built by the celebration of special days or weeks. It must be built into the day-to-day functioning of the classrooms across our country. It must be built into the social studies curriculum at scores of points. Above all it needs to be developed by teachers who really believe in respect for all human beings.

Teachers therefore need to examine their own prejudices and work on them. And they need to be well informed on how attitudes are formed, reinforced and changed. Some material on that topic is contained in this book, on pages 38–39. Many teachers will want to pursue this basic topic by referring to the reading list on page 39.

Often the oblique approach to the teaching of human relations works best. Books and plays and other literature about children can provide rich opportunities for teachers to explore the feelings of people from other groups and build positive attitudes toward themselves on the part of children of minority groups. Many such books are listed on the pages that follow. Teachers may want to develop their own reading lists on other groups, such as Indian-Americans or Jewish-Americans, depending upon the minorities in their own schools and communities.

The role-playing of situations in real life or in books also can serve as a stimulus to the examination of human relations situations and feelings. With older boys and girls, free and frank discussions in an environment of acceptance of divergent points of view can be carried on to advantage. Rumor clinics can be conducted. Films and filmstrips dealing with various groups can be shown. Visitors can be invited to class to explore stereotypes with children and young people. Sometimes they can be invited to participate in the work of a class without directly discussing the problems of their group. Then they become active participants in the life of the class rather than experts on a special group.

Attitudes are acquired over a long period of time, and changes in attitudes do not come easily or quickly. Teachers should not become discouraged if they do not see changes in attitudes immediately.

Some teachers may not feel that they can handle such controversial issues. In such cases, they may well decide not to tackle such topics until they are able to do so with real feeling and skill.

Fortunately there is a wealth of resource material for teaching about most groups in the United States. Many books are listed in the pages that follow. Lists of films, filmstrips, tapes, and other materials may be obtained from local or state organizations or the national headquarters of various groups.

Here are a few lists of some of the available books written to help boys and girls (and teachers) develop respect for human beings no matter what their backgrounds may be.

Fortunately there is now a climate of acceptance in publishing houses of manuscripts which deal with minority groups. This has been evident from the period of the late 1960s on and is a real gain for all of us. Since there are so many new books being published, you will undoubtedly want to keep track of the latest releases and try to obtain such books for your school and/or local library. Do not expect all of them to be outstanding, however. In their desire to capitalize upon the current interest in this field, some publishers are accepting materials which do not meet the highest standards.

GENERAL BOOKS ON INTERCULTURAL EDUCATION FOR BOYS AND GIRLS

1. BACMEISTER, RHODA W. *The People Downstairs*. New York: Coward-McCann (1964), 120 pp. K–3. Twenty-four stories about a variety of families of different backgrounds.
2. BEHRENS, JUNE. *Who Am I?* Encino, California: Elk Grove (1968), 21 pp. K–2. An integrated group of first grade children tell who they are and what they like to do.
3. BEIM, LORRAINE. *Carol's Side of the Street*. New York: Harcourt Brace Jovanovich (1951), 213 pp. Grades 3–5. Carol discovers that prejudice comes from not knowing people in other groups.
4. BENEDICT, RUTH and GENE WELTFISH. *In Henry's Backyard*. New York: Schuman (1948), 40 pp. Grades 4–6. Based on the pamphlet "The Races of Mankind." Old but still very useful.
5. BURCHARDT, NELLIE. *Project Cat*. New York: Watts (1966), 66 pp. Grades 4–6. A group of girls of various ethnic backgrounds discover and care for a cat and finally persuade the City Council to revise its rules about cats in city housing projects.
6. CHILD STUDY ASSOCIATION. *Round About the City: Stories You Can Read to Yourself*. New York. Crowell (1966), 116 pp. K–3. Stories about everyday life of children of various ethnic backgrounds.
7. CLIFFORD, ETH. *Your Face Is A Picture*. Indianapolis: Seale (1963), K–2. Beautiful photographs of children of various groups.
8. COHEN, MIRIAM. *Will I Have a Friend?* New York: Macmillan (1967), unpaged. K–2. A variety of children in school on a boy's first day.
9. EVANS, EVA KNOX. *All About Us*. New York: Capitol and Golden (1965), 95 pp. Grades 5–8. The variety of people in the United States.
10. GUGGENHEIM, HANS. *The World of a Wonderful Difference*. New York: Friendly (1960), 49 pp. Grades 2–4. Delightful drawings and verses about the variety of people in the world.
11. HUNT, MABEL LEIGH. *Cristy at Skipping Hills*. Philadelphia: Lippincott (1958), 139 pp. Grades 4–6. An example of a community with a variety of people in it, living together in harmony most of the time.
12. LANSDOWN, BRENDA. *Galumph*. Boston: Houghton Mifflin (1963), 48 pp. K–3. The story of a cat — and a variety of children.
13. LERNER, MARGUERITE R. *Red Man, White Man, African Chief*. Minneapolis, Minnesota: Medical Books for Children (1961), unpaged. Grades 4–6. Explains skin pigmentation scientifically.

14. SECHRIST, ELIZABETH H. and NEANETTE WOOLSEY. *It's Time for Brotherhood.* Philadelphia: Macrae (1963), 222 pp. Grades 6–9.
15. SHOWERS, PAUL. *Your Skin and Mine.* New York: Crowell (1965), unpaged. K–3. Simple and clear information on pigmentation.
16. WILSON, BETTYE D. *We Are All Americans.* New York. Friendly (1959), unpaged. Grades 4–6.

AFRO-AMERICANS

The books listed below are merely a sample of books on Afro-Americans, accenting books for younger children on families and school life. For a very complete list see Augusta Baker's *The Black Experience in Children's Books* (New York Public Library, 1971, 109 pp.).

1. ADELMAN, BOB and SUSAN HALL. *On and Off the Street.* New York: Macmillan (1970), 64 pp. Grades 4–7. Beautiful black and white pictures of Vincent and Danny and their exploits in New York City.
2. BALDWIN, ANNE N. *Sunflowers for Tina.* New York: Four Winds (1970), 42 pp. Grades 3–4. A young black girl and her desire to grow things in the city.
3. BAUM, BETTY. *Patricia Crosses Town.* New York: Knopf (1965), 178 pp. Grades 5–8. Patricia makes friends in an integrated school.
4. BEIM, JERROLD and LORRAINE. *Two Is a Team.* New York: Harcourt (1945), 61 pp. K–3. A Negro and a white boy become friends. A classic.
5. BISHOP, CURTIS K. *Little League Heroes.* Philadelphia: Lippincott (1960), 190 pp. Grades 4–6. Joel doubts he will be accepted in the Little League but he becomes a star performer and wins friends.
6. BLUE, ROSE. *Bed-Stuy Beat: Sonny's Song.* New York: Watts (1971), 48 pp. Illustrations for a song about life in the Bedford-Stuyvesant section of Brooklyn.
7. CARLSON, NATALIE S. *The Empty Schoolhouse.* New York: Harper (1965), 119 pp. Grades 5–8. Another story about an integrated school.
8. CLIFTON, LUCILLE. *The Black B C's.* New York: Dutton (1970), 46 pp. Grades 5–6. Black history and achievements told through letters of the alphabet.
9. GRAHAM, LORENZ. *North Town.* New York: Crowell (1965), 124 pp. Grades 5–8. A black boy in an integrated school. See also his earlier *South Town.*
10. JUSTUS, MAY. *A New Home for Billy.* New York: Hastings (1966), 56 pp. Grades 1–3. A six-year-old moves to a new home.
11. KEATS, EZRA K. *Whistle for Willie.* New York: Viking (1964), 83 pp. Grades 1–3. A boy learns to whistle to call his dog, as other boys do. See Ezra Keats' other books, such as *The Snowy Day.*
12. KOCH, JOHN R. *Where Did You Come From?* Eau Claire, Wisconsin. Hale (1968), 48 pp. Grades 2–4. A Negro child learns of his heritage.
13. McGOVERN, ANN. *Black Is Beautiful.* New York: Putnam (1970), 320 pp. Grades 1–3.
14. RINKOFF, BARBARA. *Headed for Trouble.* New York: Knopf (1970), 120 pp. Grades 6–8.
15. SHEARER, JOHN. *I Wish I Had An Afro.* New York: Cowles (1970), 48 pp. Grades 5–7. Dramatic photos enhance this volume about a black family.
16. SHOTWELL, LOUISA R. *Roosevelt Grady.* Cleveland: World (1963), 151 pp. Grades 4–6. The story of a son of a migrant black worker. Available also as a paperback.
17. WALTER, MILDRED P. *Lillie of Watts: A Birthday Discovery.* Los Angeles, Ward Ritchie Press (1969), 61 pp. Grades 5–7. Story of an 11-year-old girl.
18. WEINER, SANDRA. *It's Wings That Makes Birds Fly.* New York: Pantheon (1968), 57 pp. Grades 2–4. Adventures of a black boy in New York City.
19. WERSTEIN, IRVING. *A Proud People: Black Americans.* Philadelphia: Lippincott (1970), 120 pp. Grades 5–8.

ASIAN-AMERICANS

A bibliography of "Books for the Chinese-American Child" is available free from the Cooperative Children's Book Center, 411 West, State Capitol, Madison, Wisconsin. It lists and annotates books on China for children.

1. BEHRENS, JUNE. *Soo Ling Finds a Way*. San Carlos, California: Golden Gate (1965), 32 pp. K–2.
2. BROWN, JEANNETTE P. *Keiko's Birthday*. New York: Friendship (1954), 32 pp. K–3. A little girl from Japan is welcomed in an American kindergarten.
3. COATSWORTH, ELIZABETH. *Cherry Ann and the Dragon Horse*. New York: Macmillan (1955), 64 pp. Grades 5–7. A ten-year-old Chinese girl in San Francisco resolves her divided loyalties.
4. COPELAND, HELEN. *Meet Miki Takino*. New York: Lothrop (1963), 32 pp. K–3. A Japanese-American boy in the first grade turns up five "grandparents" for a school play.
5. DINES, GLEN. *The Useful Dragon of San Ling Toy*. New York: Macmillan (1956), unpaged. Grades 1–3. An amusing picture book.
6. DOANE, PELAGRIE. *Understanding Kim*. Philadelphia: Lippincott (1962), 126 pp. Grades 4–6. A Korean orphan is adopted by an American family and makes adjustments to his new life.
7. DOWDELL, DOROTHY and JOSEPH. *The Japanese Helped Build America*. New York: Messner (1970), 96 pp. For better readers only, or for teachers to summarize for use with children.
8. LATTIMORE, ELEANOR P. *The Chinese Daughter*. New York: Morrow (1960), 125 pp. Grades 3–5. A Chinese girl has problems of divided loyalties to solve.
9. LENSKI, LOIS. *San Francisco Day*. Philadelphia: Lippincott (1955), 175 pp. Grades 5–7. The family life of a Chinese-American family in San Francisco.
10. MARTIN, PATRICIA M. *The Rice Bowl Pet*. New York: Crowell (1962), unpaged. Grades 2–4. The story of Ah Jim and his pet in San Francisco.
11. MATSUNO, MASAKO. *Chie and the Sports Day*. Cleveland: World (1965), 32 pp. Grades 1–4. A younger sister, who generally is a nuisance, helps her brother win a race.
12. OAKES, VANYA. *Roy Sato: New Neighbor*. New York: Messner (1955), unpaged. Grades 4–6. A Nisei boy and his Caucasian and Negro friends discover in a trip to the laboratory that their blood is the same.
13. POLITI, LEO. *Moy Moy*. New York: Scribner (1960), 22 pp. K–2. The little sister in a Chinese-American home enjoys the Chinese New Year.
14. STRACHAN, MARGARET P. *Patience and a Mulberry Leaf*. New York: Washburn (1962), 137 pp. Grades 6–9. A girl in Seattle resolves her problems of loyalty.
15. UCHIDA, YOSHIKO. *Mik and the Prowler*. New York: Harcourt Brace Jovanovich (1960), 122 pp. Grades 2–6. Mik Watanabe, a ten-year-old, and his guest from Japan.
16. UCHIDA, YOSHIKO. *New Friends for Susan*. New York: Scribner (1951), 185 pp. Grades 2–6. Life in a Japanese-American home.
17. YASHIMA, MITSU and TARO. *Momo's Kitten*. New York: Viking (1961), 33 pp. Grades 1–3.
18. YASHIMA, TARO. *Umbrella*. New York: Viking (1958), 33 pp. Grades 1–3. A child of Japanese-American parents in New York City and her delight over a new umbrella.

CHICANOS: MEXICAN-AMERICANS

1. ACUNA, RUDOLPH. *The Story of Mexican Americans: The Men and the Land*. New York: American Book Company (1969), 140 pp. Grades 6–9.
2. BULLA, CLYDE R. *Benito*. New York: Crowell (1961), 84 pp. Grades 3–5.
3. ERDMAN, LOULA G. *My Sky Is Blue*. New York: Longmans (1963), 218 pp. Grades

6–9. The story of how a young teacher tries to work against prejudice in New Mexico.

4. ETS, MARIE HALL. *Bad Boy, Good Boy.* New York: Crowell (1967), 50 pp. Grades 1–3. Realistic story of the troubles of a Mexican American boy and his family.

5. ETS, MARIE HALL. *Gilberto and the Wind.* New York: Viking (1963), 32 pp. K–2. Outstanding illustrations of a boy and the wind.

6. FRANCHERE, RUTH. *César Chavez.* New York: Crowell (1970), 42 pp. Grades 3–6. The story of the leader of the Mexican American migrant workers.

7. FREEMAN, DOROTHY R. *Home for Memo.* Encino, California: Elk Grove Press (1969), 39 pp. Grades 3–4.

8. GALBRAITH, CLARE K. *Victor.* Boston: Little, Brown (1971), 47 pp. Grades 2–3. A Mexican American boy who dislikes school.

9. GARTHWAITE, MARION. *Mario.* Garden City, New York: Doubleday (1960), 67 pp. Grades 5–7.

10. GREENE, CARLA. *Manuel: Young Mexican American.* New York: Lantern (1969), 48 pp. Grades 3–6.

11. HARTER, HELEN. *Carmelo.* Chicago: Follett (1962), 31 pp. Grades 3–4.

12. MacLEOD, RUTH. *Buenos Dias, Teacher.* New York: Messner (1970), 192 pp. Grades 7–10.

13. MARTIN, PATRICIA M. *Chicanos: Mexicans in the United States.* New York: Parents Magazine Press (1971), 64 pp. Grades 7–10. Background.

14. MARTIN, PATRICIA M. *Trina's Boxcar.* New York: Abingdon (1967), 112 pp. Grades 3–4.

15. MULCAHY, LUCILLE. *The Blue Marshmallow Mountains.* New York: Nelson (1959), 128 pp. Grades 5–7.

16. POLITI, LEO. *Juanita.* New York: Scribner (1948), 30 pp. K–3.

17. POLITI, LEO. *Pedro: The Angel of Olvera Street.* New York: Scribner (1946), 32 pp. Grades 2–4.

18. POLITI, LEO. *Song of the Swallows.* New York: Scribner (1949), 30 pp. K–2. Beautiful illustrations enhance the value of this book on Mexican Americans.

19. SCHAEFER, JACK. *Old Ramon.* Boston: Houghton Mifflin (1963), 41 pp. K–3.

20. SCHWEITZER, BYRD B. *Amigo.* New York: Macmillan (1963), 41 pp. K–3.

21. SUMMERS, JAMES L. *You Can't Make It By Bus.* Philadelphia: Westminster (1960), 120 pp. Grades 7–10.

22. TEBBEL, JOHN and RAMOS E. RUIZ. *South by Southwest.* Garden City, New York: Doubleday (1969), 120 pp. Grades 6–9. A Zenith book on Mexican Americans.

23. WEINER, SANDRA. *Small Hands, Big Hands: Seven Profiles of Chicano Migrant Workers and Their Families.* New York: Pantheon (1970), 55 pp. Grades 2–4. Excellent photographs add to the usefulness of this volume.

24. YOUNG, BOB and JAN. *Across the Tracks.* New York: Washington Square Press (1969), 240 pp. Grades 7–9.

PUERTO RICO AND PUERTO RICAN-AMERICANS

1. BARRY, ROBERT. *The Musical Palm Tree.* New York: McGraw-Hill (1965), 32 pp. Grades 3–5. A young lad in Puerto Rico entertains visitors from a ship.

2. BARTH, EDNA. *The Day Luis Was Lost.* Boston: Little, Brown (1971), 62 pp. Grades 1–3. Small boy, new to the mainland, gets lost going to school.

3. BELPRÉ, PURA. *Juan Bob and the Queen's Necklace.* New York: Warne (1962), 48 pp. Grades 1–3. An old Puerto Rican folk tale.

4. BELPRÉ, PURA. *Santiago.* New York: Warne (1969), 31 pp. A Puerto Rican boy and his love for a pet.

5. BLUE, ROSE. *I Am Here: Yo Estoy Aqui.* New York: Watts (1971), 48 pp. K–3. A young girl adjusts to the kindergarten on the mainland.

6. BOWEN, J. DAVID. *The Island of Puerto Rico.* Philadelphia: Lippincott (1968), 136 pp. Grades 5–9. A general account in the Portraits of Nations series.

7. BRENNER, BARBARA. *Barto Takes the Subway.* New York: Knopf (1961), unpaged. A Puerto Rican boy in New York City rides the subway.

8. BUCKLEY, PETER. *I Am From Puerto Rico.* New York: Simon and Schuster (1971), 128 pp. Grades 5–6. Many black and white photographs enhance this volume.

9. BURCHARD, PETER. *Chito.* New York: Coward-McCann (1969), 46 pp. Grades 2–4.

10. COLORADO, ANTONIO J. *The First Book of Puerto Rico.* New York: Watts (1965), 74 pp. Grades 5–7.

11. CHRISTOPHER, MATTHEW F. *Baseball Flyhawks.* Boston: Little, Brown (1963), 127 pp. Grades 5–8. Stories of Puerto Ricans in New York City.

12. KURTIS, ARLENE H. *Puerto Rico: From Island to Mainland.* New York: Messner (1969), 96 pp. Grades 5–7.

13. LEWITON, MINA. *That Bad Carlos.* New York: Harper and Row (1964), 175 pp. Grades 5–6. Teacher spots a very intelligent but troublesome boy.

14. LITTLE, MARY E. *Ricardo and the Puppets.* New York: Scribner (1958), unpaged. Grades 1–3. Includes many pictures of Puerto Rican children.

15. MANN, PEGGY. *When Carlos Closed the Street.* New York: Coward-McCann (1969), 71 pp. Grades 3–4. A stick ball championship between two gangs and their difficulties with the police.

16. MANNING, JACK. *Young Puerto Rico: Children of Puerto Rico at Work and at Play.* New York: Dodd, Mead (1962), 64 pp. Grades 4–6.

17. NASH, VERONICA. *Carlito's World: A Block in Spanish Harlem.* New York: McGraw-Hill (1969), 30 pp. Grades 2–4. Lively colored illustrations.

18. ROLLINS, FRANCES. *Getting to Know Puerto Rico.* New York: Coward-McCann (1967), 64 pp. Grades 5–7.

19. SCHLOAT, G. WARREN, JR. *Maria and Ramon: A Girl and Boy of Puerto Rico.* New York: Knopf (1966), 48 pp. Grades 4–6. Profusely illustrated.

20. SONNEBORN, RUTH A. *Friday Night Is Papa Night.* New York: Viking (1970), 32 pp. Grades 2–4.

21. SPEEVACH, YETTA. *The Spider Plant.* New York: Atheneum (1966), 154 pp. Grades 3–5. Twelve-year-old Carmen comes from Puerto Rico and adjusts to life in New York City.

22. TALBOT, TOBY. *My House Is Your House.* New York: Cowles (1971), 47 pp. Grades 1–3.

31. Acting out the roles of workers

Individuals and Families in the United States

32. Getting acquainted with the father of a classmate

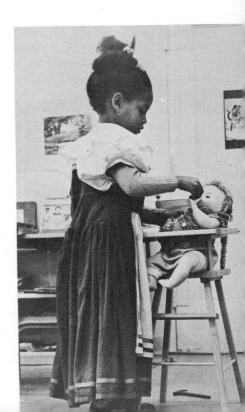

33. Trying out the role of mother

Studying Individuals
and Families

17

INDIVIDUALS AND FAMILIES IN THE LOCAL COMMUNITY
AND IN OTHER PARTS OF THE UNITED STATES

While children are learning about themselves, they should be learning about other individuals and about families. These topics are so closely interwoven that you may not want to separate them in your mind and in your teaching as they are separated in this book. In the search for self-identity, boys and girls see themselves in large part, initially, as they are seen by members of their own families. The family is like a mirror in which they see themselves.

Of course learning about families and family living should not be restricted to the early years in school. Boys and girls are concerned about their relationships with other members of this unit throughout their years in school. Therefore social studies teachers need to keep this unit of society in mind at every grade level. This is especially true of the early adolescent years when pupils often need help in making adjustments to their fathers, mothers, and siblings.

It is in the primary grades, however, that the focus ought to be on families and family living. The family is the most significant segment of society for all of us, but especially for young children. This is the unit in which they spend most of their time. This is the group in which they gain most of their satisfactions, as well as the group which often baffles them most. It is the chief unit in which they have direct experiences. It is also the sector of society which provides them with their basic needs.

Unconsciously or consciously, every child has been studying his or her own family ever since he or she was born. Under competent guidance, children in the primary grades can now begin to understand the main aspects of family life. Children can act out situations involving families, talk about episodes they have experienced, and hear stories about families. When the beginning skills in reading have been acquired, they can also read about family life. And through all kinds of pictorial materials, they can enlarge their understanding of their own families and the families of others.

Throughout this chapter the word "family" should be interpreted very broadly to include every unit from groups with only one parent to those where children are living with adults who are not their own parents. Teachers would do well to think in such terms when working with pupils in elementary schools.

Even in the kindergarten some structured study of families can be undertaken, using play situations, construction activities, pictures, songs and games, and other methods to stimulate the life of adults and children living and working together.

Introductory activities relating to this topic should certainly focus on the family of each child in the group. Considering the number of children in any kindergarten or first grade, this should provide the teacher and the children with a wide variety of experiences upon which to draw.

But no study of families should stop there. Eventually children should come into contact with many types of individuals and families in their own community and with a much wider variety of families in other parts of the United States. Some teachers in the first grade may even want to include some study of a few families in other parts of the world.

Boys and girls in these early years should begin to understand how families are alike and different. They should start to comprehend the many activities in which families engage. They should grasp in a simple way the various roles that people play in any family group. They should also learn how people work and earn a living and then make decisions about how their money should be spent. Such approaches will lead quite naturally into a study of transportation and the various means of travel that people use. A little attention can be given to the contacts of people with governments and governmental services. Boys and girls should also delve into the ways in which families spend their free time and have fun. Teachers should also help children to look at some of the problems of family or group living, with special attention to problems the children themselves face at home. This should be done with great sensitivity but it should certainly be done. Otherwise children will not be helped with the problems which baffle them and prevent learning on their part.

If you have thought about these aims in relation to the various social science disciplines, you will have realized that all or almost all of them will be included in the study of individuals and families. Much of the content will be drawn from anthropology and sociology. Some of it will be taken from the field of economics. A little will be derived from geography. A smaller amount deals with political science or government. The least will be drawn from history, but it does not need to be completely ignored.

After coping with the study of families which children have seen, they should then be ready to study some families with which they have had no direct experience.

Obviously children at this stage are not ready to study a large number of families in other parts of the United States. Ten or twelve families may be enough for them to explore in the first year in school. Otherwise they cannot study them in depth and become closely identified with them.

Therefore the selection of families to be studied needs to be approached with great care by teachers. The needs of a particular group of children should be kept uppermost in the minds of teachers in making their selection. In a segregated suburb, the children should learn about middle- and upper-class Negro families, to offset the stereotypes they have already begun to develop. City children might well study rural children, as well as one or more city families.

Before reading further in this chapter, you may want to stop and think about the variety of families in the United States that you would like to teach about with a group of children. How would you select the families to be studied? What general categories would they represent? How many would you study?

After you have made such a list, you may then want to compare it with the 10 suggestions made by the author of this book on this page. Several of the categories in that list overlap, so they need not be thought of as 10 separate families. One family can represent several of the categories included in this list. You may, however, want to think of at least these 10 general topics:

1. Families of different sizes, including some with only one parent, the usual nuclear or small, tightly knit family, and extended families, with close relationships with cousins, grandparents, and other relatives.
2. Families living in different parts of the United States, with the emphasis upon the effect of geography upon family life.
3. Families in different locations: rural, small town, city, and suburban.
4. Families of different socio-economic levels.
5. Families with different religious faiths or value systems.
6. Families representing different ethnic or racial backgrounds.
7. Families in trailer camps, migrant workers, and other families on the move because of jobs.
8. Families representing different occupations.
9. Families which represent different problems of group living.
10. Families of newcomers to this country.

There are dangers involved in such studies if they are not handled carefully by teachers. This is true in the study of families as it is in other aspects of the social studies. One danger is that of stereotyping. Another is that of superficial treatment. A third is that the class will spend so long with one family that they become bored with that cluster of persons. A fourth is that the teacher may have some kind of prejudice against the family under consideration. Possibly you can add other pitfalls.

Despite these dangers, however, the study of families seems to be a "natural" for children in the primary grades. For many years the family has been the major focus of social studies learnings at this level. The unique part of the suggestions here is that children will spend less time on their own families and on the families of their classmates and that they will begin to learn about the variety of families in the United States, including groups which are very different from their own families.

The overarching aim of all such studies should be improved behavior on the part of children in their own families. By comparison and contrast, they should learn more about the tiny unit in which they live and be better able to live in it. Coupled with this aim should be a desire on the part of teachers to foster better attitudes toward human beings of many kinds. The end result of exposure to the experiences of several families in the United States would be better behavior, better attitudes, better skills, and much more knowledge.

Another gain inherent in the program suggested here for the kindergarten and primary grades is the extension of the experiences of boys and girls beyond their own community. Still another asset is that learnings can be drawn from all or almost all of the social science disciplines.

Now let us turn to a consideration of more specific ways in which boys and girls can study their own families, the families of their classmates, and families in different parts of the United States.

As you read and think about the material in the next few pages, it might be helpful if you thought in terms of a unit you will teach on "Individuals and Families in the United States." Or you might limit yourself to teaching about one family. Then you can apply what you read to such a specific teaching situation.

ASPECTS OF A FAMILY WHICH CHILDREN SHOULD STUDY

The chart reproduced below indicates the many aspects of families which boys and girls should explore. It can apply to families anywhere in the world. The name of a family or its picture can be placed in the center of the wheel or clock and all aspects will apply to it whether it is the family of a pupil in your class, a family somewhere in the United States, or a family in some other part of the world.

This chart is intended primarily for teachers, although pupils can use it in a simplified form. But it most certainly should not be memorized by them. Nor should the order of topics be the same for every family studied. The spotlight can be pointed to different aspects of this chart with different families.

As a teacher you may want to add the word "Why" to each of these topics — or at least to most of them.

Here, then, is the chart for studying families:

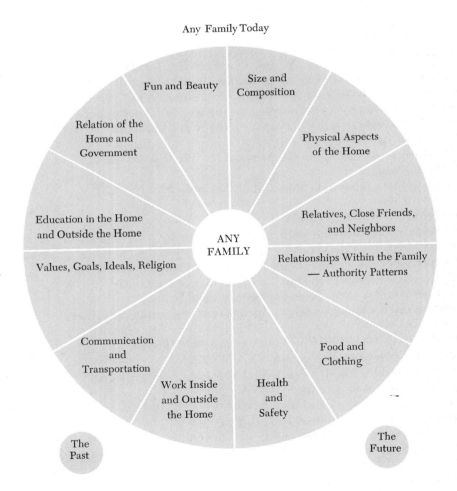

From Leonard S. Kenworthy, "Extending the Study of Families," *Childhood Education*, March 1968, p. 431. By permission of the author and the Association for Childhood Education International.

SOME POINTS TO EMPHASIZE IN STUDYING FAMILIES

Many in-service teachers will have enough background to fill in sub-points under each of the twelve categories on the foregoing chart. But some beginning teachers, and possibly a few experienced ones, may find it helpful to have a few ideas spelled out under each of the twelve headings. Here are some of the points you may want to consider in dealing with the various aspects of any family:

1. *Size and composition of the family*

   Size. Is this a nuclear family with a small, closely knit group, or an extended family (with a close relationship with relatives)? How many members are there in the family?

   Composition. Father? Mother? Stepfather or stepmother? Foster parents? Siblings? Parent substitutes? Adopted children? Other relatives living with the family?

2. *The physical aspects of the home*

   Location of the home and why it was built where it is.
   Types of materials used in the house and why.
   Layout of the house or apartment and reasons for the location of the various rooms. Number of rooms in relation to the size of the family. Why?
   Other families sharing the apartment or house. Why?
   Kitchen arrangements. Sleeping arrangements. Types of beds and why.
   Recreational facilities: play space indoors and outdoors.
   Study space.
   Lighting and heating of the house.
   Provisions for privacy.
   Location in relation to stores, churches, etc.

3. *Relatives, close friends, neighbors, and helpers*

   Other people who live with the family and why.
   Relationship of these people to the family.
   Relations with relatives, especially in extended families.
   The close friends of the family. What shared activities?
   Relationships with close neighbors or neighbors.
   Who are the home helpers — members of the family and others?

4. *Relationships within the family: authority patterns*

   What is the general "climate" of family life?
   Who are the authority figures and how is authority divided?
   Relationships of parents.
   Relationships of parents and children.
   Relationships of children and other adults.
   Relationships among the children of the family.
   To what extent do the children share in decisions? On what topics?

5. *Food and clothing*

   Food. What is eaten? Taboos?
       Climate, culture, and religion as determinants.
       Income as a determinant.
       Type of heat for cooking. Utensils used — and why.

Who prepares the food?

Does the family eat together — why or why not?

The part children play in preparing the food.

Clothing. What clothing do different members wear and why?

Effect of climate on clothing.

Effect of income on clothing.

Effect of religion and/or culture on the clothing.

Where is the clothing obtained?

What part do children play in selecting clothing?

6. *Health and safety*

General health conditions of the family. What medical help?

Any cultural determinants of health.

Taboos, superstitions.

What hazards to health and safety?

Precautions taken by the family to safeguard health and safety.

Government's role in promoting health and safety of the family.

7. *Work inside and outside the home*

Inside the home. What responsibilities does each member have? Why?

What part do the children take in jobs around the house-apartment?

What home helpers are there? Why?

Outside the home. Who works outside the home? In what kinds of jobs?

Why are the jobs located where they are?

How do the goods or services which the parents produce fit into the economy?

Dress in relation to jobs.

Transportation to and from jobs.

Training for jobs.

When paid and how?

8. *Transportation and communication*

How adults travel to and from work. Why?

How does the family travel?

Costs of transportation.

How goods and services are "delivered" to the home.

Forms of communication in the family:

Language or languages spoken.

Use of newspapers, magazines, radio, television.

Non-verbal ways of communicating in the family.

9. *Values, goals, ideals, religion*

Some of the values the family prizes most highly. Why?

Members of any formal religious group?

Worship in the home.

Conflicts of values in the home or with others in the community? Why?

10. *Education in the home and outside the home*

How the children learn the culture of the home. Informal education.

Formal education in the schools. Where? Relation to the home.

Education of the parents.

Attitude of parents to schools.

What the children study. Why?

Other places of education, such as religious schools.

Other educational activities of the family.

11. *Relation of the home and government*

Taxes the family pays to help support the government. Direct and indirect.

Services rendered by the government to the family, such as water, health, schools, post offices, recreation, highways and streets, etc.

Part the adults take in government and voting.

Attitudes of adults toward government.

Contacts of the children with government, such as teachers, policemen and safety patrols, welfare officers, health officials, etc.

12. *Fun and beauty in the home and outside the home*

Ways in which the family beautifies the home inside and outside.

Part the children play in making the home more beautiful.

Special features of the home and family life of which the family is most proud.

Special days and family fun.

   Birthdays

   Holidays

   Family trips

   Visitors

Other forms of family fun.

   Especially the children and their fun at home.

   Forms of fun in the community.

Costs of fun.

The material in the last few pages is written for teachers. From it they should be able to pick appropriate material for children and use it in a form conducive to learning. It would be overwhelming to use all of these factors in family life in the study of any one family, but pupils should be able to study something from each of these categories in their study of their own families, the families of other people in the community, and selected families in different parts of the United States.

SOME METHODS TO USE IN STUDYING FAMILIES

A wide variety of methods needs to be used by teachers in helping boys and girls learn about families and family living. The methods should vary from class to class, depending upon the ideas which the teacher decides to stress and the group of pupils. Many methods will be needed to achieve the objectives worked out by teachers. Because young children learn best through direct, firsthand experience, the stress should be upon these experiences. Here, then, are some possible activities or methods for you to consider:

1. Talking about the various aspects of family life which children have already experienced or are experiencing.
2. Reading aloud to children about various families.
3. Drawing and painting pictures of aspects of family life.
4. Role-playing and simple dramatizations of family situations and problems.
5. Constructing a play house and its equipment.

6. Weaving rugs and other materials for the play house.
7. Viewing films and filmstrips on family life.
8. Making clothes for dolls.
9. Using puppets to act out situations.
10. Observing and commenting upon pictures of families and family life.
11. Singing songs about family life.
12. Acting out safety situations.
13. Taking trips to see different types of houses and apartments.
14. Discussing the trips and making experience charts with the class.
15. Reading simple materials about houses and families.
16. Cooking simple foods.
17. Making a class scrapbook of materials on families and houses.
18. Inviting parents and other adults to talk about houses and home life.
19. Having parents or other adults take pictures for the class.
20. Learning games to play at home.
21. Working on maps of houses and their relationship to each other and to other aspects of the community, including the school.
22. Exchanging materials on families with other classes in the school or in schools in other parts of the United States.
23. Presenting a program to a kindergarten, another class, or to parents.
24. Learning how to use the telephone as a means of communication.
25. Preparing materials to beautify the homes of children in the class.
26. Using tape recordings and records about family life.
27. Making clay models of people, utensils, and furniture.
28. Having better readers read aloud to others about families.
29. Discussing books read by individual children.
30.
31.
32.
33.

SOME RESOURCES FOR STUDYING FAMILIES OF THE UNITED STATES

To do a superior job in helping children learn about a wide variety of families locally and in the United States, teachers need to be as conversant as possible with a wide range of resources on this topic.

First comes the classroom. This is an important resource for learning and one which is sometimes overlooked. It should resemble a good home as closely as possible. It should have flowers and plants. It should have curtains if the architecture permits. It may have rugs and even wall-to-wall carpeting — a feature of modern schools. The furniture should be sturdy, built for children, and colorful. There should be attractive displays and exhibits. There might even be a rocking chair for the teacher when she reads to groups — and pillows to sit on. This should be a place which children enjoy.

And it should be the children's room as much as possible. They should help plan for it and help to care for it. They should think of it as "our home" or "our room."

There should be at least one play corner, and preferably more. In it children will need small tables and chairs, a stove, a telephone, a clock, brooms and dustpans, and other equipment.

Elsewhere there should be easels, soft wood and nails and hammers, paints and brushes, sandpaper, clay, drawing paper, crayons, scissors, and other materials.

And closets and files? Yes, indeed. And books and pictures, too. Some of these can be "on loan" from the school or local library.

In the pages that follow, many more materials are listed, with the emphasis upon books, because of the space limitations in this volume.

Some Textbooks Dealing with Individuals and Families in the United States

1. ANDERSON, EDNA A. *Families and Their Needs.* Morristown, New Jersey: Silver Burdett (1966), 128 pp.
2. BRANDWEIN, PAUL and others. *The Social Sciences: Concepts and Values.* New York: Harcourt, Brace and World (1970), 176 pp. Largely individuals.
3. BUCKLEY, PETER and HORTENSE JONES. *William, Andy, and Ramon.* New York: Holt, Rinehart and Winston (1966), 70 pp. Three boys of different backgrounds — Spanish-speaking American, Negro, and Caucasian. See also the companion volume on *Five Friends At School.*
4. HANNA, PAUL and others. *Family Studies.* Chicago: Scott, Foresman (1970), 160 pp.
5. KING, FREDERICK and others. *Families and Social Needs Concepts in Social Science.* River Forest, Illinois: Laidlaw (1968), 144 pp.
6. SHACKELTON, PEGGY. *Families Are Important.* Lexington, Massachusetts: Ginn (1972), 192 pp. Nine real rather than contrived families, from different ethnic backgrounds in different parts of the United States. An interdisciplinary approach to each family. Also a Response Book for children.
7. WANN, KENNETH, EMMAS SHEEHY, and BERNARD SPODEK. *Learning About Our Families.* Rockleigh, New Jersey: Allyn and Bacon (1962), 144 pp.

Teachers are reminded of the fact that there are almost always Teachers' Editions for textbooks. In them are many useful suggestions for classroom strategies in teaching about individuals and families. Often there are lists of additional resources which teachers can use profitably with pupils.

Picture Portfolios

In the last few years several publishers have assembled portfolios or sets of large pictures for use with kindergarten and primary grade boys and girls. These are usually pictures which depict situations that can be used with children for problem-solving and for eliciting pupil comments and fostering discussion. These are printed on heavy cardboard and are large enough to use with fairly large groups of children. Sometimes they are in full color, sometimes in two colors. Most of these sets of pictures have a pamphlet for teachers with suggestions for their use. Occasionally concepts and generalizations, or other material, are printed on the back of the pictures.

Among the companies which have produced such pictures are the following:

1. THE JOHN DAY COMPANY. Several picture portfolios in spiral-bound books, including "A Family Is. . . . ." (one portfolio depicting a city family and another on a rural family).
2. GINN AND COMPANY. 48 pictures on 24 cardboard panels, one on each side, depicting problem situations. In two colors. Accompanying teacher's guide. The series is entitled "You and Me."
3. HARCOURT BRACE JOVANOVICH. "Our Changing Cities: An Early Childhood Social Studies Program." Four levels, on "All Around Me," "Living Together," "The Changing City," and "Our City and Others."
4. HOLT, RINEHART AND WINSTON. "Words and Actions: Role-playing Photo-Problems for Young Children." By Fannie and George Shaftel.

5. RAND MCNALLY. "Interaction of Man and Man." Material on two levels: "With Family and Friends" and "In the Community."

Some Films and Filmstrips on Families

"Appreciating Our Parents." Coronet, black and white, 11 minutes.

"Date With Your Family." Encyclopaedia Britannica Films, black and white, 10 minutes.

"Father Goes Away to Work." Bailey Films, color, 11 minutes. About three fathers — a construction worker, a commercial artist, and a salesman.

"Our Family Works Together." Coronet, color, 11 minutes. Ways in which children help around the house.

"Your Family." Coronet, black and white or color, 10 minutes.

An exceptionally fine set of five sound filmstrips on "Five Children" and a similar set of five sound filmstrips on "Five Families" has been issued recently by *Scholastic*. Write them for details at 50 West 44th Street, New York, N.Y. 10036.

Some Books on Families in the United States

Some teachers may be interested in the pamphlet on *Recommended Reading about Children and Family Life*, issued in 1969 by the Child Study Association of America.

1. APPEL, CLARA and MOREY. *We Are Six: The Story of a Family*. New York: Golden (1959), 49 pp. Grades 1–3. A family of five prepares for the arrival of a sixth member.
2. BAUER, HELEN. *Good Times At Home*. Chicago: Melmont (1954), 23 pp. Grades 1–3. A suburban family with two boys.
3. BAUM, BETTY. *A New Home for Theresa*. New York: Knopf (1968), 182 pp. A shy girl is placed in a foster home. The girl is black.
4. BORCHARD, RUTH. *The Children of the Old House*. Garden City, New York: Doubleday (1963), 181 pp. Grades 4–6. A family moves from a small house in town to a run-down house near a river and repairs the new home.
5. CASEY, ROSEMARY A. *The Cousinly Cousins*. New York: Dodd (1961), 175 pp. Grades 4–5. Three city children and their three country cousins spend a summer together on a farm in upper New York state.
6. CLARK, ANN NOLAN. *In My Mother's House*. New York: Viking (1960), 56 pp. Grades 2–4. The story of Tewa Indians near Santa Fe, New Mexico.
7. COLMAN, HILA. *Peter's Brownstone House*. New York: Morrow (1963), 36 pp. Grades 1–3. A comparison of life today and life at the turn of the century.
8. EVANS, EVA KNOX. *Home Is a Very Special Place*. New York: Capitol and Golden (1961), 91 pp. Grades 4–6. The joys and problems of family living.
9. GEORGE, JEAN C. *Coyote in Manhattan*. New York: Crowell (1968), 203 pp. Various racial groups are included in the story of this New York City family.
10. KRAUSS, RUTH. *The Big World and the Little House*. New York: Harper (1956), 42 pp. K–2. A family moves from one house to a little house which has been deserted. Then they turn it into a home.
11. LENSKI, LOIS. *Shoo-Fly Girl*. Philadelphia: Lippincott (1963), 176 pp. Grades 4–6. Life in a family of Amish people in Pennsylvania today.
12. LENSKI, LOIS. *We Live in the City*. Philadelphia: Lippincott (1954), 128 pp. Grades 4–6. The lives of several children, including everyone from a shoeshine boy to a girl who lives in a penthouse apartment.
13. LEXAN, JOAN M. *Striped Ice Cream*. Philadelphia: Lippincott (1969), 96 pp. Poverty and loneliness are the two main themes of this family story.

14. ROGERS, ELIZABETH. *Angela of Angel Court.* New York: Crowell (1954), 116 pp. Grades 4–6. The story of Angela Rossi who lives with her mother and younger sister in a battered apartment where everyone is poor in money but rich in affection and love for music.
15. SCHLEIN, MIRIAM. *Who.* New York: Walck (1963), 32 pp. Grades 1–3. The story of a mother and daughter and their relationship.
16. SELZ, IRMA. *Wonderful Nice.* New York: Lothrop (1960), 34 pp. Grades 1–3. An Amish family in Pennsylvania. Large, colorful drawings.
17. SHARPE, STELLA G. *Tobe.* Chapel Hill, North Carolina: University of North Carolina Press (1960), 121 pp. Grades 2–4. A Negro family on a farm in North Carolina. Black and white photographs.
18. WAKIN, EDWARD. *At the Edge of Harlem.* New York: Morrow (1965), 127 pp. Grades 6–8. A difficult text but especially useful to teachers. A family of eight in Harlem. Pictures can be shared with the children as the story is told.

Books on fathers and grandfathers

Two themes predominate in these books— the work that fathers do and the days when daddies stay home from work. Unfortunately almost all of the books are about middle- and upper-class families. Books on other classes are needed.

1. APPEL, CLARA and MOREY. *Now I Have a Daddy Haircut.* New York: Dodd (1960), 48 pp. K–2.
2. BANNON, LAURA. *Toby's Friends.* Chicago: Whitman (1963), unpaged. K–2. Toby and his grandfather play a guessing game.
3. BISHOP, CURTIS. *Little League Stepson.* Philadelphia: Lippincott (1965), 154 pp. Grades 3–5. A mother remarries and her husband is the manager of Little League clubs. The stepson wants to become a good baseball player and not make the team merely because of his stepfather.
4. BORACK, BARBARA. *Grandpa.* New York: Harper (1967), 32 pp. K–3.
5. BUCKLEY, HELEN E. *Grandfather and I.* New York: Lothrop (1959), 32 pp. K–2. Mothers, fathers, and other people are always in a hurry. Trains hurry. But grandfather and grandson take a leisurely walk and then sit in a rocking chair and enjoy each other.
6. CARTON, LONNIE C. *Daddies.* New York: Random (1963), 42 pp. Grades 1–3. Poetry which describes a whole gamut of functions of fathers.
7. GOFF, BETH. *Where Is Daddy?* Boston: Beacon Press (1970), 28 pp. Written for children without fathers, especially for very withdrawn preschoolers. Treats the bewilderment when daddy leaves and the readjustment necessary when the family goes to live with grandmother.
8. KESSLER, ETHEL and LEONARD. *The Day Daddy Stayed Home.* Garden City, New York: Doubleday (1959), 34 pp. K–2. Daddy is snowbound and the children enjoy him on this special day off.
9. LOW, ALICE. *Grandmas and Grandpas.* New York: Random (1962), unpaged. K–2. The important place of grandparents, told in verse, with large pictures to illustrate this basic theme.
10. MERRILL, JEAN. *Tell Me about the Cowbarn, Daddy.* New York: Scott (1963), 48 pp. Grades 1–3. A young boy retells, with his father's help, a familiar story of life on a dairy farm. Colorful illustrations.
11. PUNER, HELEN WALKER. *Daddies: What They Do All Day.* New York: Lothrop (1946), 34 pp. K–3. The wide variety of jobs daddies carry on, from window-cleaning to farming, from mining to working in offices. Told in rhymes.
12. RADLAUER, RUTH S. *Fathers At Work.* Chicago: Melmont (1958), 31 pp. Grades 1–3. The fathers are a steam shovel operator, a truck driver, a jet pilot, a newspaper reporter, a telephone linesman, a gardener, and a carpenter.

13. SIMON, NORMA. *The Daddy Days*. New York: Abelard-Schuman (1958), 42 pp. K–3. On "Daddy Days" Peter and Amy rake leaves, visit the library, a garage, and a barber shop, and enjoy a picnic — with Daddy.
14. WARNER, EDYTHE. *The Fishing River*. New York: Viking (1962), 64 pp. Grades 3–5. Two boys fish trout with grandfather and learn about conservation.
15. WILSON, CHRISTOPHER. *Growing Up With Daddy*. New York: Lothrop (1957), unpaged. Grades 1–3.
16. YATES, ELIZABETH. *A Place for Peter*. New York: Coward-McCann (1952), 184 pp. Grades 4–7. A farm boy in New England wins his father's confidence.

Books on mothers and on grandmothers

1. ADELBERG, DORIS. *Grandma's Holidays*. New York: Dial (1963), 32 pp. Grades 2–4. The story of a little girl and her grandmother's place on each holiday in the year, including a special "Grandmother's Day."
2. BROWNSTONE, CECILY. *All Kinds of Mothers*. New York: McKay (1969), 32 pp. Grades 1–3.
3. BUCKLEY, HELEN E. *Grandmother and I*. New York: Lothrop (1961), 27 pp. N–2. Other laps are good, but sitting in grandmother's lap in a rocking chair is especially comforting.
4. CARTON, LONNIE C. *Mommies*. New York: Random House (1960), 40 pp. K–2. Mothers are busy all day, doing a variety of things, but they are busiest "loving you." Told in rhymes. Colorful illustrations.
5. COOK, BERNADINE. *Looking for Susie*. New York: Scott (1959), 48 pp. Grades 3–4. An understanding mother finds her children in the barn with the kittens and delays supper to play with them.
6. CUSHMAN, JEAN. *We Help Mommy*. New York: Golden (1959), 24 pp. N–2. A day in the life of a boy and girl as they help their mother.
7. HARRISON, CRANE B. *The Odd One*. Boston: Little, Brown (1959), 269 pp. Grades 4–6. A forthright grandmother helps her granddaughter to grow up.
8. KAY, HELEN. *One Mitten Lewis*. New York: Lothrop (1960), 31 pp. N–3. A very human story of a boy who is always losing his mittens, and what his mother does about it.
9. LENSKI, LOIS. *Debbie and Her Grandma*. New York: Walck (1968), 48 pp. K–2.
10. LOW, ALICE. *Grandmas and Grandpas*. New York: Random House (1962), 24 pp. Grades 1–3.
11. MARINO, DOROTHY. *Where Are the Mothers?* Philadelphia: Lippincott (1950), unpaged. K–2. An easy-to-read book which correlates what children do in kindergarten with what their mothers are doing at home.
12. MERRIAM, EVE. *Mommies at Work*. New York: Knopf (1961), 39 pp. K–2. The jobs others have away from home.
13. NEVILLE, EMILY. *It's Like This, Cat*. New York: Harper (1963), 180 pp. Grades 5–7. A boy and his mother and their arrival in New York City.
14. PHELAN, MARY K. *Mother's Day*. New York: Crowell (1965), 40 pp. Grades 2–4. On the origins of Mother's Day and its purposes.
15. SCHLEIN, MIRIAM. *Who*. New York: Walck (1963), 32 pp. N–2. A mother and a make-believe game she plays with her daughter throughout the day.
16. SIBLEY, CELESTINE. *Mothers Are Always Special*. New York: Doubleday (1970), 62 pp. Grades 1–3.
17. SONNEBORN, RUTH A. *I Love Gram*. New York: Viking (1971), 32 pp. Grades 2–4. Elle, a little girl in a city apartment and her grandmother. A story of black Americans.
18. UDRY, JANICE M. *Mary Jo's Grandmother*. Chicago: Whitman (1970), 32 pp. Grades 2–4. Mary Jo's grandmother becomes ill and Mary Jo summons help. A story of black Americans.

A series of six filmstrips, each in color, on "Mothers Work," has been produced by Churchill Films. Those on mothers complement the six on fathers. The mothers include a dental assistant, a waitress, a bank clerk, and a home-maker.

Materials on Appalachian families

1. BUDD, LILLIAN. *Larry.* New York: McKay (1966), unpaged. K–2. A small boy in the Kentucky mountains runs away from home but is reunited with his family.
2. CAUDILL, REBECCA. *Did You Carry the Flag Today, Charley?* New York: Holt, Rinehart and Winston (1966), 94 pp. K–3. The first day of a mountain boy's experience in the Kentucky equivalent of Head Start.
3. CAUDILL, REBECCA. *A Pocketful of Cricket.* New York: Holt, Rinehart and Winston (1964), 48 pp. Grades 1–3. A small boy in the hills of Appalachia takes a cricket to school for show and tell, in the class of an understanding teacher. Good two-color illustrations.
4. CHASE, RICHARD. *Billy Boy.* San Carlos, California: Golden Gate (1966), unpaged. Grade K–3. An Appalachian version of the old song. Includes the music of all 17 verses.
5. JUSTUS, MAY. *Children of the Great Smoky Mountains.* New York: Dutton (1961), 158 pp. Grades 4–6. Fifteen stories of life in the Smoky Mountains. Several folk songs included.
6. LENSKI, LOIS. *Coal Camp Girl.* Philadelphia: Lippincott (1959), 192 pp. Grades 4–6. Hardships and good times in a coal-mining community in West Virginia.
7. RAYMOND, CHARLES. *Up from Appalachia.* Chicago: Follett (1966), 190 pp. Grades 6–8. The Cantrell family moves to Chicago and makes the big adjustment to life there. Grandpa and Lathe, the grandson, are the chief characters of the story.
8. SECKAR, ALVENA. *Zuska of the Burning Hills.* New York: Walck (1952), 224 pp. Grades 4–6. Life is not easy for Zuska and the family in a mining town in West Virginia, but there are good times for them, too.
9. SHULL, PEG. *Children of Appalachia.* New York: Messner (1969), 96 pp. Grades 3–4. The story of three different types of families in Bell County, Kentucky and their attachment to the mountain region. Some black and white photographs of people and the region.
10. YOUNG, LOIS H. *No Biscuits At All!* New York: Friendship Press (1966), 122 pp. Grades 2–4. Morgan, a second-grader, and the other six members of his family move to Baltimore from Appalachia and finally find friends there.

A filmstrip in black and white, or in color, on "Life in a Coal-Mining Town" is sold by Coronet. It portrays the life of a family in David, Kentucky and some of the processes of coal mining.

Teachers should find an article in *The National Geographic* for November, 1971 on "The People of the Cumberland Gap" helpful as background.

Harriette Arnow's *The Dollmaker* (Collier Books, 1954, 571 pp.) is an excellent novel of a family which moves from Appalachia to Detroit.

Teachers will need to read aloud or tell in a condensed form the story of some of the books listed above.

Materials on migrant families

Teachers will find good background on migrant farm workers in Willard A. Heaps' *Wandering Workers* (New York, Crown, 1968, 172 pp.), a volume written for young people and in Steve Allen's *The Ground Is Our Table* (New York, Doubleday, 1966, 141 pp.), on Mexican Americans in Texas. You may want to purchase "The Migrant Map of the United States" from the Friendship Press, with pictures to paste on the map accompanying it. See the reference to the novel on *The Dollmaker* under Appalachia.

1. BENEDICT, STEVE. *The Little House on Wheels.* New York: Tell-Well Press (1953), unpaged. Grades 4–6. Itinerant fruit-pickers.
2. BIESTERVELT, BETTY. *Run, Freddy, Run.* New York: Nelson (1962), 125 pp. Grades 4–6. A girl finds solace in her pets in her new home.
3. FRANCHERE, RUTY. *Cesar Chavez.* New York: Crowell (1970), 42 pp. Grades 4–6. A biography of the leader of the migrant workers in California.
4. GARTHWAITE, MARION. *Mario: A Mexican Boy's Adventures.* New York: Doubleday (1960), 167 pp. Grades 4–6. Smuggling an 11-year-old boy into the Imperial Valley in California.
5. GATES, DORIS. *Blue Willow.* New York: Viking (1940), 172 pp. Grades 5–8. An old favorite of a 10-year-old girl and her family of migrant workers.
6. HOLLAND, RUTY. *Forgotten Minority: America's Tenant Farmers and Migrant Workers.* New York: Crowell-Collier (1970), 153 pp. Grades 5–8.
7. LENSKI, LOIS. *Judy's Journey.* Philadelphia: Lippincott (1947), 212 pp. Grades 5–7. Judy tries to obtain an education despite the travels of her family.
8. MEANS, FLORENCE. *Knock at the Door, Emmy.* Boston: Houghton Mifflin (1956), 240 pp. Grades 6–9. A young migrant worker goes into social work.
9. MUSGRAVE, FLORENCE. *Robert E.* New York: Hastings (1957), 191 pp. Grades 4–6. A boy and his grandfather move from a cabin in the country to the city, and the boy gets into trouble.
10. O'DELL, SCOTT. *Journey to Jericho.* Boston: Houghton Mifflin (1969), 40 pp. Grades 1–3. The story of migrant workers in the West Virginia coalfields and in California lumbercamps.
11. ROBINSON, BENELLE H. *Citizen Pablo.* New York: John Day (1959), 128 pp. Grades 5–8. Migrant workers from Texas relocate in California.
12. SHOTWELL, LOUISA R. *Roosevelt Grady.* Cleveland: World (1963), 151 pp. Grades 4–6. Also available as a paperback. A splendid story of a black boy, his family of migrant workers, and his desire to obtain an education.
13. WALTRIP, LELA. *White Harvest.* New York: Longmans (1960), 118 pp. Grades 5–8. Susan's family follows the cotton harvest in Texas.
14. WEINER, SANDRA. *Small Hands, Big Hands: Seven Stories of Mexican-American Migrant Workers and Their Families.* New York: Pantheon (1970), 55 pp. Grades 5–8.
15. WHITNEY, PHYLLIS A. *A Long Time Coming.* New York: McKay (1954), 261 pp. Grades 5–8. 18-year-old Christie and her part in a conflict between migrant workers and the local townspeople.
16. WIER, ESTER. *The Loner.* New York: McKay (1963), 153 pp. Grades 5–8. A nameless, starving migrant child is rescued from being "a loner."

For material on Asian-Americans, Afro-Americans, Mexican-Americans, and Puerto Rican-Americans, see the bibliographies on pages 251–254.

35. Using resource persons from abroad

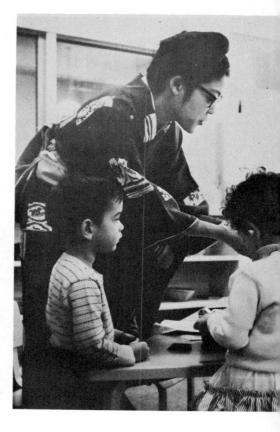

34. Encouraging pupils to report
on books they have read

Individuals and Families
in Other Parts of the World

36. Utilizing pictures of people in other parts of the world

STUDYING INDIVIDUALS AND FAMILIES IN OTHER PARTS OF THE WORLD

Having concentrated upon families in their own community and in selected parts of the United States in kindergarten and in the first grade, children should then be ready to move on to a study of a few selected individuals and families in other parts of the world at least by grade two.

Changes Which Make Such Studies Possible

This suggestion represents a definite departure in social studies curricula, but it seems warranted by changes which have already been referred to in this book. In the past, boys and girls have not been introduced to other parts of the world before the fourth or fifth grades. But that was in another period of history. Their experiences in those days were limited largely to the home and community. Today the experiences of many children are broader.

Many boys and girls today have had some experiences with people from other parts of the world, before they enter school. Some children were born abroad. Others have lived or traveled in other parts of the world. This will be true increasingly as the world grows smaller, because of transportation changes.

Many more have relatives, friends, or neighbors who have been abroad or who are now abroad. They may have older brothers or fathers who have been or are now in the armed services. They may have neighbors who have worked abroad and who talk about their experiences in other places.

Even more children have met people from other parts of the world, in their homes or in their schools. This is not an uncommon experience for second graders today.

Furthermore, almost all children have been exposed to people from other parts of our globe through the medium of television.

Therefore, a large percentage of children have had some contact with people who are "different." Children are naturally curious about such persons.

Why Concentrate upon Individuals and Families?

Readers may ask why a social studies program should concentrate upon individuals and families in other parts of the world, rather than upon larger units of society.

The answer is that it is easier for children to comprehend the life of a single individual or a family than it is to understand entire communities or countries. The concepts of community and country are much more difficult than the concept of family. Therefore it seems advisable to cope with a unit which is small enough to comprehend and yet different enough to provide some contrasts as well as some comparisons.

The study of a few selected families in other parts of the world can be the first step or stage in a spiral curriculum which later introduces them to communities, countries, and eventually cultures. One can think of this early program as a "readiness" program in international understanding.

Having studied their own families and selected families in a few parts of the United States, children in the second grade should have some "tools" or skills with which to study selected families in other parts of the world.

One of the assets of such studies is that children can get away from the local scene and learn by comparison and contrast. Thus they can learn more about their own families and families in other parts of the U.S.A. by seeing how a few families in other parts of the world live. They can begin to gain a little perspective as they widen their horizons.

Does the Study of Families of the World Have to Be Deferred until the Second Grade?

Teachers and others sometimes raise the question as to whether the study of families in other parts of the world has to be delayed to the second grade. This writer's reply would be "No." There are instances where children may be ready to examine families in other parts of the world when they are still in kindergarten or first grade. This is especially true of children with experience abroad or of children in schools where there are children from other countries. This would be true, too, of boys and girls in schools near the borders of Canada and Mexico.

An example of such a situation occurred not long ago in the demonstration nursery school at Brooklyn College. A Chinese-American girl in the class spoke English with a decided accent and did not always understand what her classmates said. One day two boys tried to persuade her to bring a wagon to them, but she did not seem to grasp what they wanted. Baffled, one of them said to the other, "I think Mei Lei speaks Spanish." To which the other replied, "No, she talks Irish." The teacher overheard the conversation and decided to take that opportunity to introduce the group to Chinese children by reading the story of *Tommy and Dee-Dee* and talking with them about children in that part of the globe.

Stories like this could be multiplied many times in a place like Miami, Florida, when the Cuban children began to arrive there. Many teachers have used such situations in our large city schools to introduce pupils to the ways of living in other parts of the world as boys and girls have entered classes from Israel, Hong Kong, and other places.

Some schools are already doing some teaching about the children of other lands in the kindergarten or first grade, but most schools will undoubtedly postpone such studies to the second grade.

The Selection of Families of the World to Be Studied

Assuming that your pupils are ready for some examination of families in places outside the United States, how should you select the families to be studied?

There are several criteria you may want to use. One would be children and families with whom the children have had some contact. This might mean the study of families in a country where some children have lived or visited. It might mean the study of families in a place where a relative or friend is now living. It might be the study of a family in a country from which the class has had a visitor, or families in a country where your school has an affiliation program.

If you, as a teacher, have had some experiences abroad, it would probably be wise to utilize them and study families whom you know well.

If your school is near the border of another country, you might want to study a family or families in the neighboring nation. Then your study of families could be enriched by a visit to such a family — or families.

You probably will also want to consider the background of the pupils in your class. If you have children of Scandinavian background, it might be well to study families in that part of the world. The same would be true of children of African descent or Polish background or Italian heritage.

The availability of resource persons might be another criterion to bear in mind. If there are people in your community who have had experiences abroad, you might like to use them. And if there are students in nearby colleges who could be persuaded to visit your class, you might want to study families from their homelands.

It is probably best to omit the study of families in places like China and the Soviet Union at this point because children are not yet ready to wrestle with the politics and economics which impinge upon the lives of families in those nations.

It is this writer's considered judgment that it is best to study first families which are not too different from those the children already know. Once they have learned to accept small differences, they can then be exposed to families which represent greater differences in ways of living.

You may want to start with a family which leads a relatively simple life. Some schools have started with an Eskimo family. There is merit in such a choice, but teachers are warned that many Eskimos today live lives that are almost the same as those of other Americans or Canadians. There is considerable value in considering Lapp or Bedouin families, especially as an introduction to the study of families in other parts of the world.

A very practical consideration in selecting families to study is the availability of materials. There are a good many books and a few pictorial materials on families in other parts of the world, but they are not so numerous yet that teachers can select any type of family and be assured that there will be teaching materials on it.

For a second grade volume in the Ginn social science series, entitled *Families Live Everywhere*, one family was selected in which the father trucks fish from Montreal to New York City. A second family is the Grice family; the father is an oceanographer in Woods Hole who goes on trips to other parts of the world. The other families are all from other parts of the world, as follows:

1. A Lapp family in northern Scandinavia (actually Norway).
2. A family in Nuremberg, Germany, where the father owns a toy factory.
3. A family in a fishing village in Japan, using modern methods; an uncle works in a camera store in Tokyo.
4. A family in São Paulo, Brazil, which has moved from the drought area in the northeastern part of that nation.
5. A cocoa-growing family in Ghana; the family goes to market in the large city of Kumasi.

In keeping with the general plan advanced earlier in this book for the selection of families, communities, and countries to be studied, this list represents several, but not all, of the major cultural regions of the world. The pupils studying these families do not need to know this basis for selection, but teachers should be aware that they are starting a process which will culminate in the high school in the study of cultural regions. Introducing children to families from various parts of the world is a part of a curriculum design which stresses the "spiraling process."

The Silver Burdett Company used a somewhat similar approach in selecting the families for their picture portfolios. They chose the following:

1. A family in France.
2. A family in Kenya.
3. A family in Japan.
4. A family in Brazil.

In addition to the pictures of families in these four countries, there is a set on a family in the United States.

Each of these picture portfolios is composed of six large pieces of cardboard, 23 inches by 19 inches in size, with pictures on either side, making a total of 12 pictures. They are excellent for group discussion.

Shall We Stress Similarities or Differences in Families?

No matter what families you decide upon, the major idea to stress is that there are interesting families everywhere and that they have common needs to meet. Those include food, clothing, shelter, jobs, and fun and recreation. In those ways the families of the world are similar.

But one should not ignore differences. It is important that boys and girls learn about differences and be led to accept differences as a part of our world — at home and abroad. As Celia Stendler points out in her volume on *Intergroup Education in Kindergarten — Primary Grades:* "We do not eliminate differences among people by ignoring them. From an early age, we ought to help children to notice differences, learn to accept them, and make use of them. There has been too much stress on the fact that such differences are only skin-deep and that underneath that skin we are all the same. While this may be true, differences continue to exist on the surface, and we must deal with them."

As a teacher you can help boys and girls to learn that people meet their basic needs in different ways. At first children may be amused by the differences they see or read about. Let them be amused. Then examine why people do what they do. Your pupils should learn that people in a warm climate often live out-of-doors much of the time and do not need much clothing. A boy in Southeast Asia may wear only a loincloth — and he doesn't need a change of clothes to go swimming. People in parts of India place their beds out-of-doors to get the cool breezes at night. They do not need thick mattresses on their beds, so they use string woven across the wooden frames. When space is at a premium in a home, people may want a bed that is like a hammock and can be taken down during the day. If you sit on the floor instead of on chairs, you probably will want good carpets and pillows and a long flowing dress for the women and girls. People do things in different ways and they usually have good reasons for what they do. Children should discover this basic generalization.

Gaining Background as a Teacher

Inasmuch as the study of families in other parts of the world is probably new to you, you may well feel at a loss when you contemplate such teaching. If that is the case, you may want to read about families in other parts of the world. Two good references would be:

MEAD, MARGARET and KEN HEYMAN. *Family*. New York: Macmillan (1965), 208 pp.
  Beautifully illustrated volume on families around the world.
WHITING, BEATRICE B. (Editor). *Six Cultures: Studies of Child Rearing*. New York: Wiley (1963), 1017 pp.

In 1967 Ginn and Company published a book based on a series of articles on families around the world. The articles had been printed originally in the *Christian Science Monitor*. The book, entitled *Fifteen Families*, should be extremely useful to you as background.

You also may want to read a volume on anthropology. Several such volumes are listed in the bibliography on pages 21–22.

You should also find it profitable to view the film by Margaret Mead on *Four Families* (National Film Board of Canada).

It would be helpful to talk with people who were born in other parts of the world, though this will not be possible for many teachers at this time. If you can arrange such talks, it might be well to get their permission to tape parts of their conversation so that you will have such material for future reference and can share it with other teachers. Notes may have to suffice of course.

You will certainly want to go through all the materials listed in the bibliography at the end of this chapter, since teachers can always learn from materials prepared for children. If you can preview the films and filmstrips, that would have many advantages.

Then undoubtedly you will want to collect some of the music which your families enjoy and play parts of those recordings to your children.

You will want to gain all the background you can, but you will always be confronted with questions from your pupils which you cannot answer immediately. You should be able to say that you do not know the answer but will try to find it somewhere.

Some Ideas to Stress in Studying Families

Before you go further, probably you will want to turn back to the chart on page 259 on how to study a family. This can be used with families in other parts of the world as well as with families in different parts of the United States. You may want to organize folders on each family or a notebook on the basis of that circle, which includes the following items: (1) size and composition of the family, (2) physical aspects of the home, (3) relatives, close friends, and neighbors, (4) relationships within the family, (5) food and clothing, (6) health and safety, (7) work inside and outside the home, (8) communication and transportation, (9) values, (10) education, (11) relation of the home and government, and (12) fun and beauty or recreation. Each of those topics is spelled out in some detail on the pages that follow.

   1. Size and composition of the family.   Many families around the world are "extended families." They include persons other than the immediate family. Often those persons live together in a compound, village, or other area. Older persons may share in the care of children, whether their own or not.

   Many families around the world are larger than American families. Until recently, families desired many children, especially boys, to produce more hands for work and more protection or help in old age.

   Such characteristics of families are being changed by industrialization.

   In many parts of the world, families live more closely knit lives than in the United States. They are at home more and entertain relatives and close friends there.

   2. Physical aspects of homes.   Houses differ greatly around the world in materials used, in size, and in arrangements. The cheapest materials are usually those from the immediate area — mud, stone, or thatch, for example. As people grow wealthier and have good transportation, materials can be brought from a distance.

   The layout of a house often indicates several things about the life of the family. Is the family indoors or outdoors most of the time? Is there one main house or a main house and several smaller houses (sometimes for older children)? What rooms are there? What provisions for climate? What provisions for pets? What means of beautifying the house? What safety precautions?

3. Relatives, close friends, and neighbors.   In many parts of the world, grand-parents, or other older relatives, live with their immediate families. Other relatives may live nearby. What are the relationships with these relatives and the rest of the family under examination?

How many friends does this family have? Who are they? What is the basis or bases of friendship? What do they do together?

How close are the neighbors? What do they do? What are their relationships with the family being studied?

4. Relationships within the family: authority.   How is authority divided in the family? Roles of mother? Roles of father? Roles of other adults?

Are the children raised largely by the mother or by both parents? (In some parts of the world, the women and children live separately from the men and older boys.)

5. Food and clothing.   What foods does the family eat? How obtained? What percentage of the family income is devoted to food? Remember that most families of the world live on subsistence diets. They may raise much of their food and use cash income for other purposes.

What utensils are used in cooking and in eating? Why? Do all members of the family eat together? Why or why not?

Does the family eat indoors or outdoors most of the time? What are the staples — rice, bread, potatoes, manioc — for example. What ceremonies in conjunction with meals?

Clothing differs tremendously around the world, depending upon such factors as climate, religion, income, and cultural tradition. What clothes do the various members of the family wear? What materials? How are they made? What part does income play in the selection of clothes? Climate? Religion? Tradition?

6. Health and safety.   A few countries, like The Netherlands, Denmark, and the U.S.A., have very high health standards and are extremely health-conscious. Many parts of the world cannot afford such standards. Consequently a lack of doctors, nurses, hospitals, clinics, and medicines. In some places superstitions play their part.

What are the health standards of this family? How are the children protected? What role does the government play in health? What hazards to safety (from rain, snow, or animals) exist?

7. Work inside and outside the home — and money.   There are thousands of jobs in various parts of the world. What work does this family do? What does the father do? The mother? The children? (Remember that millions of older boys and girls around the world care for their younger brothers and sisters and that this "work" begins often at a very early age.)

What cash income does this family have? What source or sources? How is it used? Why?

8. Transportation and communication.   How does the family travel? How do they transport goods? How does the lay of the land in their locality affect transporta-tion? How does the income affect the means of transportation used? Remember that walking and carrying goods on the backs of human beings is still very common in many parts of our planet. Bicycles are the "middle class" means of transportation in many places.

What language or languages does this family speak? Why? Are there other forms of communication in this family? What are they?

9. *Values, goals, religions.*   What religious affiliation does this family have — if any? What are the chief values of this family? What holidays are celebrated? How do the special values of this family reveal themselves in daily living?

To what extent is the religion of a given family a matter of form and ritual and to what extent is it meaningful to that family, so far as one can determine motivations?

10. *Education.*   Approximately half of the world's children still do not attend school. Schools differ tremendously throughout the world.

How do the children in this family learn the ways of living of the family and the community? From whom? Do they attend school? If so, where? For how many years? What do they study? Who are their teachers? How is their school like an American school? How does it differ?

Do the parents go to school? If so, what do they study?

11. *The home and government.*   What is the local unit of government closest to the life of this family? How are the people chosen who run this local unit?

What rules or laws are there which affect this family? How? Examples might include laws about safety, health, regulations about the location and building of homes.

What taxes are paid by this family? Are there taxes on land and buildings? Are there taxes on income? Are there taxes like our sales taxes on items purchased in markets and stores? What proportion of the cash income of a family is paid in taxes?

What does the family receive in the way of services from the local unit of government? What are the services which this family receives from other units of government (regional and/or national)?

What is the "government" of the home? Who makes the decisions about the family? Are children included in any decision-making? What rules are there in the family?

12. *Fun and beauty: recreation.*   All families have some fun and beauty in their lives. They also create beauty in different ways.

What games do the children in this family play? What pets do they have — if any? What holidays do they celebrate? Are there any special holidays for children? What recreation does the family have as a unit?

What are the costs of family fun or recreation?

You as a teacher will need to raise some of the questions in the foregoing sections as children study pictures of families, read about them, and discuss them. But children will raise most of these questions if they are stimulated to raise questions. The wording of many of the questions suggested in the last few pages will need to be "translated," of course, into the language of children.

Although separated into twelve topics, many of these items overlap. Teachers should attempt to develop the total life of any family, a concept derived largely from anthropology. You may want to add questions on changes occurring in any of the families you are studying with pupils.

Against the background of these questions and comments, let us take a quick look at some of the methods which can be used effectively in the study of selected families in other parts of the world.

SOME METHODS FOR STUDYING FAMILIES IN OTHER PARTS OF THE WORLD

In order to help children gain a feeling of identification with the families selected from various parts of the world for study, considerable time needs to be devoted to each family. Children should read about them, see pictures of them, construct models of their homes, and talk about them. Often they should role-play various aspects of their lives. These families should become friends of theirs whom they come to know intimately and to understand and respect.

The development of a feeling-tone about a few families will depend in large part upon your attitude as a teacher. If you enjoy this study and are open to differences, then your pupils will catch much of this feeling from you. Conversely, if you regard them as strange and inferior people, the children are likely to sense this and pattern their feelings after yours.

If you can do so, prevail upon people who have lived in the part of the world you are studying, to visit your class and share their experiences with you and your pupils, especially experiences pertaining to family life. Where people from that part of the world are not available, try to substitute people who have traveled there. But be sure that they have the kind of attitudes you want to develop in your pupils — and that they can communicate with children.

You will certainly want to use all the pictorial material on families that you can assemble. That will include pictures, color slides, films, and filmstrips.

Playing the roles of people from other parts of the world is about as good a teaching device as you can use for getting into their shoes.

Games and songs will help your pupils to live the lives of the families they are studying.

Probably you will want to make models of the homes of the families you are studying. This can include a good bit of "house geography." But be sure not to stereotype the people of various regions of the world. Bear in mind, for example, that all Africans do not live in mud houses. Try to develop respect for the ways in which families have learned to live in their environment.

You may want to learn and teach a simple dance to your pupils which "their" family employs.

Of course you will use the pictures in books to help you understand the family under examination. You will need to help your pupils to study rather than merely look at the pictures. For help on the many skills in studying pictures, you may want to refer to the ideas on pages 101–102 of this book.

Because your resources may be limited, you may want to develop class rather than individual scrapbooks, pooling all your resources. This should also promote cooperation rather than competition among your pupils.

Of course your pupils will want to do many drawings of the families studied, selecting their own themes in many cases.

Undoubtedly you will want to read some of the folk tales which children in other parts of the world hear. And you will want to play some of their games.

For suggestions on other methods you may want to use, refer to the ideas on pages 78–80.

Now let us examine some of the resources for such studies of families.

SOME RESOURCES FOR STUDYING INDIVIDUALS AND FAMILIES IN OTHER PARTS OF THE WORLD — GENERAL

Books

1. BRYNNER, YUL. *Bring Forth the Children*. New York: McGraw-Hill (1960), 152 pp. Grades 4–6. Pictures of refugee children in Europe and the Middle East.
2. COATSWORTH, ELIZABETH. *The Children Came Running*. New York: Golden (1963), 93 pp. Grades 4–6. UNICEF cards reproduced in color, plus a brief text.
3. McMEEKING, ISABEL M. *A First Book About Babies All Around the World*. New York: Watts (1950), 42 pp. Grades 1–3.
4. SCHARTUM, HANSEN. *Ingvild's Diary*. New York: Lothrop (1954), 142 pp. Grades 5–7. Ingvild makes friends with children from Austria, Denmark, England, France, Germany, Mexico, and the U.S.A. at an International Children's Village.
5. SOLEM, ELIZABETH F. *We Learn About Other Children*. Chicago: Encyclopaedia Britannica (1954), 316 pp. Grades 3–5.
6. UNICEF. *Children of the Developing Countries*. Cleveland: World (1963), 131 pp. Grades 5–7.
7. UNICEF. "Child of UNICEF Collection." New York: United Nations (1962). Grades 5–7. A collection of leaflets on children in several countries.

Some teachers will find helpful background in Beatrice B. Whiting (Editor). *Six Cultures: Studies of Child Rearing*. New York: Wiley (1963), 1017 pp.

Textbooks

Families in other parts of the world are mentioned in several of the newer textbooks but not dealt with in any detail in most of them. However, there are two texts which deal with them in considerable depth. One is the book by Frederick M. King, Dorothy K. Brachen, and Margaret A. Sloan, called *Families and Social Needs* (River Forest, Illinois. Laidlaw Brothers 1968, 144 pp). Families included in this volume are from Japan, Switzerland, and India. Much of the concentration is on their houses. The only textbook to date which concentrates on families around the world is the volume by Sonia Gidal and Peggy Shackelton entitled *Families Live Everywhere* (Lexington, Massachusetts: Ginn, 1972, 240 pp.) The families included for depth study are the Oskols (Lapps), the Richters (Germans), the Uezimas (Japanese), the Leal and Meneses families (Brazil), and the Donkors (Ghana). These are real families and they were photographed in their own homes and communities.

Filmstrips

UNICEF sells inexpensively, several filmstrips about children, among them "Children of the Americas," "Children of Africa," "Children of Asia," and "Children of the Developing Countries."

Encyclopaedia Britannica has a series of filmstrips on Greek, French-Canadian, Japanese, English, Irish, Italian, Mexican, Dutch, Swiss, Chinese, French, Spanish, and Norwegian children. These are in black and white. They also have a series on "Families Around the World," in color, on families of Mexico, Guatemala, Brazil, Scotland, Yugoslavia, Israel, and Jordan.

McGraw-Hill has a series of filmstrips on families in China, Czechoslovakia, Egypt, England, Equatorial Africa, Germany, Italy, Japan, Mexico, and Pakistan. The same company has a series on farm families in Turkey, Denmark, The Netherlands, Switzerland, Greece, and Canada. Their color filmstrips on the "Children of Europe" include Ireland, Norway, France, The Netherlands, Switzerland and

Italy. They also have a series on families — in England, Scandinavia, The Nether-
lands, Switzerland, France, and Italy. A similar series on "Children of Latin
America" includes life in Argentina, Brazil, Guatemala, Peru, Mexico, and Chile.
Their series on "Children of the Orient" covers the Philippines, Egypt, China,
Arabia, India, and Turkey.

The Society for Visual Education has a series on "Children of the World" which
includes Japan, The Netherlands, Norway, and Switzerland.

For current offerings see the various filmstrip catalogues.

Individuals and Families in Brazil

For an account of two families in Brazil see Sonia Gidal and Peggy Shackelton's
Everyone Lives in Families (Ginn, 1972).

Books

1. BREETVELD, JIM. *Getting to Know Brazil*. New York: Coward-McCann (1960), 64
   pp. Grades 4–6.
2. CALDWELL, JOHN C. and ELSIE F. *Our Neighbors in Brazil*. New York: John Day
   (1962), 96 pp. Grades 3–5.
3. FORMAN, LEONA. *Bico: A Brazilian Raft Fisherman's Son*. New York: Lothrop, Lee
   and Shepard (1969), 92 pp. Grades 2–4.
4. GIDAL, SONIA and TIM. *My Village in Brazil*. New York: Pantheon (1968), 78 pp.
   Grades 4–6.
5. HERMANNS, RALPH. *River Boy: Adventures on the Amazon*. New York: Harcourt
   Brace Jovanovich (1965), 48 pp. Grades 2–4.
6. McDOWELL, ELIZABETH. *Nady Goes to Market: A Story About Brazil*. New York:
   Friendship Press (1961), 32 pp. K–2.
7. MANNING, JACK. *Young Brazil*. New York: Dodd, Mead (1970), 64 pp. Grades 3–5.
8. MAZIERE, FRANCIS. *Parana: Boy of the Amazon*. Chicago: Follett (1960), 47 pp.
   Grades 3–5.
9. REIT, SEYMOUR. *Week in Bico's World*. New York: Macmillan (1970), unpaged.
   K–3.

Films and filmstrips

1. "Chico Learns to Read — Life in Brazil." McGraw-Hill, color. A filmstrip.
2. "Brazil — People of the Highlands." Encyclopaedia Britannica, black and white or
   color, 17 minutes. A film.
3. "Family of Brazil." Encyclopaedia Britannica, color. A filmstrip.

Individuals and Families in Germany

For textbook material at the second grade level see the chapter on the Richter
family in *Everyone Lives in Families* (Ginn, 1972).

Books

1. GIDAL, SONIA and TIM. *My Village in Germany*. New York: Pantheon (1964), 85
   pp. Grades 4–6. Centers on family life in a village.
2. HILLES, HELEN T. *Rainbow on the Rhine*. Philadelphia: Lippincott (1959). Grades
   3–5. An American boy and his German friend.
3. NORRIS, GRACE. *Young Germany At Work and At Play*. New York: Dodd, Mead
   (1969), 64 pp. Grades 4–6.
4. SCHIEKER, SOFIE. *The House at the City Wall*. Chicago: Follett (1955), 93 pp. Grades
   4–6. A German family adopts a World War II orphan.

Films and filmstrips
1. "Families of the World — Germany." McGraw-Hill, black and white. A filmstrip.
2. "Children of Germany." Encyclopaedia Britannica, color, 14 minutes. A film.
3. "Germany — A Family of the Industrial Ruhr." McGraw-Hill, black and white, 16 minutes. A film.
4. "West German Family." Bailey Films, 1970, color, 15 minutes.

Individuals and Families in Ghana

The only text dealing with families in Ghana is *Everyone Lives in Families* (Ginn, 1972). For material on chocolate, write the Hershey Chocolate Company in Hershey, Pennsylvania. Teachers will find some background material in the *Hi Neighbor* book, Volume 2 (U.S. Committee for UNICEF) and in *Fun and Festival in Africa* (Friendship Press).

Books
1. COURLANDER, HAROLD. *The Hat-Shaking Dance and Other Tales from Ghana.* New York: Harcourt (1957), 115 pp. Can be read by the teacher to pupils.
2. GIDAL, SONIA and TIM. *My Village in Ghana.* New York: Pantheon (1970), 96 pp. Grades 4–6. Can be used by good readers.
3. GROVE, ELIZABETH. *The Mfums.* New York: World (1967), 95 pp. Grades 2–4. Daily life in a village in Ghana today.
4. HASKETT, EDYTHE. *Some Gold, a Little Ivory: Country Tales from Ghana and the Ivory Coast.* New York: John Day (1970), 126 pp.
5. KAYE, GERALDINE. *Great Day in Ghana: Kwasi Goes to Town.* New York: Abelard-Schuman (1962), 32 pp. Grades 3–5.
6. SCHLOAT, G. WARREN, JR. *Kwaku: A Boy of Ghana.* New York: Knopf (1962), 44 pp. Grades 4–6. Photographs of the Ga tribe.
7. SUTHERLAND, EFUS. *Playtime in Africa.* New York: Atheneum (1962), 58 pp. Grades 2–4. Large black and white photographs.

Individuals and Families in Japan

Japanese families are depicted in three textbooks for the primary grades. In the volume on *Social Sciences: Concepts and Values* (Harcourt Brace Jovanovich, 1970) and in *Families and Social Needs* (Laidlaw, 1968) there are short accounts. In *Everyone Lives in Families* (Ginn, 1972) there is a fuller treatment.

Teachers will find helpful material in the *Hi Neighbor* book (Volume 2) and in *Fun and Festival from Japan* (Friendship Press, 1966).

Books
1. BANNON, LAURA. *The Other Side of the World.* Boston: Houghton Mifflin (1960), 48 pp. Grades 1–3. Japanese children and their adventures.
2. BUELL, HAL. *Young Japan.* New York: Dodd, Mead (1961), 64 pp. Grades 4–6. Young Japanese at work and at play. Younger children can use the photos.
3. COPELAND, HELEN. *Meet Miki Takimi.* New York: Lothrop (1963), 28 pp. K–2.
4. DARBOIS, DOMINIQUE. *Norika: Girl of Japan.* Chicago: Follett (1964), 48 pp. Grades 3–5.
5. GALLANT, KATHRYN. *The Flute Player of Beppu.* New York: Coward-McCann (1960), 43 pp. Grades 3–5. Music in the life of a Japanese boy.
6. GIDAL, SONIA and TIM. *My Village in Japan.* New York: Pantheon (1966), 96 pp. Grades 4–6. Many photographs of family life.

7. GLUBOK, SHIRLEY. *The Art of Japan.* New York: Macmillan (1970), 48 pp. Grades 4–6.
8. GRIFFIS, FAYE. *Lantern in the Valley.* New York: Macmillan (1956), 136 pp. Grades 3–5. Life of a farm family, with much humor included.
9. HALL, H. TOM. *The Golden Tombo.* New York: Knopf (1959), unpaged. Grades 1–3. The story of a rice farmer's son.
10. HAYES, FLORENCE. *The Boy in the 49th Seat.* New York: Random (1963), 57 pp. Grades 3–5. A lonely boy brings his uncle and his space suit to class and wins friends in the new school.
11. LEWIS, RICHARD. *There Are Two Lives: Poems by Children of Japan.* New York: Simon and Schuster (1970), 96 pp. Primary grades.
12. MATSUNO, MASAKO. *Taro and the Tofu.* Cleveland: World (1962), unpaged. Grades 1–3.
13. NUGENT, RUTH. *On Japanese Playmates.* Rutland, Vermont: Tuttle (1960), 64 pp. Grades 3–5. Jack and Judy Brown learn about Japan through their classmates.
14. SCHLOAT, G. WARREN, JR. *Junichi: A Boy of Japan.* New York: Knopf (1964), 48 pp. Grades 3–5. Well illustrated with black and white photos.
15. SHIRAKIGAWA, TOMIKO. *Children of Japan.* New York: Sterling (1969), 96 pp. Grades 5–8.
16. UCHIDO, YOSHIKO. *Sumi's Prize.* New York: Scribner (1964), 44 pp. Grades 2–4. A girl in a kite-flying contest.
17. YASHIMA, TARO. *Crow Boy.* New York: Viking (1955), 37 pp. Grades 1–3. A classic about a shy Japanese boy who finally wins his place with his peers.
18. YASHIMA, TARO. *Seashore Story.* New York: Viking (1967), unpaged. Grades 1–3.
19. YASHIMA, TARO. *The Village Tree.* New York: Viking (1953), 36 pp. K–2. Japanese children enjoy a tree by the river in their village.

One of the Match Box units sold by the Boston Children's Museum is on Japan.

Films

1. "Boy of Japan: Ito and His Kite." Coronet, color or black and white, 11 minutes.
2. "Children of Japan." Encyclopaedia Britannica, black and white, 11 minutes.
3. "Family of Toyko." Bailey Films, color, 15 minutes.
4. "A Japanese Family." International Film Foundation, black and white, 23 minutes. A family of silkweavers.

Filmstrips

1. "Families of the World — Japan." McGraw-Hill, black and white.
2. "Japanese Children." Encyclopaedia Britannica, black and white.
3. "Kimiko of Japan." Friendship Press, color.
4. "Taro and Hanako of Japan." Society for Visual Education, color. A day with a Japanese family.
   (Silver Burdett has a picture packet of colored pictures of a Japanese family.)

The Lapps

The only textbook depicting family life among the Lapps is Sonia Gidal and Peggy Shackelton, *Everyone Lives in Families* (Ginn, 1972).

Books

1. BERRY, ERICK. *Men, Moss and Reindeer: Challenge of Lapland.* New York: Coward-McCann (1959), 128 pp. Grades 5–8.
2. DARBOIS, DOMINIQUE. *Aslak: Boy of Lapland.* Chicago: Follett (1968), 48 pp. Grades 3–5.

3. GIDAL, SONIA and TIM. *Follow the Reindeer.* New York: Pantheon (1959), 79 pp. Grades 4–6. Many black and white photographs.
4. LIDE, ALICE L. and MARGARET A. JOHANSEN. *Lapland Drum.* New York: Abingdon (1955), 128 pp. Grades 3–5. Brenda, her brother Vik, and their family.
5. PALAZZO, TONY. *Jan and the Reindeer.* Champaign, Illinois: Garrard (1963), 44 pp. Grades 2–4.
6. POHLMANN, LILLIAN. *The Summer of the White Reindeer.* Philadelphia: Westminster (1965), 154 pp. Grades 4–6.
7. RIWKIN-BRICK, ANNA. *Elle Kari.* New York: Macmillan (1967), 48 pp. Grades 2–4.
8. RIWKIN-BRICK, ANNA. *Nomads of the North.* New York: Macmillan (1961), 88 pp. Grades 4–7. Colored pictures add immeasurably to this volume.

Films

1. "Lapland." Embassy of Finland, color, 15 minutes.
2. "Laplanders." Encyclopaedia Britannica, black and white, 10 minutes.
3. "People of the Reindeer." Encyclopaedia Britannica, black and white, 18 minutes.

Individuals and Families in Mexico

Books

1. AMESCUS, CAROL G. *The Story of Pablo: Mexican Boy.* Chicago: Encyclopaedia Britannica (1962), 36 pp. Grades 3–5. Colored illustrations.
2. BULLA, CLYDE. *The Poppy Seeds.* New York: Crowell (1955), unpaged. Grades 1–3. A small boy plants poppy seeds to beautify his neighborhood.
3. "A Child of UNICEF in Mexico." New York: U.S. Committee for UNICEF (1963), 16 pp. Grades 3–4.
4. CRIST, EDA and RICHARD. *Chico.* Philadelphia: Westminster (1951), 80 pp. K–2. A Mexican boy worries about the drought.
5. DARBOIS, DOMINIQUE. *Tacho: Boy of Mexico.* Chicago: Follett (1962), 47 pp. Grades 2–4. Fine black and white photographs.
6. ETS, MARIE H. *Nine Days to Christmas.* New York: Viking (1959), 48 pp. K–2. A modern Mexican family celebrates Christmas.
7. GRACE, NANCY. *Earrings for Celia.* New York: Pantheon (1962), 48 pp. Grades 3–5. A Mexican boy looks after his sister.
8. GRANT, CLARA L. and JANE W. WATSON. *Mexico: Land of the Plumed Serpent.* Champaign, Illinois: Garrard (1968), 112 pp. Grades 3–5.
9. GRIFFITH, FAY. *Hidalgo and the Gringo Train.* New York: Dutton (1958), 96 pp. Grades 3–5. Humor and suspense in the story of a Mexican family trying to make ends meet.
10. JORDAN, PHILIP. *The Burro Benedicto.* New York: Coward-McCann (1960), 92 pp. Grades 3–5. Mexican folk tales.
11. KIRN, ANN. *Two Pesos for Catalina.* Chicago: Rand McNally (1962), unpaged. K–3.
12. MARX, M. RICHARD. *About Mexico's Children.* Chicago: Melmont (1959), 47 pp. Grades 2–4.
13. MOFFITT, FREDERICK J. *The Best Burro.* Morristown, New Jersey: Silver Burdett (1967), 32 pp. Grades 1–3.
14. POLITI, LEO. *The Noble Doll.* New York: Viking (1961), 45 pp. Grades 3–5.
15. POLITI, LEO. *Rosa.* New York: Scribner (1963), 32 pp. Grades 1–3.
16. SCHLOAT, G. WARREN, JR. *Conchita and Juan: A Girl and Boy of Mexico.* New York: Knopf (1964), unpaged. Grades 3–5.
17. TRIPP, EDWARD. *The New Tuba.* New York: Oxford (1955), 103 pp. Grades 3–5. Happy family relationships in a Mexican village.
18. WIER, ESTHER. *Gift of the Mountains.* New York: McKay (1963), 128 pp. Grades 3–5. A warm family story about rural Mexico.

Films and filmstrips

1. "Boy of Mexico: Juan and His Donkey." Coronet, color or black and white, 11 minutes. A film.
2. "Mexican Children." Encyclopaedia Britannica, black and white, 11 minutes. A film.
3. "Families of the World — Mexico." McGraw-Hill, black and white. A filmstrip.
4. "Family of Mexico." Encyclopaedia Britannica, color. A filmstrip.
5. "Fiesta Day — Life in Mexico." McGraw-Hill, color. A filmstrip.
   The International Communications Foundation has a teaching kit on Mexico.

37. Using pictures to study a community

38. Using maps to study a community

Communities in the United States

39. A committee member shows data collected by his group on a community

Studying Communities 18

THE LOCAL COMMUNITY AND OTHER COMMUNITIES IN THE UNITED STATES

The second unit of society of importance to boys and girls is the community. Very early in life they begin to function as members of a community, whether they live in a rural area, a small town, or a city. All their lives they will function as workers, consumers, and citizens in some community, and, in many instances, in a series of communities.

The community should be viewed at all grade levels as the local laboratory for studying people and their activities, but it deserves special recognition in the early years in school. It is a larger unit than the family and a smaller segment of society than the nation. Therefore it seems an appropriate unit to study in grades three or four, after the study of families and before the study of nations.

By the time children reach the third or fourth grades, they should be able to make an intensive study of their own community. Such a study should cover more facets and in greater depth than their studies in the earlier grades.

The study of communities is not something new. For many years the local community has been studied in almost all schools in the United States. Often it has been examined carefully. But all too often it has been studied superficially, with the emphasis almost exclusively on community helpers.

What we are suggesting here is a study of all aspects of the local community in a way similar to the method employed in studying families. Such a comprehensive view would include the physical features, the size and subdivisions, the variety of people, the many ways of earning a living, the transportation and communications of the community, the value systems and religions, the variety of ways of living, the government, health and education, recreation and cultural opportunities, and conflicts and changes. The emphasis would be upon the present, with some projection of the pupils into the past and into the future.

Having carried on such a thorough study of the local community, boys and girls would then be ready to look at a few carefully selected communities in other parts of the United States, using the skills they had developed in viewing their own community in looking at other communities of different kinds.

In the past, most curricula have stressed the local community to the exclusion of other communities in the United States. As a result, pupils have gotten the idea that all communities are alike — and that they are like their own community. This is a very parochial view of American communities. Surely pupils should learn about a variety of communities in our nation, including small, semi-rural locales and large, metropolitan concentrations.

Then, about the fourth grade, pupils could look at a few well-selected communities in various parts of the world. This should be easier because of the skills they have acquired in examining American communities.

Some Criteria for the Selection of Communities of the United States

Since it is impossible to study more than a few communities of the United States other than the local community, it is highly important that they be chosen on the basis of some well-thought-out plan. Some criteria you might like to consider would include the following:

1. Communities from several regions of the United States.
2. Communities which represent different types of geography.
3. Communities which show different types of people.
4. Communities which show different ways of earning a living.
5. A farming community and/or county seat, with its hinterland.
6. A seat of government (the state capital or Washington, D.C.)
7. A community of people on the move, such as a trailer community or a migrant labor community.
8. A community with a long history, such as Philadelphia, Boston, Baltimore, or Charleston.
9. A community showing many changes in recent years.
10. A large city with a harbor.
11. A space age community.
12. A new planned community, such as Reston, Virginia or Columbia, Maryland.

The mention of these twelve criteria does not mean that pupils should study that number of communities. Several of the criteria above might well be combined in a single city. For example, Philadelphia represents a large city with a harbor. It is an old city and one which has undergone many changes in recent years, with the restoration of the area around Independence Hall and its midcity urban renewal. It could also represent the Middle Atlantic area.

Of course there are other types of communities you may want to consider. There are college communities, retirement communities, and military communities. There are fishing and lumbering communities, although they are not of as much importance today as they have been in the past.

The Order in Which Communities Are Studied

The order in which communities are studied does not matter too much. Usually the local community will be examined first and in considerable detail. Perhaps two months or ten weeks will be devoted to it.

Teachers of city children may then want to turn to a rural community and a county seat so that their pupils can examine those basic units in our nations and see how they relate to each other. Teachers in rural areas may want to move from their local community to a study of the nearest large city with which their pupils have some contacts.

After studying their local community, some teachers may want to select a community or city which is in the news. Others may want to move on to a community which is quite different from their own, thus enabling the pupils to learn more by contrast rather than by comparison.

In some schools with several sections of a class, all the teachers may want to study the communities of the United States in the same order so that they may use the same films, filmstrips, and speakers. In other schools it may be helpful for teachers to follow different orders so that each may use all the available materials.

General Aims in Studying Any Community in the United States

It is important that children learn about the various aspects of any community and in some depth. That will probably mean that four to five weeks are spent on each of a few communities toward the beginning of the year and shorter periods on others toward the end of the year.

There are several aspects of any community which need to be discovered by pupils. They are shown in the chart below. They should not be studied, however, in the same order for every community, with the exception of the geographic base, which may be a good starting point for most communities. Here, then, are the major features of any community to be kept in mind by the teacher:

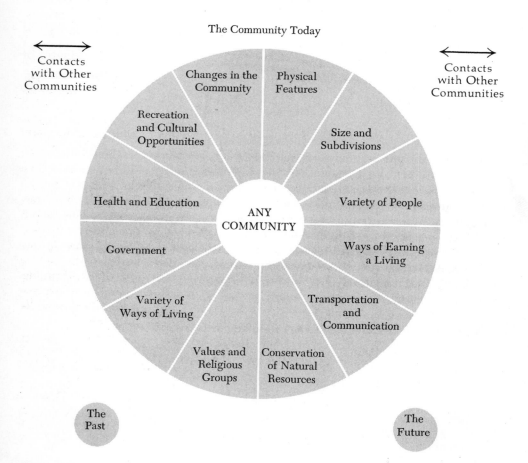

From Leonard S. Kenworthy, "Broaden Students' Study of Communities," *The Instructor*, February 1968, p. 47. By permission of the author and the publisher.

A Clarification of Terms

Some readers may be confused by the use of such terms as neighborhoods, communities, and metropolitan areas. Such confusion is understandable. It even exists to some extent among specialists in urban geography, economics, and sociology.

In general a *neighborhood* refers to a small area geographically, with a limited number of persons who have many face-to-face contacts in the course of a week, a month, or a year. A neighborhood may be a rural area or a small part of a city. Ordinarily the people in it know each other to some extent and have some mutual centers of interest. They may include the school, the church or churches, community clubs, and/or shopping centers.

The word *community* is usually used to refer to larger areas geographically. The people in those places do not need to have face-to-face contacts, shared interests, or common meeting places, although these may occur.

In this chapter and in other parts of this book the author has used the term "community" rather broadly. He has included cities in that category, too, since there is no one word which encompasses all kinds of communities, large and small.

The *larger community* or the *metropolitan area* usually refers to a large area geographically and a larger number of people than live in one town or city. For example Greater Philadelphia or Metropolitan Philadelphia includes large parts of nearby New Jersey and Delaware as well as some of the adjacent counties in Pennsylvania. It is an area which is tied to the larger city in several ways. Radio and television stations probably serve the metropolitan area. Newspapers are usually sold in the larger city. Sometimes there are governmental units which serve the larger community. Rooters for big league teams are usually found in the metropolitan area, as well as shoppers.

With some bright classes you may want to introduce the term megalopolis. In the future that is a term that they will hear more and more often as "strip cities" extend over hundreds of miles. The outstanding example of such a megalopolis, of course, is the region from Washington to Boston or even from the southern part of Virginia to Boston.

Pupils should not be burdened with these definitions, difficult as they are for adults. But they should know something about these various units of society. Some pupils still live in neighborhoods and their contacts are limited to a few blocks, especially in large cities. But most of us live in larger areas, including communities and metropolitan areas. This is especially true of suburbanites.

It is highly important that pupils begin to see the relationships between farming regions and small towns or larger cities, and vice versa. These units are increasingly dependent upon each other and many of our problems now include metropolitan areas. This is especially true of transportation and of air pollution.

It might be useful for you and your pupils to determine whether the chart or clock on page 288 applies equally well to all types of communities. Which points apply to all communities? Which apply only to the larger units? These are interesting and pertinent questions which you may want to explore as you delve into the study of various communities in many parts of the United States.

Now let us turn to a consideration of some of the general aims in studying various communities in our country.

Some Aims in Studying Communities in the United States

No doubt you will want to develop your own list of aims or generalizations for a study of your local community and other communities in the United States. You may even want to develop a master list, showing the aims that you want to highlight for each community, realizing that they may shift as you proceed with your studies with your pupils.

In the pages that follow we shall attempt to spell out in some detail some of the generalizations you will want to use. In some cases this may be information that you need to acquire; in other instances it is material you may want to use to review your knowledge of communities.

Your pupils will bring varying backgrounds to the study of American communities. Some will never have stepped out of the narrow neighborhood in which they were born. Others will have moved several times or will have traveled in various parts of the United States. Therefore it is impossible to establish a common list of aims for all pupils.

Throughout your study of communities, however, you should bear in mind the fact that you are introducing children to the various social sciences. Therefore you should help them discover pertinent data from geography, political science, anthropology, sociology, economics, and history.

Here are some concepts or generalizations which you may want to develop:

1. *There are many kinds of communities in the United States today.*

Communities differ in location, in size, and in activities. Some serve only a few purposes. Others serve a wide range of purposes. Among the types of communities in our country are the following:

    (a) Rural communities.
    (b) Small towns.
    (c) County seat communities.
    (d) Large cities.
    (e) Farming communities.
    (f) Fishing and lumbering communities.
    (g) Mining communities.
    (h) Government communities, including state capitals, and defense communities.
    (i) Resort communities.
    (j) Retirement communities.
    (k) Religious communities.
    (l) Manufacturing communities.
    (m) Communities of people on the move: Trailer camps and migrant communities.
    (n) Planned communities.
    (o) Family-owned and factory-owned communities.
    (p) Special communities, such as atomic energy centers and space centers.

Some communities fall into more than one of the above categories.

2. *Many people live on farms in our country today.*

    (a) Percentage of population of farmers decreasing. 90% at time of founding of nation. Now about 7%.

(b) Great strides in production of food, due to fertilizers, improved seeds, improved farm machinery, crop rotation and other factors.

(c) Fewer persons can produce much larger quantities of farm products. Other persons freed to work elsewhere.

(d) The size of farms differs tremendously in the U.S.A. today.
Many farms of small acreage; subsistence level living.
Some crops do not demand large acreage, such as peanuts and truck crops.
Mechanized farms of the West may include several hundred acres.

(e) Many people prefer to live on farms.
Enjoy the outdoor life and work with soil and animals.
Do not enjoy crowds of cities and city problems.
Farmers no longer isolated, with autos, radio and television.

(f) The work of farmers depends in large part on the weather.
Some efforts to control the weather, but with little success to date.
Rains a big factor on many farms.

(g) Farmers live in communities, too.
The consolidated school, a church or churches, farm organization, and/or store may be the center of community life. Clubs.
Young people often belong to farm organizations — Future Farmers of America and the 4-H club.
A farm community may be spread over a wide area.

(h) Farmers are very important people.
To live, everyone depends upon the products farmers raise.
They purchase many of the products of cities.

3. *There are many small towns in our country today.*

(a) Thousands of small towns exist across the U.S.A. today.

(b) Vary tremendously in size and population. Usually up to 5,000 inhabitants.

(c) Most small towns originally on some body of water, because of transportation and mill for grinding grain, run by water power.
Later, communities built on railroad lines.

(d) Small towns today often are incorporated into larger cities.
As cities expand, small towns are incorporated, gaining services of larger cities.

(e) Some small towns depend largely upon one industry for jobs for their inhabitants.

(f) Most small towns are shopping centers for the areas around them.

(g) Many small towns are seats of county government.

(h) Many people live in small towns today and work in large cities.
Homes are less expensive.
People prefer small town life but advantages of large city nearby.

(i) Many farmers retire to small towns.

4. *More and more people are moving to larger cities.*

(a) Most Americans today live in cities and suburbs.

(b) People have moved and are moving to the cities for a variety of reasons:
Fewer people can earn a living on farms today.
Cities demand more workers in factories, shops, businesses, services, etc.
People are attracted to the cities by the proximity to shops, stores, places of entertainment, educational opportunities, and for other reasons.

(c) There are now forty cities in the United States with a population of over 300,000. These are:

THE FORTY LARGEST U.S. CITIES

Rank	Population	Rank	Population
1. New York, N.Y.	7,895,563	21. Columbus, Ohio	540,025
2. Chicago, Ill.	3,369,359	22. Seattle, Wash.	530,831
3. Los Angeles, Cal.	2,809,596	23. Jacksonville, Fla.	528,865
4. Philadelphia, Pa.	1,950,098	24. Pittsburgh, Pa.	520,117
5. Detroit, Mich.	1,512,893	25. Denver, Colo.	514,678
6. Houston, Tex.	1,232,802	26. Kansas City, Mo.	507,330
7. Baltimore, Md.	905,759	27. Atlanta, Ga.	497,421
8. Dallas, Tex.	844,401	28. Buffalo, N.Y.	462,768
9. Washington, D.C.	756,510	29. Cincinnati, Ohio	452,524
10. Cleveland, Ohio	750,879	30. Nashville, Davidson Tenn.	447,877
11. Indianapolis, Ind.	745,739	31. San Jose, Cal.	445,779
12. Milwaukee, Wisc.	717,372	32. Minneapolis, Minn.	434,400
13. San Francisco, Cal.	715,674	33. Ft. Worth, Tex.	393,476
14. San Diego, Cal.	697,027	34. Toledo, Ohio	383,818
15. San Antonio, Tex.	654,153	35. Newark, N.J.	381,930
16. Boston, Mass.	641,071	36. Portland, Ore.	380,620
17. Memphis, Tenn.	623,530	37. Oklahoma City, Okla.	368,856
18. St. Louis, Mo.	622,236	38. Louisville, Ky.	361,958
19. New Orleans, La.	593,471	39. Oakland, Cal.	361,561
20. Phoenix, Ariz.	581,562	40. Long Beach, Cal.	358,633

Final figures 1970 census

(d) Some cities, however, are losing population in the "inner city" area. Families with children often move to the suburbs if they can afford to do so. Some families return to the cities when the children are raised.

(e) Inner cities usually are inhabited by members of lower socio-economic and minority groups.

(f) For a long period of our history, immigrants landed in the harbor cities. Often the people stayed on as workers in factories, as domestics, etc.

(g) Lived in ghettos with people of their own background. Older houses because of less expensive rents. Many such ghettos still exist in cities.

(h) With the increase of population, there has been an increase in problems. Problems include poverty, crime, juvenile delinquency, drug addiction, etc.

5. *Communities across the United States often have revealing names. These names often give us clues about the community.*

(a) Names are not always a revealing fact, but sometimes they are.

(b) Some communities are named for places in other countries.
Rome, Troy, and Syracuse, New York; New London, Connecticut; Reading, Pennsylvania.

(c) Communities are sometimes named for their founders or for explorers.
Lewiston, Idaho for Meriwether Lewis; Carson City, Nevada for Kit Carson; Baltimore, Maryland for Lord Baltimore.

(d) Communities are often named for special geographical features.
Bay City, Michigan; Great Falls, Montana; Salt Lake City, Utah.

(e) Communities are sometimes named for their products or for processes.
Galena, Illinois; Silver City, Nevada; Bessemer, Alabama; Anaconda, Montana; Gloversville, New York.

(f) Communities are often named for famous Americans.
Lincoln, Nebraska; Jefferson City, Missouri; Jackson, Tennessee.

(g) Communities are often named after Indian tribes or given the names the Indians gave localities.
Chippewa Falls, Wisconsin; Irondequoit, New York; Weehawken, New Jersey.

6. *The location of a community is very important.*

(a) Most large cities in the United States are located on rivers, on lakes, or on harbors of the oceans or the Gulf of Mexico.
Boston, New York, Philadelphia, Baltimore, Charleston, New Orleans, Pittsburgh, Cincinnati, St. Louis, Chicago, Los Angeles, San Francisco, Seattle, Portland, etc.

(b) Some of our larger cities grew because they were originally on canals.
Albany, Utica, Syracuse, Buffalo and other cities on the Erie Canal.
Might also have been on a river or on a lake — as Albany and Buffalo.

(c) Many cities have grown as railroad centers.
Chicago, St. Louis, Indianapolis, and Atlanta are four good examples.

(d) Many cities have grown because of their proximity to resources.
Kansas City and Minneapolis–St. Paul and their proximity to wheat.
Pittsburgh, Gary, Hammond, Birmingham in relation to iron, coal, and limestone — for the steel industry. More recent growth of Baltimore and Levittown, Pennsylvania as steel centers with iron brought from abroad.

(e) Some cities are growing as a result of new waterways.
Massena and other cities along the St. Lawrence Seaway.
Houston and its canal to the Gulf of Mexico.

(f) Some cities grow because of the location of new industries.
Seattle and the airplane industry. Houston and Clear Lake City, Texas and the space centers. Cape Kennedy and vicinity, and space.

(g) Any community can be studied in relation to its location and the effects of its location on its development.

7. *Communities differ in shape.*

(a) The shape of cities is determined in large part by the geography.
San Francisco with its shoreline and high hills.
Chicago in relation to Lake Michigan.
New York City as an island, with later growth of nearby areas with the building of bridges and tunnels.

(b) Many communities grow along major highways.

The development of "strip cities" along the roads connecting towns.
United States Route 1 from Boston to Washington — or into Virginia.
United States Route 40 — the National Road from east to west.
(c) Many towns grow in relation to their hinterland — the area they serve in trade.
(d) In which directions has your community grown? Why?
(e) Teachers may be interested in the four general patterns of growth which sociologists and urban geographers use, as shown in the examples below. Children can discover patterns without being told about them.

EXAMPLES OF TOWN PATTERNS

Lineal

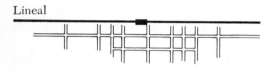

Gridiron

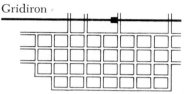

Radial

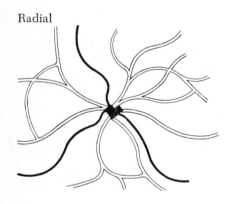

Spider Web

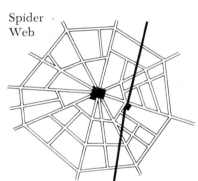

8. *Communities differ in size.*
(a) Rural communities are usually very small.
(b) Villages and small towns are limited to a few thousand people.
(c) Cities range in size from a few thousand persons to millions.
(d) Within most communities there are several fairly clearly marked sections or divisions:
A downtown shopping or commercial center.
A downtown residential area.
An industrial center.
An outer residential area.
Suburbs — within larger towns or cities.
(e) Larger cities may have several subdivisions.
A downtown shopping or commercial center.
Several industrial areas.
Several residential areas.
Several suburbs.
(f) The downtown section of a city often is called "The Inner City." This section usually was settled first and therefore begins to decay first. People

move out of the inner city to other residential areas in the city or to the suburbs.

Urban renewal projects are largely in inner city areas.

Communities within the city often are segregated on the basis of income and/or race, religion, or ethnic background.

9. *A wide variety of people usually live in any community.*
   (a) Smaller communities tend to be more homogeneous.
   (b) Some communities are inhabited by people of one general category.
       Resort communities.
       Retirement communities.
       Some suburbs — usually younger people with families.
   (c) Many smaller communities and all cities have a wide variety of inhabitants.
       Different age groups.
       Different income groups.
       Different religious groups.
       Different lengths of residence in the U.S.A. and/or in the city.
       Different educational backgrounds.
       Different recreational interests.
       Different jobs and job skills.
   (d) Differences in a community often lead to conflicts — open or submerged.
       Examples of recent conflicts, riots in cities. Watts area in Los Angeles, Chicago, Cleveland, Detroit, Newark, Rochester, etc.
   (e) Differences can serve as enrichment for a community.
   (f) Learning to live together in communities is one of the most important aspects of life in the U.S.A.

10. *The people in communities earn their living in many different ways.*
    (a) All communities have certain basic jobs:
        Retail stores.
        Service occupations.
        A few government officials, including school teachers.
        Religious personnel.
    (b) Some communities specialize in one product or industry.
        Butte, Montana as the world's largest mining town.
        Corning, New York as a glass center.
        Elkhart, Indiana and the making of musical instruments.
        Decatur, Illinois as the soybean capital of the world.
    (c) The larger the community, the more types of jobs are likely to be found.
        Among the many ways of earning a living are the following:
        Wholesale and retail trade.
        Manufacturing.
        Service occupations.
        Public administration and government.
        Entertainment and recreation.
        Finance, insurance, and real estate.
        Repair work.
        Construction.
        Transportation, communication, and public utilities.

   Mining.
   Forestry.
   Farming.
 (d) All jobs are important to the people of that community and to people in other communities.
 (e) People earn their living at different times of the day.
   Most people work during the day.
   There are many night workers — policemen, cleaners in large buildings, transportation people, telephone operators, etc.
 (f) Some people do not work at all or all the time.
   The physically handicapped may not work, although some do.
   Some people work only in certain seasons of the year.
   The work of some people depends upon the sales of products; they are sometimes out of work when business is slack.
 (g) Some people live in other communities and commute to work.
   Farmer and part-time workers in cities.
   Suburbanites.
 (h) People are paid for their work in different ways.
   Wages and salaries.
   "Fringe benefits," such as paid vacations, opportunities for investment in the business, etc.
 (i) There are laws governing the conditions under which people work and how long they can work in a week.
 (j) There are laws concerning child labor.

11. *Transportation is important in every community.*

 for getting raw materials to factories.
 for getting finished products to other parts of the country.
 for getting people to and from work.
 for getting people to and from shopping places.
 for getting people to and from schools.
 for getting people to and from other places they want to visit.
 for getting products to the homes of people in the community.
 (a) There are many types of transportation in most communities.
   Walking.
   Riding bicycles, motorcycles, and cars.
   Buses, street cars, and subways.
   Horses and wagons.
   Trains.
   Airplanes.
   Boats.
 (b) The types of transportation depend upon several factors.
  (1) The geography of the community.
   The use of boats, for example, in some places.
   Where the roads or railroads are built.
  (2) Economic and political factors.
   How much a community decides to spend on transportation.
   How much outside help the community obtains from state and/or federal aid to highways, airports, trains, etc.

    (c) Rules are made for the safety, comfort, and convenience of people.

          Rules for walking and driving.

          People are hired to enforce the laws.

    (d) Transportation is a big problem in large cities today.

          Because of the large number of people who come to a city at the same time and leave at approximately the same time, creating rush hours.

          Because of the narrowness of many downtown streets.

          Because of the large number of trucks bringing in raw materials or hauling finished products to other places.

          Because of the factor of speed required by people and the importance of moving goods quickly.

    (e) Cities are working on transportation problems.

          The development of one-way streets to speed traffic.

          The construction of new streets and highways, including superhighways.

          Improvements being made in train transportation around large cities.

          More control of traffic.

    (f) Many cities have airports, bus terminals, and railroad stations.

          Importance of railroad stations, especially in the past.

          Growing importance of bus terminals and airports.

    (g) The planning of transportation is an important part of community planning. Often a special division of the local government.

          Regional planning groups have been formed or are being formed.

12. *Communication is another important aspect of every community.*

    (a) We all need to communicate with members of our families and with neighbors and friends, as well as with people in distant places.

    (b) Much of our communication is informal — visits to the supermarkets or stores, visiting in homes, seeing people in other places.

    (c) Some communication is by telephone and by telegraph.

    (d) We write and receive letters from friends and relatives.

    (e) People in business also need to communicate.

          They use different forms of communication— from personal messengers to telegrams, telephone conversations, and advertisements. Some stores and offices have their own special delivery services.

    (f) Some types of communication must be organized.

          Postal services of the government.

          Telephone and telegraph services.

    (g) Communication is important in alerting people to impending disasters.

          Storm warnings of hurricanes and tornadoes in certain parts of the United States.

          Smog alerts in some cities.

          Weather forecasts for farmers, such as citrus fruit growers, so that they can protect their crops.

    (h) Communication is important in furnishing people with news.

          Newspapers.

          Magazines.

          Radio and television news reports.

    (i) Communication is important, too, as a form of fun.

    (j) All forms of communication are important.

13. *Values and religious groups.*
   (a) Every person and each family has a set of values.
       In our early years we learn our values from our families, chiefly.
   (b) The value systems of people differ.
   (c) Sometimes conflicts result from different value systems in a community.
   (d) It is important to learn to respect the ideas of other people even though we
       may not agree with them; freedom of speech and religion important.
   (e) There are different faiths or religions in most communities.
       The four major groups in the United States are Protestants, Roman Catho-
           lics, Orthodox Catholics, and Jews. People are free to join or not join such
           groups.
   (f) There are some commonly accepted values in communities, such as hon-
       esty, neighborliness, kindness, etc.
   (g) Some values are written into laws.
       Disregard for these laws brings punishment.
       Governments have officials who enforce these laws.
           Police officers are the major officials who enforce laws.
           Safety patrol personnel are also officers of the law.
           These are community helpers of ours.

14. *There are often a variety of ways of living in a community.*
   (a) There is often one general way of life in smaller communities.
   (b) In larger communities there are usually several ways of life.
   (c) People live in different ways because of a variety of reasons.
       (1) Because of their income.
           Different types of houses — tenements to luxury apartments.
           Different types of food.
           Different ways of having fun.
       (2) Because of religious beliefs.
           Special dress for some religious groups.
           Prohibitions against special kinds of transportation (Amish as an
               example).
           Special foods.
           Religious symbols in homes and special acts of worship; holidays.
       (3) Because of their value systems.
           Feelings about how to use their money.
           Feelings about participation in community affairs.
       (4) Because of their ethnic or national backgrounds.
           Language used in the home.
           Food eaten and how prepared.
           Membership in clubs or groups of people of their own background.
   (d) People often live in parts of a community with people of a similar back-
       ground. This may be voluntary segregation or involuntary segregation on
       the basis of religion, race, or ethnic background.

15. *Communities need rules or laws, and governments.*
   (a) People need rules or laws to maintain law and order. The more people, the
       more need for laws.
   (b) Governments are organized to do what people cannot do or do as well indi-
       vidually.

(c) There are different types of governments or units: local, state, and federal.
(d) Among the services of government to citizens are the following:
   Schools.
   Police and fire protection.
   Streets and highways.
   Parks and playgrounds.
   Welfare services.
   Inspection of houses, elevators, food, etc.
   Courts and prisons.
   Employment offices.
   Provision of clean water.
   Removal of waste, sewage, rubbish, garbage, etc.
   Planning and zoning.
(e) People support governments by paying various kinds of taxes.
(f) Each person can help observe the laws of the community and nation and make his or her community a better place in which to live.

16. *Communities help to safeguard the health of the people in them.*
   (a) With many people in a community, the health of each person depends upon the health of all.
   (b) Many people help improve the health of all of us.
      Families help.
      Doctors and nurses help.
      The personnel of hospitals and clinics help.
      School doctors and nurses help.
      People work to provide us with pure water.
      People inspect our food to make sure it is clean.
      People dispose of rubbish and garbage to protect the health of all.
   (c) There are rules for our health and safety, from requiring inoculations against disease to prohibiting spitting in public places.
   (d) Many of the health services are provided by the government and paid for by taxes. Some religious and private groups provide health services.
   (e) All the people who help with the health of individuals in the community are important.

17. *Communities are likewise concerned with the education of individuals.*
   (a) There are many reasons for schools.
      To help people to learn skills.
      To help people to get jobs.
      To help people to learn to live together in a peaceful way.
      To help people to enjoy themselves.
      To help people to become better citizens and people.
   (b) There are schools in every community.
   (c) Some schools are run by the government; some by private groups.
      Public schools.
      Religious schools.
      Private or independent schools.
   (d) Public schools are paid for by the citizens through taxes; religious and private schools are paid for by voluntary contributions.

(e) There are different types of schools in most communities.
   Preschools and kindergartens.
   Elementary schools.
   Middle schools and junior high schools.
   Senior high schools of different types.
   Junior colleges, colleges, and universities.
   Special schools for the handicapped.
(f) Some schools are for adults.
   Night schools and adult education classes.
   Schools in factories and businesses.
(g) Many people work in schools.
   Principals and assistant principals, librarians, office workers, bus drivers,
      custodians, special teachers, teachers, student teachers, etc.
(h) Many forms of education go on in homes and in the community.

18. *People in all communities like to enjoy themselves.*
   (a) Everyone likes to enjoy himself from time to time.
      Often this is done in school or on jobs. People can enjoy work.
      More often this is done in spare time.
   (b) Different people have different ways of having fun.
      This depends upon their own interests, their finances, and sometimes
         upon their religious beliefs or value systems.
   (c) There are many ways of having fun.
      (1) Some fun is unorganized.
         In homes.
         In groups of children and/or adults.
      (2) Some fun is organized commercially.
         Movies.
         Skating rinks and bowling alleys.
         Sports and plays.
      (3) Some fun is carried on by organizations.
         Community centers and settlement houses.
         Clubs like the Boy Scouts and Girl Scouts, Police Athletic Leagues,
            etc.
      (4) Some fun is carried on by churches and synagogues.
         Church and synagogue groups and clubs.
         Parties, plays, programs, outings, etc.
      (5) Some fun is carried on by schools.
         Fun in the classroom.
         Assembly programs.
         Clubs and special programs.
         After-school activities.
      (6) Some fun is carried on by governments.
         Play streets, parks and playgrounds.
         Libraries.
         City theaters, concerts, etc.
         Museums and art galleries.
   (d) We all can have fun in different ways and not hurt others in the process.
   (e) The many people who help with fun in the community are important.

19. *Communities are constantly changing.*

   (a) Communities are a little like people — they grow older.
      The central part of most cities is the oldest and becomes run down first.
   (b) There are many programs of "urban renewal" today in cities.
   (c) Different sections of communities are built up with new houses, buildings, and factories.
   (d) Increased numbers of people have brought changes.
      The need for more houses and offices and stores.
      The need for more transportation and communication facilities.
      Taller buildings are built.
   (e) There are some interesting changes in cities today.
      Underground parking spaces and special garages.
      Shopping malls, sometimes covered.
      Eliminating streets in the midcity, as in Rochester, New York, and Miami, Florida.
   (f) Some towns even disappear — "ghost towns."
   (g) Some towns change because industries change or mining is discontinued.
      Appalachia area.
      Changes in the Mesabi Range area in Minnesota.
   (h) Changes because of the movement to the suburbs.

20. *Many large cities are spreading out over a distance of many miles.*

   (a) With increases in population and the move to the cities, more space is demanded for large cities.
      Many people prefer to live in the outskirts of cities.
      Some industries move to the suburbs for more space and lower taxes.
   (b) About thirty-two persons in every one hundred now live in the twenty largest metropolitan areas of the country.
   (c) Problems of overlapping governments arise; some new metropolitan governmental units are being formed or have been formed.

21. *Communities are not self-sufficient; they depend upon each other and their hinterlands.*

   (a) A few communities are partially self-sufficient, such as Amish and some Mormon communities.
   (b) All communities depend upon the people in adjacent areas for food, a labor supply, finished products, transportation, and communication.
   (c) All communities depend to some extent upon other areas for raw materials, for home and office supplies, for entertainment (such as books, magazines, radio and television), and for markets.
   (d) Communities in the United States also depend upon communities in other countries for raw materials, finished products, and markets, as well as ideas.

22. *Most communities have a long history.*

   Every community to some extent is a museum. We need to study:
   When our community and other communities were founded, and by whom.
   Why they were founded. Reasons for the selection of their site.
   Factors which determined the growth of the community — or its eventual decline.

A few highlights of the history of a community.

Some of the important groups or individuals in the history of a community.

What life was like fifty years ago, one hundred years ago, two hundred years ago, or three hundred years ago.

Some Ways in Which Teachers Can Learn about Their Communities

The more you know about neighborhoods and communities, the better able you should be to teach about them. There is no guarantee that a great deal of knowledge will result in better teaching, but it is likely to produce good results, especially if you can organize your knowledge around basic concepts.

If you are a new teacher or are teaching for the first time in a given community, this need not bother you. You have lived all your life in at least one community and you know a great deal about it. You can use that background to advantage now. You may have lived in more than one community. If so, that should enhance your background even more.

If you are uninformed about the community in which you are to teach, you can learn along with the pupils if you have a general approach in mind. In some ways that is a distinct advantage, for you will be able to see things that you might overlook if you had lived all your life in that one place.

Even if you are a teacher who has taught in one community a long time, or you are a new teacher teaching in your home community, there is still much for you to learn about that locale.

No matter where you are teaching, it might be a good idea to start a series of folders or a note book on the community in which you are teaching. It might be organized by the twelve big ideas presented in the chart or clock on page 288, or according to the twenty-two ideas presented in the pages you have just read.

Especially if you are a new teacher in a community, there are four basic ways in which you can learn about your new community. Here are some suggestions for you.

1. *Observation*

    Take several walking trips from your home or the school to see the community at close hand.

    Make several automobile or bus trips into various parts of the community.

    On such trips note such features as:

    The houses, including new ones under construction.

    Stores, and new exhibits in stores.

    Churches and synagogues.

    Places for recreation, unorganized and organized.

    Play streets, playgrounds and parks.

    Governmental institutions.

    Transportation facilities.

    The people.

2. *Listening*

    As you take trips, listen for characteristic and special sounds.

    Listen to people on buses, trains, subways, and in stores to learn more about the community. You may want to tape record some of these sounds for later use.

3. *Talking with people*

Ask teachers, parents, and others you meet to tell you about their community. A good question is, "What should I know about . . . . as a newcomer?" Arrange to meet settlement house workers, librarians, and others to learn from them. You and/or your pupils may want to tape record some of these interviews.

4. *Reading*

Be sure to read the local newspaper for background information.
If there is a Chamber of Commerce, get its materials on the community.
Visit the public library to see if there is a history of the community or other material.
Read any material from your school and the Board of Education on the community in which you are to teach.

Learning about Other Communities in the U.S.A.

Here are some suggestions on ways to learn about other communities in the United States:

1. Write the Chamber of Commerce for materials.
2. Write the library or arrange for an interlibrary loan of materials.
3. Subscribe for a short time to a newspaper in the community you are going to study with your class.
4. Contact the Board of Education to see if they have any brochures about their community, written especially for children.
5. Check on whether there are films, filmstrips, tape recordings, or other materials about this community.
6. See if there are articles in secondhand magazines which you can clip and use.
7. Try to develop an exchange of materials with a teacher in the community you are studying, including pupil-made materials.
8. Visit that community, if at all possible, even for a day or two.
9. Develop a pen pal program with a teacher in that community.
10. Obtain as many of the materials as possible from the bibliography in this chapter.
11. Take notes on the community you are studying or are going to study, and save them for future use. You may want to assemble your notes by topics so that you can refer to them quickly.
12. Try to discover someone in your present community who has lived in the community you are studying and use him or her as a resource person for yourself and/or the class.
13. Write to the telephone company in a city you are studying and ask them for an old telephone book. Use the yellow pages to study types of industries in that community.

Some Community Resources for Studying the Local Community

It is often helpful to have an up-to-date list of community resources to use in connection with a study of a local community. You may also want to use such a list to remind yourself of features in any community you are studying.

If you are a new teacher in a community, you may want to build this resource list with the help of your pupils, their parents, other teachers, and other adults in the community.

If you are an experienced teacher in the local community, you may want to use this as a checklist to be certain that you have not overlooked some possible resources and as a checklist or reference list to save you time, instead of compiling it each year.

You may want to fill in the blank spaces in relation to the community you are working in and are studying with pupils.

A. *Government institutions and personnel*

1.

2.

3.

4.

5.

6.

B. *Business and commercial establishments*

1.

2.

3.

4.

5.

6.

C. *Cultural institutions and personnel: libraries, museums, theaters, etc.*

1.

2.

3.

4.

5.

6.

D. *Transportation and communication facilities and personnel*

1.

2.

3.

4.

5.

6.

E. *Recreational and leisure time facilities and personnel, especially for children*

1.
2.
3.
4.
5.
6.

H. *Parents and other teachers with special interests and abilities*

1.
2.
3.
4.
5.
6.

F. *Religious institutions and personnel*

1.
2.
3.
4.
5.
6.

I. *Other community resources*

1.
2.
3.
4.
5.
6.

G. *Social welfare and health institutions and personnel*

1.
2.
3.
4.
5.
6.

Some Suggested Experiences for Studying Communities

The experiences listed below apply especially to the study of the local community, although some of them can be used in conjunction with the study of other communities in the United States. They are listed somewhat in the order of difficulty. The majority of the experiences listed apply to work with pupils in the third and/or fourth grades. No doubt you and your pupils will think of other appropriate experiences.

1. *General suggestions*
   (a) Take several walking trips in your neighborhood and/or community, observing different phases of life.
   (b) Develop experience charts of what you have observed.
   (c) Draw pictures of what you saw.
   (d) Listen to the typical and unusual sounds in your neighborhood-community — especially sounds you would not hear in other communities.
   (e) Take a bus trip through your community, or several bus trips, if possible.
   (f) Write up these experiences on experience charts.
   (g) Make drawings or paintings of what you saw.

(h) Make a frieze or mural of the entire neighborhood and/or community. Be sure to include many pupils in this experience. They may do sections of the frieze or mural as groups or individuals, and then put the entire mural together.

(i) Develop a flannelboard of your neighborhood and/or community.

(j) Develop the layout of your community in a sandbox.

(k) Start a large map of your community, adding various features as you study them.

(l) Develop a series of overlay maps to use over a basic map of your community. Different pupils should make different maps. Committees can do this, too.

(m) Write poems, stories, and/or skits on some aspect of your community.

(n) Prepare a scrapbook of your community, including clippings from newspapers, drawings and pictures, maps, and brief statements about it. Plan it so that children in another community could use it as a type of "textbook" on your community. Exchange such scrapbooks with other schools.

(o) Prepare a bulletin board about your community.

(p) Make a brief list of the books about communities which seemed most helpful.

(q) Record typical and unusual sounds in your community.

(r) Take photographs of your community.

(s) Plan a simple directory of community organizations, with a little account of what they do.

(t) Discuss and/or write about "The Things I Like Best in My Community" or "The Things I Would Like to Change in My Community."

(u) Invite resources persons from the community to talk with members of the class.

2. *Activities regarding the physical aspects of communities*

(a) Take a walk or drive around the community to see its physical features.

(b) Take a trip to the highest point in the community from which you can see the "lay of the land."

(c) Make a map of the community in a sandtable.

(d) Make other types of maps — clay, flour and salt, sawdust and glue.

(e) If possible, arrange for an airplane trip over the community. Be sure to work this out with parents and with the school officials. It may have to be on out-of-school time.

3. *Activities regarding the size and subdivisions of communities*

(a) Make a map or an overlay, showing the subdivisions of the community you are studying.

(b) Make a simple bar graph, showing how your community compares in size and/or population with three or four other communities you have studied or will study.

(c) Discuss the reasons for the various subdivisions of your community or some other community.

4. *Activities regarding the variety of people in communities*

(a) Make a survey of the size of families in your class. Compare these results with results in another class.

(b) Find out how many newcomers there were in your neighborhood in the past year and some of the reasons why they moved to your neighborhood.

(c) Make a survey of the countries from which the ancestors of your pupils came. Make a world map showing the data collected.

5. *Activities regarding earning a living in communities*

(a) Make a survey of how the mothers, fathers, and other close relatives of the pupils in your class earn a living.

(b) If possible, visit two or three places where parents work.

(c) Invite a few parents to talk with the class about what they do to earn a living. Urge them to bring objects connected with their work, if possible.

(d) Add the places of work of the parents to your basic map of the community.

(e) Collect pictures representing the different type of jobs which people do in your community.

(f) Explore with the class the work of community helpers. Add relatively recent ones to the traditional list — including community planners, highway engineers, television studio operators, safety patrol personnel, etc.

(g) Make a mural of people at work in your community.

(h) Take pictures of people at work in your community. (Teachers or parents may need to take the pictures.)

(i) Make and use puppets of the people of your community at work.

6. *Activities regarding transportation in communities*

(a) Take trips to the local railroad station, bus station, and airport.

(b) Discuss what you saw on these trips and develop experience charts on them.

(c) Make drawings or paintings of some aspects of what you saw on the trips.

(d) Make a simple survey of how children get to school.

(e) Construct in a sandtable the local transportation system.

(f) Read about different methods of transportation in communities.

(g) Collect pictures of different methods of transportation.

(h) Add items on transportation to your basic map of the community.

(i) Discuss the advantages and disadvantages of different methods of transportation.

(j) Study a map of your state and its major roads. Try to determine why they were built where they were.

(k) Discuss the importance of transportation workers.

7. *Activities regarding communication in communities*

(a) Visit the local telephone office, a newspaper plant, a radio station, and a television studio. Try to get "behind the scenes."

(b) Discuss what was seen on such trips, and develop experience charts.

(c) Read about various means of communication.

(d) Make drawings and/or paintings of aspects of communication.

(e) Discuss the importance of communications workers.

(f) Make a simple survey of magazines and newspapers read at home.

(g) Make a survey of the television habits of your pupils.

8. *Activities regarding values and religions in communities*

(a) Make a map of the places of worship in your neighborhood and/or community, or place these buildings on the master map of your community.

(b) Make a collection of the "symbols" of various religious faiths, collect pictures of them.

(c) If possible, arrange for trips to places of worship, with explanations of the services. This must be handled with extreme care and may be done best as an out-of-school activity.

(d) Develop some role-playing situations involving values.

9. *Activities regarding government in communities*

(a) Discuss ways in which the government touches the lives of your pupils.

(b) Invite at least one representative of the local government to visit your class and talk with the pupils.

(c) Visit a polling booth at election time.

(d) Make a simple chart of the officers of your local government.

(e) Visit the city hall.

(f) Discuss the importance of local government workers.

10. *Activities regarding health and education*

(a) Discuss the people in the community who help keep people healthy.

(b) Interview the school doctor and/or nurse.

(c) Visit a hospital or health center.

(d) Collect pictures of health workers and discuss the importance of their work.

(e) Discuss and make a map of the different schools in your community or the largest community near you.

(f) Visit the nearest junior high school, possibly for an assembly program.

(g) Find out about the local parent-teacher organization.

11. *Activities regarding recreation and cultural opportunities*

(a) Discuss what the pupils like to do in their spare time — and why.

(b) Discuss what parents do in their spare time.

(c) Have a hobby show.

(d) Make a map of some of the cultural opportunities and/or recreational places in your community. This may be included in your master map of the community or may be an overlay on your master map.

(e) Visit one or two of the places of recreation in your community.

(f) Make a bar graph of the amount of time spent on various recreational activities by members of the class.

12. *Activities regarding changes in the community and its history*

(a) Visit an urban renewal project in the community.

(b) Visit a new building under construction and talk with the men working there.

(c) Visit an old house in the community.

(d) Visit any old landmarks and learn about their importance.

(e) Collect pictures of your community in "the olden days."

(f) Talk to older people about the changes they have seen in your community during their lifetimes.

(g) Obtain information about future plans of your community.

13. *Activities regarding contacts with other communities*
  (a) Make a map of the major roads in your community and where they lead, showing the nearest cities to which they go.
  (b) Do the same with the railroads and bus companies in your community.
  (c) Interview a local merchant as to the products he sells and where he obtains them.
  (d) Do the same for products made in your community and where they are sold.
  (e) Have pupils discover the items in their houses or apartments which came from outside your local community and make maps of those places.

BIBLIOGRAPHY ON U.S. COMMUNITIES

Some background books for teachers about U.S. communities

1. BOSKOFF, ALVIN. *The Sociology of Urban Regions.* New York: Appleton (1962), 370 pp.
2. EBERSOLE, LUKE. *American Society: An Introductory Analysis.* New York: McGraw-Hill (1955), 510 pp. Considerable material on rural communities.
3. EDITORS OF FORTUNE. *The Exploding Metropolis.* Garden City, New York: Doubleday (1958), 177 pp. A paperback.
4. GOTTMAN, JEAN. *Megalopolis: The Urbanized Northeast Seaboard of the United States.* New York: Twentieth Century Fund (1961), 810 pp.
5. HOOVER, EDGAR M. and RAYMOND VERNON. *Anatomy of a Metropolis.* Cambridge, Massachusetts: Harvard University Press (1959), 345 pp. On the New York area.
6. MUMFORD, LEWIS. *The City in History.* New York: Harcourt (1961), 576 pp. See also any of his other books on cities.
7. SCIENTIFIC AMERICAN. *Cities.* New York: Knopf (1965), 214 pp. Based on a special issue of *Scientific American.*
8. VIDICH, ARTHUR J. and JOSEPH BENSMAN. *Small Town in Mass Society: Class, Power, and Religion in a Rural Community.* Garden City, New York: Doubleday (1958), 337 pp. A paperback.

Some textbooks for children about U.S. communities

1. ANDERSON, EDNA A. *Communities and Their Needs.* Morristown, New Jersey: Silver Burdett (1966), 192 pp. Chapters on Minneapolis and a colonial village.
2. BALDWIN, ORELL and BENJAMIN STRUMPF. *The Way We Live.* New York: Noble and Noble (1965), 88 pp. Grade 3.
3. BLACK, IRMA and others. *The Bank Street Readers* (three books): *In the City, People Read,* and *Around the City.* New York: Macmillan (1965), 128 pp. each. Intended as readers for Grade 1 pupils.
4. BUCKLEY, PETER and HORTENSE JONES. *Our Growing City.* New York: Holt, Rinehart and Winston (1968), 176 pp. Based on the story of a Negro boy and his photographer father who work on an article on their city — presumably New York City. Black and white photographs.
5. EDUCATIONAL RESEARCH COUNCIL. *Communities At Home and Abroad.* Boston: Allyn and Bacon (1970), 367 pp. Intended for the first semester of a year. The second semester uses six communities, Williamsburg, Fort Bragg, Yakima–Washington, a forest products community in Arkansas, Pittsburgh, and a rural community — Webster City, Iowa.
6. HANNA, PAUL R. and others. *Metropolitan Studies.* Chicago: Scott, Foresman (1970), 304 pp. Divided into sections according to the various social science disciplines. Does not focus on individual cities.

7. JAROLIMEK, JOHN and ELIZABETH B. CAREY. *Living in Places Near and Far.* New York: Macmillan (1966), 146 pp. Grades 2–3.
8. PRESTON, RALPH and others. *Greenfield, U.S.A.* Boston: Heath (1965). Grade 2. A contrived community, dealt with from various facets.
9. QUIGLEY, CHARLES. *We Live in Communities.* Lexington, Massachusetts: Ginn (1972). Fourth grade text. An interdisciplinary approach to Mason City (Iowa), Houston and Clear Lake, Chicago, New Orleans, Minneapolis and St. Paul, Detroit, Gastonia (North Carolina), Honolulu, Philadelphia, Washington, Los Angeles, and New York City. In the final chapter pupils are asked to plan a new community on the basis of what they have learned previously, with some attention to Columbia (Maryland) and Reston (Virginia).
10. SENASH, LAWRENCE. *Our Working World: Cities at Work.* Chicago: Science Research Associates (1966), 288 pp. Short accounts of several cities, including some from history, such as Athens.

Picture portfolios

The Chandler Publishing Company has a set of pictures of city life, suitable for the early grades, arranged in a large, spiral-bound book for viewing by groups of children. Write them for details at 124 Spear Street, San Francisco, California (94105).

The John Day Company (257 Park Avenue South, New York, New York 10010) has several large portfolios on city life, stressing integrated living. One is entitled "A City Is . . ." Others are on work, urban renewal, recreation, and opportunity. Individual volumes have been issued on Chicago, Denver, Detroit, Los Angeles, San Francisco, and Washington. There is a similar series on rural living, including one on "A Rural Community Is . . ."

The Fideler Company (Grand Rapids, Michigan) has a set of pictures on "Our City."

Books for children about communities in the United States — general

1. ASSOCIATION FOR CHILDHOOD EDUCATION. *Round About the City: Stories You Can Read To Yourself.* New York: Crowell (1956), 117 pp. Grades 2–4.
2. AMES, JOCELYN and LEE. *City Street Games.* New York: Holt (1963), 25 pp. Grades 2–5. An unusual approach to city life for children.
3. BACMEISTER, RHODA W. *The People Downstairs and Other City Stories.* New York: Coward-McCann (1964), 120 pp. Grades 4–6.
4. BUDD, LILLIAN. *The People on Long Ago Streets.* Chicago: Rand McNally (1964), 47 pp. Grades 4–6.
5. CARSE, ROBERT. *Great American Harbors.* New York: Norton (1963), 110 pp. Grades 5–8. About Boston, Baltimore, Cleveland, New Orleans, New York, Los Angeles, San Francisco, and Seattle.
6. DEUSSEN, ELIZABETH. *Exploring Baltimore.* Baltimore: Board of Education (1960), 208 pp. Grades 5–8.
7. DIETTRICH, S. W. *Miami.* Garden City, New York: Doubleday (1964), 64 pp.
8. ELTING, MARY. *Water Come—Water Go.* Irvington-on-Hudson, New York: Harvey (1964), 45 pp. Grades 1–3. On water systems of communities.
9. FRIEDMAN, FRIEDA. *The Janitor's Daughter.* New York: Morrow (1956), 159 pp. Grades 5–7. Reactions of different members of the family to their life in an apartment house where their father works.
10. GRAHAM, LORENZ. *North Town.* New York: Crowell (1965), 220 pp. Grades 5–9. Prejudice in a northern community.
11. GRAHAM, LORENZ. *South Town.* Chicago: Follett (1958), 189 pp. Grades 6–9. Racial tensions on a small farm in the South.

12. HALL, DOROTHY. *Our Land of Cities.* Columbus, Ohio: Merrill (1959), 32 pp. Grades 4–6.
13. HELLER, AARON. *Let's Take a Walk: A Story Told in Pictures.* New York: Holt, Rinehart and Winston (1963), 25 pp. K–3. 12 streets depicted in drawings without text, showing different aspects of city life.
14. HOFFMAN, ELAINE and JANE HEFFLEFINGER. *About Helpers Who Work at Night.* Chicago: Melmont (1963), 32 pp. Grades 2–4. Disc jockeys, nurses, janitors, firemen, bakers, telephone operators, and others.
15. KELLY, REGINA Z. *New Orleans: Queen of the River.* New York and Chicago: Reilly and Lee (1963), 179 pp. Grades 6–9. History of its settlement, with an epilogue about the city today.
16. LAVINE, DAVID. *The Mayor and the Changing City.* New York: Random (1966), 172 pp. Grades 6–9. The story of Richard C. Lee and New Haven, Connecticut.
17. LAVINE, DAVID. *Under the City: The Wondrous World Beneath the Streets.* Garden City, New York: Doubleday (1967), 128 pp. Grades 4–6.
18. LENSKI, LOIS. *High-Rise Secret.* Philadelphia: Lippincott (1966), 160 pp. Grades 4–6.
19. MAITLAND, ANTHONY. *Ben Goes to the City.* New York: Delacorte (1967), 48 pp. Grades 3–4.
20. MARTEL, SUZANNE. *The City Under Ground.* New York: Viking (1964), 160 pp. Grades 4–6.
21. MUNZER, MARTHA E. *Planning Our Town: An Introduction to City and Regional Planning.* New York: Knopf (1964), 180 pp. Grades 6–9.
22. PEET, CREIGHTON. *The First Book of Skyscrapers.* New York: Watts (1964), 63 pp. Grades 4–6.
23. SCHATZ, LETTA. *No Lights for Brightville.* Chicago: Follett (1965), 31 pp. Grades 1–3. City lights explained as a result of a "blackout."
24. SCHNEIDER, HERMAN and NINA, *Let's Look Under the City.* New York: Scott (1950), 71 pp. Grades 4–6. Combines science and social studies in a superior way.
25. SECKAR, ALVENA. *Zuska of the Burning Hills.* New York: Walck (1952), 224 pp. Grades 4–6. A West Virginia mining community.
26. SEIFERT, SHIRLEY. *The Key to St. Louis.* Philadelphia: Lippincott (1963), 128 pp. Grades 5–8. Keys to the Cities series.
27. SHELDON, WILLIAM D. *The House Biter.* New York: Holt, Rinehart and Winston (1966), 28 pp. Grades K–2. Story of a house-wrecking derrick. A Little Owl book.
28. STANEK, MURIEL. *How People Live in the Big City.* Chicago: Benefic (1964), 48 pp. Grades 1–3.
29. THOMAS, KATRINA. *My Skyscraper City: A Child's View of New York.* New York: Doubleday (1962), 61 pp.
30. TINKLE, LON. *The Key to Dallas.* Philadelphia: Lippincott (1965), 128 pp. Grades 5–8. Keys to the Cities series.
31. TRESSELT, ALVIN. *Wake Up, City!* Chicago, Lothrop, Lee and Shepard (1956), 24 pp. Grades 2–4.
32. URELL, CATHERINE and LILLIAN GOLDMAN. *Big City Neighbors.* Chicago: Follett (1955), 95 pp. Grades 4–6. On New York City but could be any other large city and its people.
33. URELL, CATHERINE, ANNE JENNINGS, and FLORENCE R. WEINBERG. *The Big City Book of Conservation.* Chicago: Follett (1956), 96 pp. Grades 4–6.
34. URELL, CATHERINE, ANNE JENNINGS, and FLORENCE R. WEINBERG. *Big City Fun: How and Where to Find It.* Chicago: Follett (1954), 96 pp. Grades 4–6.
35. URELL, CATHERINE and ELIZABETH VREEKEN. *Big City Government.* Chicago: Follett (1957), 96 pp. Grades 4–6.
36. WELLESLEY, HOWARD R. *All Kinds of Neighbors.* New York: Holt, Rinehart and Winston (1963), 28 pp. K–2.

37. WESTON, GEORGE F., JR. and MILDRED. *The Key to Boston.* Philadelphia: Lippincott (1961), 128 pp. Grades 5–8. Keys to the Cities series.

Farms and farming communities

There are a few textbooks which include material on farms or small towns in farming regions. Among them are the Preston and Clymer volume on *Communities At Work* (Heath), the Educational Research Council's book *A Rural Community: Webster City, Iowa* (Allyn and Bacon), and Charles Quigley's text on *We Live in Communities* (Ginn), in which the opening chapter is on Mason City, Iowa and the surrounding farming area.

1. BERQUIST, GRACE. *Farm Girl.* Nashville, Tennessee: Abingdon (1963), 48 pp. Grades 2–4.
2. BUCK, MARGARET. *Country Boy.* Nashville, Tennessee: Abingdon (1963), 64 pp. Grades 3–4.
3. COLLIER, ETHEL. *I Know a Farm.* New York: Scott (1960), 72 pp. Grades 1–3. A little girl discovers fun on the farm.
4. EBERLE, IRMENGAARDE. *Apple Orchard.* New York: Walck (1962), 48 pp. Grades 4–6. Raising apples on a farm in northern New York State.
5. GRAHAM, JORY. *Children on a Farm.* Chicago: Encyclopaedia Britannica (1962), Grades 2–4. Illustrated in color.
6. IPCAR, DAHLOV. *Ten Big Farms.* New York: Knopf (1958), 36 pp. Grades 2–4. The Jordans inspect several farms before selecting an all-purpose one for purchase.
7. KNEEBONE, S. D. *Farm Machinery Works Like This.* New York: Roy (1963), 56 pp. Grades 5–8. An excellent combination of science and social studies learnings.
8. LENSKI, LOIS. *Corn-Farm Boy.* Philadelphia: Lippincott (1954), 180 pp. Grades 3–5. Life on an Iowa farm.
9. LENSKI, LOIS. *We Live in the Country.* Philadelphia: Lippincott (1960), 127 pp. Grades 2–4. A farm in Connecticut, a cotton farm in Georgia, a sheep ranch in Texas, and a tree farm in Louisiana.

Fishing and fishing communities

1. BLAUSCH, GLENN. *Who Lives at the Seashore?* New York: McGraw-Hill (1962), 32 pp. Grades 1–3.
2. BROOKS, ANITA. *The Picture Book of Fisheries.* New York: John Day (1961), 96 pp. Grades 4–6.
3. BUEHR, WALTER. *Harvest of the Sea.* New York: Morrow (1965), 96 pp. Grades 4–6. Fishing in each of the major coastal areas of the United States.
4. BURGER, CARL. *All About Fish.* New York: Random (1960), 138 pp. Grades 4–7. The fishing industry from earliest times to the present.
5. GREENE, CARLA. *I Want to Be a Fisherman.* Chicago: Children's Press (1957), unpaged. Grades 3–5.
6. HARRISON, C. WILLIAM. *The First Book of Commercial Fishing.* New York: Watts (1964), 72 pp. Grades 6–9.
7. LENT, HENRY. *Men at Work in New England.* New York: Putnam (1956), 130 pp. Grades 4–6. Includes fishing and fishing villages in New England.
8. LENT, HENRY. *Men at Work in the South.* New York: Putnam (1957), 128 pp. Grades 4–6. Includes shrimp fishing in the Gulf of Mexico.
9. RIEDMAN, SARAH. *Let's Take a Trip to a Fishery.* New York: Abelard-Schuman (1956), 127 pp. Grades 3–5.
10. SCHEIB, IDA. *The First Book of Food.* New York: Watts (1956), 65 pp. Grades 5–6.

Coronet has a filmstrip on "A Field Trip to a Fish Hatchery."

Cape Kennedy and space cities

1. CHESTER, MICHAEL. *Let's Go to a Rocket Base.* New York: Putnam (1961), 48 pp. Grades 2–5.
2. SASEK, MIROSLAV. *This Is Cape Kennedy.* New York: Macmillan (1962), 60 pp. Grades 2–4.

Chicago

1. KOHN, WALTER and MARIAN. *Chicago: Midwestern Giant.* New York: McGraw-Hill (1969), 64 pp. Grades 5–8.
2. NADEN, CORINNE J. *The Chicago Fire — October 8, 1871: The Blaze That Nearly Destroyed a City.* New York: Watts (1969), 66 pp. Grades 5–8.
3. RAY, BERT. *We Live in the City.* Chicago: Children's Press (1963), 32 pp. K–3. Life in Chicago, with colored illustrations.
4. RICE, MARY JANE JUDSON. *Chicago: Port to the World.* Chicago: Follett (1969), 160 pp. Grades 5–8.
5. SMUCKER, BARBARA C. *Wigwam in the City.* New York: Dutton (1966), 154 pp. Grades 5–7. Susan and her family leave an Indian reservation to live in Chicago. McGraw-Hill has a color filmstrip of 40 frames on Chicago.

Chicago is the subject of chapters or sections in the Follett and Nilsen text on *Exploring World Communities* (Follett), the Quigley volume on *We Live in Communities* (Ginn), and the Senash volume on *Our Working World: Cities at Work* (Science Research Associates).

Houston

Houston, Texas is developed in Paul R. Hanna's volume on *Metropolitan Studies* (Scott, Foresman), largely from the standpoint of government. Houston is also treated in Charles Quigley's text on *We Live in Communities* (Ginn), including material on Clear Lake. It is also examined in the Lawrence Senash volume on *Our Working World: Cities at Work* (Science Research Associates).

Eye Gate House has a color filmstrip on "Houston: A City Expanding."

Teachers will find good background in an article on "Houston: Prairie Dynamo" in the *National Geographic* for September, 1967.

New York City

1. BAILEY, BERNADINE. *Picture Book of New York.* Chicago: Whitman (1962), 32 pp. Grades 4–6.
2. BAMPEY, ABRAHAM. *The New York Story.* New York: Noble and Noble (1965), 87 pp. Grades 4–6.
3. BRENNER, BARBARA. *Barto Rides the Subway.* New York: Knopf (1961), 43 pp. Grades 2–4. Puerto Rican boy learns about the subway system in New York.
4. DAVIS, LAVINIA. *Island City: Adventures in Old New York.* Garden City, New York: Doubleday (1961), 256 pp. Grades 5–7. A beginning history of New York.
5. FLEMING, ALICE. *The Key to New York.* Philadelphia: Lippincott (1960), 128 pp. Grades 5–8. Keys to the Cities series.
6. GARELICK, MAY. *Manhattan Island.* New York: Crowell (1957), 56 pp. Grades 4–6. Rhythmic text and vigorous drawings.
7. HAMMOND, PENNY and KATRINA THOMAS. *My Skyscraper City: A Child's View of New York.* Garden City, New York: Doubleday (1963), 61 pp. Verses and photos.
8. MORGAN, CARD M. *A New Home for Pablo.* New York: Abelard-Schuman (1955), 144 pp. Grades 4–6. A Puerto Rican in New York City.
9. PAULL, GRACE. *Come to the City.* New York: Abelard (1959), 43 pp. Grades 1–3. A child and her family discover the variety in New York City.

10. PEET, CREIGHTON. *The First Book of Skyscrapers*. New York: Watts (1964), 63 pp. Grades 4–6.
11. SASEK, MIROSLAV. *This Is New York*. New York: Macmillan (1966), 60 pp. Grades 2–4.
12. URELL, CATHERINE and others. *The Big City and How It Grew*. Chicago: Follett (1958), 128 pp. Grades 4–6. See the other Urell books listed on page 311.

"Living in Harlem" and "Making a Living" are two filmstrips about Negroes in New York City, from Filmstrip House.

"Sounds of New York City" is a recording by Tony Schwartz which may be purchased from Folkways Records.

One of the John Day Urban Education folios is on New York City.

Philadelphia

1. CARPENTER, ALLAN. *Pennsylvania: From Its Glorious Past to the Present*. Chicago: Children's Press (1966), 95 pp. Grades 5–7.
2. CAVANNA, BETTY. *A Touch of Magic*. Philadelphia: Westminster (1961), 189 pp. Grades 5–8.
3. HALL-QUEST, OLGA. *The Bell That Rang for Freedom*. New York: Dutton (1965), 184 pp. Grades 5–7.
4. LODER, DOROTHY. *The Key to Philadelphia*. Philadelphia: Lippincott (1960), 120 pp. Grades 5–7. In the Keys to the Cities series.
5. MILHOUS, KATHERINE. *Through These Arches: The Story of Independence Hall*. Philadelphia: Lippincott (1964), 96 pp. Grades 4–7.

"Philadelphia-Colonial Shrine and Modern City" is the title of a filmstrip of McGraw-Hill, in color.

"Song of Philadelphia" is a film available from the Commercial Museum in Philadelphia. 14 minutes, in color.

San Francisco

1. FRASER, KATHLEEN and MIRIAM F. LEVY. *Adam's World: San Francisco*. Chicago: Whitman (1971), 32 pp. K–2.
2. FRITZ, JEAN. *San Francisco*. Chicago: Rand McNally (1962), 143 pp. Grades 5–8.
3. SASEK, MIROSLAV. *This Is San Francisco*. New York: Macmillan (1961), 55 pp. Grades 3–5.
4. SELVIN, DAVID P. *The Other San Francisco*. New York: Seabury Press (1969), 167 pp. Grades 5–8.

One of the volumes in the John Day Urban Education Series is on San Francisco. A color filmstrip on "San Francisco" is sold by the McGraw-Hill Company.

Washington, D.C.

1. BAILEY, BERNADINE. *Our Nation's Capital: Washington, D.C.* Chicago: Whitman (1962), 32 pp. Grades 3–5.
2. CARPENTER, FRANCES. *Holiday in Washington*. New York: Knopf (1958), 208 pp. Grades 4–6.
3. EPSTEIN, SAMUEL and BERYL. *First Book of Washington, D.C.* New York: Watts (1961), 48 pp. Grades 4–5.
4. HOLLAND, JANICE. *They Built a City*. New York: Scribner (1953), unpaged. Grades 4–6. The story of planning Washington, D.C.
5. LONG, E. JOHN. *The Real Book about Our National Capital*. Garden City, New York: Garden City Books (1959), 222 pp. Grades 5–7.
6. PETERSON, BETTINA. *Washington Is for You*. New York: Washburn (1962), 63 pp. Grades 4–6.

7. PROLMAN, MARILYN. *The Story of the Capitol.* Chicago: Children's Press (1969), 30 pp. Grades 3–5.
8. ROSENFIELD, BERNARD. *Let's Go to the Capital.* New York: Putnam (1959), 47 pp. Grades 4–6.
9. SASEK, M. *This Is Washington, D.C.* New York: Macmillan (1969), 60 pp. Grades 3–4.
10. SMITH, IRENE. *Washington, D.C.* Chicago: Rand McNally (1964), 144 pp. Grades 6–9.
11. TERRELL, JOHN U. *The Key to Washington.* Philadelphia: Lippincott (1963), 128 pp. Grades 5–8. One of the Key to the Cities series.
12. WOOD, JAMES P. *Washington, D.C.* New York: Seabury (1966), 126 pp. Grades 6–9.

See filmstrips on "Washington: City of the World" (Film Assoc. of Calif.) and "Washington: The Capitol City" (Encyclopaedia Britannica Films).

There is a short account of Washington, D.C. in the Lawrence Senash text on *Our Working World: Cities at Work* (Science Research Associates, 1966). A longer chapter appears in Charles Quigley's *We Live in Communities* (Ginn, 1972).

One of the picture portfolios of the John Day Company is on Washington.

Some films on communities in the United States

"The Changing City." Churchill Films, color, 16 minutes.
"Cities." Gateway Productions, black and white and color, 10 minutes. A history of cities.
"The City." Encyclopaedia Britannica, color, 11 minutes.
"City Highways." Bailey Films, color, 13 minutes.
"Everyone Helps in a Community." Churchill Films, color, 13½ minutes.
"Geography of Your Community." Coronet, color, 11 minutes.
"Our Community." Encyclopaedia Britannica, color, 12 minutes.
"What Is a City?" Bailey Films, color or black and white, 11 minutes.
"What Is a Neighborhood?" Coronet, color, 11 minutes.
"What Our Town Does for Us." Coronet, color, 11 minutes.

Some filmstrips on communities of the United States

Encyclopaedia Britannica has several sets of filmstrips, including:
"The City Community." Six separate filmstrips. Largely middle grades. Color.
"Community Services." Six separate filmstrips. Primary and middle grades. Color.
"The Country Community." Six separate filmstrips. Color.
"Life on the Farm." Six separate filmstrips. Primary grades. Color.
"The Neighborhood Community." Six separate filmstrips. Primary Grades. Color.
"Our Community Workers." Four filmstrips. Black and White.
"Our Public Utilities." Six separate filmstrips, each on a utility. Color.
"The Town Community." Six separate filmstrips. Middle grades. Color.
Jam Handy has sets of filmstrips on:
"My Neighborhood." Six separate filmstrips. Color.
"Our Community." Eight separate filmstrips. Color.
"Our Neighborhood Helpers." Six separate filmstrips. Color.
They also have filmstrips on "Living in a Town," color, and "Living in a Big City," color.
McGraw-Hill has filmstrips on the following:
"Farm and City Series." Four filmstrips. Color.
"Community Helpers Series." Two sets with six filmstrips each. Color.
They also have a series of six filmstrips on health in communities.
The Society for Visual Education has a series of six color filmstrips on:

"Community Helpers." They have two filmstrips on "Cities and Commerce."
A filmstrip on "Five Great Cities" includes New York, Boston, Philadelphia, Baltimore, and Washington.

The Society for Visual Education has a set of full color prints, 18" x 13" on
"Urban Life" and Picture Story Study Prints on "Community Helpers."
   For other films and filmstrips see *Educational Media Index* or other sources.
   Eye Gate House has a series of nine filmstrips on U.S. cities: Birmingham, Los
Angeles, Boston, Chicago, Detroit, Seattle, New York, and Houston, plus one on
"Why and How Cities Grow."
   The Senash volume has an accompanying set of 12 color filmstrips and six
10-inch records.

40.   Combining role-playing and map study in learning about the Bedouins

Communities in Other Parts of the World

41.   Using a transparency with the overhead
      projector to study London

42.   Reporting to the class in a study of
      the two Berlins

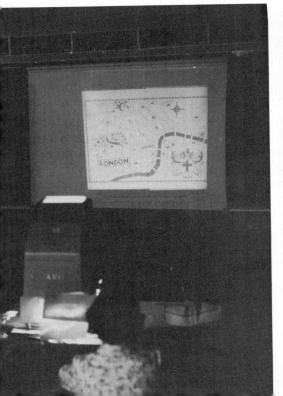

COMMUNITIES IN OTHER PARTS OF THE WORLD

In the primary grades, having studied selected families of the U.S.A. and other parts of the world and selected communities of the U.S.A., boys and girls should be ready to cope with the study of selected communities in other parts of the world by the fourth grade. Such a study could be carried on in other grades, but the fourth grade seems an especially appropriate place for such a theme. Most children at that stage in their development are not yet ready to undertake studies of countries, but they can handle the smaller and less complicated unit of the community. Having looked at several communities in our country, they should now be ready to look at a few carefully selected communities in other parts of our planet.

To such a study they should bring several skills learned in the examination of families at home and abroad, and even more pertinent skills gained in examining communities in the United States. Now they can apply and test, in new situations and new places, the concepts and generalizations they have acquired.

Since most of the people of the world still live in hamlets, compounds, or villages, pupils need to become acquainted with a variety of these smaller units of society in various parts of the globe. But people all over our planet today are moving to the large cities. Of course cities have existed for centuries, but they are growing by leaps and bounds today in every part of the world. For example, there are now more cities of one million or more inhabitants in Latin America than in North America. If this is difficult to believe, check the figures and see that this statement is true. It seems highly important, therefore, that pupils study some of the large cities of the world as well as villages and small towns.

It would be ideal if pupils could study several villages and cities in many countries. In that way they would begin to understand the variety of ways of life within any nation. However, time does not permit such an approach. Therefore it is recommended that at least one village or small town and one large city be examined in each of several nations in different parts of the world. If possible, the small town or village studied would be relatively close to the large city examined. Thus their interrelationships could be stressed. For example, your pupils might study a village or small town in Switzerland where the people live largely off the land, possibly earning their living by dairying. This community might well be near Bern or Zurich. Or you might select a small town in Finland where the people earn their living by forestry. It might be a community which is tied in many ways to the city of Helsinki.

Since the world is such a large place and the amount of time which can be spent on the study of communities is limited, it seems wise to limit a year's work on this topic to twelve to sixteen communities in different parts of the globe. Otherwise the study of communities is likely to be superficial.

In studying such communities outside the United States, you will probably want to use the "wheel" which is presented on page 288. The emphasis again would be on communities today, although the historic dimension should be included for at least some of the communities examined, such as London and Jerusalem. All twelve of the sections of this "wheel" need not be used for every community studied; you may want to emphasize three or four aspects for each village or city. During the year, concepts and generalizations would be drawn from all the social sciences.

Some Criteria for the Selection of Communities to Be Studied

If only a few communities outside the United States are to be studied, great care needs to be exercised in the choice of those places. Here are a few suggested criteria for the selection of communities in other parts of the world:

1. Communities in most of the eight cultural areas of the world.
2. Communities which represent a variety of geographic conditions.
3. Communities which represent a variety of ways of earning a living.
4. Communities which represent different value systems and/or religions.
5. Communities which represent villages and cities.
6. Communities which represent some differences in economies and types of governments, although pupils at this point probably are not yet ready to tackle communities which represent communism.
7. Communities which represent continuity *and* change. Some history can be included, although not highlighted.
8. Communities for which there are at least some resources available.

One special problem arises in the selection of communities. That is the relation of this recommended work in the fourth grade to the study of families in the primary grades and the study of countries in the sixth or seventh grades. Three possibilities present themselves. They might be charted as follows:

	Families	Communities	Countries
Plan A:	Japan	Japan	Japan
Plan B:	Korea	Hong Kong	Japan and/or China
Plan C:	Japan	Korea	Japan and China
Plan D:	Japan	Thailand	China and Japan

Under Plan A, pupils would gain considerable depth in the study of one part of the world. By the time they were through the sixth or seventh grade they would know a great deal about Japan. But this plan does not give a broad view of the region of East Asia. It includes the danger of overlapping.

Plan B gives children a broader view of East Asia and avoids overlapping. There is considerable merit in this approach.

Plan C combines the merits of Plans A and B. Children are introduced to families which are not too different from their own. Then they study a rice village in Korea and the city of Seoul. In the sixth grade they return to Japan and add the study of China. There is some concentration on Japan in this case, but pupils are exposed to a fairly broad view of East Asia.

You may find that you have selected four or more small towns and cities in Europe and that you have too many communities. In that case you may want to include Southeast Asia and East Asia as one part of the world. That is proposed in Plan D.

Some Communities Which Might Well Be Studied

In planning the volume for the new Ginn Social Science Series on *Everyone Lives in Communities*, we finally decided upon the following communities. An explanation of the choices made and some of the emphases will be found on page 321.

Cultural or geographic area	Communities in a given country	Some aspects to stress
Germanic-Scandinavian	Bueren, Switzerland	Geography, farming.
	Bern, Switzerland	Old city and new areas; economics (chocolate process).
Other European groups	Korpilahti, Finland	Economics (lumbering), geography.
	Helsinki, Finland	City planning in Tapiola.
Anglo-Saxon	London, England	Variety of people, trade and transportation, some history.
Latin	Two mining communities in Chile	Geography, economic process (copper).
	Valparaiso, Chile	World trade, poverty.
Africa: South of the Sahara	Moshi, Tanzania	Geography and economic process (coffee).
	Dar Es Salaam	Variety of people; rapid change; city planning.
Asia	Nakluya, Thailand	Economics (rice growing and fishing, religion).
	Bangkok, Thailand	Variety of people (minorities), handcraft industries, beauty.
Muslim	The Bedouins	Geography (living in a different environment), religion.
	Kuwait City	Economics (oil process) and rapid change.
Other Middle East groups	Kfar Vitkin, Israel	Pioneers in a new land and economic process (oranges)
	Jerusalem, Israel	Geography, history, government, and religion.

Obviously there are many other communities which could be added to this list. For example, it would be profitable to have a ranch in Australia and the city of Sydney. A changing village in India and Bombay or Calcutta would strengthen the list in some ways. Communities in southern Europe and the Iberian peninsula could be selected to represent the Latin culture. Communities in Canada and Mexico could be added. But then the list becomes too long. One has to limit the list to guarantee some depth.

You may have noticed that there are no communities from the communist (or left-wing socialist) countries. That omission was intentional. It is this writer's contention that such communities cannot be studied realistically without considerable attention to the special systems of economics and government under which they operate. Since that is so difficult to do with boys and girls in the fourth grade, such communities were omitted at this grade level. The first introduction of pupils to communism is held until the sixth grade where they study in considerable depth the U.S.S.R. and China. Would you agree with this point of view? Why? Why not?

The communities finally selected, however, do represent a wide range.

Pupils using such a textbook meet a variety of people from different continents and from different cultural regions. They meet people from Europe, the Middle East, Africa, Asia, and South America. Often they meet minority groups in a community. This is true, for example, in Bern (where they meet "guest workers" from Italy and Turkey), in London, in Dar Es Salaam, in Bangkok, and in Jerusalem.

Pupils will be introduced, also, to people of various religions— Christians of different backgrounds, Moslems, Buddhists, Jews, and animists.

Pupils will likewise be introduced to a large number of different types of jobs. And several economic processes are stressed— including lumbering or the production of paper, chocolate, coffee, copper, rice, and oil. By including the Bedouins and Kuwait City, a tremendous range in ways of living and in ways of earning a living is made possible.

The concept of continuity is emphasized in several communities (such as London, Bangkok, and Jerusalem). But change and planning for change are highlighted in other communities (particularly Helsinki, Dar Es Salaam, and Kuwait City).

In almost every community, there is some treatment of local government. Some special attention is also paid to the cities which are capitals. And the United Nations is brought into the book in such places as Bern and Bangkok.

Many aspects of geography are also presented. For example, the location of the small towns or villages and the cities range from deserts to mountain regions. Several of the communities are seaports; a few are located on rivers. The communities of Bueren and Bern are in valleys. So one could continue to enumerate the geographical features of these 15 communities.

Perhaps you have wondered about the order in which these communities appear. The first five are European communities. They are different in some ways from the communities in which the readers live and from the communities studied in the third grade. But the first five are not so different that they will be difficult for the readers to study early in the year. And they are communities which should not frighten teachers who have never taught about communities in different parts of the world.

YOUR LIST OF COMMUNITIES TO BE STUDIED

Perhaps you would like to compile your own list of 12 to 16 communities in different parts of the world which you would like to study with your fourth grade pupils. If so, space is provided on this page for such a list. You may use the suggested headings or disregard them.

Cultural area   *Communities*   *Major emphases*

Cultural area   *Communities*   *Major emphases*

Studying Smaller Communities Around the World

Perhaps you have already begun to wonder what you would teach about the small communities of the world if the course of study in your school called for such work. Perhaps you have already begun to make a list of ideas in your mind that you would stress or help children to discover.

Probably you know already that most of the people of the world live in villages, hamlets, compounds, or small towns. Many of them are now moving to larger cities, but the largest part of humanity still lives in relatively small clusters of people, by whatever name you call them. In India there are somewhere around 500,000 such villages; in China there are probably nearly a million villages. In other parts of the world there are additional millions of such small groups of people.

It is only in places like Canada, New Zealand, Australia, the United States, and parts of East Africa that farmers live on their own land. In most of those places such a life is possible because of the large acreage. Of course that is not true in East Africa.

In most parts of the world people have lived and still live in small communities for three basic reasons. One has been for defense. That may not still be true today in many places, but it is true in a country like Israel. For the most part this is an historic reason. A second explanation for small communities is that people want to be near others for sociability. A third reason is because most farmers own very small plots of ground and can walk to work. The average farm in the world is only four or five acres and often that acreage is divided into several small plots in different areas because of the way in which land has been divided among several sons. Lack of adequate transportation might be considered a fourth factor. Because of the expense of cars or trucks or tractors, most farmers cannot possibly own them.

Over the period of a year you may want to help your pupils learn about the wide variety of small towns and villages in the world. Some of them consist of only a few persons. They may be related to each other or may belong to the same tribe. Often these persons share a common faith.

You may also want to help your pupils discover the differences in the ages of communities. There are a few new communities around the world. They may be villages and towns in a country like Israel where newcomers have staked out new communities. They may be new places in the Philippines or Vietnam or some other Southeast Asian country where refugees have been assisted by the government to start new communities. But most communities around the world are very old. Some have existed for centuries.

Undoubtedly you will want to help children find out how various communities govern themselves. It may be important for you to remember in this regard that some form of democracy is practiced in local communities in large parts of the world, even though their national governments may not be democratic. In the past such local governments were often councils of older men. In a few communities women were selected, especially in Africa. With your pupils you can develop a list of the services which local governments render citizens and see if the list applies to all communities.

In most small communities, your pupils will not discover a great variety of people, but there will be different types of people in some villages or towns.

Certainly you will want to explore with the boys and girls in your class the variety of ways in which people earn their living. We already have indicated that the choice of communities proposed was made partially upon that basis. This aspect of community life should lead often to a consideration of specialization, one-crop economics and their effect on standards of living, the exchange of goods, and transportation as it relates to moving of crops and goods.

You may not want to go very deeply into a study of the various religions or value systems, but you can hardly ignore this topic. Children will be confronted with strange customs and ways of living. They should learn that there are reasons for such customs and modes of life, based usually upon a value system or a religion. Children may not agree that the ways are good ways, but they should learn to respect people with other ways of living.

You undoubtedly will want to explore education in its broadest sense. The basic question in most instances will be, "How do children learn the ways of their families and their communities?" In many parts of the world children do not attend school; they learn their ways of living solely in their homes and in their communities. In other places you will find special emphasis upon education. This would be true, for example, in Switzerland, Israel, and Finland. In all three countries the basic resources are few, and the people have banked heavily upon human resources to raise their standards of living. Children can understand something of the importance of education in such places.

Health is another factor which needs to be studied. Pupils will be amazed to learn how few medical people and how few medical services there are in large parts of the world. This would be true, for example, in the villages of India and Africa and large parts of East Asia. No doubt your brighter pupils will inquire why this is true. One reason is the low standard of living, with two-thirds of the people of the world earning less than $200 per person per year. Another reason is the lack of education. A third is the prevalence of superstitions. But in other places health ranks high, as in The Netherlands.

Your pupils surely should discover how clever people are all over the world in learning to adapt to their natural surroundings. Boys and girls should be amazed at how well Bedouins have adjusted to life in the desert. They should be impressed that members of the Chagga tribe learned a long time ago to build irrigation ditches to make the best use of their limited water supply for raising crops. They should be impressed with the ingenuity of villagers in rice-growing areas in using that grain for such a wide variety of purposes.

They also should be confronted with the fact of change in communities. This is occurring rapidly in many parts of the world, especially in the so-called economically underdeveloped areas.

Your pupils should also enjoy some of the activities of the people who live in small communities. This may be their music. It may be their dances. It may be their holidays and fun-days.

There are many aspects of life for you and your pupils to explore. Such studies will include many aspects of the social sciences, from economics to history and from anthropology to political science.

Here are a few notes which may be helpful to you in the study of small communities in various parts of the world. From them you can cull the ideas that you feel are most important to you — and to your pupils.

1. *Small communities exist in all parts of the world.*

   Villages or small communities exist in every part of the world.
   A million such villages in China; 500,000 in India.

2. *There are reasons for the location and physical pattern of communities.*
   (a) Often they are located on or near bodies of water.
       For water supply and for transportation.
   (b) Sometimes they are at important junctions, such as camel caravan junction points — for example Kano, Nigeria.
   (c) In the past, communities often were built on hills, for defense.
   (d) The shape of communities depends in large part on geography.
       Paths leading to the water.
       Paths or streets along the waterfront.
   (e) Sometimes communities are built around a central plaza.
       This may be a market, a grazing spot, or a place of worship.

3. *Many communities are small in area and in population.*
   (a) There may be from fifty to five hundred persons in villages.
   (b) Often these persons are related.
   (c) Small communities are not likely to have subdivisions.

4. *People in small communities earn their living in varied ways.*
   (a) In some places most or all of the laborers are engaged in one major activity.
   (b) This is especially true in countries with one-crop economies or one major occupation.
       A community may specialize in raising coffee or cacao or sisal.
       A community may specialize in rugmaking or producing cheese.
       A community may devote itself largely to mining.
   (c) In all communities there are some special jobs.
       From barbers to persons performing religious rites.
   (d) The caste system of India may have developed because of specialization in villages.
   (e) Some communities are self-sufficient with a variety of jobs.

5. *Transportation is important in most communities.*
   (a) There is a tremendous variety of modes of transportation in communities around the world.
       In many places the backs of human beings are still used for transportation.
       Animals used for transportation vary from buffaloes to llamas, depending upon the locality.
       Bicycles are increasingly important in many parts of the world.

Many more modern means of transportation are being introduced all over the world.

(b) Paths or dirt roads are still very common throughout the world.

(c) With modernization, hard-surfaced roads are increasingly important to transport goods from one place to another.

(d) Lack of good transportation isolates a community.

6. *Means of communication often are limited in small communities.*

(a) Some small communities have little contact with other communities. Some communities want to be isolated — often for religious reasons or to protect their way of life.

(b) Small communities usually cannot afford their own newspapers, radio stations, and other mass media. However, many villages have one radio.

(c) Yet, some small communities have much contact with the outside world. Switzerland is an example of a country with a highly developed communication system — telephones.

7. *Value systems in small communities are important.*

(a) Small communities may be organized around families or religions or tribes.

(b) People tend to be more homogeneous in their beliefs — and more conservative in upholding the status quo.

(c) However, small communities may be more supportive of individuals than larger ones. People know and care for each other.

(d) Some small communities, nevertheless, may be torn by divisions along religious lines or differences in other values.

8. *Small communities have their own organizational patterns.*

(a) Small communities do not always have formal, organized governments. The pattern of control may be informal. Nevertheless, it exists.

(b) There may be much cooperation in economic activities.
The cooperatives in the Scandinavian countries are examples of such cooperation economically.
In communist countries there is control and central planning.

(c) There may be some political organization of small communities.
In many parts of the world the local governmental organization depends upon age — usually older men in a council.
Democracy at this level is much more common than at the national level.

(d) In most smaller communities there is more participation by the people today than in times past.

9. *Health and educational activities are carried on in small communities.*

(a) Health is closely related to standards of living.
Health in small communities less likely to be good than in larger places.
Fewer facilities for health — doctors, nurses, hospitals.

(b) Superstitions may exist in smaller, isolated communities.

(c) Children learn the ways of their communities from adults and older boys and girls. Some societies provide for special training by "age groups" — as in African tribal society.

(d) Schools may be the center of activities in smaller communities.
Often they are controlled by a religious group.

Teachers often among the "elite" in small communities; leaders.
  (e) The introduction of schools may radically change local communities.

10. *There are always some means of recreation in small communities.*
  (a) Children always have some special games and pastimes in any community in the world.
       These activities often imitate the work of adults.
       Such activities depend in part on the geography of the locality.
  (b) Children are included in adult activities more often than in larger communities.
       Attend puppet shows with their parents (in Indonesia), take part in community dancing (throughout Southeast Asia and other places).
  (c) Some means of recreation are different from leisure-time activities in the United States.
       Important place of the dance throughout most of the world.
       Storytelling by older people in many parts of the world.
       Vast amount of reading during long winters in far north communities.

11. *Small communities may have some contacts with other communities.*
  (a) Some small communities are still isolated, sometimes by choice.
  (b) Small communities today have increasing contacts with other places, because of improved transportation and communication.
  (c) Markets and market days provide contact with the people of other communities in many parts of the world.

12. *Changes are taking place in many small communities today around the world.*
  (a) The building of roads, dams, airports and other construction is bringing about change in many places.
  (b) The introduction of radio and television is facilitating change, too.
  (c) Many nations are concentrating upon community development programs in small communities.
  (d) Schools are another factor in fostering change.

13. *Most small communities have long histories.*
  (a) Many new communities are being built today, but most small communities have long histories.
  (b) People in most places are proud of their past and try to perpetuate this pride.
       Local celebrations help to perpetuate pride in the past.
       Monuments often are erected to remind people of the past.
       Stories of the history of the locality are told to children by adults.
       Place of the storytellers in some localities.
       Sagas of the Norsemen told by adults to children.
  (c) Hostilities between communities also may be perpetuated by adults.

14. *The future of many small communities is uncertain.*
  (a) Many people are leaving small communities for larger cities.
       Usually men go first, without their families.
       Often young people move for better education, better jobs, or for the "bright lights" of cities.

(b) Changing agricultural practices demand more land for farms, affecting life in small communities.
(c) Conflicts often occur between the proponents of change and those who support the status quo.

Studying Some Larger Cities Around the World

Boys and girls in the middle grades, however, should not be confined to the study of several villages or small towns. Such an approach would be misleading. The study of such small communities needs to be supplemented by the examination of several larger cities. It is especially important that these be included, because cities will be increasingly the centers of change. Also, pupils should not think of the people of the world as living solely in small towns and villages when so many of them live in large cities.

Sometimes teachers are unaware of the rapidity with which cities in many parts of the world are growing. For example, there are now over 10 cities in the U.S.S.R. with a population of over one million. There are various estimates of Mexico City; some recent "guesses" place it at a figure of over nine million. A city like Seoul, Korea is now over three million.

There are many cities from which you may select. As we have indicated earlier, you may want to choose such places as London, Bern, Helsinki, Jerusalem, Kuwait, Dar Es Salaam, Bangkok, and Valparaiso. You may want to select other cities, using other criteria.

To this writer it seems important to show the relationships between the villages or small towns studied and the larger cities. It would be wise, therefore, to study the small town and the larger city consecutively rather than making a study of all the small towns and then all the larger cities. In most instances you will want to start with the smaller locality and then move on to the larger one. But sometimes the order can be reversed. The city can be studied and then the question raised as to how it exists — thus introducing its dependence upon other smaller places. Such a plan provides for variety, an important factor in learning.

In the study of cities, all the factors listed on page 288 on the wheel or chart of how to study a community, should be included. They should not be considered, however, in the same order so that the chart becomes a kind of ritual. Nor should the emphasis be the same for each city. You will want to focus on different factors for each locality. For example, in studying Jerusalem you might want to concentrate on its history and religious significance. When studying Bangkok you might want to stress its arts and crafts and its importance as a center of United Nations activities for Southeast Asia.

There are a good many similarities among the cities of the world. These need to be discovered by boys and girls and underlined by teachers. As a matter of fact, the cities of the world are much more alike than are the villages and small towns on our planet.

But there are differences, too, and these should be brought out in any study of cities. There are differences in location, in the types of jobs held, in architecture, in transportation, in dress, in government and in other characteristics.

You will need to bear in mind also the fact that population figures for cities have to be examined closely. Sometimes they are for the city itself and often they are for the entire metropolitan area. This accounts for what are apparently discrepancies in the figures for various cities.

Here are some statements which may help you as you think about the study of some of the major cities of the world:

1. *There are large cities in almost all parts of the world, and most of them are growing rapidly today.*
   (a) The fifteen largest cities of the world today are Tokyo, New York, Shanghai, Moscow, Bombay, Peking, Chicago, Cairo, Rio de Janeiro, Tientsin, Leningrad, Osaka, London, São Paulo, and Mexico City.
   (b) The highest percentage of city dwellers to total population for any region is in Oceania — Australia and New Zealand.
   (c) Asia and Africa have large cities, despite popular belief to the contrary.
   (d) About one person in every nine or ten lives in a large city today.
   (e) Around the world, most large cities are growing rapidly.

2. *People are moving to large cities for a variety of reasons.*
   (a) Lack of work on farms and in small towns as specialization and mechanization cut down on the number of people needed in villages and small towns.
   (b) Special conditions on farms, such as the drought area in northeastern Brazil, are driving people to the cities.
   (c) The cities need workers for factories and other jobs.
   (d) Hope of a better standard of living in the cities.
   (e) Hope for a better education for children.
   (f) The success of relatives or friends in cities.
   (g) The lure of the "bright lights" in cities.

3. *Many cities were rebuilt after World War II.*
   (a) This was especially true in Europe and parts of Asia (especially in Japan) Rotterdam, Berlin, and Hiroshima are three prime examples.
   (b) Some old cities are being rebuilt with the aid of famous architects, such as Baghdad in Iraq.

4. *Some new cities are being built around the world.*
   (a) Capitals of regions or countries, such as Brasilia in Brazil, Islamabad in Pakistan, and Chandigarh in India.
   (b) New towns for industry and trade, like Tema in Ghana, as a result of the building of the Volta Dam. Many such cities in the Soviet Union.

5. *There are special reasons for the location of large cities.*
   (a) Almost always they are on some large body of water.
       Tokyo on the Sumida river where it meets the sea.
       New York on the Hudson where it meets the sea.
       London on the Thames.
   (b) They need a large "hinterland" to exist, especially for food.
   (c) A few large cities of the world are on camel caravan routes or originally were established for that purpose.
       Kano in northern Nigeria and Teheran in Iran.

6. *Cities tend to have a wide variety of people.*
   (a) Cities draw from large parts of a nation with diverse peoples.

(b) Often they draw from other countries.
Australia, for example, is drawing its labor supply from southern Europe today.
(c) Usually there is great diversity in religions represented.
For instance, Tokyo with Shintoists, Buddhists, and Christians.
(d) Value systems may bring conflicts between groups in large cities.
(e) There is almost always a great diversity of income in large cities.
Great wealth and great poverty, plus a middle class.
Bombay and Calcutta examples of tremendous poverty.

7. *The people in large cities earn their living in many ways.*
(a) With so many people, cities demand a great variety of services and jobs.
(b) Most cities are manufacturing centers.
(c) Most large cities are financial centers.
(d) Most large cities are cultural centers.
(e) Some cities are the capitals of countries, although the largest city is not necessarily the capital.

8. *Cities have a variety of means of transportation.*
(a) Water transportation in almost all large cities.
(b) Cities are railroad centers. Also bus centers today.
(c) Cities today have airports.
(d) Oxcarts or camel carts can still be seen in many cities. Bicycles are seen in large numbers in many cities. Motorcycles, pedicabs and taxis are common. Streetcars frequently are found in large cities.
(e) Several large cities have subway systems.

9. *Cities have similar means of communication today.*
(a) A variety of languages are usually spoken in any large city.
(b) Telephones, telegraph, radio, and television common to all major cities in the world today.
(c) In some cities are found means of communication which we use infrequently or not at all in the United States.
Kiosks for communication in many cities.
Wall newspapers in some cities.
(d) Many newspapers in most large cities.
Often special language papers for groups of people.
(e) Governments are relying heavily on radio and television to reach large numbers of persons.
Radio and television are government-owned in many large cities.

10. *Cities are usually centers of higher education.*
(a) There are usually several colleges or universities in large cities.
Some of these are private institutions in certain countries.
Some of these are government-run institutions, especially in the communist countries where there are only government institutions.
(b) Often there are research institutes in large cities.
(c) There are almost always centers for the study of science, technology, and the arts.

(d) Throughout history, cities have been the seedbeds of most new ideas and changes.
This is due in part to capital to support special activities.
This is also due to the ability of people to exchange ideas.
Another factor has been (and still is) contacts through trade with people in other parts of the world.

11. *Cities usually have long and interesting histories.*
    (a) Many cities are hundreds or even thousands of years old.
    Rome, Cairo, Shanghai, Jerusalem as examples.
    (b) The people of a city often take great pride in their history.
    Monuments honoring heroes and special holidays of the city.
    (c) Some cities attract many tourists because of their history.
    London, Rome, Jerusalem, or Mexico City as examples.

12. *Cities have their own governments.*
    (a) The type of government depends usually upon the type of government of the country in which the city is located.
    Moscow and Leningrad and communist governments; London and Paris with democratic governments.
    (b) Most large cities of the world have some type of elected government with a mayor and a council.
    (c) City governments are plagued with fairly common problems of large cities.

13. *Cities have problems.*
    (a) Any large concentration of people causes problems.
    (b) The rapid growth of large cities causes special problems.
    (c) Among the common problems of large cities are:
    Health measures.
    Care of the unemployed and the poor.
    Group conflicts based on race, religion, caste, language, etc.
    Crime.
    Housing.
    Transportation and communication.
    Education.
    Planning and zoning.
    Care of refugees.
    Graft and corruption.
    Pollution.

14. *Most large cities spread out to include large areas outside the city.*
    (a) The wealthy and the middle class often move to the suburbs.
    (b) The nature of the transportation system determines in part whether people move to the suburbs.
    Inner cities growing in a place like São Paulo with poor transportation.
    (c) Some cities in the past have spread out because of the terrain.
    Inability to build tall buildings in the center of the city in the past in a city like London. Lacked hard rock base, which New York City has.

15. *Opportunities for a variety of leisure time activities exist in cities.*

   (a) Some of this is due to aggregations of wealth.
   (b) Some of this is due to the variety of people and their varied interests.
   (c) Some of this is due to government assistance.
   (d) Some of this is due to the fact that cities are more likely to be centers of change — new interests.

METHODS AND MATERIALS FOR STUDYING COMMUNITIES IN OTHER PARTS OF THE WORLD

The study of communities in various parts of the world from many angles is a new approach in the social studies programs of elementary schools. Heretofore communities have been studied almost exclusively from a geographical point of view. Therefore there is a great deal of room for experimentation on the part of teachers. What communities should be studied? How many can be studied in depth in a year? Is there a real carryover or transfer of skills learned in studying U.S. communities when pupils attack communities in other parts of the world? To what extent can boys and girls understand the different ways people live in places they have not seen? These are some of the questions which need considerable examination.

Teachers should know, however, that there are several places in the United States where such full-fledged studies of foreign communities are underway. The schools of Chicago, for example, are studying Chicago, New York, and London in the third grade. The schools of Contra Costa County, California, are studying several communities in different parts of the world. Recently the schools of Ridgewood and Fair Lawn, New Jersey and Great Neck, Locust Valley, and Manhasset, New York have embarked on a study of communities of the world in the fourth grade. This is an important and interesting development and one in which you may want to take part, either with special permission as an experimental project or as a part of a schoolwide curriculum revision.

A large number of teachers have been ready to teach about communities outside the United States, providing that they got the "green light" from their schools or school systems and/or had adequate materials. Within the last three or four years several textbooks have appeared with this emphasis. It seems likely that this new approach will gain considerable ground in the next few years. For a list of such textbooks, see pages 334–335.

The approaches to the study of communities in various parts of the world should be similar to those used in studying communities in the United States. However, since few pupils will have visited any of these places, there should be even more visualizations used than in the study of U.S. communities. All the pictures you can assemble will be helpful for children to view individually and in small groups, as well as in class groups, using opaque projectors. You will want to use many films and filmstrips, too, to help boys and girls visualize these places. If the captions are difficult, you can read them. Or you can cut off the sound track on films and arrange the filmstrip projector so the captions are not visible, letting children discover for themselves the communities they are studying, without being told by some adult what they are to see.

Much use should be made of maps in the study of communities in various parts of the world. Every class undertaking such studies should have a large vacant space on the floor where a community can be constructed. Lacking such space, a class should have a table or sandtable where the community under examination can be reproduced. It can be developed as the pupils read about it, see pictures of it, interview people who have been there, and listen to accounts read to them by their teachers. Such an activity will make learning more real and the development of skills more effective than merely reading about communities in other parts of the world.

Wherever possible, resource persons should be used. This may mean the use of a student from the country in which your "current community" is located. It may be a teacher or an adult in the community who has visited in that part of the world. In a few instances it can be a governmental representative from that country. Since such persons cannot be asked to visit several classes or make repeated visits, their visits should be taped or even filmed for use by other classes and in future years. In some instances such resource material should be made available to other school systems.

Recordings of the music and films of the dances done by people in the various communities being studied should be other methods used by teachers.

Sometimes posters and charts can be obtained from governmental information bureaus, travel agencies, or business firms.

If you know of people who are living now in a community you are studying, or who are traveling there, you should ask them to help you with materials. If they can take pictures, fine. If teachers are traveling to such communities, the school board should be asked if they will contribute film, provided the pictures are to become the property of the school system.

Magazines like *Holiday, Life,* and the *National Geographic* should be "mined" for possible pictures.

Files of clippings and articles should be kept on the communities you are studying for your use, and often for use by pupils.

Since the number of books and other materials is still somewhat limited on many communities you want to study, it is recommended that all classes in a given school not study the same community at the same time. In that way all the resources will be available at one time to a class.

On villages the best source for books is the series by Sonia and Tim Gidal, published by Pantheon Press. Several of those volumes are listed in the bibliography at the end of this section.

As we mentioned earlier in this section, there are now several textbooks with material in them on communities outside the United States. Some deal exclusively with communities outside our nation. Others combine a study of communities in the United States with some from other parts of the world.

There is a splendid opportunity for potential writers and audio-visual persons to prepare supplementary material on various communities of the world. This is an area of the curriculum in which new materials are sorely needed.

Some teachers and/or school systems may want to explore the possibility of developing affiliation programs in one or two parts of the world, exchanging materials with those schools over a period of years. Parents as well as teachers and pupils can be involved in such programs. The materials sent from abroad should be kept carefully in the school library or in some other place so that they can be used many times by many classes. Such materials can be invaluable to a school or school system.

With some groups you may want to develop dioramas or murals of the community you are studying.

There are numerous ways to introduce the study of a community in another part of the world. A film or a filmstrip may be a good introduction. A set of pictures may be shown to pupils and a discussion may ensue on how they would live and earn a living in such a locality. For example, a set of five or six pictures might be shown of a lumbering community in Finland and the pupils asked how they would build a house, what they would do for a living, and how they would transport their products. With such pictures of water and trees, real discovery learning can be carried on.

Occasionally, visitors can be used to tell about their communities to answer questions. In some cases you will want to start a study of a community in this way. At other times you will want to use your visitor after the class has done some studying and has questions for the visitor.

Stories of children in different places can be read aloud as a way of motivating the study of a given community. After you have read a short section, you can engage the class in summarizing what they have learned and then turn to questions which remain unanswered, for which they will need to find answers.

Since some of the materials on communities are too difficult for slower readers, you may want to do considerable reading aloud to pupils. Undoubtedly you will want to use the better readers in your class and have them read to small groups doing research or to the entire class. Some schools may want to have good readers tape-record some materials and have slower readers listen to the recordings.

Teachers should develop a good background on the countries in which the communities they study with pupils are located. But they should *not* attempt at this point to study countries. It is highly important that they limit the study to communities. For most pupils in the fourth grade the concept of a country is too difficult and involved. Also, there is considerable danger of overlap between the study of communities and countries unless some sharp lines are drawn.

Since most teachers do not have a very good background on various communities of the world, they may want to team up with other teachers. Each teacher in a school with several sections of one grade could become the expert on one, two, or three sets of communities, sharing his or her information and materials with colleagues. This would not necessarily be "team teaching," but it would be a form of cooperative planning.

Professors in schools of education or departments of education would enhance the background of their students and would contribute to the teaching profession if they worked on units on various communities of the world and made them available to teachers throughout the nation.

In the pages that follow are many books and some films and filmstrips on specific communities. A general reading list for teachers is also included so that teachers may enhance their own store of information.

Materials on other communities are also available but are not included here because of the limitations of space in this volume.

You may want to start a notebook or folder on each of the communities you plan to study, placing in it everything you can lay your hands on about those localities, adding to your storehouse as you find new materials in newspapers and magazines. Such notebooks or folders should be invaluable to you now and in the foreseeable future as a source of background information, teaching methods, and pictorial materials.

BIBLIOGRAPHY FOR TEACHERS ON COMMUNITIES OF THE WORLD

1. ANDERSON, NELS. *The Urban Community: A World Perspective.* New York: Holt, Rinehart and Winston (1959), 500 pp.

2. BOOTH, ESMA R. *The Village, the City, and the World.* New York: McKay (1966), 282 pp.
3. DEYOUNG,, JOHN E. *Village Life in Modern Thailand.* Berkeley, California: University of California Press (1955), 225 pp.
4. DORE, R. P. *City Life in Japan.* Berkeley, California: University of California Press (1958), 412 pp.
5. DUBE, S. C. *Indian Village.* Ithaca, New York: Cornell University Press (1955), 248 pp.
6. GOODFRIEND, ARTHUR. *Rice Roots: An American in Asia.* New York: Simon and Schuster (1958), 209 pp. The author and his family spend a year in a village in Indonesia. Very readable.
7. Gunther, John. *Twelve Cities.* New York: Harper and Row (1969), 370 pp. On London, Paris, Brussels, Rome, Hamburg, Vienna, Warsaw, Moscow, Jerusalem, Beirut, Amman, and Tokyo.
8. HALL, PETER. *The World Cities.* New York: McGraw-Hill (1966), 256 pp. On London, Paris, Randstad (Holland), Rhine-Ruhr, Moscow, New York, and Tokyo.
9. International Urban Research. *The World's Metropolitan Areas.* Berkeley, California: University of California Press (1959), 115 pp.
10. LEWIS, OSCAR. *Tepoztlan: Village in Mexico.* New York: Holt, Rinehart and Winston (1960), 104 pp.
11. LEWIS, OSCAR. *Village Life in Northern India.* New York: Random (1965), 246 pp.
12. MORRIS, JAMES. *Cities.* New York: Harcourt Brace Jovanovich (1964), 375 pp. Short, vivid accounts of many of the major cities of the world. Highly readable.
13. MYRDAL, JAN. *Report from a Chinese Village.* New York: Random (1965), 374 pp.
14. NAIR, KUSUM. *Blossoms in the Dust.* New York: Praeger (1962), 201 pp. A novel on the impact of change on an Indian village. Also a paperback.
15. POWDERMAKER, HORTENSE. *Copper Town.* New York: Harper and Row (1962), 391 pp. A study of a town in Southern Africa.
16. *Scientific American* staff. *Cities.* New York: Knopf (1968), 211 pp. Includes Calcutta, Stockholm, Ciudad Guayana, and New York.
17. WAGLEY, CHARLES. *Amazon Town: A Study of Man in the Tropics.* New York: Macmillan (1953), 305 pp. A Brazilian village.
18. WILIE, LAURENCE. *Village in the Vaucluse: An Account of Life in a French Village.* New York: Harper and Row (1964), 375 pp.
19. WISER, WILLIAM and CHARLOTTE. *Behind Mud Walls: 1930–1960.* Berkeley, California: University of California Press (1963), 249 pp. On India.
20. *World Health Magazine.* (February–March, 1966 issue). On cities of the world, with many black and white photographs.

The Pantheon Press is issuing a series of books on communities of the world, written for adults. Write them for a list of current titles, if interested.

BIBLIOGRAPHY FOR PUPILS ON COMMUNITIES IN OTHER PARTS OF THE WORLD

General

NEUENDORFFER, MARY JANE. *Journey to Nine Villages.* New York: Ungar (1962), 138 pp. Grades 5–7. Villages in India, Argentina, Japan, Germany, Nigeria, Saudi Arabia, Finland, Vietnam, and the Navajos of the United States.

Textbooks

1. *Around Our World: A Study of Communities.* Toronto: Ginn (1965), 216 pp. A geographical approach, including communities in a tundra region, a tropical rain forest, a desert, a high plateau, a sheep station, a mountainous coast, a river delta, and below sea level.

2. *At Home Around the World.* Lexington, Massachusetts: Ginn (1965), 336 pp. A geographical approach. Communities in an island in the Pacific, in the mountains of Ecuador, on the tundra in Alaska, on a fjord in Norway, on a desert in Saudi Arabia, by a river in India, in a forest in Nigeria, on a ranch in Argentina, and on a polder in The Netherlands.

3. *Communities Around the World.* New York: Sadlier (1971), 320 pp. Life in Lapland, in a Portuguese fishing village, in the Sahara, in an African village, on a kibbutz, in an Indian village, on an Argentine cattle ranch, and in a Peruvian mountain valley.

4. *Everyone Lives in Communities.* Lexington, Massachusetts: Ginn (1972), 416 pp. An interdisciplinary approach. All photographs taken on the spot by the author of the text. Communities include Bueren and Bern in Switzerland; Korpilahti and Helsinki in Finland; London; Sewell and La Africana and Valparaiso in Chile; Moshi and Dar Es Salaam in Tanzania; Nakluya and Bangkok in Thailand; the Bedouins and Kuwait City in Kuwait; and Kfar Vitkin and Jerusalem in Israel.

5. *Exploring World Communities.* Chicago: Follett (1969), 256 pp. Includes an oil city in Canada, Brasilia in Brazil, a town under the snow in the Antarctic, Moscow, London, Tokyo, and Ibadan in Nigeria.

6. *Metropolitan Studies.* Glenview, Illinois: Scott, Foresman (1970), 304 pp. Largely a study of communities in the United States, but the book does include material on Nairobi, Kenya and Tokyo.

7. *Our Working World: Cities at Work.* Chicago: Science Research Associates (1966), 288 pp. Considered a third grade book in the Senash program, but very difficult reading for third grade. Depicts largely communities in the U.S.A. Emphasis shifts quickly back and forth from communities today to communities in the past. Included are the cities of Athens, London, Venice, Calcutta, and Singapore.

8. *People in Communities* and *People in States.* Menlo Park, California: Addison–Wesley. *People in Communities* is the third grade book in the Hilda Taba program in social science. It includes chapters on "The Bedouin of the Negev," "The Yoruba of Ife (Nigeria)," "The Thai of Bangkok," and the Norwegians of Hemmensberget." The fourth grade book is on people in "states" and includes "The People of Mysore (India) State," "The People of Osaka (Japan) State," "The People of Serbia," and "The People of Nova Scotia."

The Field Educational Publications (Palo Alto, California) has a set of filmstrips and accompanying records on people and communities in Ghana, India, Japan, Mexico, Norway, and the United States.

Communities in Chile

Some material on communities in Chile may be obtained from:

Anaconda, American Brass Company, 25 Broadway, New York, New York 10006.

Consulado General de Chile, 61 Broadway, New York, New York 10006.

Pamphlets on the Araucanians, Chile, Copper, and Introduction to Chile may be purchased at a low cost from the Organization of American States, Washington, D.C. 20006.

A kit on copper is available from the White Pine Chamber of Commerce and Mines, Box 239, Ely, Nevada 89301.

Two copper mining towns and Valparaiso are examined in the fourth grade textbook by Tim Gidal, *Everyone Lives in Communities* (Ginn, 1972). An account of a town in Chile appears in Edna Anderson's *Communities and Their Needs* (Silver Burdett).

Books

1. BOWEN, DAVID. *The Land and People of Chile.* Philadelphia: Lippincott (1966), 154 pp. Grades 6–9.
2. BREETVELD, JIM. *Getting to Know Chile.* New York: Coward–McCann (1960), 64 pp. Grades 4–6.
3. CALDWELL, JOHN C. *Let's Visit Chile.* New York: John Day (1963), 96 pp. Grades 4–6.
4. "Chile, Long and Narrow Land" in *National Geographic* (February, 1960), pp. 185–235. Pictures can be used by all pupils; text is difficult for most.
5. FENTON, CARROL L. *Riches from the Earth.* New York: John Day (1953), 159 pp. Pp. 45–63 on copper. Grades 5–7.
6. FIDELER, RAYMOND and CAROL KVANDE. *South America.* Grand Rapids, Michigan: Fideler (1965), 224 pp. Grades 4–7. Section on Chile.
7. GOETZ, DELIA. *Neighbors to the South.* New York: Harcourt Brace Jovanovich (1956), 179 pp. Grades 4–6. Some material on Chile.
8. KEPPLE, ELLA H. *Three Children of Chile.* New York: Friendship Press (1961), 127 pp. Grades 4–6.
9. LAMBERT, C. D. *The Copper Nail: A Story of Atacama.* Chicago: Follett (1965), 174 pp. Grades 5–8. Story of an Incan boy.
10. PENDLE, GEORGE. *The Land and Peoples of Chile.* New York: Macmillan (1960), 96 pp. Grades 5–7.
11. PRUDEN, DURWARD. *The Story of Chile.* Cincinnati: McCormick–Mathers (1966), 96 pp. Grades 5–8.
12. TRACY, EDWARD B. *The New World of Copper.* New York: Dodd, Mead (1964), 80 pp. Grades 5–7.

Films

1. "Chile's Copper." International Film Bureau, color, 10 minutes.
2. "Copper Mining." Pat Dowling Pictures, color, 14 minutes.
3. "People of Chile." International Film Bureau, color, 22 minutes.
4. "Valparaiso." McGraw-Hill.

Communities in Finland

Some free material is available from:

The Consulate General of Finland, 200 East 42nd Street, New York, New York 10017.

Finnish National Travel Office, 505 Fifth Avenue, New York, New York 10017.
Korpilahti, a lumbering community in northern Finland, and Helsinki are described in Tim Gidal's volume for the fourth grade on *Everyone Lives in Communities* (Ginn, 1972).

A free booklet on "How You Can Make Paper" may be obtained from The American Paper Institute, 260 Madison Avenue, New York, New York 10016.

Several charts on lumber and paper may be obtained free from The American Forest Products Industries, 1835 K Street, N.W., Washington, D.C. 20006.

Books

1. BERRY, ERICK. *Land and People of Finland.* Philadelphia: Lippincott (1959), 128 pp. Grades 6–9.
2. BOWMAN, JAMES C. and MARGERY BIANCO. *Tales from a Finnish Tupa.* Chicago: Whitman (1970), 48 pp. Finnish folk tales.
3. BROOKS, ANITA. *The Picture Book of Timber.* New York: John Day (1967), Grades 4–6.

4. CLARKE, MOLLIE. *Aldar the Trickster*. Chicago: Follett (1967), 32 pp. Humorous and easy-to-read Finnish folk tale.
5. GIDAL, SONIA and TIM. *My Village in Finland*. New York: Pantheon (1966), 81 pp. Grades 3–5. Large black and white photographs add to this volume.
6. JOUTSEN, BRITTA-LISA. *Lingonberries in the Snow*. Chicago: Follett (1968), 190 pp. Grades 5–7. An American family lives in Turku for a year. Fiction.
7. LISLE, ALICE A. *Magic World of Elin*. New York: Abingdon (1958), 160 pp.
8. NUENDORFFER, MARY JANE. *Journey to Nine Villages*. New York: Ungar (1962), 138 pp. Grades 4–6. Pp. 101–112 on "The Snowbound Cooperatives" of Finland.
9. RIWKIN–BRICK, ANNA and ASTRID LINDGREN. *Matti Lives in Finland*. New York: Macmillan (1969), 46 pp. Grades 2–4. Many black and white photographs.

Films

The Consulate General of Finland has several films available for transportation charges only. These include the following films on communities:

1. "Helsinki." 20 minutes, color.
2. "Lappeenranta: A Lake District Town." 10 minutes, color.
3. "Tampere: Finland's Second Largest City." 12 minutes, color.
4. "Tapiola: Garden City of the North." 15 minutes, color. This is the famous planned suburb of Helsinki and one of the outstanding communities of the world.

Filmstrips

The Society for Visual Education has a filmstrip on "How People Live in Stavanger, Norway and Turku, Finland." 51 frames, color.

Eye Gate House has a color filmstrip on Helsinki.

Communities in India

Some free material may be obtained from the India Information Services, 3 East 64th Street, New York, New York 10021.

Several textbooks now include communities of India. There are many books on that nation; those listed below deal primarily with village and city life.

Books

1. AHMED, ZAHIR. *Dusk and Dawn in Village India*. New York: Praeger (1966), 144 pp. Grades 5–7. Change in an Indian village.
2. ARORA, SHIRLEY L. *What Then, Raman?* Chicago: Follett (1960), 176 pp. Grades 4–6. The first boy in a village learns to read.
3. BOTHWELL, JEAN. *The First Book of India*. New York: Watts (1966), 88 pp. Grades 4–6.
4. BOTHWELL, JEAN. *The Holy Man's Secret: A Story of India*. New York: Abelard (1967), 160 pp. Tensions in a family over changes in the village.
5. CHANDAVARKAR, SAMANA. *Children of India*. New York: Lothrop, Lee and Shepard (1971), 128 pp. Grades 2–4.
6. COOKE, DAVID C. *Dera: A Village in India*. New York: Norton (1967), 128 pp. Grades 5–7.
7. DARBOIS, DOMINIQUE. *Lakhmi: Girl of India*. Chicago: Follett (1964), 47 pp. The story of a girl and her brother in the city of Benares. Grades 4–5.
8. GIDAL, SONIA and TIM. *My Village in India*. New York: Pantheon (1958), 75 pp. Grades 4–6.
9. GOBHAI, MEHLLI. *Ramu and the Kite*. Englewood Cliffs, New Jersey: Prentice-Hall (1968), 32 pp. A story of the Indian Festival of Kites.
10. HAMILTON, LEE D. *Let's Go to a Dam*. New York: Putnam (1963), 47 pp. A visit to a new hydroelectric dam. Grades 4–6.

11. Joy, Charles R. *Taming India's Indus River: The Challenge of Desert, Drought and Flood.* New York: Coward-McCann (1964), 119 pp. Grades 5–8.
12. Lewis, Richard (Editor). *Moon, for What Do You Wait?* New York: Atheneum (1967), 32 pp. A selection of Tagore's two-line nature poems.
13. Naylor, Phyllis. *The New Schoolmaster.* Morristown, New Jersey: Silver Burdett (1967), 32 pp. Grades 3–5.
14. Nirodi, Hira. *Chikka.* Chicago: Reilly and Lee (1962), 154 pp. Grades 4–6. An eleven-year-old and life in Bangalore.
15. Norris, Marianna. *Young India: Children at Work and at Play.* New York: Dodd, Mead (1965), 64 pp. Grades 4–6.
16. Polk, Emily. *Delhi: Old and New.* Chicago: Rand McNally (1963), 144 pp. Grades 6–9.
17. Schloat, G. Warren, Jr. *Uttam: A Boy of India.* New York: Knopf (1963), 46 pp. Grades 3–5.
18. Silverstone, Marilyn and Lauree Miller. *Bala: Child of India.* New York: Hastings (1968), 48 pp. Life in a village.
19. Thampi, Parvathi. *Geeta and the Village School.* Garden City, New York: Doubleday (1960), 64 pp. Grades 3–5. A South India village.
20. Thoeger, Marie. *Shanta.* Chicago: Follett (1968), 160 pp. Grades 3–5. A girl in a village and her visit to the big city for a festival.
21. Wyckoff, Charlotte C. *Kumar.* New York: Norton (1965), 192 pp. Grades 4–6. Story of a boy in the city of Madras; India in transition.

Communities in Israel

Much free material may be obtained from the Israel Office of Information, 11 East 70th Street, New York, New York 10021.

The new cooperative community of Kfar Vitkin and the city of Jerusalem are examined in the text *Everyone Lives in Communities* (Ginn, 1972) by Tim Gidal. "At Home in Beersheba" is the title of a chapter in the Carls, Templin and Sorenson volume on *Knowing Our Neighbors Around the World* (Holt, Rinehart and Winston).

Books

1. American Geographical Society. *Israel.* Garden City, New York: Doubleday (1970), 64 pp. Text for adults but colored illustrations can be used with pupils.
2. Braverman, Libbie L. *Children of the Emek.* Brooklyn, New York: Furrow (1950), 120 pp. Grades 5–7.
3. Comay, Jean and Moshe Pearlman. *Israel.* New York: Macmillan (1964), 120 pp. Grades 5–7.
4. Edelman, Lily. *Israel: New Life in an Old Land.* Camden, New Jersey: Nelson (1969), 223 pp. Grades 5–8.
5. Edwardson, Cordella. *Miriam Lives in a Kibbutz.* New York: Lothrop, Lee and Shepard (1971), 48 pp. Grades 2–4.
6. Gidal, Sonia. *Meier Shfeya: A Children's Village in Israel.* New York: Behrman (1950), 44 pp. Grades 4–6.
7. Gidal, Sonia and Tim. *My Village in Israel.* New York: Pantheon (1959), 76 pp. Grades 4–6. A boy in a new village in a fertile area.
8. Gillsater, Sven and Pia. *Pia's Journey to the Holy Land.* New York: Harcourt Brace Jovanovich (1961), unpaged. K–3. Full color photographs.
9. Hamori, Lazzlo. *Flight to the Promised Land.* New York: Harcourt Brace Jovanovich (1963), 189 pp. Grades 6–9. A Jewish family moves from Yemen to Israel.
10. Holisher, Desider. *Growing Up in Israel.* New York: Viking (1963), 180 pp. Grades 5–8.
11. Joy, Charles R. *Getting to Know Israel.* New York: Coward-McCann (1961), 64 pp. Grades 4–6.

12. KUBIE, NORA B. *Israel*. New York: Watts (1968), 96 pp. Grades 5–8.
13. LOTAN, JOEL (Editor). *A Kibbutz Adventure*. New York: Warne (1963), 64 pp. Grades 4–6.
14. PINNEY, ROY. *Young Israel: Children of Israel at Work and at Play*. New York: Dodd, Mead (1963), 64 pp. Grades 4–6.
15. RIWKIN-BRICK, ANNA and LEA GOLDBERG. *Eli Lives in Israel*. London: Methuen, (1964), unpaged. Grades 3–5.
16. SASEK, MIROSLAV. *This Is Israel*. New York: Macmillan (1962), 60 pp. Grades 1–3.

Films and filmstrips

1. "Family of Israel." Encyclopaedia Britannica, color, 47 frames. A filmstrip.
2. "Jerusalem — The Holy City." Encyclopaedia Britannica, color, 10 minutes.
3. "Life in Israel." Curriculum, color, 25 frames. A filmstrip.
4. "Other Hearts in Other Lands." Anti-Defamation League, color, 15 minutes. Children learn about life in an Israeli kibbutz.

Communities in Kuwait and Saudi Arabia

Some material may be obtained from the following sources:

Embassy of the State of Kuwait, 2940 Tilden Street, N.W., Washington, D.C. 20008.

Embassy of Saudi Arabia, 2333 Wisconsin Avenue, N.W., Washington, D.C. 20007.

Public Relations Department, Arabian-American Oil Company, 505 Park Avenue, New York, New York 10022. Ask for posters and copies of *Aramco World* magazine, as well as a kit which includes stamps, coins, pictures, "worry beads," and a book.

Textbooks

The life of the Bedouins and the very modern city of Kuwait comprise two chapters in *Everyone Lives in Communities* (Ginn, 1972). There are chapters on life in the desert in Delia Goetz's *At Home Around the World* (Ginn) and D. L. Massey's *Around Our World: A Study of Communities* (Toronto: Ginn).

Books

1. DARBOIS, DOMINIQUE. *Hassan: Boy of the Desert*. Chicago: Follett (1961), 47 pp. Grades 2–4.
2. GIDAL, SONIA and TIM. *Sons of the Desert*. New York: Pantheon (1960), 80 pp. The story of a son of a desert tribal chief. Grades 4–6.
3. NEUENDORFFER, MARY JANE. *Journey to Nine Villages*. New York: Ungar (1962), 139 pp. Grades 4–6. Chapter 6, "My Home Is Your Home," is about Saudi Arabia.
4. NEVIL, SUSAN R. *The Picture Story of the Middle East*. New York: McKay (1956), unpaged. Grades 4–6. Has information on "City Life" and other topics.
5. PHILLIPS, TED. *Getting to Know Saudi Arabia*. New York: Coward-McCann (1963), 64 pp. Grades 4–6. As seen through the eyes of two children. Well illustrated.
6. PONT, CLARICE. *No School on Friday: A Story of Saudi Arabia*. New York: McKay (1953), 213 pp. Grades 5–7. Three American children spend a year in Saudi Arabia and share experiences with children there.
7. WIMMER, HEDWIG. *Maha and Her Donkey*. Chicago: Rand McNally (1965), 78 pp. Grades 3–5.

Films

1. "Ali and His Baby Camel." Atlantis, color or black and white, 11 minutes. A nine-year-old and his camel.

2. "Flowering Desert." Bailey, color, 11 minutes. On deserts in general.
3. "Life of a Nomad People: Desert Dwellers." Coronet, color or black and white, 11 minutes.

Filmstrips

1. "Ahmed and Adah of the Desert." Eye Gate House, color, 30 frames.
2. "Ali of Saudi Arabia." McGraw-Hill, color, 40 frames.
3. "Living in Saudi Arabia." Visual Education Consultants, black and white, 30 frames.
4. "Saudi Arabia." Disney, color, 40 frames.
5. "Saudi Arabia." Eye Gate House, color, 30 frames.

Communities in Switzerland

Some material may be obtained free from the Consulate General of Switzerland, 444 Madison Avenue, New York, New York 10022.

The communities of Bueren and Bern are treated in *Everyone Lives in Communities* (Ginn, 1972).

Books

1. BENARY-ISBERT, MARGOT. *The Shooting Star*. New York: Harcourt Brace Jovanovich (1954), 118 pp. Day-to-day activities in the Swiss mountains.
2. BEMELMANS, LUDWIG. *The High World*. New York: Harper and Row (1954), 114 pp. Grades 4–6. Changes in Lech when the government builds a hydroelectric plant nearby.
3. BRAGDON, LILLIAN J. *Land and People of Switzerland*. Philadelphia: Lippincott (1961), 128 pp. Grades 6–9.
4. CARIGIET, ALOIS. *The Pear Tree, The Birchtree, and the Barberry Bush*. New York: Walck (1967), 32 pp. Change of seasons.
5. EPSTEIN, SAM and BERYL. *The First Book of Switzerland*. New York: Watts (1964), 85 pp. Grades 4–6.
6. GEIS, DARLENE (Editor). *Let's Travel in Switzerland*. Chicago: Children's Press (1964), 64 pp. 32 full color photos.
7. GIDAL, SONIA and TIM. *My Village in Switzerland*. New York: Pantheon (1961), 82 pp. Grades 4–6. A cheesemaking village and warm family life.
8. HOFFMAN, GEORGE W. *Life in Europe: Switzerland*. Grand Rapids, Michigan: Fideler (1961), 160 pp. Grades 5–7.
9. HULDSCHINER, ROBERT E. *The Cow That Spoke for Sepple and Other Alpine Tales*. New York: Doubleday (1968), 160 pp. Twelve stories of village life in the Alps.
10. LAUBER, PATRICIA. *Getting to Know Switzerland*. New York: Coward-McCann (1960), 64 pp. Grades 4–6.
11. MARITHNER, MARIA. *Christiane Lives in the Alps*. New York: Hastings House (1967), 48 pp. Family life in the highest village in Switzerland.
12. MILLEN, NINA. *Children's Games from Many Lands*. New York: Friendship Press (1960), 214 pp. Some Swiss games are included.
13. PEATTIE, MARGARET R. *Switzerland*. Columbus, Ohio: Merrill (1957), 32 pp. Grades 3–5.
14. *Swiss Alpine Songs*. Delaware, Ohio: Cooperative Recreation Service (1950), 32 pp. Texts in English.

Films and filmstrips

1. "Alpine Village." McGraw-Hill, black and white or color, 31 minutes.
2. "Anthony and Marie of Switzerland." Society for Visual Education, color, 48 frames. A filmstrip on the Eastern Alps region.

3. "Children of Switzerland." Encyclopaedia Britannica, black and white, 11 minutes. Home life, tending of cattle, and a visit to a nearby village.
4. "Village of Switzerland." Churchill Films, color, 16 minutes.

Other materials

Some material may be obtained from The Swiss Industries Group, The Swiss Center, 608 Fifth Avenue, New York, New York 10020 and from The Watchmakers of Switzerland, 730 Fifth Avenue, New York, New York 10019.

Communities in Tanzania

The coffee growing Chagga tribe, the town of Moshi on the slopes of Kilimanjaro, and the capital of Dar Es Salaam are described in *Everyone Lives in Communities* (Ginn, 1972).

Books

1. BLEEKER, SONIA. *The Masai: Herders of East Africa.* New York: Morrow (1963), 155 pp. Grades 4–6. An anthropologist's account for children.
2. CALDWELL, JOHN C. *Let's Visit Middle Africa.* New York: John Day (1958), 96 pp. Pp. 69–72 on Tanzania. Grades 5–7.
3. CARPENTER, FRANCES. *The Story of East Africa.* Cincinnati: McCormick–Mathers (1967), 140 pp. Grades 5–8.
4. DAVIS, RUSSELL and BRENT ASHABRANNER. *The Lion's Whisker: Tales of High Africa.* Boston: Little, Brown (1959), 191 pp. Grades 5–8.
5. DONNA, NATALIE. *Boy of the Masai.* New York: Dodd, Mead (1964), 64 pp. Grades 2–4.
6. HALL, GEORGE. *Come Along to Tanganyika.* Minneapolis: Denison (1963), 175 pp. A general account. Grades 4–7.
7. HALMI, ROBERT. *Visit to a Chief's Son.* New York: Holt, Rinehart and Winston (1963), 95 pp. Adventures of a Masai boy near Kilimanjaro and his American friend.
8. HEADY, ELEANOR B. *When the Stones Were Soft: East African Fireside Tales.* New York: Funk and Wagnalls (1964), 94 pp. Grades 4–6.
9. JOHNSON, JAMES R. *Animal Paradise.* New York: McKay (1969), 218 pp. Grades 4–7. The story of the Ngorogoro Conservation Area.
10. JOY, CHARLES R. *Getting to Know Tanzania.* New York: Coward-McCann (1966), 64 pp. Grades 5–7.
11. KAULA, EDNA M. *The Land and People of Tanganyika.* Philadelphia: Lippincott (1963), 160 pp. Grades 6–9. In the Portraits of Nations series.
12. LOBSENZ, NORMAN M. *The First Book of East Africa.* New York: Dodd, Mead (1964), 88 pp. Grades 5–7.
13. NEVINS, ALBERT J. *Away to East Africa.* New York: Dodd, Mead (1959), 96 pp. Grades 6–9.
14. PERKINS, CAROL M. *The Shattered Skull.* New York: Atheneum (1965), 64 pp. Grades 4–6. The work of the archaeologist Dr. Leakey in East Africa.
15. RIWKIN-BRICK, ANNA and ASTRID LINDGREN. *Sia Lives on Kilimanjaro.* New York: Macmillan (1959), 48 pp. Grades 1–4. A delightful story of a girl of the Chagga tribe. Wonderful illustrations.
16. UNITED STATES COMMITTEE FOR UNICEF. *Hi Neighbor: Book Eight.* New York: Hastings House (1965), 64 pp. Pp. 37–47 on Tanzania. Includes some background information, recipes, two songs, games, a folk tale, and suggestions on what to make from gourds.

Films and filmstrips

1. "Tanganyika." Common Ground filmstrips. People and products in various regions. Color.
2. "Tanganyika: Industries, Products, and Cities." Eye Gate, 1962, 46 frames, color.
3. "Tanzania." International Film Bureau. Color, 11 minutes.

Communities in Thailand

Some material may be obtained from the Office of the Public Relations Attaché, Royal Thai Embassy, 2300 Kalorama Road, N.W., Washington, D.C. 20008.

Textbooks

A town in Thailand is described in Edna Anderson's *Communities and Their Needs*, published by Silver Burdett.

In Tim Gidal's *Everyone Lives in Communities* (Ginn, 1972), the village of Nakluya, where the people earn their living by growing rice and fishing, is described. Another chapter is on big, booming, beautiful Bangkok.

Books

1. AYER, JACQUELINE. *The Paper-Flower Tree.* New York: Harcourt Brace Jovanovich (1962), 32 pp. Grades 1–4.
2. AYER, JACQUELINE. *Getting to Know Thailand.* New York: Coward-McCann (1959), 64 pp. Grades 4–6. Daily life of the Thais is stressed.
3. AYER, MARGARET. *Made in Thailand.* New York: Knopf (1964), 237 pp. Grades 5–7. On the arts and crafts of Thailand.
4. BERRY, ERICK. *The Spring of the Rice: A Story of Thailand.* New York: Macmillan (1966), 85 pp. Grades 3–4. The ritual of rice-planting.
5. CALDWELL, JOHN C. *Let's Visit Southeast Asia.* New York: John Day (1967), 96 pp. Grades 4–6.
6. COOKE, DAVID C. *Thailand: The Land of Smiles.* New York: Grosset and Dunlap (1969), 150 pp. Grades 5–7.
7. EYRE, JOHN D. *Thailand.* Lexington, Massachusetts: Ginn (1964), 122 pp. For better readers.
8. EXELL, F. K. *The Land and the People of Thailand.* New York: Macmillan (1961), 96 pp. Grades 5–8.
9. FLOETHE, LOUISE and RICHARD. *Floating Market.* New York: Farrar, Straus and Giroux (1969), 32 pp. Grades 3–5. The markets of Bangkok.
10. GARLAN, PATRICIA W. and MARYJANE DUNSTAN. *Orange-Robed Boy.* New York: Viking (1967), 90 pp. Life of a young Burmese koyin or monk in a temple.
11. GEISS, DARLENE. *Let's Travel in Thailand.* Chicago: Children's Press (1964), 32 pp. Grades 4–6.
12. HAMORI, LAZLO. *Adventure in Bangkok.* New York: Macmillan (1966), 188 pp. Grades 5–8. A young Swedish boy and his adventures in Bangkok.
13. MATTHEW, EUNICE S. *The Land and People of Thailand.* Philadelphia: Lippincott (1964), 160 pp. For better readers.
14. RIWKIN-BRICK, ANNA and ASTRID LINDGREN. *Noy Lives in Thailand.* New York: Macmillan (1967), 48 pp. Grades 3–5.
15. SCHLOAT, G. WARREN, JR. *Prapan: A Boy of Thailand.* New York. Knopf (1963), 48 pp. Grades 4–6. A 10-year-old in Bangkok.

16. SPIEGELMAN, JUDITH M. *Galong: River Boy of Thailand.* New York: Messner (1970), 64 pp. Grades 4–6.
17. WATSON, JANE W. *Thailand: Rice Bowl of Asia.* Champaign, Illinois: Garrard (1966), 112 pp. Grades 4–6.

Teachers will find some background in Peter T. White's article in the July 1967 *National Geographic* on Thailand and in the *Hi Neighbor— Book 3* on music, folk tales, games, and recipes.

43. A teacher uses the overhead projector in a study of the U.S.A.

44. A pupil-made graph illustrates the growth of population

The United States — Today and Yesterday

45. Boys explain their illustrated time-line of the U.S.A.

46. Girls exhibit
    models they have constructed

47. Pupils prepare a flannelboard map of the United States

48. The teacher uses a large income tax form to study taxation

Studying the
United States Today 19

One of the strangest situations in the social studies curricula of most American schools is the lack of attention given to the contemporary scene in our own nation. Boys and girls are expected to learn about life in many other countries, concentrating upon the present. But they are not asked to learn about their own country except through its history. How curious — and how sad.

Sometimes boys and girls make a study of the United States through regional geography. Usually that is in the fourth grade. Such an approach, however, is likely to be limited. They are not likely to gain a comprehensive view of many aspects of the U.S.A. today.

There does not seem to be any valid explanation for this situation. Perhaps it is assumed that children will learn about the United States today through current events. Possibly curriculum-makers have felt that pupils will pick up bits and pieces of information about the contemporary scene and fit them together into a meaningful whole. More probably the lack of attention to life in our own nation today is explained by the fact that geography has occupied such an important place in social studies programs to the neglect of other social science disciplines.

The Importance of Studying about the United States Today

It seems to this writer extremely important for boys and girls to learn about their own country as it is now. It is here that they live now and will continue to live throughout their lives, with the exception of some who will live abroad. It is here that they will earn a living and participate in various groups. It is with the problems of our society now, rather than in the past, that they will have to wrestle in the foreseeable future. Why, then, should they not learn about the U.S.A. today, even before they study its history?

Furthermore, it is easier for children in their early years in school to learn about the present than it is to learn about the past. Their lack of a sense of time militates against too early an introduction of the history of our country in a formal way. But they certainly can profit from a study of various aspects of our nation at the moment.

It is also easier to study other countries if pupils have first studied their own. The skills they have acquired, even in an elementary way, can then be applied to places with which they have had no direct contact.

Certainly a comprehensive picture of the United States cannot be obtained solely through current events, no matter how well they are taught.

What we need at some point in the elementary school is a lengthy period of time in which boys and girls can explore their own nation today in a comprehensive fashion, taking up such varied topics as the land, the people, ways of earning a living, the government, health, the schools, values, transportation and communication, and our dealings with other nations.

Grade Placement of a Study of the U.S.A. Today

At many points in the early grades, pupils should be gaining a partial view of the United States through their study of families and communities. By the time they have reached the middle grades, they should be ready for a full-fledged study of our contemporary life. This should be built on what they have learned about families and communities in the United States, but should go far beyond those topics.

The fifth grade seems to be an appropriate place for such a study, although it can be made at other grade levels.

Schools will need to determine how much time should be devoted to such a study in depth. Some may want to devote as much as a half-year to this theme. Others may decide upon less time. Surely one-third of a year should be the minimum time devoted to a study of the U.S.A. today.

What a Comprehensive Study of the U.S.A. Might Include

There are a great many aspects of our nation to which children should be exposed. By the end of their study of the contemporary scene in the U.S.A., they should have discovered a good many facets in our national life.

They should certainly learn much more about the land than they have in previous grades and how it has shaped our life as well as how men have changed it. They should learn about the great variety of people in our nation and how they earn their living in many, many ways. Those two topics might lead quite naturally into a study of transportation and communication, with particular reference to the importance of transportation in obtaining raw materials and in shipping finished products throughout the U.S.A. Some study of the various units of government might well precede a study of health and education, which depend in large part upon the work of governments. An elementary introduction might well be made, too, to the relations of the United States with other nations of the world. During this study, children should learn something, too, about the various religious groups in our country. Such a study would include some of the factors which bind us together and make us feel united as Americans, as well as some of the factors which divide us and create problems.

Probably you have noticed already that these various topics include all of the social sciences and some related areas. Geography is highlighted in the study of land. Psychology, social psychology, anthropology, and sociology are emphasized in the study of people. Economics and geography should be stressed in the theme of earning a living. Political science comes to the fore in the section on governmental units. International relations would be the pivot of the study of our relations with other nations. Elementary philosophy and religion form the basis of the study of values or beliefs. Education would take the center of the stage in the consideration of schools. Some history would be included in each of these topics, although it would not be highlighted in any of them.

Of course various disciplines would be needed for each topic, but one would be emphasized in each unit or theme.

Models or Constructs for a Study of the U.S.A. Today

Teachers may want to develop their own approaches to a study of the United States today. For those not so inclined, two suggestions are made here. One of them would start with the land and people and their interaction. It would highlight the geographical distribution of people, and pupils would be encouraged to discover why people live where they do today.

This would be followed by a study of basic values which we hold in common and some of the conflicts over values.

Following these two approaches, a study might well be made of the various institutions of the United States, including the family, the educational system, the economy, government, and the mass media. The various religious groups in our country might well be considered at this point, too.

The various creative ideas and expressions which have so much to do with the future of our country could then be examined.

Finally, the contacts of our country with other countries could be explored.

All of these topics would be considered in their current dimensions. The emphasis would be upon the here and now.

Teachers could then turn to a study of the United States in the past.

A model or social construct for such an approach might look something like this:

THE FUTURE

		Creative Ideas and Expressions		
	T H E	Institutions Family Economy	T H E	
Contacts with Other Countries ←——→	P R E S E N T	Education Government Religion Mass Media	P R E S E N T	Contacts with Other Countries ←——→
		Values, Beliefs, Goals		
		The Land and the People		

THE PAST

A less structured approach is suggested in the following model in which the people of the United States are placed in the center. They are shown as being affected by many factors, from land to values and beliefs. In this approach a teacher would begin with people and relate them to the other factors presented. No fixed order is absolutely necessary.

The entire class could work on all or carefully selected topics chosen from the 12 topics listed in this chart. Or the class could be divided into groups or committees, with each group working on one of the 12 subjects. A compromise between these two

Based on chart appearing in *Social Education*, April 1959, p. 161. By permission of the author and the National Council for the Social Studies.

plans would be for the class to work on certain topics as a whole, with committees working on other topics and reporting back to the entire group.

The focus would again be on the present, as in the plan suggested on the previous page. In some topics, it would be wise to dip back into the past in order to explain the present. The historical dimension, however, would be left to a half year's study of our development as a nation, told in chronological fashion, probably picking a few decades or eras for concentration.

THE FUTURE

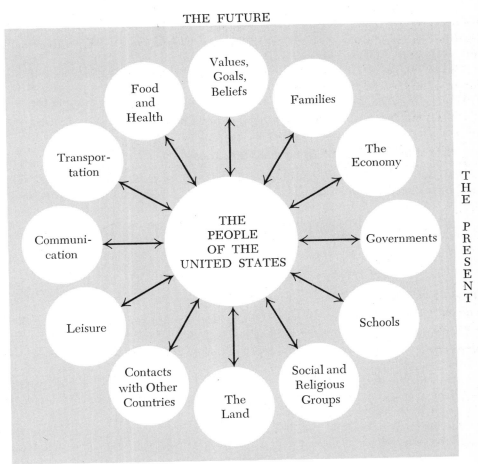

THE PAST

The Role of History in Studying the U.S.A. Today

In either of the alternatives to the approach suggested in this chapter, history would be minimized. Some explanations would be necessary to clarify aspects of our country today, but history would not be stressed. The history of our nation would be postponed until a little later in the year.

You may well ask, Why follow this approach? There are three main reasons. One is because boys and girls can cope best with the here and now. It is the world around them and they can experience some of it firsthand. The newspapers, radio, television, and other media are exposing them to this world.

Secondly, it seems wise to postpone the study of history as long as one can in the middle grades. This is based upon the assumption that the time sense does not develop very much until children are nine or ten — or older. Thus, history is postponed until the latter part of the fifth grade.

A third reason is because there are so many concepts and generalizations to discover, that children should be exposed to them without having to add the historical dimension at the same time. Thus various aspects of government are treated in this first part of the fifth grade year. When they are fairly well mastered, pupils can go on to the historical approach to government, bringing some basic learning to this study. Otherwise they would have to combine learning concepts and generalizations about government with the historical approach to government, thus making the study even more difficult.

Of course teachers can approach a study of the United States first through its history and then through the contemporary scene. Historically-minded teachers will find this more satisfying. But it probably will be much more difficult for children.

The Role of the Future in Studying the U.S.A. Today

Boys and girls need to learn something, too, about the foreseeable future. They can be exposed to possible changes in transportation and communication, in schools, in jobs, and to similar topics. They are likely to find a certain amount of fascination in projecting themselves into the future — "when they grow up."

But again, the present seems a more feasible focus than the future. It should be brought in, but not stressed at very many points.

The Role of Problems

Many children in the fifth grade should find certain problems of our society today of interest. This approach may well be included, spotlighting some of the current events in our nation. Such an approach should lend reality to a study of the United States.

However, this study of the United States should not be essentially a course in the problems of our society. The writer has already suggested that this be the focus for one grade in the junior high school years..

This writer believes that the first study of the U.S.A. today should accent the strengths rather than the weaknesses in our nation. Problems should be included at several points but they should not be the major focus in this particular grade.

The Major Aims in a Study of the U.S.A. Today

The major aims of a study of the contemporary scene in the United States have been outlined implicitly in what has already been said. Now let us attempt to state them more explicitly. They might include the following:

1. A further understanding of the importance of our land and how its main features affect living today in the U.S.A.
2. An understanding of some of the basic resources of our nation and how they are used.

3. Some understanding of the variety of people in our nation.
4. An introduction to geographic distribution of our population.
5. An understanding of climate and its effects on people.
6. Some understanding of the ways in which our land is being changed.
7. Considerable background on the location and role of rivers and other bodies of water and how they affect life in our country.
8. Further understanding of the place of family life in the U.S.A.
9. An elementary understanding of our economy and how it works.
10. Acquaintance with a fairly wide variety of ways of earning a living in the United States today.
11. An overview of the various units of government.
12. An overview of the school system.
13. Some acquaintance with the variety of groups in the U.S.A. today — including religious groups, ethnic groups, and political groups.
14. An introduction to the themes of food and health.
15. A basic understanding of the importance of transportation.
16. An elementary knowledge of various forms of communication in our country.
17. Some knowledge of the use of leisure by children and by adults.
18. An introduction to the contacts we have with the people and governments of other nations, stressing interdependence in today's world.
19. An introduction to the various religious groups in our nation and the importance of values-beliefs-goals.
20. Illustrative material from the history of our country in order to understand the present and function in the future as citizens.
21. Occasional glimpses into the future of our nation.
22. Considerable attention to current events and the reading of newspapers and current events publications.
23. The refinement of a sense of identification with our nation — patriotism.
24. Concentration on skills in the social studies that pertain to the study of the U.S.A. today.
25. An increasing respect for the variety of people in our country, no matter to what group they belong.
26. Some attention to problems which face our nation now.
27. An appreciation of the rights and responsibilities of all Americans.

Background for Teachers on Studying the U.S.A. Today

Perhaps you already have said to yourself that you do not really know much about the United States and therefore cannot teach about it. This is likely to be true of any teacher and particularly of a young teacher.

Probably you are too modest. You certainly know much more about the U.S.A. than your pupils. You have lived longer. You may have traveled in a few states. You certainly have seen television programs, read books, and followed the newspapers and magazines. That is more than most of your pupils have done.

You can learn along with them if you know what you are going to teach about. That has been suggested in this chapter already.

Furthermore, there are a good many things you can do at once to give you even more background. Here are a few of them. Provision is made for you to test yourself on what you have already done and what you can do now.

Gaining background on the U.S.A. today some suggestions	*Have already begun*	*Should start*
1. Start folders on each of the topics on the U.S.A. which you expect to cover.		
2. Save current events magazines with articles on the U.S.A.		
3. Save clippings from current newspapers and magazines which will help you.		
4. Start a collection of pictures on the U.S.A. Browse in secondhand bookstores for inexpensive magazines, back issues, from which you can cut such pictures.		
5. Start a collection of maps of the U.S.A. (See pages 133–140 on maps.)		
6. Take at least one short trip within the next year to a place in the U.S.A. which you never have visited before.		
7. Send to Chambers of Commerce of states and cities for information.		
8. Send for materials on the U.S.A. listed in various booklets on free and inexpensive materials. (See pages 151–152.)		
9. Make a list of parents, teachers, and other people in the community who might be used as resource people on various parts of the U.S.A. and on various themes.		
10. Develop a small collection of background books on the United States today.		
11. Make a large map of the U.S.A. today on cardboard or oaktag. Do not fill in. Use this in your class as you learn about the U.S.A. (You also can use this to draw the U.S.A. on the chalkboard if you are not very good at drawing maps.)		
12. Make a large flannelboard of the U.S.A. which you can use for many lessons and activities. Get the pupils to prepare materials to use on the flannelboard.		
13. Make two or three large charts on the U.S.A. which can be used for several years.		
14. Start a collection of 3″ x 5″ cards on books for children on the U.S.A. today. Include annotations on the backs of the cards as you use the books.		
15. Start a similar collection of cards on films and filmstrips and other audio-visual materials on the U.S.A. today.		
16. Make a collection of phonograph records on the U.S.A. today.		
17. Do some background reading. Place a few magazines, pamphlets, and books in places where you might pick them up at home to read them from time to time.		

Gaining background on the U.S.A. today *some suggestions: continued*	*Have* *already* *begun*	*Should* *start*
18. Talk to people who know about various phases of our life today or parts of the U.S.A., and make brief notes of what they have said.		
19. Obtain a copy of a good reference book, such as *Information Please Almanac, World Almanac, Reader's Digest Almanac,* or the *Desk Book of Facts and Figures,* and place on your desk at school.		

Background Reading on the U.S.A. for Teachers and Pupils

If you are planning to visit or even study India, Israel, or Indonesia, it is possible to find several books on those countries as they are today. The same is true of almost every nation in the world. Strange as it may seem, however, this does not apply to the United States. There are numerous books for adults and for children on various regions of our country and on innumerable topics, from art to zoos. But there are very few comprehensive accounts about our nation as a whole today.

One of the best single volumes is the Life World Library book on *The United States,* published by Time Incorporated in 1965. In it there are chapters on nine phases of our national life, each written by a commentator from some other part of the world.

Probably the best brief account is a 64-page book by the American Geographical Society called *The United States of America* (Doubleday, 1962). It is one of their Around the World series and has beautiful colored pictures, like postage stamps, to be pasted onto the pages of the book. A very short pamphlet is published by the U.S. Government Printing Office, called "Facts about the United States."

A book which is almost like an encyclopedia is the volume produced in 1968 by the Reader's Digest Association in Pleasantville, New York on *These United States: Our Nation's Geography, History, and People.* Many colored maps are included in this remarkable publication. Somewhat analogous, but with more emphasis upon maps, charts, and graphs is the book by Philip Schreier and George Vuicich on *A Guide to Understanding the United States,* printed by Guiness in 1970. That book contains a wealth of information. These two volumes are fairly expensive but should be in all school libraries as references.

Over the last few years several organizations have produced books of pictures of our nation today. One of these is the paperback entitled *This America,* published by Random House in 1965. The photographs are in black and white and are the results of the artistic work of Ken Heyman. A much larger and more expensive volume is Charles E. Rotkin's *The U.S.A. An Aerial Close-Up,* published by Crown Publishers in 1968, featuring largely black and white photographs. A third volume is John Steinbeck's *America and the Americans* (Viking, 1966), with superb photographs.

Sometimes the account of a person's travels across our country gives us new insights. Two such volumes are:

ARNOLD, ALAN. *How to Visit America and Enjoy It.* London: Putnam (1964), 187 pp.
MOYERS, BILL. *Listening to America: A Traveler Re-Discovers His Country.* New York: Harper and Row (1971), 342 pp.

A slightly different type of tour was taken by the Teales, the naturalists, and reported in four books published by Dodd, Mead. They are *North With the Spring, Journey Into Summer, Autumn Across America,* and *Wandering Through Winter.*

Sometimes the anthropologist brings special perspective or skills to seeing a nation. Three anthropologists have given us accounts of the United States which are unusually provocative. One is Geoffrey Gorer's *The American People: A Study of National Character* (Norton, 1964). That volume may make you angry at times, but it will certainly make you stop and think about what he says. Another account is by the late Clyde Kluckhohn in his book *Mirror for Man* (a Mentor paperback), in which he included an incisive chapter on the U.S.A. The third is *The American Way of Life,* authored by Ashley Montagu (Putnam, 1967). Again, this may irk you but it is worth being irked from time to time by such an anthropologist.

Two of Max Lerner's volumes should prove helpful, too. They are *The Unfinished Country* (Simon and Schuster, 1959) and *America As a Civilization: Life and Thought in the United States Today* (Simon and Schuster, 1957).

A good geography would be an invaluable reference volume on the U.S.A. for most teachers. One of the best is G. Langdon White, Edwin J. Foscue, and Tom L. McKnight's *Regional Geography of Anglo-America* (Prentice-Hall, 1964). Of course there are several similar volumes by geographers.

If you are interested in looking ahead, you may want to use the summer issue of *Daedalus* magazine, 1967, on *Toward the Year 2000: Work in Progress,* with a wide range of articles by outstanding persons in their respective fields.

If you have the time, a quick reading of current magazines and newspapers will yield much background, particularly if you read a wide variety of periodicals and papers so that you are exposed to a variety of points of view.

It is possible that you will want to take notes on whatever reading you do in order to retain much of the information for future use. It would be wise to clip articles and place them in appropriate folders on many aspects of life in the United States, or to take notes on small pieces of paper or cards, so that you can quickly file these materials and not have to re-read the articles in the future. In the long run this can be a great time saver. When you are ready to start a new topic on the United States, you will already have a reference folder at hand to refresh your memory on points that struck you at some time in the past.

Another way to gain background is to read the books written for children. Again, there are very few on the United States today written for elementary and middle school boys and girls.

One of the few such accounts is entitled *Let's Visit the U.S.A.* and was written by Noel Barber and published by the John Day Company in 1968. It is a volume of only 95 pages, with several black and white photographs to illustrate the text.

Another volume is entitled *My Country the U.S.A.* and is outstanding, probably the best single account for elementary school pupils of our nation today. It is a Golden Book published in 1968. Remarkable photographs add greatly to the value of this volume.

Strangely enough, there are very few textbook accounts of our nation today. The most extensive treatment is found in the author's volume *One Nation: The United States* (Ginn, 1972). The first half is devoted to an analysis of our nation today. There are chapters in it on The Land, The People, Our Resources, Earning a Living, Transportation and Communication, Religion, Government, and Our World Neighbors. Together, these should give pupils at the fifth grade level a relatively comprehensive view of the U.S.A. today.

Of the other textbooks, the Patterson, Patterson, Hunnicutt, Grambs, and Smith volume, *This Is Our Land* (Singer) includes about 100 pages about the United States today. The Eibling, King and Harlow volume on *Our Country's Story* (Laidlaw) has a final section on "Our Country Today," with sections on "The Geography of Our Country," "A Land of Great Wealth," "Our Transportation System," "Sending Messages in America," "A Land of Great Inventions," and "Our America" — a chapter on our national symbols and our way of life. The Allen-Howland book on *The United States of America* (Prentice-Hall) has a very brief introductory section on the geography of the United States today.

SOME "BIG IDEAS" TO STRESS ABOUT THE UNITED STATES TODAY

There are a few topics which should certainly be included in any study of the United States today. Mention has already been made of them. They are the land, the people, the economy, transportation and communication, food and health, governments, values, leisure time, religions, and contacts with other countries. The family might be included in such a list if it has not already been treated fully in the schools.

Let us look at some major considerations for study in conjunction with a few of these basic themes. You may want to develop statements on some of the topics not spelled out here.

The Geography of the U.S.A.

No one can really understand and interpret the United States without a fairly extensive knowledge of its geography. That is basic, fundamental, essential. As Robert Frost once said, "What makes a nation in the beginning is a good piece of geography." With a poet's insight he was expressing a very important truth.

Children need to discover the importance of our land base and what has been done to it and with it. There are scores of basic ideas and features which might be included in gaining a deep understanding of our geography. Cut to the barest minimum, there might be seven which should be granted the highest priority. They would be the Atlantic and Pacific Oceans, the Great Lakes, the Gulf of Mexico, the Mississippi River and its tributaries, and the Appalachian and Rocky Mountains. Given a real understanding of these features of our land base, almost anyone can explain much of our contemporary life as well as our history.

But these facts must be tied to many generalizations and concepts. Otherwise they will be a mere list of a few places the teacher thinks are important. Very early in a study of our geography, boys and girls might be confronted with a map of our population distribution and set to work on finding out why people live where they do today. They could discover where all the food for these people is grown and where the many minerals are found which supply our factories with raw materials. They could learn how these raw materials and our finished products are transported, thus seeing the importance of our various bodies of water.

On a giant map of the United States a class should begin to locate these seven important features of our land, as well as others which the teacher decides upon. In the process of developing such a map, children should learn a great many map skills. In a similar way a flannelboard of the U.S.A. today might well be used.

Quite early in a study of the geographic base of our country, boys and girls should see the U.S.A. in a very broad frame of reference. This would include its place in

North America, in the Western Hemisphere, on planet Earth, and in the Solar System.

A study of the grid system would certainly develop from the approach just mentioned. Longitude and latitude would be taught but without the great amount of time which is so often devoted to it. Teachers will probably ensure much better learning if they liken the grid system to a nearby football field with the goalposts as the North and South Poles.

Very early in such a study of our geography, pupils should learn how large our country is in area. Bar graphs of the largest four or five nations will help considerably in obtaining good learning. In this way pupils can find out that our country is outstripped in size only by the Soviet Union, Canada, and China.

The location of the United States and its size also mean that we have a great variety of climatic conditions. What that means in terms of living conditions, houses, and food products can be explored with profit by the pupils.

Boys and girls also need to learn what a well-watered land we have, in general and just what that means. They might well study at least the largest drainage systems of our country. These include the Eastern Coastal area, the St. Lawrence, the Hudson Bay, the Missouri-Mississippi, the Gulf and Rio Grande regions, the valleys of the Colorado and the Columbia Rivers, and the Western Coastal region.

At this point they might well be introduced to the problem of water, without any depth study of that vast contemporary problem at this point.

Our soil is another important feature. Few spots in the world compare favorably with the productive soil of the Great Plains, making it the Bread Basket and Colossal Meat Plant of our nation.

Location? Yes, that should be included as a major aim. The fact of our location on the globe has been and still is an important explanation of our nation.

While the favorable aspects of our land base should be discovered, children should also learn that a great deal of ingenuity, thinking, planning, and action has gone into our development of this "good piece of geography." We have widened and deepened and tamed some of our rivers. We have made dams on them. We have built the world's largest irrigation system. We have learned to build bridges, tunnels, and canals as well as highways and railroads.

Children should also learn something about our mineral resources. Perhaps it is enough for them to know about our petroleum and natural gas deposits, our coal and iron resources, and one or two other major minerals. Certainly there is not time enough to develop all of our resources at this point in school.

Some attention needs to be given, too, to the story of conservation and its importance. This should not be a historical development at this point. But it can grow out of the importance of keeping what we have, improving the resources we have, and adding new resources.

You might decide that a study of our land should include some attention to the current interest in ecology and pollution. This would give added relevance to the study of this aspect of our nation. Pupils should learn by the fifth grade that we have polluted our streams, poisoned our lakes, and begun to destroy our oceans. They should learn that smoke and fog have curtailed our use and enjoyment of clean air and that it is high time we began to work to restore these important aspects of nature.

You can get so involved, however, in such studies that you fail to cover other important topics during the year. So you will have to work out some time limits for such a study.

There is no lack of important data with which pupils can work. The biggest task of teachers in such a study is to limit and concentrate on essentials and to help children to discover rather than to memorize these "big ideas."

Some Suggested Methods in Developing Geographic Learnings About the U.S.A.

Map study should be at the heart of any unit on the geographic base of the United States today. We have already suggested that a very large map of the United States be used. We have proposed that a flannelboard be developed of the United States. These should provide learnings for the pupils. A large relief map of the United States might well be bought. (See the section on maps for suggestions.)

But children should develop their own small maps whether they are made from sawdust and glue, salt and flour, clay, or some other materials. A few pupils may want to experiment with battery fact maps.

Pupils should also make maps for use in the opaque projector, and in the overhead projector, and maps on frosted glass slides.

Many children at this point are beginning to use cameras. They should be encouraged to take pictures of geographic features in the area in which they live or in other parts of the country where they travel.

You and your pupils should be constantly on the lookout for pictures which can help them to visualize the land of the United States.

Of course films and filmstrips will be invaluable in developing the learnings on which you want to concentrate. Many such materials are available and can be used most effectively. They can take children to places and help them understand processes they would learn about in no other way.

A good many pupils can be encouraged to pursue special topics in conjunction with a study of the land of the United States. One pupil can learn where sugar is raised and why. At the same time, another pupil can be learning about the conditions under which rice is grown. Every pupil does not need to learn about these and other products. There can be some individualization in learnings.

Many of the encyclopedias are especially strong in presenting vividly the land base of the United States. Pupils can use a variety of encyclopedias, and teachers can profit by turning to them at this point. They should be utilized to the fullest extent possible.

Many of the pictorial materials developed by pupils should be exhibited in the classroom and used as frequently as possible. A few of the best ones can be saved to use in other classes as examples of the interesting and valuable types of materials pupils can develop.

If there are parents who work in the weather bureau, in the soil conservation office, or in similar jobs, they should be invited to meet and talk to and work with the class. Often their remarks can be taped, because they should not be asked back to the school over and over. Their time is too precious for continuous invitations.

These are just a few ways in which learning about the geographic base of the United States can be enhanced. Undoubtedly you will think of many other methods you should use. You might want to use the checklist on general methods on pages 78–80 to stimulate your thinking.

BOOKS FOR PUPILS ON THE LAND OF THE U.S.A.

(For books on conservation in the United States, see pages 484–485.)

1. CARMER, ELIZABETH and CARL. *The Susquehanna: From New York to the Chesapeake.* Champaign, Illinois: Garrard (1964), 96 pp. Grades 5–8.
2. CARSE, ROBERT. *Great American Harbors.* New York: Norton (1963), 128 pp. Grades 4–6.
3. COON, MARTHA S. *Oake Dam: Master of the Missouri.* Irvington-on-Hudson, New York: Hastings (1969), 124 pp. Grades 5–8.
4. CRAZ, ALBERT G. *Getting to Know the Mississippi River.* New York: Coward-McCann (1965), 64 pp. Grades 4–7.
5. CROSBY, ALEXANDER L. *The Colorado: Mover of Mountains.* Champaign, Illinois: Garrard (1961), 96 pp. Grades 5–8.
6. CROSBY, ALEXANDER L. and NANCY LARRICK. *Rivers: What They Do.* Racine, Wisconsin: Whitman (1961), 57 pp. Grades 4–6. In color.
7. EPSTEIN, SAMUEL and BERYL W. *All About the Desert.* New York: Random (1957), 148 pp. Grades 5–8.
8. FARB, PETER. *Face of North America.* New York: Harper and Row (1963), 254 pp. Grades 5–8. A young people's edition of his book for adults.
9. HAMILTON, EDWARD A. and CHARLES PRESTON. *Our Land, Our People.* Englewood Cliffs, New Jersey: Prentice-Hall (1963), 216 pp. Grades 5–8.
10. HAMMOND, DIANA. *Let's Go to a Harbor.* New York: Putnam (1959), 44 pp. Grades 3–5.
11. HARMER, MABEL. *About Dams.* Chicago: Melmont (1963), 63 pp. Grades 4–6.
12. HAVIGHURST, WALTER. *The Long Ships Passing: The Story of the Great Lakes.* New York: Macmillan (1961), 291 pp. Grades 5–8.
13. HELFMAN, ELIZABETH S. *Land, People and History.* New York: McKay (1962), 271 pp. Grades 6–9.
14. HELFMAN, ELIZABETH S. *Water for the World.* New York: Longmans (1960), 213 pp. Grades 5–8.
15. HIRSCH, S. CARL. *Mapmakers of America: From the Age of Discovery to the Space Era.* New York: Viking (1970), 176 pp. Grades 6–9.
16. LAUBER, PATRICIA. *Changing the Face of North America: The Challenge of the St. Lawrence Seaway.* New York: Coward-McCann (1959), 94 pp. Grades 5–7.
17. LAUBER, PATRICIA. *The Mississippi: Giant at Work.* Champaign, Illinois: Garrard (1961), 95 pp. Grades 5–8. Vivid and comprehensive.
18. McNEER, MAY. *The Hudson: River of History.* Champaign, Illinois: Garrard (1962), 96 pp. Grades 5–8.
19. MEYER, JEROME S. *Water at Work.* Cleveland: World (1963), 92 pp. Grades 4–6.
20. NEURATH, MARIE. *The Wonder World of Land and Water.* New York: Lothrop (1958), 36 pp. Grades 4–6. Excellent diagrams and illustrations.
21. PILKINGTON, ROGER. *The River.* New York: Walck (1965), 64 pp. Grades 4–6.
22. SANDERSON, IVAN T. *The Continent We Live On.* New York: Random (1962), 208 pp. Grades 5–8. Beautifully illustrated children's edition.
23. WATTENBERG, BEN. *Busy Waterways: The Story of America's Inland Water Transportation.* New York: John Day (1964), 127 pp. Grades 6–9.
24. WHITE, ANNE TERRY. *The St. Lawrence: Seaway of North America.* Champaign, Illinois, Garrard (1961), 96 pp. Grades 5–8.
25. WIESENTHAL, ELEANOR and TED. *Let's Find Out About Rivers.* New York: Watts (1971), 46 pp. Grades 1–3.

The People of the U.S.A. Today

At many points in their social studies programs in elementary schools, pupils should meet a wide variety of people in the United States, both literally and figuratively. But at some point these experiences should be tied together in some kind of comprehensive picture. How better than in a short unit on the people of the United States today? This need not be a long unit, although it could be. But it should help children to arrive at certain generalizations about their fellow-Americans.

One of these would be the large number of people in our country. Actually we are the fourth largest nation in the world in terms of population, being surpassed only by China, the Soviet Union, and India. But children also should learn that we represent only a small percentage of the world's people.

Pupils should meet as wide a variety of these people as possible. This should not be crowded into the space of a few days, but they should certainly have some contact, during a unit on people, with a number of different kinds of people. This is especially true of boys and girls in segregated schools, whether they are in the mid-city or in the suburbs.

The "big idea" that is being stressed is the pluralistic nature of American society. Children probably cannot cope with that word but they can be challenged by the idea behind it. We represent different races. We represent different religious groups. We represent different occupations. We represent different backgrounds in terms of ethnic groups and nationalities. We represent different political parties and views. We represent different socio-economic levels or groups.

However, we do have much in common. Some of those ideas can be spelled out in a unit on the people of the United States. Some of them can be handled when the beliefs, values, and goals of Americans are discussed.

The story of immigration should be reserved in detail for the study of United States history, but it can be touched upon in this unit. Likewise, children should learn that people are still coming to the U.S.A., although not in large numbers. They are coming from Cuba, from Europe, from Mexico, from Hong Kong, and from many other places, for a variety of reasons.

Much of the material about the people of the United States can be dealt with on a basis of their study of geography. The basic question would be, where do all these people live? The depth of such a study will have to be determined by every teacher according to the time available and by the nature of his or her class. But it might well include some of the major shifts in population in our country— to the Southeast, to the Southwest, and to the West. It might well include a look at the large cities where people live, including some study of the ten or fifteen largest cities. They are, according to the 1970 census: New York, Chicago, Los Angeles, Philadelphia, Detroit, Houston, Baltimore, Dallas, Washington, Indianapolis, Cleveland, Milwaukee, San Francisco, San Diego, and San Antonio. With some pupils it would be interesting to compare that list with the population of the same cities in 1960, letting them try to figure out why changes occurred in the list.

With all the talk about how crowded we are, it might be interesting for boys and girls (as well as adults) to know that 70% of our people live on 2% of our land.

Boys and girls need to know, too, that the people of our country are moving. Perhaps they already have learned something about migrant workers and people on the move. Perhaps some of them have experienced moving. In any case, people on the move should be another topic with which they deal.

Some of them may be interested in the fact that some groups are gaining in population. The group which is gaining most rapidly is the Indian group.

Wherever people come together there is likely to be tension and conflict, sometimes under the surface, sometimes in the open. Children know this and this idea should not be neglected. Teachers need to help them see the rights and responsibilities of minority groups as well as majority groups and to begin to feel as a minority group person might feel or does feel, recognizing that there are many points of view even in minority groups. Teachers need to bear in mind that the minority groups are many, including Japanese-Americans, Chinese-Americans, Mexican-Americans, Jews, and others, as well as Afro-Americans. Today many people think largely of the Afro-American when they think of minorities. We need to keep many groups in mind while not neglecting the largest group — the Afro-Americans.

At the same time, children should begin to understand the similarities of all people, starting with the fact that we are not different in our basic needs and in our intellectual potential.

In our understandable concentration on narrowing the gap between the opportunities available to all groups, we tend to neglect the fact that we have done well as a nation in learning to live together. Much remains to be done, but much already has been done. Children need to know this and to have examples of it in their minds.

Some teachers may also want to teach a little history in connection with this unit, showing how people have come to the United States from all parts of the world, but especially from Europe, over a long period of time. If this is done, teachers need to keep in mind the fact that all people did not come with equal amounts of skills and capital, and that some have been handicapped. This is most certainly true of the Negroes brought here as slaves from Africa. It was true of the Irish who came in such large numbers in the nineteenth century. It has been true of other groups, too, throughout our history.

Biography may be utilized to advantage in this unit on the people of the United States, with the emphasis upon the men and women who have made outstanding contributions to our nation. But the "little men and women" of history also should be given credit for their contributions, whether their names are known or not.

Such a unit ought to be a thrilling one for boys and girls. It is a people-centered unit and therefore at the heart of social studies learnings. By the use of pictures, films and filmstrips, and real live human beings, it ought to be made a memorable experience for boys and girls as well as for teachers. Through this unit every child should become more proud of his past and feel a stronger identification with the group or groups of which he or she is a part. This is of paramount importance.

At the same time, they should begin to identify with other persons and groups in the United States and to respect others and their rights. It is not too early for them to sense some concern about the welfare of all their fellow-Americans. Such values are worth striving for as a teacher of boys and girls.

Some Methods of Introducing Children to the People of the United States Today

Mention already has been made of the importance of having children meet a variety of people from different backgrounds during a unit on the people of the United States. Such visitors need not talk about their own group, whether it is a religious group, a racial group, or some other cluster of people. They should be invited to work with your class on some topic of mutual interest. Boys and girls should come to know them as human beings, not just as members of some minority. Good resource persons are essential in any unit on the people of the United States. You can teach about education in human relations without such persons, but you cannot carry on education *in* human relations without such people.

Then, of course, there are pictures. They should be used in many ways. They should be used on bulletin boards. They should be passed around the class for examination. They should be used in the opaque projector. You will never have enough good pictures of people. Some of them should be collected by the pupils from current newspapers and magazines. In this connection, for pictures of Negroes, you will probably want to bear in mind *Ebony* magazine as a good source which is too often overlooked. Many of your pictures should be mounted by the pupils and used in future classes.

Film and filmstrips on the people of the United States are not plentiful, but there are a few which you can use to advantage.

Books of biography should also be utilized fully. There are several collections of biographies of famous American immigrants suitable for use by elementary school pupils, written on different reading levels. There are also several volumes which deal with one person rather than a group.

If you are teaching this topic at about the time of a national census, there should be many articles about it in the newspapers and in magazines. Such materials will include many charts and graphs. This will give you an opportunity to do considerable work in the teaching of social studies skills.

It would be fine if there were some fairly recent immigrants in your community who would be willing to talk to your pupils about life in their homelands and why they came to this country. This would be a memorable experience for your pupils. You might even want to tape such a discussion, because you cannot do this every year. This is invaluable "source material" for classes.

In this unit, undoubtedly you will want to develop a list of words which are new to your pupils.

Some of the data on people of the United States lend themselves to the use of charts and graphs. You can use some of these in an overhead projector or an opaque projector; others can be made by the pupils, especially gifted ones. With a few pupils you will probably want to introduce population distribution maps.

Role-playing is highly recommended in conjunction with this unit.

The preparation of a montage on "We the People of the U.S.A. — Today," or the creation of a mural would be a good culminating activity for this unit.

Current conflicts between groups in the United States should be handled with care but certainly not neglected. They are the "stuff" of this unit. Children need to know that there are conflicts, but that many of them are being resolved. They should know how this is being done.

BOOKS FOR BOYS AND GIRLS ON THE PEOPLE OF THE U.S.A.

General

1. FRIEDMAN, FRIEDA. *A Sundae with Judy.* New York: Morrow (1950), 192 pp. Grades 4–6. A community group of various backgrounds puts on a show.
2. HOFF, RHODA. *America's Immigrants: Adventures in Eyewitness History.* New York: Walck (1967), 156 pp. Grades 6–9. An anthology.
3. HUNT, MABEL LEIGH. *Cristy of Skipping Hills.* Philadelphia: Lippincott (1958), 139 pp. Grades 4–6. People of different backgrounds live together in a community in harmony.
4. WILSON, BETTY D. *We Are All Americans.* New York: Friendly (1959), 32 pp. Grades 2–4.

Indians

(For books on Indians in our history, see pages 396–397. The books below are about Indians today.)

1. BREWSTER, BENJAMIN. *The First Book of Indians.* New York: Watts (1950), 69 pp. Grades 4–6. Includes some material on Indians today.
2. BULLA, CLYDE R. *Indian Hill.* New York: Crowell (1964), 80 pp. Grades 3–5. An Indian family moves from the reservation to the city.
3. CLARK, ANN NOLAN. *Medicine Man's Daughter.* New York: Farrar (1963), 178 pp. Grades 5–8. Dilemma of a girl trying to decide whether to follow ancient or modern ways.
4. FLOETHE, LOUISE L. *The Indian and His Pueblo.* New York: Scribner (1960), unpaged. Grades 3–5. Rio Grande Indians today and yesterday.
5. HOFSINDE, ROBERT. *Indians at Home.* New York: Morrow (1964), 96 pp. Grades 4–7. Seven tribes today and yesterday.

Asian–Americans (see also page 252)

1. DOWDELL, DOROTHY and JOSEPH. *The Japanese Helped Build America.* New York: Messner (1970), 96 pp. Grades 5–8.
2. HOLLAND, RUTH. *The Oriental Immigrants in America.* New York: Grosset and Dunlap (1969), 61 pp. Grades 5–8.

German–Americans

Care should be exercised not to lean unduly on books on the "Pennsylvania Dutch," as most books on German–Americans are on this unusual group.

1. BENARY-ISBERT, MARGOT. *The Long Way Home.* New York: Harcourt Brace Jovanovich (1959), 280 pp. Grades 7–10. An orphan during World War II goes to the U.S.A.
2. CUNZ, DIETER. *They Came from Germany.* New York: Dodd, Mead (1966), 178 pp. Grades 6–9.
3. HARK, ANN. *The Story of the Pennsylvania Dutch.* New York: Harper and Row (1957), 84 pp. Grades 4–6. Beautifully illustrated volume.
4. HOLLAND, RUTH. *The German Immigrants in America.* New York: Grosset and Dunlap (1969), 61 pp. Grades 5–8.
5. LEITON, MINA. *Elizabeth and the Young Stranger.* New York: McKay (1961), 133 pp. Grades 5–8. A girl who has lived in Nazi Germany and in a concentration camp emigrates to the United States.
6. LENSKI, LOIS. *Shoo-Fly Girl.* Philadelphia: Lippincott (1963), 176 pp. Grades 4–6. A "Pennsylvania Dutch" family's activities.

Italian–Americans

1. MANGIONE, JERRE. *America Is Also Italian.* New York: Putnam's (1969), 126 pp. Grades 6–9.
2. PELLEGRINI, ANGELO. *Americans by Choice.* New York: Macmillan (1956), 240 pp. Grades 5–8. Six brief biographies.
3. POLITI, LEO. *A Boat for Peppe.* New York: Scribner (1950), 30 pp. Grades 3–4. Life in the Sicilian section of Monterey, California.
4. SILVERMAN, MEL. *Ciri-biri-bin.* Cleveland: World (1957), 40 pp. Grades 1–3. An Italian boy in New York City at festival time.

Jewish–Americans

1. KURTIS, ARLENE H. *The Jews Helped Build America.* New York: Messner (1970), 75 pp. Grades 5–8.

Scandinavian–Americans

1. CARR, HARRIETT H. *Young Viking of Brooklyn.* New York: Viking (1961), 72 pp. Grades 3–5. Young Eric in a home for children of Norwegian background.
2. CROWLEY, MAUDE. *Tor and Azor.* New York: Oxford (1955), 123 pp. Grades 4–6. A Norwegian boy's adjustment to a new home in Marblehead, Massachusetts.
3. KINGMAN, LEE. *Quarry Adventure.* Garden City, New York: Doubleday (1951), 209 pp. Grades 4–6. A Finnish family in Massachusetts, with the father working in a quarry.
4. LINDQUIST, JENNIE D. *The Little Silver House.* New York: Harper and Row (1959), 213 pp. Grades 4–6. Life in a rural Swedish-American community.
5. MORGAN, NINA H. *Prairie Star.* New York: Viking (1955), 189 pp. Grades 5–7. A Norwegian boy and his grandparents in South Dakota.

Americans from western Europe

See the "They Came from . . . . . ." series of books published by Follett. The series includes:

> *Pierre's Lucky Bough: They Came from France.*
> *Bruce Carries the Flag: They Came from Scotland.*
> *The Lost Violin: They Came from Bohemia.*
> *Michael's Victory: They Came from Ireland.*
> *Sod-House Winter: They Came from Sweden.*
> *Peter's Treasure: They Came from Dalmatia.*

Americans from eastern Europe

1. EICHELBERGER, ROSA K. *Bienko.* New York: Morrow (1955), 192 pp. Grades 5–8. A Polish refugee boy and his difficulties in the U.S.A.
2. HAYES, FLORENCE. *Joe Pole: New American.* Boston: Houghton Mifflin (1952), 244 pp. Grades 6–9. A Polish boy wins out in his new land.
3. LANSING, ELIZABETH H. *A House for Henrietta.* New York: Crowell (1958), 196 pp. Grades 4–6. A Hungarian immigrant girl.
4. SECKAR, A. V. *Zuska of the Burning Hills.* New York: Walck (1952), 222 pp. Grades 4–7. The daughter in a Czechoslovakian family in West Virginia.

The Economy of the United States Today

The boys and girls in our elementary schools should certainly learn much about our economy. Perhaps this can be done best by a study of how people earn a living in the United States now. That seems an appropriate unit in any study of life in the contemporary U.S.A.

Pupils might well think in terms of the United States as a giant workshop in which approximately 100 million people are at work day and night. From our land, from our factories, and from other places come a constant flow of goods. From human beings there is likewise a constant flow of services.

Altogether, the workers of the United States produce nearly half of all the wealth of the world in any given year. That is a startling fact. It is a significant one, too, in understanding world affairs. Its importance cannot be grasped fully by young pupils, but at least they can be introduced to it, especially in pictorial and in chart form.

Boys and girls could assemble a list of all the different kinds of jobs they know about. Such a list is likely to be fairly long if enough time is given to it by teachers. Then pupils could learn that there are many more jobs than they have listed. Actually there are about 25,000 different types of jobs in our nation today.

Then a class might well be divided into groups, with each one representing a category of jobs. To obtain a realistic ratio in a class of 30, there should be seven persons who are manufacturers, six who are tradesmen, four who are government workers, four professional people, three service workers, two farmers, one transportation and one communication worker, one finance, real estate, and insurance person, and one miner. Each group may want to develop a placard showing its designation and some of its jobs. Then each group can do simple research on its category.

This might lead quite naturally into a study of plants and factories and the organization of business. If kept at a simple level, this should not be too difficult for most fifth graders.

Another important idea which can be taught in connection with a unit on earning a living is the idea of specialization. Some such specialization in our nation today is by regions. But not all of it is so organized. Most of the large companies, for example, now have regional plants instead of one large establishment in one locality. The importance of transportation and the need to have factories near the large concentrations of labor can be developed with pupils.

Pupils can then discover two other major concepts in economics. One is the idea of markets, which grow out of specialization and the need for a place to exchange goods. The other is money as a medium of exchange or demands for goods and/or services.

Role-playing can continue as pupils ask about what they receive in return for their work. This takes a class almost automatically into the topic of the rewards of labor in the form of wages and fringe benefits. Neither of these two topics should be too difficult for children if handled in a simple fashion. It may be helpful for members of a class to negotiate with employers for the amount of wages they receive. This will make their discussions much more realistic. They also can decide where to build their factories, shops, or other concerns, and raise capital with which to establish them.

A natural development from this would be the organization of labor unions and the formation of groups of employers. Children can learn about groups such as the American Federation of Labor–Congress of Industrial Organizations (AFL–CIO) and the National Association of Manufacturers (NAM).

Growing out of this would be a study of the division of income among families in the United States today. Teachers should find the diamond-shaped diagram below helpful in developing this concept.

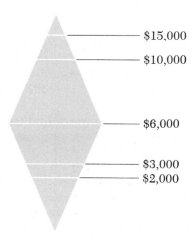

From an elementary study of what rent, food, and other costs are, children can see how important it is to use money wisely. Here the basic economic concept of unlimited wants and limited resources can be reintroduced. Some teachers will want to touch lightly upon the use of budgets by families at this point in their study.

As a result of this information, many children will ask how families exist on the small incomes that many of them have. At this juncture, you can deal with the work of governments, of organizations, and of individuals in the constant war on poverty.

Does all this material seem too difficult to you as a teacher? A few years ago it might have seemed so, but there has been enough experimentation in the teaching of economics to children to know that it is not too difficult if approached dramatically and realistically. The use of many charts and graphs and the use of role-playing will make it possible for you to develop in a simple way all of the concepts and generalizations suggested here.

This unit is primarily an economics unit and it provides a wonderful opportunity to build on previous learnings in that field.

Teachers may need to do some background reading in economics in order to feel comfortable in teaching such a unit. If teachers keep in mind a half dozen basic concepts in economics and relate them to the question of jobs in the U.S.A. today, this should not be too difficult a study for almost any pupil in the fifth grade.

Now let us turn to the question of methodology in teaching this unit and see what ways can be found to enhance learning for boys and girls in the field of economics as related to jobs in the U.S.A. today.

Some Methods of Studying the Economy of the United States Today

There are a number of methods which can be used to promote learning about our economy today. Perhaps the best of these is role-playing. As suggested already, the pupils might well be divided into several groups, to represent the various types of workers in our country today. Placards could be developed for each group. Such a dramatization of the army of workers in our nation should help to promote learning.

Interviewing parents and other adults about their jobs is another excellent approach and particularly applicable to this type of study.

Pictures, films, and filmstrips should enhance such a study, too.

In this unit there is a chance to develop further the ability to read and interpret charts and graphs. They should be very simple, like the diamond-shaped one on page 364. But this unit lends itself to charts and graphs. Some of them can be developed with drawings by the pupils. In this connection, the large charts of the Industrial Relations Center of the University of Chicago should be used. They may be purchased through the Allied Education Council (5533 South Woodlawn Avenue, Chicago, Illinois 60637). Teachers will find some helpful suggestions in the paperback book on *The U.S.A. and Its Economic Future* by Arnold B. Barach (Macmillan).

Pupils should have some experience in interpreting and making flow charts or process charts. An example would be the steps in producing oil, tracing the process from the discovery of petroleum to the final stages which result in oil for consumers. Pupils can also see the steps in our free enterprise system through a chart showing the various stages. Such a chart is included in the unit on "Our People At Work" in the author's text on *One Nation: The United States* (Ginn, 1972).

An unusual and provocative approach to the variations in jobs in different parts of the United States is given in the volume on *Social Study: Inquiry in Elementary*

Classrooms by M. Millard Clements, William R. Fielder, and B. Robert Tabach-
nick (Bobbs-Merrill). In it the authors depict a classroom where a teacher had collected telephone books from various parts of the United States. The children used
these as source materials to discover how many listings there were for snowplows,
air conditioners, grain companies, and other firms dependent upon certain geographical conditions.

If there is time, there are many trips which can be taken profitably in conjunction with this unit. They will depend, of course, upon the types of industry in your
locality.

Considerable map study also can be included in such a unit. The location of
resources, the means of transportation of resources to plants, the location of plants,
and the means of taking finished products to many parts of the U.S.A. are topics
which lend themselves to mapmaking and map reading.

There are many books for children about economic activities. Several of them are
listed in the next pages. Among the best are the volumes in the Men at Work
series of Putnam, written by Henry B. Lent and Harry C. Rubicam, Jr.

Approached in the ways suggested here, a unit on earning a living in the United
States today should give elementary school pupils a good introduction to the study of
economics.

BIBLIOGRAPHY FOR PUPILS ON EARNING A LIVING

There is much material of value in the *Men at Work* series, published by
Putnam even though some of the books were written in the late 1950s. There are
books by Henry Lent on *Men at Work in the South, Men at Work in New England,
Men at Work in the Mid-Atlantic States, Men at Work in the Great Lakes States,*
and with Harry C. Rubicam *Men at Work in the Mountain States.* Grades 5–7.
About 125 pages each.

General: Stressing machines, tools, computers, and automation

1. ADLER, IRVING and RUTH. *Machines.* New York: John Day (1964), 48 pp. Grades
   5–8.
2. ARNOLD, PAULINE and PERCIVAL WHITE. *The Automation Age.* New York: Holiday
   (1963), 197 pp. Grades 7–10.
3. BORTON, MILDRED. *Men and Machines Work Together.* Chicago: Scott, Foresman
   (1961), 56 pp. Grades 3–5.
4. JACKSON, DAVID. *The Wonderful World of Engineering.* New York: Doubleday
   (1969), 96 pp. Grades 7–10. Colored illustrations add to this volume.
5. JONES, WEYMAN. *Computer: The Mind Stretcher.* New York: Dial (1969), 120 pp.
   Grades 5–8.
6. LIBERTY, GENE. *The First Book of Tools.* New York: Watts (1960), 62 pp. Grades
   6–9.
7. MEADOW, CHARLES T. *The Story of Computers.* Irvington-on-Hudson, New York:
   Hudson House (1970), 124 pp. Grades 6–9.
8. MUNVES, JAMES. *We Were There at the Opening of the Atomic Era.* New York:
   Grosset and Dunlap (1960), 175 pp. Grades 4–6.
9. NEAL, HARRY E. *From Spinning Wheel to Spacecraft: The Story of the Industrial
   Revolution.* New York: Messner (1964), 191 pp. Grades 7–10.
10. POLING, JAMES. *The Story of Tools: How They Built Our World and Shaped Man's
    Life.* New York: Norton (1969), 150 pp. Grades 6–9.
11. RUSCH, RICHARD B. *Man's Marvelous Computer: The Next Quarter Century.* New
    York: Simon and Schuster (1970), 128 pp. Grades 6–9.

12. SOULE, GARDNER. *Tomorrow's World of Science: The Challenge of Today's Experiments.* New York: Coward-McCann (1963), 120 pp. Grades 6–9.
13. STEINBERG, FRED J. *Computers.* New York: Watts (1969), 89 pp. Grades 6–9.
14. SULLIVAN, GEORGE. *Rise of the Robots.* New York: Dodd, Mead (1971), 114 pp. Grades 5–8.
15. VORWALD, ALAN and FRANK CLARK. *Computers: From Sand Table to Electronic Brain.* New York: McGraw-Hill (1970), 191 pp. Grades 7–10.
16. WALTER, LESLIE. *American Inventions: A Book to Begin On.* New York: Holt, Rinehart and Winston (1963), 45 pp. Grades 3–4.
17. ZIM, HERBERT and JAMES R. SKELLY. *Hoists, Cranes and Derricks.* New York: Morrow (1969), 64 pp. Grades 4–6. Sketches with this volume.
18. ZIM, HERBERT S. and JAMES R. SKELLY. *Machine Tools.* New York: Morrow (1969), 64 pp. Grades 4–6. Many black and white sketches included.

Farming and food

1. BUEHR, WALTER. *Food from Farm to Home.* New York: Morrow (1970), 96 pp. Grades 4–6.
2. EBERLE, IRMENGARDE. *Apple Orchard.* New York: Walck (1962), 39 pp. Grades 5–8. Apple growing in upper New York state.
3. EDWARDS, D. J. *Growing Food.* New York: John Day (1969), 48 pp. Grades 4–6.
4. LIMBURG, PETER. *The Story of Corn.* New York: Messner (1971), 96 pp. Grades 4–6.
5. RUSSELL, SOLVEIG P. *The Farm.* New York: Parent's Magazine Press (1970), 64 pp. Grades 3–4.
6. UHL, MELVIN. *About Grasses, Grains, and Canes.* Chicago: Melmont (1963), 48 pp. Grades 4–6.

Fishing

1. BROOKS, ANITA. *Picture Book of Fisheries.* New York: John Day (1961), 96 pp. Grades 5–8.
2. FENTON, D. K. *Harvesting the Sea.* Philadelphia: Lippincott (1970), 64 pp. Grades 4–6.

Glass

1. BUEHR, WALTER. *The Marvel of Glass.* New York: Morrow (1963), 95 pp. Grades 5–7.
2. EBERLE, IRMENGARDE. *The New World of Glass.* New York: Dodd, Mead (1963), 79 pp. Grades 7–10.
3. WYMER, NORMAN. *Glass.* New York: Roy (1969), 62 pp. Grades 7–10.

Lumbering and lumber products

1. BROOKS, ANITA. *Picture Books of Timber.* New York: John Day (1967), 96 pp. Grades 4–7.
2. FLOETHE, LOUISE L. *The Story of Lumber.* New York: Scribner (1962), 32 pp. Grades 3–5.
3. HARRISON, C. WILLIAM. *Forests: Riches of the Earth.* New York: Messner (1969), 191 pp. Grades 6–9.
4. RICH, LOUISE D. *First Book of Lumbering.* New York: Watts (1967), 72 pp. Grades 4–7.
5. TAYLOR, ARTHUR S. and others. *Logging: The Story of an Industry.* Philadelphia: Lippincott (1962), 64 pp. Grades 5–8.

Mining and mining products

1. COLBY, CARROLL B. *Aluminum: The Miracle Metal.* New York: Coward-McCann (1958), 48 pp. Grades 5–8.

2. CROSS, CLARK I. *Mines and Mining*. Chicago: Melmont (1964), 64 pp. Grades 5–8.
3. GREEN, ERNA. *Let's Go to a Steel Mill*. New York: Putnam (1961), 48 pp. Grades 3–5.
4. TRACY, EDWARD B. *The New World of Iron and Steel*. New York: Dodd, Mead (1971), 96 pp. Grades 6–9.

Money and banking

1. BRAUDE, MICHAEL. *Shelby Goes to Wall Street*. Minneapolis: Denison (1965), 24 pp. Grades 4–6.
2. COBB, VICKI. *Making Sense of Money*. New York: Parent's Magazine Press (1971), 64 pp. Grades 3–4.
3. FRIEDLANDER, JOANNE J. and JEAN NEAL. *Stock Market ABC*. Chicago: Follett (1969), 96 pp. Grades 4–6.
4. MAHER, JOHN E. *Ideas About Money*. New York: Watts (1970), 47 pp. Grades 4–6.
5. MITSCHE, ROLAND. *Money*. New York: McGraw-Hill (1970), 128 pp. Grades 6–9.
6. ROSENFIELD, BERNARD. *Let's Go to the U.S. Mint*. New York: Putnam (1960), 48 pp. Grades 4–6. Very detailed account.
7. SHAY, ARTHUR. *What Happens When You Put Money in the Bank*. Chicago: Reilly and Lee (1967), 32 pp. Grades 1–3.

Petroleum and oil

1. BUEHR, WALTER. *Oil: Today's Black Magic*. New York: Morrow (1957), 96 pp. Grades 5–8.
2. NEAL, HARRY E. *Oil*. New York: Messner (1970), 192 pp. Grades 6–8.
3. SHILSTONE, BEATRICE. *The First Book of Oil*. New York: Watts (1969), 90 pp. Grades 5–8.

Power

1. DeCAMP, L. SPRAGUE. *Engines: Man's Use of Power from the Water Wheel to the Atomic Pile*. New York: Golden (1959), 56 pp. Grades 4–6.
2. SCHNEIDER, HERMAN and NINA. *More Power to You: A Short History of Power from the Windmill to the Atom*. New York: Scott (1963), 128 pp. Grades 5–8.

Rubber and rubber products

1. BUEHR, WALTER. *Rubber: Natural and Synthetic*. New York: Morrow (1964), 96 pp. Grades 5–8.
2. DRESNY, E. JOSEPH. *The Magic of Rubber*. New York: Putnam (1960), 93 pp. Grades 5–8.
3. WILSON, MARILYN. *Let's Go to a Rubber Plant*. New York: Putnam (1960), 47 pp. Grades 4–6.

Textiles and clothing

1. ARNOLD, PAULINE and PERCIVAL WHITE. *Clothes and Cloth: America's Apparel Business*. New York: Holiday (1961), 355 pp. Grades 4–6.
2. BUEHR, WALTER. *Cloth: From Fiber to Fabric*. New York: Morrow (1965), 96 pp. Grades 4–6.
3. LAZARUS, HARRY. *Let's Go to a Clothing Factory*. New York: Putnam (1961), 48 pp. Grades 4–6.
4. WALLER, LESLIE. *Clothing: A Book to Begin On*. New York: Holt, Rinehart and Winston (1969), 48 pp. Grades 3–4.

Transportation and Communication

In order to help boys and girls understand the importance of improved transportation and communication in the United States today, teachers will need to bear in mind the economic bases of these two systems. Historically the various regions of the

United States attempted to specialize in the products they could develop best. To do so, they needed good, fast transportation to move their goods. This, in turn, led to faster ships and then to canals. Later the same forces brought about railroads and highways. Still later came the invention of airplanes. In a similar way new forms of communication have been brought about, including telephones, the telegraph, radio, and television, and now communication by satellites.

Two methods should help pupils to gain an understanding of these basic concepts. One is role-playing and the other is the preparation of time-lines. Boys and girls can be "given" farms or factories in various part of the country and can be asked how they can sell their products in other sections of the United States. Almost immediately they will grasp the importance of transportation. By the use of time-lines, pupils can begin to see how various inventions have appeared over a period of many years — some of them produced in our country and some in other parts of the world.

Boys, in particular, are likely to be interested in the inventions and the inventors behind the new discoveries. Biography should therefore be stressed in any concentration upon transportation and communication in our national life today.

To most children the marvels of our present transportation and communication systems will seem ordinary, commonplace, and expected. They will need to discover that this was not always true. By placing themselves in time, one hundred years or so, they can learn how different life was then and how many changes have been made in the last one hundred or even fifty years. This is one of the places where history seems most relevant in understanding the contemporary scene in our nation.

After a general introduction to these two topics, teachers may want to have their pupils work in committees. These topics lend themselves easily to such group work. There can be committees on air transportation, railroads, water transportation, and transportation by autos, buses, and trucks, as well as committees on various forms of communication.

Maps showing the major highways, the invisible network of pipelines, and the major railroad systems can help children to learn geography and at the same time assist them in discovering how our nation is bound together today by a vast system of transportation facilities.

Some teachers may want to delve into the place of governments — local, state, and national — in our transportation and communications system, although many teachers find this too involved for most pupils.

Another topic of importance and interest that you may want to explore is the future of transportation and communication in the U.S.A.

Here are some of the books which should be useful in this unit:

BOOKS FOR PUPILS ON TRANSPORTATION AND COMMUNICATION

Transportation — General

1. BIENVENU, HAROLD J. *Transportation — Lifeline of America.* Chicago: Scott Foresman (1961), 36 pp. Grades 4–6. Stresses the economic aspects.
2. CAIN, WILLIAM W. *Story of American Transportation.* Grand Rapids, Michigan: Fideler (1964), 128 pp. Grades 5–8. Includes many large black and white photos.
3. DE LEEUW, HENDRIK. *From Flying Horse to Man in the Moon.* New York: St. Martin's (1963), 310 pp. Grades 4–7.
4. FRISKEY, MARGARET. *Caveman to Spaceman: A Picture History of Transportation.* Chicago: Children's Press (1961), 54 pp. Grades 2–4.
5. RECK, FRANKLIN M. *The Romance of American Transportation.* New York: Crowell (1962), 312 pp. Grades 5–8.

6. RESS, ETTA S. *Transportation in Today's World.* Mankato, Minnesota: Creative Educational Society (1965), 176 pp. Grades 4–6.

Air transportation

1. CAGLE, MALCOLM W. *Flying Ships: Hovercraft and Hydrofoils.* New York: Dodd, Mead (1970), 142 pp. Grades 6–9.
2. CANBY, COURLANDT. *A History of Flight.* New York: Hawthorn (1963), 113 pp. Grades 7–10.
3. DALLISON, KEN. *When Zeppelins Flew.* New York: Time-Life (1969), 57 pp. Grades 5–8.
4. DIETRICH, FRED and SEYMOUR REIT. *Wheels, Sails and Wings: The Story of Transportation.* New York: Golden (1959), 96 pp. Grades 6–9.
5. EDITORS of *The American Heritage. The History of Flight.* New York: Golden (1964), 105 pp. Grades 6–9.
6. ELTING, MARY. *Aircraft at Work.* Irvington-on-Hudson, New York: Harvey (1964), 92 pp. Grades 5–8.
7. FENTEN, D. X. *Aviation Careers.* Philadelphia: Lippincott (1969), 208 pp. Grades 6–9.
8. HELLMAN, HAL. *Helicopters and Other VTOL's.* Garden City, New York: Doubleday (1970), 140 pp. Grades 7–10.
9. HIGHLAND, HAROLD J. *The How and Why Wonder Book of Flight.* Columbus, Ohio: Merrill (1961), 48 pp. Grades 4–6.
10. HOOD, JOSEPH F. *Skyway Round the World: The Story of the First Global Airway.* New York: Scribner (1968), 192 pp. Grades 6–9.
11. HYDE, MARGARET O. *Flight Today and Tomorrow.* New York: Whittlesey (1962), 140 pp. Grades 6–9.
12. LOOMIS, ROBERT D. *All About Aviation.* New York: Random House (1964), 139 pp. Grades 6–9.
13. MAY, CHARLES. *Women in Aeronautics.* New York: Nelson (1962), 260 pp. Grades 7–10.
14. SHAY, ARTHUR. *What Happens When You Travel by Plane?* Chicago: Reilly and Lee (1968), 32 pp. Grades 1–3.

Autos, buses, trucks, and roads

1. BUTTERWORTH, W. E. *Wheels and Pistons: The Story of the Automobile.* New York: Four Winds (1971), 191 pp. Grades 7–10.
2. CARLISLE, NORMAN and MADELYN. *About Roads.* Chicago: Melmont (1965), 46 pp. Grades 2–4.
3. COOKE, DAVID C. *How Superhighways Are Made.* New York: Dodd, Mead (1958), 64 pp. Grades 4–6. Many excellent photographs.
4. LENT, HENRY B. *What Car Is This?* New York: Dutton (1969), 128 pp. Grades 5–8.
5. LORD, BEMAN. *Look at Cars.* New York: Walck (1970) revised edition, 47 pp. Cars of today and of the past, with many black and white illustrations.
6. MORRIS, GWYNN. *The Story of Cars.* New York: Putnam (1963), 92 pp. Grades 5–8.
7. ZIM, HERBERT S. and JAMES SKELLY. *Trucks.* New York: Morrow (1970), 64 pp. Grades 4–6.

Railroads

1. BURLEIGH, DAVID R. *Piggyback.* Chicago: Follett (1962), 32 pp. Grades 3–5. Transporting goods by piggyback on railroads.
2. ELTIN, MARY. *All Aboard: The Railroad Trains That Built America.* New York: Four Winds (1971), 125 pp. Grades 4–6.

3. HOLBROOK, STEWART H. *The Golden Age of Railroads.* New York: Random House (1960), 182 pp. Grades 6–9. A Landmark book.
4. HOWARD, ROBERT W. *The Great Iron Trail: The Story of the First Transcontinental Railroad.* New York (Putnam), 1962, 376 pp. Grades 5–8.
5. McCALL, EDITH. *Men on Iron Horses.* Chicago: Melmont (1960), 127 pp. Grades 4–6.
6. SCHARFF, ROBERT. *The How and Why Wonder Books of Railroads.* New York: Grosset and Dunlap (1964), 48 pp. Grades 4–6.
7. SIMMONS, MORTIMER. *The Story of Trains.* New York: Putnam (1963), 78 pp. Grades 5–7.

Water transportation

1. BENDICK, JEANNE. *Sea So Big, Ship So Small.* Chicago: Rand McNally (1963), 80 pp. Grades 4–6.
2. BILLINGS, HENRY. *Bridges.* New York: Viking (1963), 153 pp. Grades 5–8.
3. CARSE, ROBERT. *Great American Harbors.* New York: Norton (1963), 110 pp. Grades 5–8.
4. CARTER, KATHERINE. *True Book of Ships and Seaports.* Chicago: Children's Press (1963), 47 pp. Grades 2–4.
5. CORBETT, SCOTT. *What Makes a Boat Float?* Boston: Little, Brown (1970), 42 pp. Grades 4–5.
6. DOHERTY, C. H. *Bridges.* New York: Meredith (1969), 120 pp. Grades 5–8.
7. SCHARFF, ROBERT. *The Why and How Wonder Book of Ships.* Columbus, Ohio: Merrill (1963), 48 pp. Grades 5–8.
8. WATTENBERG, BEN. *Busy Waterways: The Story of America's Inland Water Transportation.* New York: John Day (1964), 127 pp. Grades 5–8.
9. ZIM, HERBERT S. and JAMES S. SKELLY. *Cargo Ships.* New York: Morrow (1970), 64 pp. Grades 4–6.

Communication

1. ARNOLD, OREN. *Marvels of the United States Mail.* New York: Abelard-Schuman (1964), 125 pp. Grades 5–8.
2. BENDICK, JEANNE. *Television Works Like This.* New York: Whittlesey (1959), 64 pp. Grades 4–6.
3. BUEHR, WALTER. *Sending the Word: The Story of Communication.* New York: Putnam (1959), 95 pp. Grades 4–6.
4. BUSBY, EDITH. *Behind the Scenes at the Library.* New York: Dodd, Mead (1960), 62 pp. Grades 4–6.
5. CAHN, WILLIAM and RHODA. *The Story of Writing.* Irvington-on-Hudson, New York: Harvey (1963), 128 pp. Grades 6–8.
6. COLBY, C. B. *Communications: How Man Talks to Man Across Land, Sea and Space.* New York: Coward-McCann (1964), 48 pp. Grades 6–8.
7. COOMBS, CHARLES I. *Window on the World: The Story of Television Production.* Cleveland: World (1965), 122 pp. Grades 6–9.
8. DIETZ, BETTY W. *You Can Work in the Communications Industry.* New York: John Day (1970), 96 pp. Grades 7–10.
9. DUDLEY, NANCY. *Linda Goes to a TV Studio.* New York: Coward-McCann (1959), 46 pp. Grades 3–5.
10. FABER, DORIS. *Behind the Headlines: The Story of Newspapers.* New York: Pantheon (1963), 151 pp. Grades 6–9.
11. GALLANT, ROY A. *Man Must Speak: The Story of Language and How We Use It.* New York: Random House (1969), 177 pp.
12. GRAHAM, CLARENCE R. *The First Book of Public Libraries.* New York: Watts (1959), 59 pp. Grades 4–6.

13. JOHNSON, NICHOLAS. *How to Talk Back to Your Television Set.* New York: Bantam (1970), 245 pp. A paperback. For better junior high readers.
14. JUPO, FRANK. *Any Mail for Me? 5000 Years of Postal Service.* New York: Dodd, Mead (1964), 64 pp. Grades 5–8.
15. MINER, IRENE S. *The True Book of Communication.* Chicago: Children's Press (1960), 47 pp. Grades 3–5.
16. NURNBERG, MAXWELL. *Wonders in Words.* Englewood Cliffs, New Jersey: Prentice-Hall (1969), 90 pp. Introduction to word derivation.
17. OSMOND, EDWARD. *From Drumbeat to Tickertape.* New York: Criterion (1961), 128 pp. Grades 7–10.
18. RESS, ETTA S. *Signals to Satellites.* Mankato, Minnesota: Creative Educational Society (1965), 176 pp. Grades 4–6.
19. SOLOMON, LOUIS. *America Goes to Press: The Story of Newspapers from Colonial Times to the Present.* New York: Crowell-Collier (1970), 106 pp. Grades 5–8.
20. SOLOMON, LOUIS. *Telstar: Communication Break-Through by Satellite.* New York: McGraw-Hill (1962), 64 pp. Grades 6–9.
21. STEIN, M. L. *Freedom of the Press: The Continuing Struggle.* New York: Messner (1966), 190 pp. Grades 5–8.
22. SODDARD, EDWARD. *Television.* New York: Watts (1970), 64 pp. Grades 5–7.
23. TOOZE, RUTY. *Telephone Wires Up!* Chicago: Melmont (1964), 64 pp. Grades 4–6.
24. WEISBERGER, BERNARD A. *The American Newspaperman.* Chicago: University of Chicago Press (1961), 225 pp. For good readers only.

Government in the United States Today

Recent research has indicated that children know something about government before the third grade. They often identify with the President of the United States and sometimes with the mayor of their city. They sometimes know that governments make rules or laws and help people.

By the time children are in the fourth, fifth, or sixth grades, they should be ready for a relatively deep study of governmental units in our nation. Such studies, however, should stress the reasons for government, the lawmaking and service functions, and the financing of our governmental units rather than concentrating upon the structure of political units.

Perhaps the best way to start a study of government is to imagine a community or city without laws or rules. It might be fun for a short time. Pupils may like the idea. But they can discover what would happen. Eventually there would be havoc. The idea can then be introduced that government is established by groups of people to do the things they cannot do alone or think they cannot do alone. This is certainly one of the central concepts of political science.

What our government does can be tackled next. It can be illustrated best by the local unit of government, whether that is a city, a township, a district, a county, or some other unit. Perhaps all these names need to be lumped together as local government, without burdening pupils with the differences among them.

Children can find out some of the tasks of local government, and teachers can add others. As background for such a study, teachers need to know about their local government. Usually a city has such divisions as SAFETY, HEALTH, EDUCATION, PARKS AND RECREATION, STREETS AND HIGHWAYS, FINANCE, and CITY PLANNING.

As a natural outgrowth of this learning, pupils should find out how the people are selected to carry out these tasks. This can lead into an elementary study of elections. Pupils also can learn about two or three types of city government. With the city manager movement gaining slowly in our country, they might well be introduced to that form, as well as the mayor-council form of local government. The emphasis should be upon the local unit, however, no matter what it is.

Children probably will inquire what it costs to run all these offices, and a beginning study of the costs of government can be dealt with next, including elementary material on how the money is obtained.

Some brief study can also be made of local courts.

After the completion of this work, pupils can learn that state governments are similar. A great deal of time need not be spent on them.

With this background, pupils can learn something, too, about the national government, following the same pattern. It is doubtful whether an extensive study of our federal system is useful to boys and girls at this stage in their development.

A few fifth and sixth grade pupils may be interested in the idea of different forms of government. Some teachers may want to touch briefly on this topic at this point.

The conclusion of a short unit on Our Government Today might include some attention to the rights and the responsibilities of its citizens.

Sometimes there is merit in trying to select a word or phrase to cover a single social science discipline. In this way one is likely to think in terms of the bull's eye of the target of teaching. If there is one word to represent political science, it is the word "power." With more able upper school or junior high school pupils, you may want to explore that idea. Draw a circle on the chalkboard or on a transparency and mark it power. Then discuss how that circle is to be divided. One way would be to divide the circle among the various units of government — local, state, and federal. Another way would be to use the circle to represent the national government. How, then, would it be divided among the executive, legislative, and judicial branches? A device like this can help pupils to visualize the central concept of this discipline.

Some Methods of Studying Government Units

As with other topics, there are many ways of studying governmental units or divisions. One way has been suggested already, through an imaginary city without rules or laws. Another is to discover the many services to citizens which taxes pay for. Still another is to take a current newspaper and find all the references to government in it.

An interesting approach is to study the duties of a mayor, a governor, or President through a typical day in their lives.

Pie charts can be developed on the income of various governmental units, and pictorial charts of the services of government can be produced by pupils.

There are a number of films and filmstrips on government which can be used to advantage in studies of government by elementary school pupils.

The writer once saw a teacher in the sixth grade of rather slow pupils develop the idea of state government by drawing three "houses" in which she placed the governor, the legislators, and the judges, developing the concepts of their jobs through this pictorial device.

In this unit, as in others, pupils should meet a resource person, if at all possible. It need not be a high-ranking individual. It might well be someone from the local health office or the city planner's office. If he or she can be asked to bring "realia" for a meeting with the class, illustrating what he or she does, it would be much more meaningful to children.

There are several books for boys and girls on the functions of government. These are listed in the bibliography on pages 374–375. Three of them, for example, are *First Books* — on Congress, the President, and the Supreme Court. Another is on *What the President Does All Day*. Four are in a series on *Let's Go to the Capitol, . . . to the White House, . . . to the Supreme Court,* and *. . . to the U.S. Mint.* A similar approach is used in Louis Wolfe's *Let's Go to a City Hall.* There are individual books on special features of our government, too. One of them is Josephine Hemphill's *Fruitcake and Arsenic,* the story of the Food and Drug Administration. Another is on the work of the United States Public Health Service, titled *To Save Your Life.* Altogether, teachers will find that there are more books on various aspects of government than on some other topics about the U.S.A. Today.

This is not the easiest unit to teach to boys and girls, but with careful planning it can be done effectively and in an interesting manner.

BOOKS FOR BOYS AND GIRLS ON GOVERNMENT IN THE U.S.A.

1. ACHESON, PATRICIA C. *Our Federal Government: How It Works.* New York: Dodd, Mead (1962), 168 pp. Grades 6–9.
2. ARCHER, JULES. *Revolution in Our Time.* New York: Messner (1971), 192 pp. Grades 6–9. Revolutions around the world in our time and dissent in the U.S.A.
3. ARNOLD, OREN. *Marvels of the U.S. Mail.* New York: Abelard-Schuman (1964), 125 pp. Grades 3–5.
4. BEECH, LINDA. *On the Campaign Trail: The Story of Elections.* New York: Messner (1971), 96 pp. Grades 5–8.
5. BORIE, MARCIA. *Famous Presidents of the United States.* New York: Dodd, Mead (1963), 159 pp. Grades 6–9.
6. BOTTER, DAVID. *Politicians and What They Do.* New York: Watts (1960), 213 pp. Grades 7–10.
7. BRADLEY, DUANE. *Electing a President.* Princeton, New Jersey: Van Nostrand (1963), 156 pp. Illustrations from several campaigns.
8. BRINDZE, RUTH. *All About Courts and the Law.* New York: Random House (1964), 138 pp. Grades 5–8.
9. COY, HAROLD. *The First Book of Presidents.* New York: Watts (1961), 69 pp. Grades 5–7. Brief introduction and one page on each President.
10. COY, HAROLD. *The First Book of the Supreme Court.* New York: Watts (1968), 59 pp. Grades 5–8.
11. DAVIDSON, BILL. *President Kennedy Selects Six Brave Presidents.* New York: Harper and Row (1962), 96 pp. Washington, John Quincy Adams, Lincoln, Andrew Johnson, Chester A. Arthur and Theodore Roosevelt.
12. EICHNER, JAMES A. *The Cabinet of the President of the United States.* New York: Watts (1969), 62 pp. Grades 5–8.
13. EICHNER, JAMES A. *The First Book of Local Government.* New York: Watts (1964), 61 pp. Grades 5–8.
14. ESKIE, SUNNY. *A Land Full of Freedom.* New York: Friendly (1963), 32 pp. Grades 2–4.
15. FRIBOURG, MARJORIE G. *Ports of Entry: U.S.A.* Boston: Little, Brown (1962), 240 pp. Grades 7–10. The U.S. Customs Service.
16. GRAY, LEE L. *How We Choose a President.* New York: St. Martin's (1968), 175 pp. Grades 5–9.
17. GREEN, WILLIAM. *The Congressman.* New York: McGraw-Hill (1969), 128 pp. Grades 5–8. A week in the life of a Philadelphia congressman.
18. HABENSTREIT, BARBARA. *Changing America and the Supreme Court.* New York: Messner (1970), 192 pp. Grades 6–9.
19. HEAPS, WILLARD A. *Taxation: U.S.A.* New York: Seabury Press (1971), 208 pp. Grades 6–9.

20. HEMPHILL, JOSEPHINE. *Fruitcake and Arsenic*. Boston: Little, Brown (1962), 144 pp. Grades 6–9. The story of the Pure Food and Drug Administration.
21. HOOPES, ROY. *What the President Does All Day*. New York: John Day (1962), 64 pp. Grades 4–6.
22. HOOPES, ROY. *What a United States Senator Does*. New York: John Day (1970), 128 pp. Grades 4–6.
23. JEFFERS, H. PAUL. *How the U.S. Senate Works: The A.B.M. Debate*. New York: McGraw-Hill (1970), 95 pp. Grades 5–8.
24. JOHNSON, GERALD W. *The Congress*. New York: Morrow (1963), 128 pp. Grades 5–8.
25. JOHNSON, GERALD W. *The Presidency*. New York: Morrow (1963), 127 pp. Grades 5–8.
26. JOHNSON, GERALD W. *The Supreme Court*. New York: Morrow (1963), 127 pp. Grades 5–8.
27. KELLY, FRANK K. *Your Freedom: The Bill of Rights*. New York: Putnam (1964), 192 pp. Grades 7–10.
28. LAVINE, DAVID. *The Mayor and the Changing City*. New York: Random House (1966), 64 pp. Grades 6–9.
29. LAVINE, DAVID. *What Does a Congressman Do?* New York: Dodd, Mead (1965), 64 pp. Grades 4–6.
30. McCARTHY, AGNES. *Let's Go to Vote*. New York: Putnam (1962), 48 pp. Grades 3–5. A visit to a polling place with adults.
31. MIERS, EARL S. *The Capitol and Our Lawmakers*. New York: Grosset and Dunlap (1965), 48 pp. Grades 6–9.
32. MIERS, EARL S. *The White House and the Presidency*. New York: Grosset and Dunlap (1965), 48 pp. Grades 6–8.
33. NEAL, HARRY E. *Diary of Democracy: The Story of Political Parties in America*. New York: Messner (1970), 191 pp. Revised edition. Grades 6–9.
34. NEWMAN, SHIRLEE R. and DIANA F. *About People Who Run Your City*. Chicago: Melmont (1963), 48 pp. Grades 5–7.
35. PARADIS, ADRIAN. *Government in Action*. New York: Messner (1965), 192 pp. Grades 7–9.
36. PHELAN, MARY K. *The White House: A Book to Begin On*. New York: Holt, Rinehart and Winston (1962), 44 pp. Grades 2–4.
37. POLKING, KIRK. *Let's Go to See Congress at Work*. New York: Putnam (1966), 48 pp. Grades 4–6.
38. ROSENFIELD, BERNARD. *Let's Go to the Capitol*. New York: Putnam (1959), 47 pp. Grades 4–7.
39. ROSENFIELD, BERNARD. *Let's Go to the Supreme Court*. New York: Putnam (1960), 47 pp. Grades 4–7. Mostly on the building.
40. ROSENFIELD, BERNARD. *Let's Go to the White House*. New York: Putnam (1959), 48 pp. Grades 4–7.
41. SPRINGARN, NATALIE D. *To Save Your Life*. Boston: Little, Brown (1963), 213 pp. Grades 7–10. The work of the Public Health Service.
42. STEVENS, LEONARD A. *How a Law Is Made: The Story of a Bill Against Pollution*. New York: Crowell (1970), 109 pp. Grades 5–8.
43. SULLIVAN, MARY B. *Careers in Government*. New York: Walck (1964), 128 pp. Grades 8–10.
44. TERRELL, JOHN U. *The United States Department of Agriculture: A Story of Foods, Farms, and Forests*. New York: Duell (1966), 130 pp. Grades 6–9.
45. TERRELL, JOHN U. *The United States Department of the Interior: A Story of Rangeland, Wildlife, and Dams*. New York: Duell (1963), 117 pp. Grades 6–9.
46. TERRELL, JOHN U. *The United States Department of the Treasury: A Story of Dollars, Customs, and Secret Agents*. New York: Duell (1963), 121 pp. Grades 6–9.

47. WEAVER, WARREN, JR. *Making Our Government Work: The Challenge of American Citizenship.* New York: Coward-McCann (1964), 119 pp. Grades 6–9.
48. WEINGAST, DAVID. *We Elect a President.* New York: Messner (1968), 190 pp. Revised edition. Grades 8–10.

Living with Our World Neighbors

Certainly one of the basic themes throughout school should be the interdependence of the modern world, the relations of the United States and its people with other countries and their people, and the growth of regional and international organizations. The children who are in our schools today will live well into the twenty-first century and theirs will be an even more interdependent world than the one that exists today.

Possibly the best place for such a study would be the end of the period spent on the United States Today. That does not exclude many references to the world in other units. But it would provide an opportunity to concentrate upon this theme for several days and thus promote better learning. A unit on "Living with Our World Neighbors," could be taught at other points. Some teachers may want to postpone such a unit until the end of the year, letting it serve as a bridge to the study of other lands and peoples in the sixth grade. What is taught and how it is taught are far more important than when such a unit is developed.

Such a unit can be introduced in many ways. It might be approached through a study of the daily newspapers, the current news over radio, or contemporary affairs as reported by television programs. It might be through a good film or filmstrip. Or it might be through a pooling of information about the billions of neighbors we have. A creative teacher could well start with the word "neighbor" and develop the concept that our neighbors in today's world include all the people on our globe.

Then the pupils could see how these people are distributed around our planet and organized in approximately 150 territories and countries.

This should lead into the topic of how people of the United States and the government of our nation are affected by their world neighbors. This can begin with simple illustrations or examples which affect children and expand to relations between nations. Different pupils might be asked to find out about the different parts of a baseball and where they come from. Others might be asked to discover many products from other parts of the world, go into a candy bar, into a telephone, or into an automobile or airplane. Research will soon reveal how many materials from different parts of the world go into these common products.

In a similar way individuals or groups of pupils might well interview merchants as to the raw materials they obtain from abroad or the places where they sell their finished products. In this way, too, boys and girls would begin to see the interdependence of today's world.

Eventually the class should certainly make an introductory study of our relations with other countries and of our cooperative efforts in regional and international relations, including the Organization of American States and the United Nations and its various agencies.

Any study of the O.A.S. and/or the United Nations should focus, however, on the functions of such organizations rather than on structure. Pupils should learn how regional and international organizations are limited in power, but nevertheless wrestle with some of the big problems of our contemporary world, such as health, education, and economic development. If some study of the U.N. is included, it should probably emphasize the work of some of the specialized agencies.

Some Ways of Developing the World Neighbor Theme

Teachers will need to be clear on their aims in such a study and be careful to avoid too deep a study. This should not be a course in international relations or foreign policy. It should be an introduction to the relations of the United States with other countries and their people.

Again, people should be the central focus. If you can invite one or more persons from other parts of the world to discuss their homelands and the interdependence of their nations and ours, it should be helpful.

You may want to encourage pupils to gather information about the products in their homes which come from abroad. Or you may want to have them interview local merchants about the products they obtain from abroad or those which they sell abroad.

Occasionally cartoons on current affairs can be used to show interdependence and international relations. A few simple cartoons are used in the writer's volume *One Nation: The United States*. But cartoons are usually rather sophisticated devices and should be used sparingly.

Large posters about other countries can be dramatic and can be utilized effectively in this unit.

You may want to encourage some pupils to make a mural or a montage on the theme of interdependence or world neighbors. A few might be interested in trying their hand at simple cartoons.

Of course you will want to use television programs and current newspapers to relate this theme to on-going current events. In that sense this is a unit which will extend throughout the year.

As a part of this unit, you may want to encourage your pupils to take part in some action project involving the children of some other part of the world. If the pupils can earn the money they send, it will be a better learning experience for them. They may want to take part in a UNESCO or UNICEF project, raise money for Meals for Millions, prepare a Red Cross project, contribute to the aid of a child abroad through the Save the Children Federation, or send equipment to an economically underdeveloped area through CARE.

Unfortunately there are not too many books for elementary school children as yet which stress interdependence and international relations. A few that do exist are listed on pages 462–463 in the chapter on Studying the World — General. One of the best of those volumes is Marie Neurath's *Living With One Another*. Two other volumes which you may want to use are Eva Knox Evans's *People Are Important* and the writer's *Three Billion Neighbors*. You may want to use some of the pictures, too, in the famous photographic volume *The Family of Man*. There is certainly room for many publishers to solicit books on the general theme of the U.S. and its world neighbors, written for this age group.

A few films and filmstrips also are available. They may be found in the *Educational Media Index*.

BOOKS FOR BOYS AND GIRLS ON
LIVING WITH OUR WORLD NEIGHBORS

1. ABERNATHY, ROBERT G. *Introduction to Tomorrow: The U.S. and the Wider World.* 1945–1965. New York: Harcourt Brace Jovanovich (1966), 286 pp. Grades 8 and up.
2. CALDWELL, JOHN. *Let's Visit Americans Overseas.* New York: John Day (1958), 96 pp. Grades 5–8. Americans working abroad.

3. EVANS, EVA KNOX. *People Are Important.* New York: Capitol and Golden (1960), 86 pp. Grades 5–7. Anthropology with a humorous touch.
4. FISHER, ROGER. *International Conflict: For Beginners.* New York: Harper and Row (1969), 231 pp. Grades 7–10.
5. GALT, ROM. *Peace and War: Man Made.* Boston: Beacon (1962), 138 pp. Grades 6–9. A history of international conflict and cooperation.
6. GREENE, WADE. *Disarmament: The Challenge of Civilization.* New York: Coward-McCann (1966), 128 pp. Grades 6–9.
7. HANFF, HELEN. *Good Neighbors: The Peace Corps in Latin America.* New York: Grosset and Dunlap (1966), 64 pp. Grades 5–7.
8. HYDE, MARGARET O. *This Crowded Planet.* New York: Whittlesey (1961), 159 pp. Grades 5–8.
9. JOYCE, JAMES A. *Decade of Development: The Challenge of the Underdeveloped Nations.* New York: Coward-McCann (1969), 118 pp. Grades 6–9.
10. KAREN, RUTH. *Neighbors in a New World.* Cleveland: World (1966), 188 pp. Grades 6–9. Focuses on the U.S.A. and Latin America.
11. KENWORTHY, LEONARD S. *Three Billion Neighbors.* Lexington, Massachusetts: Ginn (1965), 160 pp. Grades 4–7. 450 black and white photographs.
12. LARSEN, PETER. *The United Nations at Work Throughout the World.* New York: Lothrop, Lee and Shepard (1971). Grades 3–7. Largely pictorial.
13. LAVINE, DAVID. *Outposts of Adventure: The Story of the Foreign Service.* Garden City, New York: Doubleday (1966), 84 pp. Grades 6–9.
14. LAVINE, DAVID, and IRA MANDELBAUM. *What Does a Peace Corps Volunteer Do?* New York: Dodd, Mead (1964), 64 pp. Grades 6–9.
15. McGUIRE, EDNA. *The Peace Corps.* New York: Macmillan (1966), 224 pp. Grades 6–9.
16. NEURATH, MARIE. *Living With One Another.* New York: Watts (1966), 128 pp. With interesting black and white pictographs.
17. PARADIS, ADRIAN A. *You and the Next Decade.* New York: McKay (1965), 179 pp. Grades 7–10.
18. PRINGLE, LAURENCE. *One Earth, Many People: The Challenge of Human Population Growth.* New York: Macmillan, 1971. Grades 5–8.
19. SAVAGE, KATHERINE. *The Story of the United Nations.* New York: Walck (1970). Grades 6–9.
20. SCOTT, JOHN. *Hunger: Man's Struggle to Feed Himself.* New York: Parents' Magazine Press (1969), 181 pp. Grades 6–9.
21. TOR, REGINA and ELEANOR ROOSEVELT. *Growing Toward Peace.* New York: Random House (1960), 84 pp. Grades 5–8. A history of international cooperation.
22. WHITTLESEY, SUSAN. *Good Will Toward Men: The Challenge of the Peace Corps.* New York: Coward-McCann (1963), 128 pp. Grades 5–8.

20

Studying
United States History

To what extent should United States history be taught in elementary schools or even in junior high schools? That is a question on which you will find many conflicting answers.

Many authorities have maintained for years that we introduce children too early to the formal study of our nation's history. Their contention is that we teach "too much history and too soon." They point to research which indicates that children do not really develop the time sense until the fourth, fifth, or sixth grade levels. Even in the seventh and eighth grades boys and girls are not too acute in their handling of time concepts. Since children do not have direct experience with the past, they find it difficult to comprehend.

On the other hand, many concerned persons have felt that we do not teach enough history to children. Many, but not all, think of history as indoctrination in certain values and the development of their brand of patriotism. They would have us separate history from the other social sciences and teach a great deal of it early and often. This would include much local and state history as well as the history of our nation.

In recent years some researchers and persons interested in cognitive learning have begun to maintain that children can learn history earlier than it has been taught in the past, just as they can learn more complicated mathematics and science than we assumed. They do agree, however, that the way in which history is taught is as important as the history which they learn.

This writer believes that much informal teaching of history can go on in the primary grades if it is simple in nature and related to events in which boys and girls participate. They can be holiday celebrations, elections, trips to places of historic interest in the community, and similar dramatic experiences. Most of the history children learn at that level will be through role-playing, pictorial materials, dramatics, stories, and problem-solving. Most of it will be outside of books, although some of it will be learned through stories and biographies.

Probably there should be some more formal introduction to history by about the fifth grade. It need not last for more than a half-year and it should be tied to the present as much as possible. But this does not seem too early to start on a dramatic, simple study of the drama of our national history. That initial involvement should concentrate upon people and their daily lives in selected periods of history. Later in this chapter we will spell out in detail just what such a study might entail.

Dr. Linwood Chase spelled out the kind of a history this should be, in his chapter on "American History in the Middle Grades" in *Interpreting and Teaching American History*, the National Council for the Social Studies 1961 yearbook. In that chapter he wrote:

> To a ten-year-old, history can be tramping through the wilderness with Daniel Boone, raising the flag with the Marines at Iwo Jima, living with seventy-five others 'tween decks on the *Mayflower* in the long voyage across the Atlantic, riding with Paul Revere through the night to Concord town, and floating down the Mississippi with Père Marquette. It can be an Indian watching the *Santa Maria* and her sister ships drop anchor, climbing on the heights with Balboa, surveying with Washington in virgin territory, imprisonment by Indians with Captain John Smith, and traveling under the North Pole in a nuclear submarine. It can be seeing new homes rise one by one in Jamestown, tasting buffalo meat on the trek by covered wagon across the western plains, smelling exploding gunpowder in many a French and Indian battle, feeling cold water rise to one's armpits in Arnold's march to Quebec, and hearing the great peal of the Liberty Bell on that first Fourth of July.

In that passage he has captured some of the dramatic moments of our history as young children should relive them. Would that history were taught widely in that way. Then children would capture some of the feelings as well as the facts of our national story.

DETERMINING AIMS IN THE STUDY OF UNITED STATES HISTORY

Perhaps you would like to stop at this point and make a list of the major objectives which you think should be attained with boys and girls in the middle grades through the study of our history. Then you can compare your list with the one compiled by this author.

Certainly you will want to develop facts, but facts organized around major concepts, or "big ideas." Those facts certainly will be better retained and internalized if they are discovered by pupils rather than told to them or read to them.

Doubtless you will agree, too, that much attention should be given to the development of a whole range of skills, most of which are spelled out in the chapter of this book on that topic. These would include a wide range, from reading to oral reporting and from reading maps and graphs to participation in panel discussions. Acquiring many skills in the social studies field is important to every pupil. Each boy and girl will use them as individuals and as citizens in a democracy. Long after many of the facts learned in a course in United States history are forgotten, today's pupils will be using (or not using) skills learned (or not learned) in their social studies classes. Which ones do you think will be most useful to them?

In a similar way values, attitudes, and appreciations need to be stressed. They, too, will remain with pupils for the rest of their lives, probably without much change. What values do you hope to instill? What attitudes? What appreciations? For example, many middle school pupils today have already acquired a large degree of cynicism about the functioning of governments. Do you feel that should be combatted? If not, why not? If so, how? It would be well for you to consider making a list of the values, attitudes, and appreciations you want to develop.

Following is a list of some of the aims this writer would like to see stressed in the teaching of United States history in the middle grades and in junior high schools. How does it compare with your list?

Aims in Studying United States History

1. to gain an understanding of the relationship between geography and history, with many examples of this relationship.
2. to understand the rich resources found in our country and how they were developed over a period of many years.
3. to realize that a variety of people came to this country and to understand why they came here.
4. to discover how people adapted their ways of living to the land where they lived and changed the land eventually.
5. to learn about the different ways of life in this land and the conflicts which arose out of them as well as some of the means of resolving these conflicts.
6. to understand why governments were developed and how the idea of democracy has expanded over a long period of time.
7. to discover how men have earned a living at various points in our history and how specialization has developed.
8. to begin to understand the role of religion in the life of our nation.
9. to grasp in an elementary way the importance of education.
10. to see how people moved west and how that affected our nation.
11. to delve into the story of our relations with some other countries.
12. to see how people have found ways to enjoy themselves in different activities and in different periods.
13. to discover how an American way of life has developed and what values Americans have held in common.
14. to understand in an elementary way how industries developed and how agriculture became more scientific.
15. to realize that health standards have been raised greatly.
16. to become acquainted with a few of the outstanding men and women in our national history and to realize that many unknown people have contributed to our wellbeing today.
17. to feel some of the pressures on people in moving to this country and developing it.
18. to take part, even indirectly, in the struggle for equal rights for all Americans in different times and places in our history.
19. to examine ways of resolving conflicts, seeing the role of compromise as not necessarily bad.
20. to develop and refine a sense of patriotism and a desire to contribute to the progress of the United States.
21. to improve one's reading skills in the social studies.
22. to improve one's vocabulary in the social studies.
23. to develop further skills in reading maps, charts, graphs, and related materials.
24. to develop an increasing ability to understand "time."
25. to develop further research skills and reporting skills.
26. to stimulate an interest in history and the reading of history.
27. to appreciate the rights and responsibilities of citizens.
28. to examine several values now held by Americans and to understand that people do not always agree on values.
29. to develop skills in working with others.

VARIOUS APPROACHES TO THE STUDY OF
UNITED STATES HISTORY

With these different aims in view, how should one organize the study of United States history? That is a question with which many textbook writers and curriculum construction personnel have wrestled. Many classroom teachers also have been confronted with this problem.

At the present time, some schools devote four or five years to the study of United States history, with at least one year given to this subject in the elementary, junior high, and senior high grades. Often there are two years in either the junior or senior high programs devoted to United States history. The more common pattern is to have one year of American history in the elementary school (usually at the fifth grade), another at the junior high school level (usually at the eighth grade), and still another at the senior high school level (usually at the eleventh grade).

Often there are state requirements about teaching United States history. Many states require at least one year of such history in the first eight grades and one more year in the next four years. Such requirements are not likely to change radically in the next few years. The pressures on state legislators are great for such requirements and they are not likely to modify existing laws lest they be accused of lack of patriotism. There we must all think in terms of the best use of the two or three years of United States history which will be required.

Perhaps the best way to organize courses is to think in terms of variety of approaches, depending upon the age of pupils and the major learnings which they can handle. Here are some of the ways in which United States history may be organized:

Plan one: Strict chronological order.   Most courses of study today at all levels are organized on the basis of strict chronology. The pupils start with prehistoric times or with Columbus and move slowly through the story of our history to the present time. This plan has the merit of using the main method of history-chronology or time. Certainly at some point in their school careers, pupils should be exposed to the drama of our history in this form or in a modified approach to it. The disadvantage of this approach for children is that it places undue emphasis upon chronology. Children do not have a well-developed sense of time even in the fifth grade. This militates against the success of this approach as the sole or even chief method with elementary school pupils.

Chronology helps people to see the interrelated nature of many events in a given period of history, but they do not always see the development of topics (such as labor unions, schools, or foreign relations).

Plan two: Chronological and topical approaches combined.   A variation of the strict chronological approach is often used at the junior or senior high school levels — or both. That is to treat American history to the Civil War in strict chronological fashion. From that point on, the topical and chronological approaches are combined. Thus, for example, the theme of labor unions would be developed from the Civil War to the present as a topic, treated chronologically. The role of farmers in our society would be handled in a similar way from the Civil War to the present. Other topics would be treated the same way.

This plan has merit, too. It establishes a chronological base but modifies it to show the development of movements in recent times. One may ask, however, why this cannot be done throughout the entire history of our country instead of from the Civil War on. This idea leads to a third approach.

Plan three: The topical approach. Might there not be merit in taking a few themes in United States history and developing them as topics? Probably one would want to begin with the present and then go back in history and see how these ideas or themes have developed. Here are some basic themes this writer feels might be handled in that way:

Our Land
Our People
Our Economy: From Farms to Factories
Our Government: The Development of Democracy
Our Foreign Relations: Isolationism and Internationalism
Our Leisure Time: Fun and Beauty
Our Great Men and Women

Starting with the present would provide more motivation than starting with the past. Then pupils would go back and trace the development of such themes as our ways of living and our idea of democracy in different eras.

The drawback of such an approach is that it does not show the interrelated nature of events. Theoretically, the approach is fine; practically it has some drawbacks.

The writer is not alone in thinking that strict chronology is not the only way or even the best way to deal with the development of our nation with younger pupils. Professor C. Benjamin Cox, for example, wrote recently in a chapter on "American History: Chronology and the Inquiry Process" in a book on *Social Studies in the United States: A Critical Appraisal* to the effect that "Still needed on the United States history textbook scene is a book that rejects the narrative chronological approach."

Plan four: Dividing United States history. Some schools have decided to concentrate on the early periods of our history in the elementary school and on the later periods in the junior high school. Or, they have asked junior high school teachers to focus on the periods prior to the Civil War and the senior high school teachers to deal in depth with the post-Civil War periods.

This is a convenient way to divide history, but little more can be said for it. There is no evidence to indicate that younger children are more interested in the earlier periods or to show that they can handle these periods better than more recent times. This plan of studying United States history seems to this writer merely an easy way out of a difficult situation.

Plan five: "Postholing" United States history. Is it really necessary for children to know about every President of the United States? Is it essential in their elementary school years for them to study the War of 1812 or the tariff issue or to learn about every territorial acquisition of our nation? Many people today are saying "No" to these questions.

What is needed in those years is a dramatic story of the development of our nation, touching merely upon the highlights. Some people now are calling this "postholing." Postholing is somewhat like placing several posts around a large field and then stringing wire between them. In this case the "posts" are in-depth treatments of major periods or events in our history; transitional material is the wire.

For a fifth grade class this seems like an excellent approach. It should be possible to give pupils such a panoramic view even in a half-year. Thus, the first half of the fifth grade might well be devoted to a comprehensive picture of the United States today and the second half of the year spent on a quick survey of our history. In such an approach, social history should be stressed. The accent might well be on how people lived at various times in our history. Other events would be mentioned, but the focus would be on social history.

A plan of study developed in this fashion might look something like the following outline. After a brief time spent on the prehistoric period, the Indians, and the explorers, children would concentrate on these periods:

Life in the Colonies: 1700
Life in the Colonies and in the New Nation: the 1770s and 1780s
Life in the U.S.A. in the 1830s
Life in the U.S.A. in the 1860s
Life in the U.S.A. in the 1900s
Life in the U.S.A. in the 1930s

The author of this book used this approach in writing the text for fifth graders on *One Nation: The United States* (Ginn, 1972). Such an approach has difficulties, but it can result in an exciting story of our national history for boys and girls.

Plan six: "Postholing" by important decisions. In a somewhat similar fashion, boys and girls can study American history by decisions. Such a method gives focus to a course and also enables the pupils to identify with problems with which our people have wrestled. It is also a means of promoting problem-solving, if well used.

Of course it is not easy to determine what decisions should be stressed. Here is the list of decisions which George Shaftel used in writing *Decisions in United States History* (Ginn, 1972), for use in junior high schools.

1. Can We Find a New Route to Asia?
2. What Type of Life Should the Spanish Develop in the New World?
3. What Type of Life Should the English Develop in the New World?
4. Who Shall Control North America?
5. English Colonies or a New Nation?
6. What Kind of Government Shall the New Nation Have?
7. Shall We Add the Louisiana Territory?
8. Shall We Become Friends with France or Great Britain?
9. How Shall We React to Europe's Interest in the Americas?
10. Shall We Extend Democracy?
11. Shall We Push to the Pacific?
12. What Shall We Do About Slavery?
13. United States or Divided States?
14. How Can We Bind Up the Wounds of the Civil War?
15. Shall Laborers and Farmers Share More Fully in the Wealth of Our Nation?
16. Shall We Acquire More Territories in Other Parts of the World?
17. Can We Establish an Enduring Peace?
18. Shall We Limit Immigration?
19. Can We Preserve Democracy and Capitalism?
20. What Part Should the United States Play in World War II?
21. Shall We Drop the Atomic Bomb?
22. What Decisions Are Needed Now to Strengthen Our Democracy?

It would be interesting for you to make up your own list of the great decisions in United States history. How many of these would you include? Which ones would you eliminate? Why? What ones would you add? Why?

Such an approach has the distinct advantage of encouraging discovery learning. Pupils are likely to feel as if they are taking part in decisions of a given period of our national history. Of course role-playing is one of the chief methods teachers will utilize in using such an approach.

Summary

Given a choice for three years of American history, this writer would opt for a postholing approach in the elementary school, with special attention to social history in selected periods. In the junior high school years he would like to see something along the lines of Plan Six more widely adopted, with considerable use of source materials in the solving of problems organized around basic decisions in various periods of our history. In the senior high school this writer would be willing to use the more traditional approach of chronology to the Civil War, and a combination of topics and chronology since that time. But he would prefer a two-year course in grades eleven and twelve, placing United States history in its worldwide setting and combining it with attention in grade twelve to the overriding interests and issues of our time. Such a plan would provide for a variety of approaches to our history at different grade levels and should enhance learning.

METHODS IN STUDYING UNITED STATES HISTORY

In the study of United States history, teachers should use a wide variety of methods. This is necessary to achieve the aims set forth earlier in this section of the book. It is also important in order to meet the individual needs of pupils. It is essential, too, merely for the sake of variety in order to retain or to further motivate interest in the subjects at hand.

You may want to turn back to the section on "The Variety of Methods" and see which ones you can use. This can also be done in connection with the development of each unit.

However, there are certain methods which need to be stressed in the study of history. These will be merely mentioned; teachers will need to decide how best to develop them in connection with specific content.

Time-Lines. Since the sense of time has to be developed in all pupils and especially with those in elementary schools, time-lines should be widely used. The number of dates should be limited to a very few with beginners. Drawings and pictures should be used with events and dates on many of these time-lines, especially in the first study of United States history.

Recordings. Recordings can be especially helpful in giving pupils a feeling that they actually are participating in the making of history. The recordings of the Landmark books, published by Random House, are the type of material teachers certainly will find invaluable.

Source Materials. Pupils should begin to use source materials even in the first course in United States history. They should read letters, diary accounts, old newspapers and other materials which will give them a "feel" for the period under study. Much of this material can be used to promote problem-solving in social studies classes.

Trips. In many communities there are historic places to visit. For a few schools, there will be landmarks of history nearby, whether the school is in Massachusetts,

Pennsylvania, Virginia, California, Texas, Illinois, or some other state. For others, it will be an old house which helps children to gain the feeling of life in a given period.

Museums. Some museums can help children develop a feeling of our national history, especially where the exhibits are well arranged. A museum may not, however, encourage children to enjoy history. Teachers need to know whether they can use the local museum or not.

Music. In connection with every period of American history some music should be used. The pupils can sing the songs as a group. Individuals can play or sing some of the music of a period. Recordings can be used effectively. This, too, can give children a feeling for the times they are studying. Almost any good book of music for elementary school children will have some such songs in it. Probably you can find others, with the help of a music teacher or some other teacher especially interested in this aspect of our national history. Sometimes an individual pupil can find songs or a committee can work on the music of a period as a group project.

Films and Filmstrips. These visualizations can do more than almost anything else to give boys and girls a "live" approach to the study of history. Fortunately there are hundreds of films and filmstrips which can be used in conjunction with studies of American history.

Hobbies. Often the interest of pupils can be aroused in the study of history by encouraging them to use their collections of stamps, coins, or dolls to illustrate what the class is examining. Some children also can be encouraged to begin such collections. Postcards and pictures also can be saved and used.

Interviews. Some pupils enjoy interviewing older people about the events of history through which they have lived, such as the Depression and World War II. This is well worth considering as a special method.

Map Study. Map study is a "must" in connection with the study of our national history. Map activities can range from the use of atlases to the construction of various types of maps. For more ideas along this line, see the section on maps and globes in this book.

Pictures. The local library and the school library should have collections of mounted pictures which can be used in history classes. If not, librarians should be encouraged to develop such collections. Teachers may want to assemble their own sets of pictures, often with the help of pupils.

Role-Playing. Throughout this book the author has stressed the importance of role-playing or sociodrama. This is a method which helps pupils feel as if they actually were living in a given place and at a given time. It should be used extensively in studies of American history.

Reading. It should go without saying that a wide variety of books should be used in conjunction with any study of American history. Textbooks should be compared and the illustrations in various books utilized. Encyclopedias can be very helpful. Books of plays and poetry have a place in such studies. So do biographies and autobiographies, of which there are hundreds on leaders in our history and on less well-known people. The references at the end of this chapter contain such volumes. Sometimes teachers will want to read aloud to pupils, especially with slow learners.

Models.   Some pupils like to construct. If they are urged to read about the models they are making, more learning will take place and more history taught. An excellent paperback on *Historic Models of Early America,* written by C. J. Maginley (Harcourt Brace Jovanovich) is the type of reference that every school should have for work along this line.

Newspapers.   Old newspapers often help children understand what life was like in a given period. Headlines of important events on the day they happened can create excitement in a class. And the prices of newspapers at various times and prices of everyday items in different periods can help pupils to relive the events of a period they are studying.

Television.   Occasionally there are television programs which re-create events of history. These can sometimes be viewed as homework assignments.

Charts.   Children often can develop their own charts of a period and, in so doing, learn a great deal of history. Some charts also can be obtained from various companies, such as the charts on the Revolutionary War, the Civil War, and the Presidents of the United States, published by the Civic Education Service. Textbooks usually carry some charts as illustrations, too. They should be studied by pupils, under the direction of teachers.

The History of a Local Community.   In some classes, teachers may want to develop a time-line on the history of their own community, running it parallel with a similar time-line of the history of our nation. Trips can be made to local points of historic interest in connection with such work. Many communities are veritable museums in themselves.

Cartoons.   Sometimes an apt cartoon can help children to learn about a person or an event. These can be passed around the class or used in an opaque projector for the entire class to view. Teachers need to remember, however, that cartoons often are relatively sophisticated and are not easily understood by all pupils.

Plays.   Often children can learn about a person or an event or period if they prepare their own short plays and act them out. At other times you will want to use plays which already have been written for pupils. The best single source for printed plays is Plays, Inc., in Boston.

The Rewriting of Documents.   Adults often take for granted that children understand the language of famous documents like the Declaration of Independence and the Constitution. But this is seldom so. Children often find the words in them extremely difficult and the sentences long and complicated. From time to time it is helpful to have boys and girls write their own versions of such famous documents.

Halls of Fame.   The selection by pupils of a limited number of persons to go into a classroom or school Hall of Fame for a given period also can be stimulating. It matters less who is chosen than whether children debate the merits of including certain persons.

Open Textbook Studies.   There should be many periods in which pupils study their textbooks with the teacher. This is especially important in the development of skills, such as reading, studying pictures, and interpreting maps and charts. Open textbook work is recommended especially for slower readers and pupils.

Puppets.   Some pupils in elementary schools get a great deal of enjoyment and learning from the use of puppets. This approach can be used in American history classes, with the puppets representing famous men and women and some of the inconspicuous persons in our history.

Every available method in studying American history has not been mentioned here, but certainly enough have been cited to show what a wide range of possibilities there are for alert and creative teachers.

LIVING IN THE U.S.A.

	In the 1700s	In the 1700s and 1780s	In the 1830s	In the 1860s	In the 1900s	In the 1930s
Population						
Number of Colonies or States						
Issues to Stress						
Outstanding Persons						
City to Emphasize						
The People						
New Materials for Building						
New Materials for Clothes						
Economy						
New Forms of Energy						
New Means of Transportation and Communication						
New Tools and Major Inventions						
Food and Health						
Government						
Changes in Education						
Major Developments in Religion						
Some Famous Songs						
Outstanding Artists						
Major Events in World Affairs						

PREPARING YOURSELF FOR THE TEACHING OF UNITED STATES HISTORY

As a reader you probably fall into one of two general categories. You are either a young person who has never taught, or you are a person of a few years' experience as a teacher of United States history. What can you do in either case to prepare yourself for teaching this subject?

Rethinking Aims

In either case you undoubtedly have given some thought to the reasons why boys and girls should learn a great deal about the history of their own nation. But you may have thought in rather general terms. Probably you should think more specifically about the aims or objectives of such a course in the light of specific pupils you are teaching now or are about to teach.

Perhaps you will want to think about a simple, short test to administer which will help you to determine what they already know and what their general attitudes actually are. How else can you know specifically what you should teach to this particular group?

Since you probably adhere to the philosophy that they should be learning in order to change or improve their behavior and attitudes, you may want to write down the specific changes you hope will occur in the time you spend with them on this subject. What changes do you want? What changes can you reasonably expect in the period of time at your disposal?

After you have done these things, you should be ready to write out a realistic list of aims for your class. It should be brief enough to be very practical and should include changes in behavior, skills to be learned, attitudes to be developed, and knowledge to be acquired — grouped around specific concepts or generalizations. You may want to draw upon the list of aims on page 381.

Retooling Your Knowledge of United States History

You certainly have some background in American history already. You could not have lived in this country without acquiring some such background. But it may not be sufficient to make you feel comfortable or secure in teaching. You will need to study along with your pupils — or ahead of them as the year moves along. And you probably will need to do some general reviewing of American history now.

If you are an experienced teacher, you may not have read too much in recent years on this subject because of the pressures to learn the new math, the new science, or some other subject. Perhaps you are now ready to take a summer or several weekends and holidays to retool yourself in the history of the United States.

Perhaps you would like to purchase a half-dozen books on our history or borrow them from some library. Some of them might be paperbacks which you could carry with you or leave in convenient places around the house where you can pick them up at your leisure. You may want to take notes about these books on cards or on small sheets of paper, dropping them into folders on persons and places and periods.

Here is a list of outstanding adult books on U.S. history to help you:

BOOKS FOR TEACHERS ON UNITED STATES HISTORY

1. ALLEN, FREDERICK LEWIS. *The Big Change: America Transforms Itself: 1900–1950.* New York: Harper (1953), 308 pp.

2. BARACH, ARNOLD S. *U.S.A. and Its Economic Future.* New York: Macmillan (1964), 143 pp. A graphic representation based on a Twentieth Century Fund report.
3. BEARD, CHARLES and MARY. *The Rise of American Civilization.* New York: Macmillan (1933), 865 pp. Numerous editions. An economic interpretation of our history.
4. BROGAN, DENIS W. *The American Character.* New York: Knopf (1944), 200 pp. An informed and astute British observer interprets American life.
5. COMMAGER, HENRY STEELE. *The American Mind: An Interpretation of American Thought and Character Since the 1880s.* New Haven, Connecticut: Yale University Press (1957), 476 pp. Also available as a paperback.
6. FAULKNER, HAROLD U. and HERBERT C. ROSENTHAL. *A Visual History of the United States.* New York: McGraw-Hill (1961), 188 pp. Each period presented in charts, graphs, and maps.
7. GORER, GEOFFREY. *The American People: A Study in National Character.* New York: Norton (1948), 267 pp. An English anthropologist dissects American life.
8. HOFSTADTER, RICHARD and MICHAEL WALLACE (Editors). *American Violence: A Documentary History.* New York: Knopf (1970), 478 pp. Also a paperback.
9. KRAMER, PAUL and FREDERICK L. HOLBORN (Editors). *The City in American Life: From Colonial Times to the Present.* New York: Capricorn Books (1970), 384 pp. A book of readings. Paperback.
10. LERNER, MAX. *America As a Civilization: Life and Thought in the United States Today.* New York: Simon and Schuster (1957), 1036 pp.
11. MIERS, EARL S. (Editor). *The American Story: The Age of Exploration to the Age of the Atom.* Great Neck, New York: Channel Press (1956), 352 pp.
12. MILLER, WILLIAM. *A New History of the United States.* New York: Delta (1962), 466 pp. Also available as a paperback.
13. MORISON, SAMUEL ELIOT. *The Oxford History of the American People.* New York: Oxford University Press (1965), 1150 pp. A superb volume, especially valuable on the early periods of our history.
14. NEVINS, ALLAN and HENRY STEELE COMMAGER. *A Short History of the United States.* New York: Pocket Books (1966 edition), 502 pp. Brief and highly readable account by two eminent historians.
15. PARRINGTON, VERNON L. *Main Currents in American Thought.* New York: Harcourt Brace Jovanovich (1930), 1300 pp. Also available as three paperbacks.
16. WEYL, WALTER. *The New Democracy.: An Essay on Certain Political and Economic Tendencies in the U.S.A.* New York: Harper (1964), 369 pp. Also available as a paperback.
17. WILLIAMS, WILLIAM A. *The Roots of the Modern American Empire: A Study of the Growth and Shaping of Social Consciousness in a Marketplace Society.* New York: Vintage Books (1969), 546 pp.
18. WOODWARD, WILLIAM F. *The Way Our People Lived.* New York: Washington Square Press (1970 printing), 390 pp. A paperback on social history.
19. WRIGHT, KATHLEEN. *The Other Americans: Minorities in American History.* Greenwich, Connecticut: Fawcett (1969), 256 pp. A paperback.

Building a Personal Library of Reference Books

Certainly you will want to start a small collection of reference books or augment those you already have. Some of them should be on your desk in the classroom for ready reference by yourself and/or your pupils. You may want to include:

1. An *Information Please Almanac, World Almanac,* or similar volume.
2. An American history volume, such as the Nevins–Commager volume.
3. An atlas of American history.

4. A review book on American history.
5. A quick reference volume, such as Richard B. Morris's *Encyclopedia of American History.*

Interpretations of United States History

You probably will want to review or learn about the various interpretations of United States history. Here are a few of them. If you do not know what ideas they represent and their significance, you certainly will want to find out immediately.

1. The influence of the frontier (Turner and others).
2. The economic interpretation of our history (Beard and others).
3. The influence of ideas (Gabriel, Curti, and others).
4. The urban influence (Bridenbaugh and others).
5. The influence of immigrants (Handlin and others).
6. The influence of leaders; the biographical approach.
7. The anti-intellectual aspects of our history (Hofstadter and others).
8. Violence in American society (See the Hofstadter and Wallace volume listed on page 390).

You certainly will want to include all of these ideas in your teaching of our history, without citing the names of the men mentioned above or going into detail about their theories.

Three books which should help you to understand these various approaches are:

1. HIGHAM, JOHN (Editor). *The Reconstruction of American History.* New York: Harper and Row (1962), 244 pp. Eleven sections by eleven different authors, suggesting the different interpretations of those periods.
2. KRAUS, MICHAEL. *The Writing of American History.* Norman, Oklahoma: University of Oklahoma Press (1953), 387 pp.
3. WISH, HARVEY. *The American Historian.* New York: Oxford University Press (1960), 366 pp.

Of those three volumes, the paperback edition of the Higham book is especially recommended.

You also may want to obtain the list of booklets on United States history published and sold by the Service Center for Teachers of History (400 A Street, S.E., Washington, D.C. 20003). They are inexpensive and cover over fifty titles or themes. Some of them are on the interpretations of periods in our national history. Each of the booklets is written by a highly qualified person in the field. They are very brief but cover a wide range of background for teachers of United States history.

Thinking About Explanations of Our History

Closely allied with the interpretations of our history is the topic of explanations of how we grew and developed. You should be thinking often about these explanations in order to give focus to your teaching. Here for example, is a list of explanations which the late Professor Donald Tewksbury used with students from abroad who were taking a course on American Civilization at Teachers College, Columbia University. He maintained that ours has been:

1. A transplanted civilization.
2. A frontier civilization.

3. An experimental civilization.
4. A scientific, technological civilization.
5. A young civilization.
6. A fluid civilization.
7. An isolated civilization.
8. A "grass roots" civilization.
9. A Protestant civilization.
10. A capitalistic civilization.

A slightly longer list of explanations of the development of our nation, made by the author, includes the following points:

1. Our rich soil and valuable mineral resources and their development.
2. Our generally favorable climate, with great diversity.
3. Our good geographical location.
4. Our generally wise land policy, with no system of feudal estates to contend with.
5. Our inquiring minds and inventors, leading to industrialization.
6. Our agricultural revolution.
7. Our expansion from the Atlantic to the Pacific without undue foreign difficulties.
8. Our single language.
9. Our emphasis upon hard work as a virtue, probably stemming from Puritanism.
10. Our remarkable system of transportation.
11. Our expanding concept of democracy — political, social, religious, educational, and economic.
12. Our free, public school system.
13. Our common culture from a relatively early period.
14. Our expanding population and adequate labor supply, aided by millions of immigrants.
15. Our able leaders in the period of formation of our nation.
16. Our lack of devastation by major wars, except the Civil War.
17. Our philosophy of optimism and our drive for a higher standard of living.
18. Our health and sanitation revolution.
19. Our social mobility and the early development of a middle class.
20. Our release of the capacities of women more than in most societies.

Starting a Collection of Files

Enormous amounts of time and energy will be saved if you start a collection of different periods of our history. Into such folders you can drop magazine and newspaper articles, articles from current events magazines, pictures, notes that you take on books and booklets, and anything else that you collect.

Collecting Maps, Posters, and Related Materials

In a similar way you undoubtedly will want to collect all the maps, posters, charts, and exhibit materials that you can find. Some of these have been mentioned already in this chapter. Others are mentioned in other parts of this volume. You also will want to save outstanding materials which your pupils have made so that they can be used for teaching other classes, and as examples of what pupils can do.

Taking a Trip or Trips

If you can find the time — and the money — why not plan a trip to a famous historic spot that you have never visited, or to one that you have not visited for a long time? There are such spots in every part of the United States. Several of them may be near your home or place of work. Such trips should not be expensive.

Starting a Large Time-Line

You should also plan to have a large but very simple time-line in your classroom on the history of our country. If possible, hang it in the room and keep it there throughout the year, perhaps above the chalkboard (see the illustration on page 344). You can start such a time-line at the beginning of a course and fill it in as the year goes along. You might want to indicate certain large periods of history yourself and then have the pupils fill in more details about each of those periods as they study them.

Examining Textbooks and Trade Books

Undoubtedly you will want to examine several up-to-date textbooks, choosing two or three you want to use. These may well represent different approaches or strengths or even different reading levels. You may want to start reading quickly a number of trade books, making notes on them. Then you can drop these notes into your folder which covers that period of history. By using such a plan, you can develop a great deal of background on resources over a period of several semesters or years.

GENERAL BOOKS ON UNITED STATES HISTORY FOR BOYS AND GIRLS

1. ASSOCIATION FOR CHILDHOOD EDUCATION. *Told Under Stars and Stripes.* New York: Macmillan (1962), 346 pp. Grades 4–7. Stories of regions in the U.S.A.
2. COMMAGER, HENRY STEELE. *The First Book of American History.* New York: Watts (1957), 63 pp. Grades 4–6. With sketches.
3. COMMAGER, HENRY STEELE. *A Picture History of the United States of America.* New York: Watts (1958), 61 pp. Grades 5–8.
4. COOKE, DONALD E. *Marvels of American History.* Maplewood, New Jersey: Hammond (1962), 275 pp. Grades 5–8.
5. COY, HAROLD. *The Americans.* Boston: Little, Brown (1958), 320 pp. Grades 6–9. Selected incidents in our history enliven this book.
6. EDITORS OF THE AMERICAN HERITAGE. *Golden Book of America.* New York: Golden (1957), 216 pp. Grades 5–8.
7. FAULKNER, HAROLD U. and HERBERT C. ROSENTHAL. *A Visual History of the United States.* New York: McGraw-Hill (1961), 188 pp. For better readers only.
8. FERGUSON, CHARLES W. *Getting to Know the U.S.A.* New York: Coward-McCann (1963), 96 pp. Grades 6–9. Emphasizes the size of the United States and its effects on history. Exciting.
9. HEAL, EDITH. *The First Book of America.* New York: Watts (1952), 93 pp. Grades 3–5.
10. HOFF, RHODA. *America: Adventures in Eyewitness History.* New York: Walck (1962), 167 pp. Grades 6–9. Source materials.
11. JOHNSON, GERALD W. *America Is Born: A History for Peter.* New York: Morrow (1959), 254 pp. Grades 4–7. Early beginnings to the Revolutionary War.
12. JOHNSON, GERALD W. *America Grows Up: A History for Peter.* New York: Morrow (1960), 223 pp. Grades 4–7. From 1776 through World War I.

13. JOHNSON, GERALD W. *America Moves Forward: A History for Peter.* New York: Morrow (1960), 256 pp. Grades 4–7. World War I to the present.
14. MEADOWCRAFT, ENID A. *Land of the Free.* New York: Crowell (1961), 160 pp. Grades 4–6.
15. PARKER, GORDON. *Great Moments in American History.* Chicago: Follett (1961), 43 pp. Grades 4–7. Dramatic stories about eight episodes.
16. ROBBIN, IRVING. *The How and Why Wonder Book of North America.* Columbus, Ohio: Merrill (1962), 48 pp. Grades 5–7. Easy reading.
17. TOBIAS, JOHN and SAVIN HOFFECKER (Editors). *The Adventure of America.* New York: Random (1962), 258 pp. Grades 5–8. A fine source book of poetry, letters, diary accounts and other data. Beautifully illustrated.

Anthologies of Poetry for Boys and Girls on United States History
1. BREWTON, SARA and JOHN F. *America Forever New: A Book of Poems.* New York: Thomas Y. Crowell (1968), 269 pp.
2. GINIGER, KENNETH S. *America, America, America: Prose and Poetry About the Land, the People, and the Promise.* New York: Watts (1957), 231 pp. Grades 5–8.
3. HINE, AL. *This Land Is Mine.* Philadelphia: Lippincott (1965), 244 pp. Grades 5–8. An anthology of American verse.

VARIOUS PERIODS IN UNITED STATES HISTORY

Prehistoric Times in North America

Two or three days should be sufficient time to spend on the millions of years which represent prehistoric times in what is now North America. Children may have studied something about this period in their earlier years without realizing what it represents. By the fifth grade, or at least by the junior high school years, they should comprehend a little of this lengthy period. Teachers can help to develop the time sense by using a long time-line in which the last inch or so represents the length of time that man has been on the earth.

For younger children there is an unexplained fascination about the animals which roamed the earth in those days. They ought to gain a little understanding, too, of the ice sheets which once covered large parts of our continent and left in their wake the Great Lakes and the rocks of New England and other geographic features as they retreated northward.

A fine chart of the various periods represented in prehistoric times appears on pages 18 and 19 of Alice Dickinson's *The First Book of Prehistoric Animals* and might well be used by teachers. A trip to a nearby museum which has a collection of prehistoric animals and fish should yield rich dividends. The pictures in the books listed below should help to arouse interest in this period and enhance learning. Some of the books available on this period include the following:

1. ADRIAN, MARY. *Mystery of the Dinosaur Bones.* New York: Hastings (1965), 136 pp. Grades 3–5.
2. BAINTY, ELIZABETH C. *America Before Man.* New York: Viking (1953), 224 pp. Grades 7–10. A fairly detailed account.
3. BALDWIN, GORDON C. *America's Buried Past: The Story of North American Archaeology.* New York: Putnam (1962), 192 pp. Grades 6–9.
4. COLBERT, EDWIN. *Millions of Years Ago: Prehistoric Life in North America.* New York: Crowell (1958), 153 pp. Grades 6–9. The text is superior to the illustrations.
5. DICKINSON, ALICE. *The First Book of Prehistoric Animals.* New York: Watts (1954),

93 pp. Grades 5–8. Fairly simple text and striking illustrations. An excellent chart of periods and changes in land and animals on pages 18–19. Highly recommended.
6. Fox, William and Samuel Welles. *From Bones to Bodies: A Story of Paleontology.* New York: Walck (1959), 118 pp. Grades 6–9. Largely on the methods of paleontologists, so of interest only to a small number of pupils, especially in the junior high school grades.
7. Greene, Carla. *How to Know Dinosaurs.* Indianapolis: Bobbs-Merrill (1965), 64 pp. Grades 3–5.
8. Johnson, Gaylord. *The Story of Animals: Mammals Around the World.* Irvington-on-Hudson, New York: Harvey (1958), 120 pp. Grades 6–9.
9. Watson, Hane W. *The Giant Book of Dinosaurs and Other Prehistoric Reptiles.* New York: Golden (1960), 60 pp. Grades 3–5. Large, beautiful, striking illustrations and a good text. Highly recommended.
10. White, Anne Terry. *Prehistoric America.* New York: Random (1951), 182 pp. Grades 5–8. A Landmark book. The most readable of all these books. The illustrations are not superior.

Indians in What Is Now the U.S.A.

In many schools too much time is devoted to the study of the Indians of hundreds of years ago. They were interesting people and certainly deserve a place in the social studies curriculum. But in view of the pressures of other themes, they surely do not warrant more than ten days or two weeks in a year devoted to the study of our history or the history of our country and present-day affairs.

With such a short time available, teachers will need to think carefully about the aims of studying the Indians in our early history. They might well be treated as our first immigrants, coming across the Bering Strait over a period of thousands of years and scattering across what is now the United States and into parts of Central and South America.

The heart of the study of Indians in that period should be on the ways that they adapted themselves to their locations. For example, the Indians of the Northwest lived in an area where there was plenty of water. Food was not a major problem with them. Fish became their major food. Whales were caught by some of the Indians in that locale, and various parts of those mammals were used for food, for seasoning of salmon and berries and clams. The oil from the whales was used for lights. Those Indians were surrounded by the forests and their homes were made of wood. Some of them developed great skill in carving totem poles to tell the stories of their families. Children need to discover how these men, women, and children developed a way of life suitable to that environment.

Two or three other groups of Indians can be studied with profit in a similar way. These might be the Indians of the Southwest, the Indians of the Plains, and some of the Eastern Indians — perhaps the Iroquois. Stereotypes need to be shattered and respect developed for the cleverness with which most Indians adjusted to their environment.

Most boys and girls think of the Plains Indians when they hear the word "Indians." They think of tepees and feather bonnets. Can you extend their thinking to a variety of Indian tribes, living in quite different environments but adjusting to them, often in ingenious ways?

With limitations of time, you may want to divide the class into five or six groups, studying different tribes of Indians. You could prepare for these groups an outline of topics or questions, or the class as a whole could develop such a common list of questions. You might like to have a large wall chart which the various committees then fill in as they report on their Indian groups. This would help the pupils to compare and to contrast.

Pictures, films, and filmstrips of the Indians of long ago are available in quantity. A trip to a nearby museum is certainly recommended if the children have not been there frequently before this study. Time will not permit a great deal of arts and crafts activity during this short unit, desirable as this might be.

A list of some of the many books on Indians follows. By working with the school and/or local librarian, you may be able to arrange for a collection of volumes on Indians in the classroom. Pupils could use them for general browsing and for their committee reports. One committee group might work solely on illustrations of Indian life in North America. They could be the resource or research committee for the other groups.

1. BAKER, BETTY. *Little Runner of the Longhouse.* New York: Harper and Row (1962), 63 pp. Grades 3–4. The Iroquois through the eyes of a child.
2. BLEEKER, SONIA. *The Cherokee.* New York: Morrow (1952), 154 pp. Grades 4–6. See also her volumes on *The Chippewa Indians, The Crow Indians, Horsemen of the Western Plains, The Nez Perce Indians, Indians of the Longhouse, The Story of the Iroquois, The Pueblo Indians, Farmers of the Rio Grande, The Sea Hunters: Indians of the Northwest Coast, The Seminole Indians,* and *The Sioux Indians: Hunters and Warriors of the Plains.* All Morrow books and all for grades 4–6.
3. BOYCE, GEORGE A. *Some People Are Indians.* New York: Vanguard (forthcoming).
4. BRANDON, WILLIAM. *The American Indian.* New York: Random (1963), 216 pp. Grades 5–8. Richly illustrated. Adapted from *The American Heritage Books of Indians.*
5. BULLA, CLYDE R. *Pocahontas and the Strangers.* New York: Crowell (1971), 128 pp. Grades 5–8.
6. CLARK, ANN NOLAN. *The Desert People.* New York (Viking), 1962. 59 pp. Grades 1–5.
7. COATSWORTH, ELIZABETH. *Indian Encounters: An Anthology of Stories and Poems.* New York: Macmillan (1960), 264 pp. Grades 5–8.
8. COMPTON, MARGARET. *American Indian Fairy Tales.* New York: Dodd, Mead (1971), 159 pp. Grades 5–10.
9. ESTEP, IRENE. *Seminoles.* Chicago: Melmont (1963), 31 pp. Grades 4–6.
10. FARQUHAR, MARGARET C. *Indian Children of America.* New York: Holt, Rinehart and Winston (1964), 48 pp. Grades 3–5. Indians in colonial times.
11. HARRIS, CHRISTIE. *Once Upon a Totem Pole.* New York: Atheneum (1963), 148 pp. Grades 4–6. Five stories of Indians of the Northwest.
12. HOFSINDE, ROBERT. *Indian Games and Crafts.* New York: Morrow (1957), 126 pp. Grades 4–6. See also his books on *Indian Sign Language, Indian Picture Language,* and *Indian's Secret World.*
13. HUNT, W. BEN. *The Golden Book of Indian Crafts and Lore.* New York: Simon and Schuster (1954), 112 pp. Grades 4–6. Excellent illustrations.
14. JONES, HETTIE. *The Trees Stand Shining: Poetry of the North American Indians.* New York: Dial (1971), 32 pp. Grades 5–8.
15. LaFARGE, OLIVER. *The American Indian.* New York: Golden (1960), 213 pp. Grades 4–6.
16. LaFARGE, OLIVER. *A Pictorial History of the American Indian.* New York: Crown (1956), 272 pp. Grades 6–9.
17. MARRIOTT, ALICE LEE. *The First Comers: Indians of America's Dawn.* New York: Longmans (1960), 246 pp. Grades 6–9.
18. MORRIS, LAVERNE. *The American Indian as Farmer.* Chicago: Melmont (1963), 48 pp. Grades 4–6.
19. PINE, TILLIE S. *The Indians Knew.* New York: Whittlesey (1957), 32 pp. Grades 4–6.

20. Rachlis, Eugene. *Indians of the Plains.* New York: Golden (1960), 152 pp. Grades 5–7.
21. Scheele, William E. *The Earliest Americans.* Cleveland: World (1963), 64 pp. Grades 5–8.
22. Scheele, William E. *The Mound Builders.* Cleveland: World (1960), 61 pp. Grades 5–8.
23. Shapp, Charles and Martha. *Let's Find Out About Indians.* New York: Watts (1962), 48 pp. Grades 2–4.

The Explorers

The efforts of most of us to teach the period of exploration have not been very successful. Pupils have been confronted with a great many names which meant little or nothing to them and they have been confused by them and by the places of their origin and the countries they set sail for. As a result, there has been little meaningful learning from studies of this period of world history.

Can we do better? This writer believes so. To do a more effective job we need to rethink our aims and our methods. Here are five suggestions for your consideration in teaching about the Explorers of the New World.

1. The Known World of That Day.   It is difficult but not impossible for pupils to imagine a time when people did not know about the Americas and parts of Africa and Asia. Why not bring in a large block of wood and on it mark the known world, and discuss the ideas of people on the world of that day? Or you can do the same thing with a flannelboard, including on it only those parts of the world known in 1450 or thereabout.

2. The Significance of Asia.   Again, it is difficult for boys and girls to think in terms of Asia as the most highly developed area of the world. Yet that was the situation in the fifteenth century. Europe was, for the most part, the underdeveloped part of the world of that day. With children it would be wise to discuss why Asia was so advanced and why Europeans wanted to go there, remembering that it was not solely because of spices, as is so often said in textbooks.

3. The Rivalry of Nations.   The Age of Exploration was a period of rivalry among city-states and nations. Why not liken it to the children's game of King of the Mountain and see who was on top at various times? Perhaps you will want to make a "totem pole" showing the Italian city-states on top. Then you can erase them and place Portugal on top and see who was challenging that nation. In a similar way you can show the rivalry of Spain, England, France, and The Netherlands for the leadership of the world.

4. A World Treasure Hunt.   The nations of that day were engaged in a world treasure hunt for new materials and new lands. Perhaps this idea can be discovered best by pupils by assigning them to the various countries of that time and having them discover their claims to the new areas of the globe, using a flannelboard or a large map on oaktag.

5. A Changed World.   You may want to explore with your pupils, especially older ones, the radical changes brought about in the world by this period of exploration, with nations replacing city-states in Europe in importance, new nations coming to the fore, and wars eventually occurring between them. Imperialism has its roots, too, in the period of exploration.

Here are some books for children to help you in this task:

BOOKS FOR BOYS AND GIRLS ON THE EXPLORERS

1. ANTHONY, BARBARA K. and MARCILLENE BARNES. *Explorers All.* Grand Rapids, Michigan: Fideler (1952), 112 pp. Grades 5–8. Short accounts of several explorers, with black and white illustrations.
2. AVERILL, ESTHER H. *Cartier Sails the St. Lawrence.* New York: Harper and Row (1956), 108 pp. Grades 5–7. Much material from Cartier's logs and letters.
3. BAKELESS, KATHERINE and JOHN. *They Saw America First.* Philadelphia: Lippincott (1957), 22 pp. Grades 9–11.
4. BAKER, NINA BROWN. *Amerigo Vespucci.* New York: Knopf (1956), 148 pp. Grades 5–8.
5. BAKER, NINA BROWN. *Juan Ponce de Leon.* New York: Knopf (1957), 160 pp. Grades 4–6.
6. BAKER, NINA BROWN. *Henry Hudson.* New York: Knopf (1958), 142 pp. Grades 4–6. A compact life of the four years from 1607 to 1611 and his four trips.
7. BUTCHER, THOMAS K. *Africa: The Great Explorations.* New York: Roy (1959), 191 pp. Grades 5–8.
8. DALGLEISH, ALICE. *America Begins: The Story of the Founding of the World.* New York: Scribner (1958), 63 pp. Grades 3–5.
9. DALGLEISH, ALICE. *The Columbus Story.* New York: Scribner (1955), unpaged. Grades 3–5.
10. D'AULAIRE, INGRI and EDGAR. *Columbus.* Garden City, New York: Doubleday (1955), 56 pp. Grades 3–5. Beautifully illustrated.
11. FOLSOM, FRANKLIN. *The Explorations of America.* New York: Grosset and Dunlap (1958), 149 pp. Grades 5–8. Very good for above-average readers. Enhanced by maps and black and white illustrations.
12. GRAFF, STEWART. *World Explorer: Hernando Cortes.* Champaign, Illinois: Garrard (1970), 96 pp. Grades 4–6.
13. GRAH, LYNN. *Ferdinand Magellan.* Champaign, Illinois: Garrard (1963), 96 pp. Grades 4–6.
14. HEWES, AGNES D. *Spice and the Devil's Cave.* New York: Knopf (1942), 331 pp. Grades 5–8. Rivalry of Spain and Portugal. See also *Spice Ho,* by the same author.
15. JOHNSON, GERALD W. *America Is Born: A History for Peter.* New York: Morrow (1959), 254 pp. Grades 4–7. Chapter 2 "The People Who Won It."
16. KAUFMAN, MERVYN D. *A World Explorer: Christopher Columbus.* Champaign, Illinois: Garrard (1963), 96 pp. Grades 4–6.
17. LAMBERT, R. S. *The World's Most Daring Explorers: 38 Men Who Opened Up the New World.* New York: Sterling (1956), 176 pp. Grades 6–9.
18. NOBLE, IRIS. *Honor of Balboa.* New York: Messner (1970), 192 pp. Grades 7–10.
19. SANDERLIN, GEORGE. *Across the Ocean Sea: A Journal of Columbus's Voyage.* New York: Harper and Row (1967), 275 pp. Grades 6–9.
20. SCOTT, JAMES M. *Hudson of Hudson Bay.* New York: Abelard-Schuman (1951), 176 pp. Grades 5–7.
21. SYME, RONALD. *De Soto: Finder of the Mississippi.* New York: Morrow (1957), 96 pp. Grades 4–6.
22. SYME, RONALD. *First Man to Cross America: The Story of Cabeza de Vaca.* New York: Morrow (1961), 190 pp. Grades 5–8.
23. TERRELL, JOHN U. *Search for the Seven Cities: The Opening of the American Southwest.* New York: Harcourt Brace Jovanovich (1970), 151 pp. Grades 5–8.

The Colonial Period

Few people realize the length of our colonial period. From 1607 to 1776 is 169 years. From 1776 to 1970 is 194 years. Therefore almost half of the history of Europeans in the New World took place in what we refer to briefly as The Colonial

Period. Probably the only way to get this idea across with anyone, including children, is to develop a simple time-line and to work out the point where the Declaration of Independence was proclaimed or the Revolutionary War began.

Despite that considerable length of time, the colonial period need not take more than two to four weeks of time in most elementary and junior high school classes. It is important, but it should not receive undue attention in relation to other more recent periods of our history.

In the middle grades the approach to that period probably should be on the settlement of the eastern seaboard and life in some of the colonies. This can be done by tracing the settlement of the various colonies from the beginning through the establishment of Georgia. Or it can be done by taking a date such as 1700 and seeing what life was like at that time. The date 1700 is suggested as eleven of the thirteen colonies were established by that time and some specialization in work had begun. (Georgia had not yet been established, and Delaware was still a part of the colony of Pennsylvania at that time.)

Boys and girls need to do considerable problem-solving in connection with the study of the colonial period. They need to discover why people decided to come to this country (or were forced to come as was the case with the African immigrants). They need to decide what they should take on a trip in those days, and consult source materials to determine what they would take. They should decide where to settle and how they would earn their living. All of these activities will help them to gain an understanding of the period and to learn the techniques of the social sciences.

Probably the best way to do this is to select a few colonies and concentrate on life in those few places. Massachusetts would be one. Pennsylvania or New York would be the second. Virginia would be the third. In the study of life in each of those places, children should see the total life of the people — their work, their food, their education, their religious activities, their recreation, their clothes, their government, their transportation and communication, and their health.

If you are teaching in a community which was in one of the other original colonies, you may want to substitute the history of this period in your own locale. This would be especially worthwhile if there is a local museum which deals with the colonial period in your region, where pupils could see (and hopefully touch) some of the objects used by people in that era.

If the land base of each place is determined and well developed in the minds of children, they can discover for themselves what the people would do. This is the essential approach to the colonial period.

Teachers need to think about the attitudes they want to develop in a study of the colonial period. One would be admiration for the people who braved the hardships of travel to the New World and the hazards of life in their new homeland. Another would be admiration for their hard work.

Visits to the homes of some of the people in the colonies can help children to relive life in that period. They can work in the fields with the men and boys, prepare meals with the women and girls, sit by the fireplaces in New England and pare apples or card wool. They can attend church and take part in town meetings. Much of this can be done in role-playing situations.

With older pupils, teachers may want to organize their study of this period around such key questions as "What Way of Life Is Best Suited to the New World?" The first topic would help pupils to see that the Indians had one way of life, the French and Spanish another, and the English at least two ways of life — in the northern colonies and in the southern colonies. This is primarily an anthropological approach. In regard to the second basic question or decision, pupils can see the clash of interests in the New World among the English, the French, the Spanish — and the Indians, leading to minor skirmishes and to wars.

In handling this period, it is important for teachers to realize the role that beliefs or values hold in the life of any group. Thus the people who were religiously motivated developed a different type of life than those whose major emphasis was not upon religion. In Puritan New England, therefore, the Church was the center of activities and the form of government depended upon church membership in the early days. The people who did not agree with the religious views of the dominant group were driven out of the colonies.

Another example of a way of life would be represented by the colonies of Rhode Island and Pennsylvania, where more tolerant views prevailed.

However, teachers should not portray the Puritans as people who failed to contribute to our nation. Recent historians are beginning to restore them to a higher place on the American totem pole than they have held for some years. These historians point to their practicality, their moral stamina, their intellectual interests, and even their zest for life.

The study of the colonial period gives pupils some opportunity to stress biography. Among the men who should be treated fully are William Penn, Roger Williams, and Lord Baltimore. Fortunately there are several good biographies of these men. Of course there are others you may want to stress. Your list should certainly include any famous persons from this period locally. At the junior high school level you may want to expand the use of biography, with various pupils reading about an outstanding person.

You may want to pick out a few places for emphasis, in order to help pupils see how people lived in cities. Colonial Boston and colonial Williamsburg would be two very good selections, although other spots could be chosen.

There is certainly no need to bother younger children with the various forms of colonies — self-governing, proprietary, and royal, although junior high school pupils can wrestle with these differences.

Some of the books on colonial times which will be helpful to pupils are listed on the pages which follow. Your school and/or local library will undoubtedly have some of the volumes listed. These libraries may have others which can be substituted for some of the books on this list.

BOOKS FOR BOYS AND GIRLS ON THE COLONIAL PERIOD

1. ALTER, ROBERT. *Listen, the Drum.* New York: Putnam (1963), 190 pp. Grades 6–9.
2. BERTHEIM, MARC and EVELYNE. *Growing Up in Old New England.* New York: Macmillan (1971), 96 pp. Grades 5–8.
3. BORRESON, MARY JO. *Let's Go to Plymouth with the Pilgrims.* New York: Putnam (1963), 48 pp. Grades 4–6.
4. CARSE, ROBERT. *Hudson River Hayride.* New York: Graphic (1962), 89 pp. Grades 5–8.
5. CLARK, ALLEN P. *Growing Up in Colonial America.* New York: Hill and Wang (1963), 96 pp. Grades 4–7.
6. COLBY, JEAN P. *Plimoth Plantation: Then and Now.* New York: Hastings House (1970), 128 pp. Grades 5–8.
7. DALGLEISH, ALICE. *The Thanksgiving Story.* New York: Scribner (1954), unpaged. Grades 3–5.
8. DAUGHERTY, JAMES. *The Landing of the Pilgrims.* New York: Random House (1950), 185 pp. Grades 5–8.

9. DE ANGELL, MARGUERITE. *Jared's Island*. Garden City, New York: Doubleday (1947), 95 pp. Grades 4–6. A small boy's adventures in New Jersey in the 1760s.

10. Editors of *The American Heritage*. *Jamestown: First English Colony*. New York: Harper and Row (1965), 122 pp. Grades 6–9.

11. Editors of *The American Heritage*. *The Pilgrims and Plymouth Colony*. New York: Harper and Row (1961), 128 pp. Grades 5–8.

12. EMERSON, CAROLINE D. *Pioneer Children of America*. Boston: Heath (1959), 318 pp. Grades 4–6.

13. FARQUHAR, MARGARET. *Colonial Life in America: A Book to Begin On*. New York: Holt, Rinehart and Winston (1962), unpaged. Grades 2–4.

14. FISHER, LEONARD E. *The Glassmakers*. New York: Watts (1965), 46 pp. Grades 5–8.

15. FISHER, LEONARD E. *The Papermakers*. New York: Watts (1965), 46 pp. Grades 5–8.

16. FISHER, LEONARD E. *The Printers*. New York: Watts (1964), 46 pp. Grades 5–8.

17. FISHER, LEONARD E. *The Silversmiths*. New York: Watts (1965), 46 pp. Grades 5–8.

18. FORBES, ESTHER. *Johnny Tremain*. Boston: Houghton Mifflin (1945), 256 pp. Grades 5–8. Old, but a stirring account of American patriotism.

19. FOSTER, GENEVIEVE. *The World of Captain John Smith: 1580–1631*. New York: Scribner (1959), 406 pp. Grades 6–9.

20. FOWLER, MARY JANE and MARGARET FISHER. *Colonial America*. Grand Rapids, Michigan: Fideler (1960), 128 pp. Grades 5–8.

21. HAYS, WILMA P. *Christmas on the Mayflower*. New York: Coward-McCann (1956), unpaged. Grades 2–4.

22. HAYS, WILMA P. *Naughty Little Pilgrim*. New York: Ives Washburn (1969), 48 pp. Grades 2–4.

23. HAYS, WILMA P. *Pilgrim to the Rescue*. New York: McKay (1971), 48 pp. Grades 3–5.

24. HAYS, WILMA P. *Rebel Pilgrim*. Philadelphia: Westminster (1969), 96 pp. Grades 4–6. A biography of William Bradford.

25. HULTS, DOROTHY N. *New Amsterdam Days and Ways*. New York: Harcourt Brace Jovanovich (1963), 224 pp. Grades 4–6.

26. JOHNSON, GERALD W. *America Is Born*. New York: Morrow (1959), 254 pp. Grades 4–7.

27. LATHAM, FRANK B. *The Trials of John Peter Zenger*. New York: Watts (1970), 64 pp. Grades 5–6.

28. LAWRENCE, ISABELLE. *Drumbeats in Williamsburg*. Chicago: Rand McNally (1965), 224 pp. Grades 5–8.

29. LORD, BEMAN. *On the Banks of the Delaware: A View of Its History and Folklore*. New York: Walck (1971), 63 pp. Grades 5–7.

30. MCGOVERN, ANN. *If You Lived in Colonial Times*. New York: Scholastic (1964), 79 pp. Grades 3–5. An excellent paperback.

31. MILLER, SHANE. *Peter Stuyvesant's Drummer*. New York: Coward-McCann (1959), 88 pp. Grades 3–5.

32. NEAL, HARRY E. *The Virginia Colony*. New York: Hawthorn (1969), 86 pp. Grades 4–6.

33. OGLE, NAN H. and FRANCES A. BACON. *The Lords Baltimore*. New York: Holt, Rinehart and Winston (1962), 134 pp. Grades 6–9.

34. PETERSHAM, MAUD and MISKA. *The Silver Mace: The Story of Williamsburg*. Eau Claire, Wisconsin: Hale (1961), 40 pp. Grades 5–8. Beautifully illustrated.

35. PILKINGTON, ROGER. *I Sailed on the Mayflower*. New York: St. Martin's (1966), 216 pp. Grades 4–6.

36. PINE, TILLIE S. and JOSEPH LEVINE. *The Pilgrims Knew*. New York: Whittlesey (1957), 32 pp. Grades 3–5. Contributions of the Pilgrims to today.

37. PROHLMAN, MARILY. *The Story of Jamestown*. Chicago: Children's Press (1969), 31 pp. Grades 4–5.

38. RICH, LOUISE D. *The First Book of Early Settlers.* New York: Watts (1959), 81 pp. Grades 5–8. On Plymouth, New Amsterdam, Jamestown, and Fort Christian.
39. ROGERS, LOU. *The First Thanksgiving.* Chicago: Follett (1962), 29 pp. Grades 3–5.
40. SIMPSON, WILMA and JOHN. *About Pioneers: Yesterday, Today, and Tomorrow.* Chicago: Melmont (1963), 63 pp. Grades 4–6.
41. SLOANE, ERIC. *ABC Book of Early Americana: A Sketchbook of Antiquities and Americana Firsts.* Garden City, New York: Doubleday (1963), 62 pp. Grades 5–8. Includes many detailed sketches of items used in colonial times.
42. SMITH, E. BROOKS and ROBERT MEREDITH. *The Coming of the Pilgrims.* Boston: Little, Brown (1964), 60 pp. Grades 4–6.
43. SPEARE, ELIZABETH G. *Life in Colonial America.* New York: Random House (1963), 172 pp. Grades 6–9. A Landmark book.
44. TUNIS, EDWIN. *Colonial Living.* Cleveland: World (1957), 112 pp. Grades 6–9. With 230 pictures.
45. VIPONT, ELFRIDA. *Children of the Mayflower.* New York: Watts (1969), 48 pp. Grades 2–4.
46. WEBB, ROBERT N. *We Were There with the Mayflower Pilgrims.* New York: Grosset and Dunlap (1956), 178 pp. Grades 5–8.
47. WILLIAMS, BARRY. *The Struggle for North America.* New York: McGraw-Hill (1967), 96 pp. Grades 5–8.

The Revolutionary War Period and the New Nation

There are at least three major emphases for the period of the Revolutionary War and the formation of the United States of America. One is life in the colonies in that period. Another is the Revolution and its causes. The third is the establishment of the new government after the Articles of Confederation. The first two of these might well be stressed in the middle grades, with little attention to the Constitutional Convention and the formation of the new federal government. In the junior high school years the Articles of Confederation period and the formation of the federal government can be treated in much more detail.

If you are approaching the study of history from the standpoint of decisions, there are two that stand out. One of them might be phrased in this fashion: "Colonies — or a New Nation?" The other might be "What Kind of Government for Our New Nation?" In the latter, pupils should come to realize that there were people in favor of a king. There were those who wanted an oligarchy. And there were those who preferred a limited democracy. No one at that time wanted the type of democracy we have today.

Particularly with middle grade children, the people and their ways of life need to be stressed. The pupils can meet them again in their homes and at their places of business, as they did in the study of life in the colonial period, noting the changes that occurred in a period of seventy-five to one hundred years. They can see the development of cities, with New York, Philadelphia, Boston, Charleston, Baltimore, Salem, Newport, Providence, Marblehead, and Gloucester as the ten leading localities. You will want to "move upstream" with your pupils and into the "back country," as the people did in the period before the Revolution.

Considerable economics can be included in the study of this period, even with middle grade pupils. Further specialization had taken place since the 1700 era, with more rice grown, more tobacco produced, and more cattle raised. Fishing and whaling had grown rapidly (see the list of cities above, which includes several fishing and whaling centers). Iron had developed an industry in places like Connecticut, Pennsylvania, and New Jersey. So had the glass industry. Some textiles also were produced in this country. As a result of these and other economic changes, a wealthy landowning class had sprung up and there were many wealthy merchants, including such men as George Washington and John Randolph, to cite a wealthy farmer and a wealthy merchant.

Of course you will want to develop the theme of the causes of the American Revolution, keeping in mind the fact that the reasons for it were numerous and not just limited to "taxation without representation." The distance from England was a prime factor and with it the development of an American way of life. Experience in local and even state government and the desire to maintain local autonomy were two other reasons. Reaction against the mercantile system was still another reason, coupled with the acts of England to punish the colonies for their lack of cooperation. It is highly important for pupils to realize that the Revolution did not come quickly and that even those who favored it in the end, did so reluctantly. You may want to liken the events prior to the Revolution to a tug-of-war between England and the colonies or to a volcano which finally erupted, or to steps of a house. You may even draw up a list of actions which England took, and, in a parallel column, list what the colonies did to resist. Such a listing will be found in the writer's text: *One Nation: the United States*, if you want to refer to it. With older pupils, you may want to point out that about a third of the colonists were in favor of maintaining ties with England, about a third in favor of independence, and the other third wavering at the beginning of the Revolution. You may want to role-play the feelings of these different groups.

In the middle grades, little attention need be given to the battles, although a few of them may be included. Certainly Saratoga and Yorktown must be mentioned. With older pupils you may want to include a little more of the military history, but it should not overshadow other aspects of the period. If you are not already steeped in the American Revolution, it may help you (and then your pupils) to realize that the war was fought initially in New England. Then the conflict turned to the Middle Colonies. Toward the end of the war the major battles were in the South.

In the earliest study of American history, little attention needs to be given to the Articles of Confederation and the formation of the new government, except for a passing reference to them. Pupils can get lost in the intricacies of the various compromises. All they need to know is that a government was agreed upon. Older pupils should make some study of the Constitutional Convention without dissecting the Constitution line by line. The importance of the Northwest Ordinance should certainly be brought out with junior high school pupils.

There are some colorful, wonderful people to deal with in this period, including Franklin, John and Sam Adams, Jefferson, Madison, Patrick Henry, Tom Paine, and several others. These can be a rich source for teaching about this period. For girls the personality of Abigail Adams may be of special interest. If the musical *1776* is still being shown, pupils will certainly enjoy that play, with Abigail and John Adams as central figures.

As a teacher you may learn more about this period from a novel or a biography than by reading history books. For example, the volume by Catherine Drinker Bowen on John Adams and the American Revolution (a Grosset and Dunlap paperback) is fascinating, and the story of Abigail and John Adams as told by Irving Stone in *Those Who Love* (Doubleday) is equally thrilling. Such accounts give readers a feeling for that period and intimate insights into the lives of some of the people as they were affected by events of the Revolution.

There are many accounts of the Revolution to which you may want to refer. One of the best from the standpoint of illustrations is the volume in the Life World Library on *The Making of a Nation: 1775–1789.* Lawrence Gipson's *The Coming of the Revolution: 1763–1775* (Harper and Row) is a splendid volume. So is Bruce Lancaster's *From Lexington to Liberty* (Doubleday). Some of the books for children on this period are listed on the next two pages. Teachers will find that they also throw light on the period and are a good source of background for teachers as well as for pupils.

If you are devoting only approximately a half year to the study of United States history, perhaps you will need to be reminded that you cannot go into tremendous detail on this or on any other period. After all, this is not the last time your pupils will learn about our national history.

BOOKS FOR BOYS AND GIRLS ON THE AMERICAN REVOLUTION

1. ALBRECHT, LILLIE V. *The Grist Mill Secret.* New York: Hastings (1962), 126 pp. Grades 4–6. New England during this period.
2. ALDERMAN, CLIFFORD L. *The Story of the Thirteen Colonies.* New York: Random (1966), 192 pp. Grades 5–8.
3. ALTER, ROBERT E. *Listen, the Drum: A Novel of Washington's First Command.* New York: Putnam (1963), 190 pp. Grades 6–9.
4. BARNES, ERIC W. *Free Men Must Stand: The American War of Independence.* New York: Whittlesey (1962), 159 pp. Grades 6–9.
5. BLIVEN, BRUCE, JR. *The American Revolution: 1760–1783.* New York: Random (1958), 182 pp. Grades 6–9.
6. BORRESON, MARY JO. *Let's Go to the First Independence Day.* New York: Putnam (1962), 48 pp. Grades 3–5.
7. CLARKE, MARY S. *Petticoat Rebel.* New York: Viking (1964), 255 pp. Grades 6–9. A seventeen-year-old in the time of the Revolution.
8. COMMAGER, HENRY STEELE. *The Great Declaration: A Book for Young Americans.* Indianapolis: Bobbs-Merrill (1958), 112 pp. Grades 4–7. A vivid account using documents to tell the story.
9. COOK, F. *Golden Book of the American Revolution.* New York: Golden (1959), 193 pp. Grades 5–8. Profusely illustrated.
10. DA CRUZ, DANIEL, JR. *Men Who Made America: The Founders of a Nation.* New York: Crowell (1962), 143 pp. Grades 6–9.
11. DOUTY, ESTHER. *Under the New Roof: Five Patriots of the Young Republic.* Chicago: Rand McNally (1965), 288 pp. Grades 5–8.
12. FINDLAY, BRUCE A. and ESTHER B. *Your Magnificent Declaration.* New York: Holt, Rinehart and Winston (1961), 128 pp. Grades 7–10.
13. FISHER, DOROTHY CANFIELD. *U. S. Revolution and Constitution.* New York: Random (1950), 187 pp. Grades 6–9. A Landmark book.
14. FISHER, LEONARD E. *The First Book Edition of the Declaration of Independence.* New York: Watts (1960), 27 pp. Grades 5–8.
15. FISHER, LEONARD EVERETT. *Two If By Sea.* New York: Random House (1970), 64 pp. Grades 4–6.
16. FLEMING, ALICE. *A Son of Liberty.* New York: St. Martin's (1961), 182 pp. Grades 7–10.
17. GERSON, NEAL B. *Nathan Hale: Espionage Agent.* Garden City, New York: Doubleday (1960), 55 pp. Grades 4–7.
18. GREEN, MARGARET. *Radical of the Revolution: Samuel Adams.* New York: Messner (1971), 192 pp. Grades 6–9.
19. LANCASTER, BRUCE. *The American Revolution.* Garden City, New York: Doubleday (1967), 61 pp. Grades 6–9.

20. LANCASTER, BRUCE. *The First Book of the American Revolution.* New York: Watts (1959), 70 pp. Grades 5–8.

21. LANCASTER, BRUCE. *Ticonderoga: The Story of a Fort.* Boston: Houghton Mifflin (1959), 181 pp. Grades 7–10.

22. LECKIE, ROBERT. *Great American Battles.* New York: Random House (1968), 128 pp. Grades 6–9.

23. LENS, SIDNEY. *A Country Is Born: The Story of the American Revolution.* New York: Putnam (1964), 191 pp. Grades 5–8.

24. MCKOWN, ROBIN. *Washington's America.* New York: Grosset and Dunlap (1960), 94 pp. Grades 5–8.

25. MIERS, EARL S. *We Were There When Washington Won at Yorktown.* New York: Grosset and Dunlap (1959), 176 pp. Grades 4–6.

26. MIERS, EARL S. and FELIX SUTTON. *America During Four Wars.* New York: Grosset and Dunlap (1965), 192 pp. Grades 5–8.

27. MORRIS, RICHARD B. *The First Book of the American Revolution.* New York: Watts (1956), 64 pp. Grades 4–6.

28. RIPLEY, SHELDON N. *Ethan Allen: Green Mountain Hero.* Boston: Houghton Mifflin (1961), 191 pp. Grades 5–8.

29. ROSS, GEORGE F. *Know Your Declaration of Independence and the 50 Signers.* Chicago: Rand McNally (1963), 72 pp. Grades 5–7.

30. SOBOL, DONALD J. *Lock, Stock, and Barrel.* Philadelphia: Westminster (1965), 256 pp. Grades 5–8.

31. SUTTON, FELIX. *The How and Why Wonder Book of the American Revolution.* Columbus, Ohio: Merrill (1963), 48 pp. Grades 5–7.

32. SUTTON, FELIX. *Sons of Liberty.* New York: Messner (1969), 90 pp. Grades 5–6.

33. SUTTON, FELIX. *We Were There at the Battle of Lexington and Concord.* New York: Grosset and Dunlap (1958), 182 pp. Grades 4–6.

34. VIERECK, PHILLIP. *Independence Must Be Won.* New York: John Day (1964), 158 pp. Grades 5–8.

35. WEBB, ROBERT N. *We Were There at the Boston Tea Party.* New York: Grosset and Dunlap (1956), 173 pp. Grades 4–6.

36. WOOD, JAMES P. *The People of Concord.* New York: Seabury Press (1969), 152 pp. Grades 5–8.

The United States as a new nation

1. BRUSH, HAMILTON and LEE S. PATTISON. *The Declaration of Independence and the Constitution.* Cambridge, Massachusetts: Educators Publishing Service (1961), 33 pp. Grades 5–8.

2. BUELL, CHARLES C. *They Made a Nation.* Medford, Massachusetts: Filene Center (1952), 48 pp. Grades 6–9. Simple story of the writing of the Constitution.

3. COMMAGER, HENRY STEELE. *The Great Constitution: A Book for Young Americans.* Indianapolis: Bobbs-Merrill (1961), 128 pp. Grades 6–8.

4. COMMAGER, HENRY STEELE. *James Madison.* New York: Messner (1963), 191 pp. Grades 6–9.

5. FISHER, DOROTHY CANFIELD. *Our Independence and the Constitution.* Indianapolis: Bobbs-Merrill (1961), 128 pp. Grades 6–8.

6. MILHAUS, KATHERINE. *Through These Arches: The Story of Independence Hall.* Philadelphia: Lippincott (1964), 96 pp. Grades 5–8.

7. PHELAN, MARY K. *Four Days in Philadelphia: 1776.* New York: Crowell (1967), 189 pp. Grades 5–8.

8. WILKIE, KATHERINE E. and ELIZABETH R. MOSELEY. *Father of the Constitution: James Madison.* New York: Messner (1963), 191 pp. Grades 6–9.

9. WILLIAMS, SELMA R. *Fifty-Five Fathers: The Story of the Constitutional Convention.* New York: Dodd, Mead (1970), 179 pp. Grades 5–8.

From the Formation of the New Nation to the Civil War

From the formation of the new federal government to the Civil War is a period of approximately seventy years. In that time many important events took place. The United States expanded vastly in territory and in population, industrialization began in the East and "King Cotton" began his reign in the South, the Mexican War took place, and a network of canals and highways was built, sectional issues came to the fore, and the idea of democracy was applied in new fields.

All this means that teachers need to be clear on their aims in teaching about this period with middle grade and junior high school pupils. One way to give focus to this span of years with middle grade children is to select the 1830s and concentrate upon them. Then the pupils can do the same with the 1860s. Some events can be ignored or minimized, such as the War of 1812 and the Mexican War. Pupils can learn about them later in their school years.

In the junior high school years, teachers who want to approach the study of our history through major decisions might well think in terms of ten great decisions, as follows: (1) Shall We Purchase the Louisiana Territory? (2) How Shall the Old Southwest and the Old Northwest Be Organized? (3) Shall We Be Friends with England or France? (4) How Shall We Handle Europe's Interest in Latin America? (5) How Can We Extend Democracy? (6) Can We Develop Further the American Way of Life? (7) Shall We Push On to the Pacific? (8) Shall We Fight Mexico? (9) How Can We Unify This Vast Nation? and (10) How Can We Solve the Problem of Slavery?

In the middle grades, teachers may want to concentrate on the acquisition of territory and the westward migrations, stressing the ways of living of people on the frontier in this period. Pupils can become involved in this period if they role-play the people who went west and the people who stayed at home, figuring out why they would take either action. They can do problem-solving on the materials they would take on a journey west. More point will be given to such an exercise if the pupils decide which two or three tools would be most important to them. They also can read diaries and novels of the men, women, and children who moved west. And they can write their own diaries of journeys west with the pioneers.

Pupils can discover how we grew from a nation of four million to a nation of thirteen million persons and from a country of thirteen states to a country of twenty-six states. Some study can be made with profit of the topic of immigration in this regard.

Then teachers can turn the attention of pupils to what was happening back in the East, with the invention of new tools and processes by men like Samuel Slater and Eli Whitney. The same can be done for the South, with the changes wrought by the invention of the cotton gin.

With such a vast country, the problem arose of how to unify it. This could lead very naturally into the story of the building of canals and railroads. It may be difficult for boys and girls to imagine a time when there were few roads and no highways, but you need to try to develop such a feeling on their part. If you have developed the idea of specialization, they will be better able to see why new forms of transportation were necessary to move goods. Hence, people began to develop canals and later highways and railroads between the major points in our nation.

Sectionalism existed prior to this period, but it was certainly accentuated in the years we are discussing. Children can see some of the problems which arose in regard to internal improvements, the tariff, and slavery. Dividing the class into the

East, the South, and the West can help them to understand how the people of those various sections felt on those issues. The tariff can be dealt with in passing in the middle grades and in slightly more detail in the junior high grades.

The election of Jackson can climax a study of this period, for most of the issues of the period appear in his election, with the possible exception of slavery — which was being postponed until a later time. There are numerous interpretations of Jackson, of which teachers should be aware, but for children he can be primarily a frontiersman who became President. He was the embodiment of the western pioneer, despite the fact that he also received support from the laboring classes in the East.

Some teachers may want to introduce a few of the western cities of this period, such as Cincinnati, St. Louis, and New Orleans, but this is not essential.

Biography can also have a prominent place in this unit, with Horace Mann, Dorothea Dix, Susan B. Anthony, and others stressed.

There is considerable room in this study for dramatization. The story of the westward movement lends itself to short skits, plays, and art work. The writing of diaries and travel accounts can enhance learning, too.

Geography also can be included in this unit, with the addition of new territories and the formation of several states. Children also can discover the routes the people traveled in moving west and the connection between the regions in which they settled, and the new states which came into the Union.

Many teachers will want to bring out the place of the little known, the unknown, or the inconspicuous persons who made history, whether they were the Irish who did so much work on the Erie Canal, or the pioneers who built up the great Midwest in that period. Children also can "visit" people in their homes as they lived in eastern cities, in the South, and in the West in those times.

Comparing the growth of democracy to the growth of a tree may prove helpful to boys and girls in seeing the expansion of the democratic idea in education, in political life, and in social and economic areas. In this period several movements arose or gathered strength. Among these were the women's rights movement, the peace movement, the anti-slavery movement (especially in the South), public education, and temperance. Some more able pupils may want to try to figure out why this was so.

With older boys and girls more attention can be given to the figures of Clay, Calhoun, Webster, and others.

Teachers will find a great many accounts of this period in our national history. In addition to chapters in the books cited on the history of the United States (see pages 393–394), they will find helpful such volumes as Paul Angle *The New Nation Grows* (Premier paperback), the Life history volume on *The Sweep Westward*, and Douglas North *The Economic Growth of the United States 1790–1860* (a Norton paperback). Biographies of Jackson by Arthur Schlesinger, Jr. (*The Age of Jackson*) and Leonard S. White (*The Jacksonians*) will provide further background.

Here are some books which should help pupils to understand this era.

BOOKS FOR BOYS AND GIRLS ON THE PERIOD FROM THE FORMATION OF THE UNITED STATES TO THE CIVIL WAR

1. ADAMS, SAMUEL E. *The Erie Canal*. New York: Random House (1953), 170 pp. Grades 6–9. A Landmark book. Excellent description.
2. ALLEN, EDWARD. *Heroes of Texas*. New York: Messner (1970), 94 pp. Grades 4–6.
3. ANDRIST, RALPH K. *The Erie Canal*. New York: Harper and Row (1964), 156 pp. Grades 5–7.

4. BAILEY, RALPH E. *Wagons Westward: The Story of Alexander Majors*. New York: Morrow (1970), 132 pp. Grades 5–8.

5. BAKELESS, JOHN. *The Adventures of Lewis and Clark*. Boston: Houghton Mifflin (1962), 128 pp. Grades 4–6.

6. BAUER, HELEN. *California Gold Rush Days*. Garden City, New York: Doubleday (1956), 128 pp. Grades 4–6.

7. BERRY, WILLIAM D. *Buffalo Land: The Untamed Wilderness of the High Plains*. New York: Macmillan (1961), 48 pp. Grades 5–8.

8. BUEHR, WALTER. *Westward with American Explorers*. New York: Putnam (1963), 95 pp. Grades 4–7.

9. CARR, MARY J. *Children of the Covered Wagon: A Story of the Oregon Trail*. New York: Crowell (1957), 303 pp. Grades 4–6.

10. CASTOR, HENRY. *The First Book of the Spanish-American West*. New York: Watts (1963), 72 pp. Grades 5–8. From the period of Indian domination until this area became a part of the United States.

11. CASTOR, HENRY. *The First Book of the War with Mexico*. New York: Watts (1964), 87 pp. Grades 5–8.

12. CHURCH, ALBERT C. *Whale Ships and Whaling*. New York: Norton (1958), 179 pp. Grades 6–9.

13. COUSINS, MARGARET. *We Were There at the Battle of the Alamo*. New York: Grosset and Dunlap (1958), 180 pp. Grades 4–6.

14. DOBIE, J. FRANK. *Up the Trail from Texas*. New York: Random House (1955), 192 pp. Grades 6–9. A Landmark book.

15. DUNLAP, EUGENE. *The Trail to Santa Fe*. Chicago: Encyclopaedia Britannica (1963), 32 pp. Grades 4–6. Colored illustrations.

16. Editors of *The American Heritage*. *Adventures in the Wilderness*. New York: Harper (1963), 153 pp. Grades 6–8.

17. Editors of *The American Heritage*. *Clipper Ships and Captains*. New York: American Heritage (1963), 153 pp. Grades 6–8.

18. FABER, DORIS. *I Will Be Heard: The Life of William Lloyd Garrison*. New York: Lothrop, Lee and Shepard (1970), 127 pp. Grades 5–8.

19. GERSON, NOEL B. *Mr. Madison's War*. New York: Messner (1966), 192 pp. Grades 6–9.

20. HARTER, WALTER. *Four Flags Over Florida*. New York: Messner (1971), 96 pp. Grades 5–8.

21. HAVIGHURST, WALTER. *The First Book of the Oregon Trail*. New York: Watts (1960), 58 pp. Grades 4–7.

22. HEIDERSTADT, DOROTHY. *Frontier Leaders and Pioneers*. New York: McKay (1962), 181 pp. Grades 5–8. Boone, Clark, Lewis, Pike, Houston, and others.

23. HIRSCHFIELD, BURT. *After the Alamo: The Story of the Mexican War*. New York: Messner (1966), 192 pp. Grades 7–10.

24. HIRSCHFIELD, BURT. *Four Cents an Acre: The Story of the Louisiana Purchase*. New York: Messner (1965), 191 pp. Grades 5–8.

25. HOLBROOK, STEWART. *Davy Crockett: From the Backwoods of Tennessee to the Alamo*. New York: Random House (1955), 192 pp. Grades 5–8.

26. HOLT, STEPHEN. *We Were There with the California Forty-Niners*. New York: Grosset and Dunlap (1956), 175 pp. Grades 4–6.

27. HONIG, DONALD. *In the Days of Cowboys*. New York: Random House (1970), 98 pp. Grades 4–6.

28. HOUGH, HENRY B. *Great Days of Whaling*. Boston: Houghton Mifflin (1958), 184 pp. Grades 7–10.

29. JOHNSON, WILL. *The Birth of Texas*. Boston: Houghton Mifflin (1960), 183 pp. Grades 4–6. A North Star book.

30. JONES, HELEN N. *Over the Mormon Trail*. Chicago: Children's Press (1963), 128 pp. Grades 4–7.

31. LAMPMAN, EVELYN. *Wheels West: The Story of Tabitha Brown.* Garden City, New York: Doubleday (1965), 226 pp. Grades 5–8. The story of a grandmother who moved to Oregon and founded an orphans' home.

32. LAVENDAR, DAVID. *Westward Vision: The Story of the Oregon Trail.* New York: McGraw-Hill (1963), 424 pp. Grades 8–10.

33. LAWSON, DON. *The War of 1812: America's Second War for Independence.* New York: Abelard-Schuman (1966), 160 pp. Grades 6–9.

34. LESTER, JULIUS. *To Be a Slave.* New York: Dial (1968), 160 pp. Grades 5–8.

35. McSPADDEN, J. WALKER. *Pioneer Heroes.* New York: Crowell (1957), 215 pp. Grades 5–8. Clark, Boone, Pike, Fremont, and others.

36. MEADOWCRAFT, ENID. *By Wagon and Flatboat.* New York: Crowell (1956), 170 pp. Grades 4–7.

37. MEADOWCRAFT, ENID. *We Were There at the Opening of the Erie Canal.* New York: Grosset and Dunlap (1958), 182 pp. Grades 4–6.

38. MONTGOMERY, ELIZABETH R. *When Pioneers Rushed West to Oregon.* Champaign, Illinois: Garrard (1970), 128 pp. Grades 5–8.

39. NORTH, STERLING. *The First Steamboat on the Mississippi.* Boston: Houghton Mifflin (1962), 184 pp. Grades 7–9.

40. POLKING, KIRK. *Let's Go With Lewis and Clark.* New York: Putnam (1963), 46 pp. Grades 5–8.

41. REINFERD, FRED. *The Real Book of Whales and Whaling.* Garden City, New York: Doubleday (1960), 214 pp. Grades 5–8.

42. RUIZ, RAMON E. *The Mexican War: Was It Manifest Destiny?* New York: Holt, Rinehart and Winston (1963), 118 pp. Grades 5–8.

43. SHAFTEL, GEORGE and others. *Westward the Nation.* Pasadena, California: Franklin (1965), 256 pp. Grades 5–8.

44. STEELE, WILLIAM O. *We Were There on the Oregon Trail.* New York: Grosset and Dunlap (1955), 177 pp., Grades 4–6.

45. SUTTON, FELIX. *The How and Why Wonder Book of Winning the West.* Columbus, Ohio: Merrill (1963), 43 pp. Grades 4–6.

46. TAYLOR, ROSS M. *We Were There on the Santa Fe Trail.* New York: Grosset and Dunlap (1960), Grades 4–6.

47. WIBBERLEY, LEONARD. *Zebulon Pike.* New York: Funk and Wagnalls (1961), 179 pp. Grades 5–8.

48. WOLFE, LOUIS. *Let's Go to the Louisiana Purchase.* New York: Putnam (1963), 48 pp. Grades 5–7.

The Civil War Period

Everyone, or nearly everyone, would certainly agree that the Civil War period should receive some attention in any story of the history of the United States. It is one of the great "watersheds" in our development. From that point on, however, teachers may not agree. They are likely to have varying views on what should be taught or even on what the period includes in point of time. Some would even challenge the wording of the title above, preferring that it be called "The Period of the War Between the States."

If the "postholing" approach is used, the 1860s might well be included as one of the major periods to be emphasized. The prior date to be accented would probably be the 1830s and the subsequent date the 1900s.

If this approach is used, the Civil War would be only one point of emphasis. For middle grade children, other events and movements should be included. There is the further expansion of our country around that time, and the settlement of parts of the West. There is the story of the development of agriculture, affected greatly by the increased use of the reaper. There is the history of immigration and the rise of industry. There is the story of the cotton gin and its effect on the economy of the South. And there is the important development of railroads.

In the middle grades the Civil War needs to be studied and its effect on our national history examined in an elementary way. But it need not be pursued in depth by pupils at this point in school. And it should certainly be seen against the background of these other developments mentioned.

To children, the Civil War and slavery should probably be seen as a big problem or a series of problems which adults were not able to solve. As a result, they went to war. But the war did not solve their problems. Children can learn that we are still wrestling with these problems. The Civil Rights movement and the efforts to provide equal opportunities for all our citizens today are very much with us. They affect children in many parts of our nation. This "human relations" aspect of the period should be developed in considerable detail with middle grade children.

With children in the junior high school years, more attention can be given to the causes of the Civil War and the war itself, without fighting every battle of that long and costly conflict. If the "decisions" approach is used, two questions can be highlighted. One would be, "Can We Solve the Differences Between the North and South?" The other would be "How Can We Bind the Wounds of the War Between the North and the South?"

As students wrestle with the problems connected with this second decision, they can learn for themselves how difficult it is to develop unity and friendship between sections of a nation which have fought against each other in civil war. They can also learn about the compromises which were made in the treatment of blacks and how those compromises postponed for a century or more the winning of equal rights. Handled well, a study of this period should have great relevance for today's problems.

Teachers need to handle this period with great sensitivity, trying to interpret the views of southerners and northerners, pro-slavery and anti-slavery advocates, Negroes and whites; that can be a real test of objective teaching.

Many teachers would do well to try to role-play the views of a variety of individuals, even as they encourage their pupils to do this. They might think in terms of the high hopes of Negroes over the Emancipation Proclamation—and their disillusionment when their condition reverted, between 1885 and 1900, to almost the same condition they were in, prior to that conflict. It might well include the feelings of southern plantation owners who treated their slaves humanely, judged by the standards of that time.

As background for teaching about that period, teachers would do well to read one or more accounts of the explanations of the Civil War and of Reconstruction. One such account is found in John Higham's volume on *The Reconstruction of American History* (Harper and Row). Another is in the chapter on "The Civil War and Reconstruction" in the 31st yearbook of the National Council for the Social Studies on *Interpreting and Teaching American History*. In both accounts, the writers point out that few historians today accept the economic interpretation of that period as the sole explanation of the Civil War. Sectionalism, emotional factors, slavery as a moral issue, and states' rights all played their part in the complicated causes of war.

There are many methods which can be used to advantage with middle grade and junior high school pupils. One is the use of newspaper accounts of the period. Another is the comparison of textbook accounts.

The period under discussion was rich in music, and pupils should see and feel the impact of music on the people of that time, taking part themselves in singing some of the songs of those times. Then they should examine the words of the songs they have sung to see how they express the feelings of people about the times in which they were living.

Again, biography has a part to play in learning about the 1860s and the years prior to and following the Civil War. The story of Abraham Lincoln is a must and can prove exciting to almost every pupil. The biographies of Robert E. Lee, Cyrus McCormick, Frederick Douglass, Harriet Tubman, and others need to be read and/or told.

For older pupils the story of the various compromises on slavery and sectionalism can be made vivid by building two parallel columns of the states with colored paper on which the names of the states are written or printed. In that way pupils can see how Congress tried to keep a balance between the free and slave states.

Role-playing, after the reading of some materials on this period, has already been suggested as a very important method.

Some groups, especially in the junior high school years, will profit from developing a pictorial time-line of this era, drawing their own representations of the most important men and events. In the junior high school grades some pupils may also want to develop parallel charts of the strengths and weaknesses of the North and South at the beginning of the war.

This is not an easy unit to teach, but it is an important and exciting era for pupils to discover. By studying that important era in our nation's story, they can gain a better understanding of many issues today and begin to see that history can make sense by shedding light on the present.

There are scores of books on this period for boys and girls. Those which follow are merely a sample.

BOOKS FOR BOYS AND GIRLS ON THE CIVIL WAR PERIOD

1. BROWIN, FRANCES W. *Looking for Orlando.* New York: Criterion (1961), 159 pp. Grades 7–9. About the Underground Railroad.
2. BUCKMASTER, HENRIETTA. *Flight to Freedom: The Story of the Underground Railroad.* New York: Crowell (1958), 217 pp. Grades 7–9.
3. COIT, MARGARET L. *The Fight for Union.* Boston: Houghton Mifflin (1961), 136 pp. Grades 5–8.
4. COMMAGER, HENRY STEELE. *The Great Proclamation: A Book for Young People.* Indianapolis: Bobbs-Merrill (1960), 112 pp. Grades 4–6.
5. CROSS, HELEN R. *Life in Lincoln's America.* New York: Random (1964), 171 pp. Grades 7–10. A tremendous compendium of information. Illustrated.
6. DANIELS, JONATHAN. *Stonewall Jackson.* New York: Random (1959), 192 pp. Grades 7–9.
7. DAUGHERTY, MICHAEL. *Diary of a Civil War Hero.* New York: Pyramid (1960), 128 pp. Grades 7–10.
8. DONOVAN, FRANK R. *Ironclads of the Civil War.* New York: American Heritage (1964), 153 pp. Grades 6–9.
9. DUPUY, TREVOR. *The First Book of Civil War Land Battles.* New York: Watts (1960), 96 pp. Grades 4–7.
10. FLATO, CHARLES. *The Golden Book of the Civil War.* New York: Golden (1960), 290 pp. Grades 5–8. Profusely illustrated.
11. FOSTER, C. A. *The Eyes and Ears of the Civil War.* New York: Criterion (1964), 168 pp. Grades 7–10.
12. FOSTER, GENEVIEVE. *Abraham Lincoln's World.* New York: Scribner (1944), 347 pp. Grades 5–8. An old book but still extremely useful on life in those times.

13. HAGLER, MARGARET. *Larry and the Freedom Man.* New York: Lothrop (1959), 160 pp. Grades 5–8. Two children of different races live through the Civil War period as friends.

14. HALL, ANNA G. *Cyrus Holt and the Civil War.* New York: Viking (1964), 128 pp. Grades 5–8. A small boy lives through the Civil War in an upper New York state town.

15. HEAPS, WILLARD A. *Bravest Teenage Yankee.* New York: Dell (1963), 150 pp. Grades 7–10.

16. HENKLE, HENRIETTA. *Flight to Freedom.* New York: Crowell (1958), 217 pp. Grades 7–10. The Underground Railroad.

17. KANTOR, MACKINLAY. *Gettysburg.* New York: Random (1952), 183 pp. A Landmark book.

18. KANTOR, MACKINLAY. *Lee and Grant at Appomattox.* New York: Random (1950), 192 pp. Grades 5–8. A Landmark book.

19. KEITH, HAROLD. *Rifles for Watie.* New York: Crowell (1957), 382 pp. Grades 7–10. Fighting in the Western States during the Civil War.

20. MCCARTHY, AGNES and LAWRENCE REDDICK. *Worth Fighting For.* Garden City, New York: Doubleday (1965), 118 pp. Grades 5–8. Negroes in the Civil War.

21. MCCONNELL, JANE T. *Cornelia: The Story of a Civil War Nurse.* New York: Crowell (1959), 192 pp. Grades 6–9. A twenty-year-old Quaker as a nurse in the Civil War.

22. MALKUS, ALIDA S. *We Were There at the Battle of Gettysburg.* New York: Grosset and Dunlap (1955), 176 pp. Grades 7–9. Johnny and his sister are caught between the Union and Confederate armies at this decisive battle.

23. MERIWETHER, LOUISE. *The Freedom Ship of Robert Smalls.* Englewood Cliffs, New Jersey: Prentice-Hall (1971), 32 pp. Grades 5–8. A slave takes over a Confederate ship and transports 15 slaves to freedom.

24. MIERS, EARL S. and FELIX SUTTON. *America During Four Wars.* New York: Grosset and Dunlap (1965), 192 pp. Grades 6–9. One chapter on the Civil War.

25. MIERS, EARL S. *Billy Yank and Johnny Reb: How They Fought and Made Up.* New York: Grosset and Dunlap (1960), 256 pp. Grades 4–6.

26. MIERS, EARL S. *The How and Why Wonder Book of the Civil War.* Columbus, Ohio: Merrill (1961), 48 pp. Grades 5–8.

27. MIERS, EARL S. *We Were There When Grant Met Lee at Appomattox.* New York: Grosset and Dunlap (1960), 176 pp. Grades 4–6.

28. NATHAN, ADELE G. *Lincoln's America.* New York: Grosset and Dunlap (1961), 93 pp. Grades 5–8.

29. PRATT, FLETCHER. *The Civil War.* Garden City, New York: Doubleday (1955), 63 pp. Grades 5–8.

30. PRATT, FLETCHER. *The Monitor and the Merrimac.* New York: Random (1951), 185 pp. Grades 5–8. A Landmark book.

31. REEDER, RED. *The Southern Generals.* New York: Duell (1965), 237 pp. Grades 4–6.

32. SAYRE, ANNE. *Never Call Retreat.* New York: Crowell (1957), 164 pp. Grades 4–6. A Quaker family and the seventeen-year-old daughter face difficulties with the Ku Klux Klan.

33. STEELE, WILLIAM O. *The Perilous Road.* New York: Harcourt Brace Jovanovich (1958), 191 pp. Grades 4–7.

34. STERLING, DOROTHY. *Forever Free: The Story of the Emancipation Proclamation.* Garden City, New York: Doubleday (1963), 208 pp. Grades 6–9.

35. WACKIN, EDWARD. *Black Fighting Men in U.S. History.* New York: Lothrop, Lee and Shepard (1971), 192 pp. Grades 5–8.

36. YOUNG, BOB and JOAN. *Ulysses S. Grant.* New York: Messner (1970), 192 pp. Grades 5–8.

For poetry on this period, see the three anthologies listed on page 394.
Any book of music of the United States for children will have songs from this

period which you may want to use. Folkways/Scholastic has a record on "Ballads of the Civil War" which you may want to use with your pupils.

There is a wealth of films and filmstrips on the Civil War period. Inquire of your audio-visual department, the state library, or the state university.

From the Civil War to World War I

From the Civil War to World War I is a period of a little over fifty years. In the history of our country that is not a long time, but there are numerous movements during that period which should be discovered by elementary and junior high school pupils.

First there is the period of reconstruction, which can be tied very closely with the present movement for civil rights and equal opportunity which looms so large at present.

Then comes the industrialization of our nation, with the rise of big business and the holders of great wealth. Some have referred to this as "The Gilded Age." Although the titans of business were long known as "The Robber Barons," recent interpretations are less harsh. The men who accumulated great wealth are seen by some historians today as men who realized the importance of capital for the industrialization of our nation and had the acumen to collect such capital.

Closely linked with this facet of our national life are the stories of the new inventions and the part they played in turning the United States into a giant workshop. In this period the typewriter, the telephone, electric lights, movie projectors, linotype, and new means of producing power came into being.

Middle grade children and upper grade pupils — and even teachers? — may find it difficult to imagine a time when horses did most of the work in the fields and in transporting people and goods. But this period belongs to the horses. However, the Wright Brothers flew their strange contraption known as an airplane at Kitty Hawk in 1903, and it was around the turn of the century that "horseless carriages" appeared on the scene.

To man the new factories, millions of immigrants came to our nation between the Civil War and World War I, emigrating largely from eastern and southern Europe, and adding immeasurably to our growth.

Obviously there was new wealth and new power. The question then arose as to who would share in this wealth and power. Would the farmers? Would the workers? The growth of labor unions is an important aspect of this period. So, too, is the organization of farmers into a pressure group or into various pressure groups.

Meanwhile, the United States was becoming more involved in world affairs. Our land was extended to include Alaska, Hawaii, the Philippines, Puerto Rico, Guam, American Samoa, and, to some extent, the Canal Zone.

There were many important political figures in this period, but Theodore Roosevelt certainly stands head and shoulders above the other Presidents. Children should get to know this flamboyant character and his contributions, especially in the field of conservation.

Two significant changes which are seldom highlighted are the revolution in agriculture and the revolution in health and sanitation. Both occurred in large part between the Civil War and World War I.

In this half-century, our population more than doubled, and the number of states increased from thirty-six to forty-eight. The importance of those figures needs to be examined by pupils.

With middle grade pupils, much emphasis can be placed on the contributions of inventors to the process of industrialization. They should be thrilled by many of the stories of inventors of that time.

Groups of "unhappy people" should be encountered. They should include the farmers, the laborers, the immigrants, the women, and the members of several minority groups. Children can discover why these groups were unhappy about their place in our national life in those times, what they did about their unhappiness, and what the results of their actions were. They may want to role-play the feelings of these different minorities.

In conjunction with the acquisition of new territory, some map study can be carried on. Pupils usually find it difficult, for example, to locate the two areas in which the Spanish-American War was fought and to gain an understanding of where the new possessions were.

Life in the 1900s can also be stressed, especially by visits to the homes of three or four "typical" families. One might well be the home of a Vanderbilt, a Gould, a Carnegie, or one of the other wealthy industrialists. A second might be to the home of a small-town family. A third would certainly be to a farm in the Middle West. A fourth might be to the "flat" of an immigrant family in a large city.

There are many individuals whose biographies can be studied with profit in any survey of this period. Among them would be Theodore Roosevelt, Thomas Edison, Jane Addams, Samuel Gompers, Luther Burbank, and Alexander Graham Bell.

There is plenty of opportunity in this period to explore some of the music. This might include the band music of John Philip Sousa and the compositions of Edward MacDowell as well as such popular songs as *Clementine, A Bicycle Built for Two,* and *When Irish Eyes Are Smiling.*

Boys in particular will be interested in the "invention" of basketball and the formation of Big Leagues in baseball.

If the "postholing" approach is used, the year 1900 (or the decade of the 1900s) is recommended as one possible point.

If the "decisions" approach is used in the junior high grades, these six big decisions might well be stressed: (1) How Can We Bind Up the Wounds of the Civil War? (2) How Can We Revolutionize Agriculture? (3) How Big Should Big Business Become? (4) Should Farmers and Laborers Share in the New Wealth and Power? (5) Shall We Add Land in Other Parts of the World? and (6) What Further Reforms Are Needed to Improve Democracy? Some teachers may want to add decisions regarding conservation and the treatment of immigrants.

Some of you may feel that you need more background about this period than you now have. If so, there are plenty of books to read. You may want to start with the general histories we have mentioned several times before, especially the Nevins-Commager volume. Then four paperbacks might be considered. They are George E. Mowry *The Era of Theodore Roosevelt and the Birth of Modern America: 1900–1912* (Harper), Harold U. Faulkner *Politics, Reform and Expansion 1890–1900* (Harper), Paul M. Angle *The Making of a World Power* (Premier), and Frederick L. Allen *The Big Change* (Harper). For a vivid account of the years around the turn of the century, you may want to turn to Mark Sullivan's volumes on *Our Times.*

There are numerous books for children which can be recommended for the years between the Civil War and World War I. They include the following:

BOOKS FOR BOYS AND GIRLS FROM THE CIVIL WAR TO WORLD WAR I

1. BLOW, MICHAEL. *Men of Science and Invention.* New York: American Heritage (1961), 153 pp. Grades 7–9.
2. BURLINGAME, ROGER. *Scientists Behind the Inventors.* New York: Harcourt Brace Jovanovich (1960), 211 pp. Grades 7–10.
3. CONSIDINE, BOB. *The Panama Canal.* New York: Random (1951), 192 pp. Grades 6–9. The story of the building of the Canal. A Landmark book.
4. FULLER, EDMUND. *Tinkers and Genius.* New York: Hastings (1955), 308 pp. Grades 6–9.
5. HARTMAN, GERTRUDE. *Machines and the Men Who Made the World of Industry.* New York: Macmillan (1961), 278 pp. Grades 7–10.
6. LAVINE, SIGMUND. *Famous Industrialists.* New York: Dodd, Mead (1961), 157 pp. Grades 7–10. Carnegie, Firestone, Eastman, and ten others.
7. LENSKI, LOIS. *Texas Tomboy.* Philadelphia: Lippincott (1950), 180 pp. Grades 5–7.
8. LOWREY, JANETTE. *Margaret.* New York: Harper and Row (1950), 277 pp. Grades 7–10. Growing up in a small town in Texas in the 1900s.
9. McKOWN, ROBIN. *Roosevelt's America.* New York: Grosset and Dunlap (1962), 92 pp. Grades 5–8. A panoramic view of the period of Theodore Roosevelt.
10. MANCHESTER, HARLAND. *Trail Blazers of Technology: The Story of Nine Inventors.* New York: Scribner (1962), 256 pp. Grades 7–10.
11. MELTZER, MILTON. *Bread and Roses: The Struggle of American Labor, 1865–1915.* New York: Knopf (1967), 224 pp. Grades 6–9.
12. MIERS, EARL S. and FELIX SUTTON. *America During Four Wars.* New York: Grosset and Dunlap (1965), 192 pp. Grades 6–9.
13. NEAL, HARRY E. *From Spinning Wheel to Spacecraft: The Story of the Industrial Revolution.* New York: Messner (1964), 191 pp. Grades 5–8.
14. PARADIS, ADRIAN A. *Labor in Action: The Story of the American Labor Movement.* New York: Messner (1963), 191 pp. Grades 5–8.
15. SANDBURG, CARL. *Prairie-Town Boy.* New York: Harcourt Brace Jovanovich (1955), 179 pp. Grades 5–8. An autobiographical account of Sandburg as a boy in an Illinois town.
16. SHEEHY, EMMA D. *Molly and the Golden Wedding.* New York: Holt, Rinehart and Winston (1956), 159 pp. Grades 4–6. Molly and her parents in a small town in Pennsylvania in the 1900s.
17. SHIPPEN, KATHERINE B. *Andrew Carnegie and the Age of Steel.* New York: Random (1958), 183 pp. Grades 6–9. A Landmark book.
18. SPENCER, CORNELIA. *More Hands for Man: A Brief History of the Industrial Revolution.* New York: John Day (1960), 192 pp. Grades 6–9.
19. STRUIK, DERK J. *Yankee Science in the Making.* New York: Collier (1962), 544 pp. Grades 7–10.
20. WERSTEIN, IRVING. *The Great Struggle: Labor in America.* New York: Scribner (1965), 190 pp. Grades 6–9.
21. WERSTEIN, IRVING. *Turning Point for America: The Story of the Spanish-American War.* New York: Messner (1964), 191 pp. Grades 6–9.

    You may want to use the color filmstrips on Bell, Edison, and Wright from the Jam Handy Company.

    McGraw-Hill has two films on Theodore Roosevelt which may be useful to you. The Society for Visual Education has a filmstrip on "The Road to World Power and Responsibility: 1876–1900."

From World War I to the Present

From World War I to the present is another period of approximately fifty years. Tremendous changes have taken place in our country in that period, and teachers will need to think through carefully the points they want to stress with middle grade and junior high school pupils in that span of time.

If the "postholing" approach is used, the year 1933 or the decade of the 1930s may be the best years to choose. Certainly the Depression, the New Deal, and the era of Franklin D. Roosevelt represent another great watershed in the story of our country.

If the "decisions" approach is used in the junior high school grades, several topics can be stressed. They might include the following: (1) Can We Make the World Safe for Democracy? (2) What Do We Want for the United States in This Postwar Period? (3) Can We Preserve Democracy and Capitalism? (4) Can We Prevent the Spread of Totalitarianism? (5) Shall We Drop the Atomic Bomb? (6) How Should We Cope With Urbanization and Its Problems? and (7) What "Unfinished Business" Is There in Our Democracy?

World War I looms large on the international horizon in this span of half a century, but it need not be developed in detail with children. The 1920s also do not warrant a great deal of time in a quick survey of our national history.

But it does seem to this writer that the Depression and the changes which occurred during the New Deal days need to be emphasized. Economics can be stressed in the story of the Depression and changes in government and the attitudes of people toward government in the 1930s can be emphasized in treating the Roosevelt administration. Children can begin to understand such measures as Social Security, the T.V.A., and the insuring of bank deposits. The figure of Franklin Delano Roosevelt is an interesting one which can be highlighted here.

As in other treatments of our history, people should be kept in the center of the picture of this period. During the 1917–1970 period, immigration declined to a trickle, but the population of the nation increased greatly. Most noticeable have been the shifts of population to the West, the Southwest, and the Southeast, and the movement of people to the cities and the suburbs. With this has been another migration, that of southerners to the northern cities.

In some parts of our country, boys and girls will be especially interested in the movement of Puerto Ricans to the mainland in the 1950s, largely to do the hard work in our big cities. Later came the Cubans, after the take-over of that island by Castro. In more recent times there has been some movement of other Latin Americans, especially those from the Caribbean area, to the United States. Whether you go into any detail on special movements will depend upon your class.

With the shift of population to urban centers have come new problems, such as water supplies, transportation, air pollution, and intergroup conflicts. Pupils can begin to understand these and other problems as themes which affect their lives.

Industrialization also needs to be highlighted, for it was during these years that big business grew and supergiants became commonplace. The appearance of computers and automation also can be emphasized as becoming of prime importance in this period.

Equally striking have been the changes in transportation and communication. The century was ushered in with transportation by horses and boats. Soon automobiles and airplanes changed the means of transportation radically. In the field of communications, radio, the movies, and television have markedly changed the picture in every part of our land.

Probably the greatest change, however, has been in our involvement in international affairs. This might well serve as a subunit, tracing the story of our relations with other nations from World War I through World War II and on down to the present, with such additional themes as foreign aid, "The Cold War," and the United Nations.

This need not be done, however, if pupils have spent considerable time earlier in the year in a discussion of "Our Relations With Our World Neighbors" in a study of the U.S.A. Today.

The story of the rights of minorities, and the changes in the status of Negroes in particular, should form another part of the work on the period from World War I to the present.

Some time and attention should be given to the topic of leisure and recreation as an increasingly important aspect of life in our nation today.

In the study of this period, teachers may want to develop a time-line, encouraging pupils to decide upon the major events that should go on it. They might also prepare illustrations for each of the topics to go on such a time-line.

In the field of biography, who should be stressed? Certainly Woodrow Wilson and Franklin D. Roosevelt and possibly other Presidents since those two men. But what other persons? Perhaps Eleanor Roosevelt, Marian Anderson, John Glenn, and Ralph Bunche. Who else would you include from a long list of possibilities? Why would you stress them?

For pupils interested in science, there can be special reports on the invention of nylon and plastics, on changes in automobiles, and on atomic energy.

There is a wealth of music in this period, with special emphasis upon the development of symphony orchestras, the rise of musicals like *Oklahoma!* and *West Side Story*, and the tremendous increase of phonograph record purchasing.

There is plenty of material to consider. The big question is one of focus — selecting the few outstanding persons and movements to highlight.

A unique approach to depicting a period of history is used in the author's fifth grade textbook on *One Nation — The U.S.A.* (Ginn, 1972). In it the period from 1945 to the present is chronicled in a photo-essay so that pupils can survey some of the major events in that 25-year period quickly and can do some real discovery learning by using it, supplementing it when needed by encyclopedias and trade books.

Teachers may want to turn to the general histories of the United States written for adults in order to review the events of this half-century. In addition, they may want to consult such specialized accounts of the period as Frederick Lewis Allen *The Big Change: America Transforms Itself: 1900–1950*, his *Only Yesterday*, and his volume *Since Yesterday*. A good survey of the last twenty-five years in our history is John Brook *The Great Leap*. Individual volumes on other aspects of this period will not be difficult to find.

There are a few books on this period for boys and girls, although their number is not yet great. Among those available for middle grade and junior high school pupils are the following:

BOOKS FOR BOYS AND GIRLS ON THE PERIOD FROM WORLD WAR I TO THE PRESENT

World War I

1. CONGDON, DON (Editor). *Combat: World War I.* New York: Delacorte Press (1965), 448 pp. Grades 6–9.

2. LAWSON, DON. *The United States in World War I.* New York: Abelard-Schuman (1963), 157 pp. Grades 6–9.
3. LECKIE, ROBERT. *The Story of World War I.* New York: Random (1965), 190 pp. Grades 4–6. A Landmark book.
4. SELLMAN, ROGER R. *The First World War.* New York: Criterion (1962), 160 pp. Grades 7–9.
5. SNYDER, LOUIS L. *The First Book of World War I.* New York: Watts (1958), 96 pp. Grades 5–8.
6. WERSTEIN, IRVING. *The Many Faces of World War I.* New York: Messner (1963), 191 pp. Grades 6–9.

The 1920s and 1930s

1. BOARDMAN, FON W., JR. *The Thirties: America and the Great Depression.* New York: Walck (1967), 160 pp. Grades 7–9.
2. VAN DOREN, CHARLES. *Growing Up in the Great Depression.* New York: Hill and Wang (1963), 128 pp. An eleven-year-old boy and his thirteen-year-old sister and their parents in the Depression.
3. WERSTEIN, IRVING. *A Nation Fights Back: The Depression and Its Aftermath.* New York: Messner (1963), 191 pp. Grades 5–8.

Perhaps you will want to use a filmstrip on the Depression. The Society for Visual Education and Ealing are two companies which have produced such filmstrips.

World War II and the Korean War

1. ABERNETHY, ROBERT G. *Introduction to Tomorrow: The United States and the Wider World.* New York: Harcourt Brace Jovanovich (1966), 286 pp. Grades 7–9.
2. AMERICAN HERITAGE. *The Battle of the Bulge.* New York: Harper and Row (1969), 153 pp. Grades 7–10.
3. BLASSINGAME, WYATT. *Medical Corps Heroes of World War II.* New York: Random House (1969), 177 pp. Grades 6–9.
4. CARTER, HODDING. *The Commandos of World War II.* New York: Random House (1966), 192 pp. Grades 5–8.
5. COWAN, LORE. *Children of the Resistance.* New York: Meredith (1969), 179 pp. Grades 7–10.
6. LAWSON, DON. *The United States in World War II.* New York: Abelard-Schuman (1963), 224 pp. Grades 6–9.
7. LECKIE, ROBERT. *The Story of World War II.* New York: Random House (1963), 172 pp. Grades 6–9. A Landmark book.
8. MIERS, EARL S. and FELIX SUTTON. *America During Four Wars.* New York: Grosset and Dunlap (1965), 192 pp. Grades 5–8.
9. MIERS, EARL S. *Men of Valor: The Story of World War II.* Chicago: Rand McNally (1965), 256 pp. Grades 6–9.
10. REEDER, COLONEL RED. *The Story of the Second World War.* New York: Meredith (1969), 266 pp. Grades 7–10.
11. SAVAGE, KATHERINE. *The Story of the Second World War.* New York: Walck (1958), 271 pp. Grades 7–10.
12. SELLMAN, ROGER R. *The Second World War.* New York: Roy (1966), 111 pp. Grades 6–9.
13. SHEPARD, DAVID. *We Were There at the Battle of the Bulge.* New York: Grosset and Dunlap (1961), 181 pp. Grades 5–8.
14. SNYDER, LOUIS. *The First Book of World War II.* New York: Watts (1958), 94 pp. Grades 5–8.
15. SUTTON, FELIX. *We Were There at Pearl Harbor.* New York: Grosset and Dunlap. (1957), 177 pp. Grades 4–6.

16. TAYLOR, THEODORE. *Air Raid — Pearl Harbor! The Story of December 7, 1941.* New York: Crowell (1971), 185 pp. Grades 5–8.
17. WEYR, THOMAS. *World War II.* New York: Messner (1969), 224 pp. Grades 7–10.

Atoms and atomic energy

1. ADLER, IRVING. *Atomic Energy.* New York: John Day (1971), 48 pp. Grades 4–6.
2. ASIMOV, ISAAC. *Inside the Atom.* New York: Abelard-Schuman (1966), revised edition, 224 pp. Grades 6–8.
3. BARR, DONALD. *The How and Why Wonder Book of Atomic Energy.* Columbus, Ohio: Merrill (1961), 48 pp. Grades 5–8.
4. COLBY, CARROLL B. *Atoms at Work.* New York: Coward-McCann (1968), 48 pp. Grades 5–7.
5. GAINES, MATTHEW. *Atomic Energy.* New York: Grosset and Dunlap (1970), 128 pp. Grades 5–8.
6. HYDE, MARGARET O. *Atoms Today and Tomorrow.* New York: McGraw-Hill (1966), third edition, 141 pp. Grades 5–8.
7. KOHN, BEATRICE. *The Peaceful Atom.* New York: John Day (1963), 48 pp. Grades 4–6.
8. LARSON, EGON. *Atoms and Atomic Energy.* New York: John Day (1963), 48 pp. Grades 4–6.
9. MOORE, WILLIAM. *Atomic Pioneers.* New York: Putnam (1970), 160 pp. Grades 5–8.
10. MUNUES, JAMES. *We Were There at the Opening of the Atomic Era.* New York: Grosset and Dunlap (1960), 175 pp. Grades 5–8.
11. POLKING, KIRK. *Let's Go to an Atomic Energy Town.* New York: Putnam (1968), Grades 3–5.
12. ROWLAND, JOHN. *Atoms Work Like This.* New York: Roy (1966), 64 pp. Grades 7–10.
13. SILVERBERG, ROBERT. *Men Who Mastered the Atom.* New York: Putnam (1966), 193 pp. Grades 6–9.
14. WEBB, ROBERT N. *We Were There on the Nautilus.* New York: Grosset and Dunlap (1961), 178 pp. Grades 5–7.

Space

For books on space see pages 469–470.

49. A committee reports on per capita income using a bar graph they made

50. Using the Word Association device to learn what pupils know about China

Studying Selected Nations in Depth

51. Learning from charts prepared by commercial companies

52. Studying pictures to get acquainted with aspects of a nation

Studying Other Nations 21

By the time boys and girls have reached the sixth grade, they should be ready to study several carefully selected nations of the world. This is especially true if they have already explored in fifth grade the many facets of our country and have dealt with the complicated concept of a country.

If they have studied, in the primary grades, selected families of the world, and, in the middle grades, selected communities of the world, the task of analyzing and understanding other nations should be much easier.

The Concept of a Nation

To many adults the concept of a country or nation seems relatively simple. They have known about countries so long that they assume this is an easy concept to grasp. But that is not so. Children find it extremely difficult to comprehend what a nation or a country really is.

Perhaps that is because nations differ so markedly and seem to fit no common definition. Usually the land of any nation is contiguous. But that is not always so. Pakistan and the United States are two examples of nations which do not conform to that pattern. Nations often have a common language. But all nations are not so blessed. In fact many of them today have several tongues. Switzerland and India are two such examples. Sometimes there is a common religion (often a state religion). But there are only a few such nations. Even in a nation like Israel, not everyone is bound by a common faith.

What, then, makes a nation? Usually there is a common territory which is undivided. Often there is a common faith or religion. Often there is a common language. Even more important, there usually is a common past in which all the people take pride. This is not true of all the new nations, but it is true of most countries. This common history may be even a myth, but people may believe it. For example, Nkrumah named his nation "Ghana" in order that the people might have a common history dating back to the famous and advanced Ghana Empire, even though the country of Ghana does not lie exactly where the Ghana Empire lay. Often there are common ways of living, which are the ways of a common or similar culture. This helps people feel that they belong together. Then there is the feeling of facing the future together. This is an important psychological consideration in the building of nationhood.

Basically the concept of a nation is a political science concept but there are psychological considerations which are extremely important, as we have just tried to point out.

In international law a nation is recognized by other nations if it has three characteristics. It must control the territory over which it governs. It must control its own foreign affairs. And it must recognize the debts of any previous government. These three criteria should prove useful to teachers but not to children.

Louis Fischer, an international journalist, has written of nations or countries in *This Is Our World*, in these terms:

> A country is a complicated tapestry or rather a symphony of politics and poetry, economics and religion, law and language, tradition and imported isms, and of prejudice, pride, hope, hate, fear, faith, geography, natural resources, natural catastrophes, and many other things. One must not listen only to the brass and big drums, and although the conductor — the Prime Minister or the President — occupies the center of the stage, he would be nothing more than an acrobat without the orchestra or the composer. Moreover the people always come for the performance and consequently should always remain in focus.

This brief account catches some of the complicated nature of nations and perhaps helps us to understand why they are difficult to teach about to children without care and planning on the part of teachers.

Perhaps you have noticed that the writer has used the words "nation" and "country" interchangeably so far in this chapter. Actually, to be correct, the word "nation" should be used. But the two words are almost synonymous in common usage. Teachers should not fret about using the word "country."

Why Study Countries?

If nations or countries are so complicated, why should one study them early in the years in schools? There are many answers to that pertinent question, but the most important reply is that nations matter greatly in today's world. In some respects, the day of nations and nationalism is over in our interdependent world. Yet this is increasingly a world of nations and even of increasing nationalism. To understand the world today, one must cope with the concept of nations and their interactions, interdependence, cooperation, and conflicts. Alert pupils will certainly be confronted with the concept of nations almost every day. On the television screen they see stories about events in various countries. The headlines and pictures in daily newspapers and in magazines constantly refer to the nations of the world. Pupils hear adults frequently referring to nations. Boys and girls therefore need to know something about several countries.

Furthermore, children can understand their own nation better if they study other nations, as well as their own. One gains perspective by standing away from a picture and looking at it. Something of the same experience can come to boys and girls who leave their own country for a time and see how other nations have developed and how they work. As Ralph Linton stated in *The Study of Man*, "Those who know no other culture than their own, cannot know their own." Substitute the word nation for culture and the statement still holds true.

In the third place, it is extremely important to study other nations in order to help boys and girls to learn to live with the wide variety of human beings on our planet. In the study of other nations, the focus should be on the people of those places and how and why they live as they do. By learning how other people live and why they live as they do, children should develop knowledge and attitudes which will help them to live in our closely knit, interdependent world. This is certainly one of the major objectives of the social studies programs of elementary as well as of secondary schools today.

A fourth reason for studying other lands and peoples is to gain new ideas and new insights from them. There are Americans who feel that we do everything in our country better than people do anywhere in the world. But an objective look at other parts of the world soon will reveal that we excel in many respects but not in everything. No nation has a monopoly on creativity, organization, or social concern. Each nation does some things well and some things poorly. We should be able to learn from each other.

Margaret Mead, the anthropologist, illustrates this point by drawing a circle and then stating that no nation on earth represents this perfect shape. Each is "rounded" in some aspects of its national life.

Likewise, she maintains that some nations or cultures are sharp or pointed and almost impervious to new ideas. Others have been porous for centuries, taking in other ideas than their own and absorbing invaders.

These ideas might be represented in the following fashion:

Boys and girls need to learn something about the outstanding contributions to the world of each of the nations they study, whether those contributions have been in art and music, in architecture, in the sciences, or in social welfare.

In the fifth place, it is fun to learn about other nations and their inhabitants. Such studies can be interesting, intriguing, and enriching.

On the surface, other people may appear to conduct themselves in "crazy" ways. But children should learn that people have developed their ways of life for reasons. Often these have been geographical or climatic reasons. Sometimes they have been for purposes of defense. Often they have been rooted in beliefs or value systems in a given part of the world. Teachers need to explore with children WHY people do as they do, rather than dismissing their ways as strange merely because they differ from the way in which we do things in the United States.

Moreover, the study of other nations is essential in understanding current events. Today we need to understand Cuba, Egypt, and Vietnam. They occupy the headlines and the television screens. Yesterday it was Indonesia, India, and Iran. Tomorrow it may well be Cambodia, Chile, or Canada — or any number of other nations around our globe. Without some background on these places, current events will have little meaning to pupils — or older people.

For many children in today's world there is another reason for studying other nations. Many of today's children will be traveling to other nations some time in their lives. A few may be studying or working in them in the years to come. Now is the time to begin preparation for such experiences.

Can you add other reasons for studying nations?

WHICH NATIONS TO STUDY

If you are convinced that boys and girls should study other nations by at least the sixth grade, then the question obviously arises as to which countries they should study.

One answer is that they should study all of them. Each of them is important. Each is in the news now or will be in the future. Each has features which pupils should know about.

But such thinking leads in two directions. It means that too much time has to be devoted to the study of the nations of the world. Or it leads to a kind of hop-skip-jump approach. That means that one tries to study all or most of the 150 nations and territories of the world within an 180-day school year. Just how much can one learn about a nation in two or three days, or even in two or three weeks?

It seems to this writer infinitely better to study a few countries in depth rather than to study a great many in a superficial way.

A broad overview of an area or region of the world can be gained in a fairly short time and then one, two, or three nations in that region can be studied in some depth.

All this means, however, that there need to be some criteria for the selection of the nations to be examined by children. Here are a few suggestions on that point:

1. *World Powers.*   Certainly countries which are world powers today should be studied. But which nations are world powers? Certainly everyone will agree that the United States and the Soviet Union are. Perhaps mainland China should be added.

2. *Countries of the Future.*   It is difficult to determine which nations will play increasingly important roles in the future. But here are some which may well be more important tomorrow than they are today — Brazil, Canada, China, India, Indonesia, and possibly Nigeria.

3. *Neighboring Nations.*   Our nearest nations certainly should be given a high priority in any list of nations to be studied in elementary school. This is in part because they are neighbors. It is in part because some teachers and children can visit them. But it is also because Canada and Mexico are important nations in and of themselves. So those two nations should be given a good many points in any system of deciding which nations to study.

4. *Countries Which Represent Cultural Areas.*   Another approach is to take the eight major cultural regions of the world and be certain that at least one country from each of them is selected for study by elementary school boys and girls. In some of the areas there is one obvious choice. For example, the Soviet Union must represent the Slavic area. In other regions, it is more difficult to decide. For Africa, south of the Sahara, Kenya, Nigeria, or Tanzania seem likely choices. The Union of South Africa seems too difficult for boys and girls of this age to handle. It might better be left for pupils in the secondary school years. In Southeast Asia, Burma, Thailand, or Indonesia might be chosen. It matters very little which one is decided upon. Of course a class might do two countries for that region. In the Sinitic group, China and Japan should be studied. In the Anglo-Saxon group, the United Kingdom is the obvious choice. In the Latin group, it is very difficult to decide. France, Italy, Brazil,

and Argentina come to mind quickly. In the Germanic-Scandinavian region, certainly Germany should be considered first. It is almost impossible to select one nation in the Moslem area, but Egypt, Iran, and Turkey are three strong contenders. Teachers might be given their choice of one of these three. Obviously India should represent the Indic area.

5. *Countries of the Ancestors of the Pupils.*   A fifth criterion might well be the countries of the ancestors of the children in your class. If they are largely of English and German descent, those two countries should certainly be considered for your list. If a large proportion of your pupils come from Italy or Poland, then those places might be given high priority. The same would be true of other nations represented by a large majority of the class — or even by a significant minority.

6. *Countries Representing Emerging Nations.*   At least a couple of the nations finally selected for study should represent the many emerging nations of the Middle East, Africa, Asia, and Latin America.

7. *Countries of Our Western Heritage.*   Since so much of our heritage comes from the United Kingdom, that nation certainly should represent our Western heritage. But there are other nations that helped to form our nation, as we pointed out under the fifth criterion. They may be northern and western European nations or they may be eastern and southern European countries.

8. *Countries Against Which the Pupils Have Prejudice or Little Up-to-Date Information.*   Inasmuch as the purpose of education is to develop intelligent, informed persons, it might be well to discover the nations against which your pupils have the most prejudice and consider including them in your master list. The Soviet Union and Communist China are likely to be high on such a list. What others would you include?

9. *Countries Representing Different Forms of Government and Economy and Religions.*   Since there are various religions in the world and several forms of government and economies, some nations might be selected which represent this diversity. The countries which are finally selected on the basis of the foregoing criteria probably will include ones with different religions and different economic and governmental forms, but these points should not be overlooked. Obviously the Soviet Union and/or China will take care of communism. Any nation from the Moslem world can represent that major religion. India will provide for a brief introduction to Hinduism. Any of the Southeast Asia nations will provide for a study of Buddhism. Israel might well be considered for several reasons, one of them being Judaism.

10. *Countries on Which Adequate Materials Are Available.*   A very practical criterion is the availability of adequate materials for boys and girls. There is, however, material now on almost every nation in the world which you are likely to select. China is probably the one exception and, even on it, there are some relatively recent and satisfactory resources.

To encourage you to think about the important topic of selection, this section is provided for you to develop your own list, with any criteria added which you think are important to include.

1. World Powers

2. Countries of
   the Future

3. Neighboring
   Nations

4. Countries
   Representing
   Major Cultural
   Regions

5. Countries of
   Ancestors of
   Your Pupils

6. Countries
   Representing
   Emerging Nations

7. Countries
   Representing
   Our Western
   Heritage

8. Countries Against
   Which Pupils Have
   Strong Prejudices

9. Countries
   Representing
   Varieties of
   Religions,
   Economic and
   Governmental Forms

10. Countries for Which
    Adequate Materials
    Are Available

11.

12.

A Working List of Countries to Be Studied

Using the various criteria mentioned earlier in this chapter, here is a list of twenty countries to be considered. If the four choices are used, then the list is cut to sixteen.

1. Canada	9. Israel
2. The United Kingdom	10. India
3. France	11. Indonesia or Thailand
4. Germany	12. Japan
5. Italy or Greece	13. China
6. The U.S.S.R.	14. Mexico
7. Nigeria or Kenya	15. Brazil
8. The United Arab Republic (Egypt) or Turkey	16. Argentina

Of course there are other countries which you would like to include. Several come to mind at once. The Scandinavian countries are not represented here at all. Africa has only one, or at best, two representatives on the list. The countries of Latin America with large Indian populations are not included. One might like to include Peru or Guatemala. And there is no tiny nation on the list to represent the many small countries and their problems in living in a world of bigger powers.

However, the job is to limit rather than enlarge. How can this be done? There are several answers:

One is to devote two years to a study of the nations of the world, in grades five and six or six and seven.

Another way is to include the study of Mexico and Canada in the fifth grade year and eliminate them from the sixth grade list.

A third way is to allow teachers to select their own eight to ten nations, provided that there is at least one nation from each of the cultural areas of the world.

Closely related to this method is a fourth approach. That is to have a required list and a choice list. Teachers could concentrate solely on the required list and ignore the choice list if they so desired. On the required list would be perhaps eight nations.

Another way is to have a list of a few nations and not worry about the others, realizing that they can be studied in the secondary school, especially if there is a two-year course in world cultural areas.

A sixth approach would be to have all pupils study the one or two examples of an area and then have individuals and/or groups study one or two others, not necessarily reporting their findings at length to the entire class.

A seventh possibility is to do some limited studies of nations in conjunction with the study of current affairs or current events.

It is very easy for people to write about the importance of selectivity and depth today in the social studies but quite another thing to tackle the difficult and delicate task of making choices to carry out such a theory. How did you do in narrowing your choice? Was it easy or difficult?

In the D. C. Heath series of social studies textbooks for elementary schools, the choice for the sixth grade volume was limited to four nations. Those four are Egypt, Switzerland, India, and Brazil. How do you react to limiting the study of nations to only four? If you agree with that idea, what four nations would you select? Why would you choose those four? In what order would you want to study the four you have selected?

For the sixth grade book in the new Ginn social science series, written by Bani

Shorter and Nancy Starr, eleven nations were selected. They are Great Britain, Germany, the U.S.S.R., Egypt, Israel, Nigeria, India, China, Japan, Guatemala, and Brazil. Supplementary volumes are planned on France, Italy, Turkey, Kenya, and possibly two or three others, to make a flexible curriculum for that grade possible. How do you react to the idea of eleven nations for a year? How do you feel about the eleven nations chosen? If you had material available on the fifteen nations listed, how many nations would you teach? Why? Which of the fifteen would you select? Why?

A Basic Theme or Themes for Each Country

Each of the countries finally selected for study should then be examined from a variety of points of view. We shall discuss such foci in a little while. But it is highly recommended that there be one, two, or three major themes which are stressed for each nation. Such themes would be a little like the themes in a symphony, recurring in different forms. They might also be framed as questions or as problems for problem-solving learning. Here are some possible basic themes for nations you probably will decide to study:

Nation	Possible basic theme or themes
1. Canada	1. Achieving unity in a diversified country, ethnically and geographically.
2. France	2. Her many contributions to world culture.
3. Germany	3. A nation which developed "late," and some of the resulting problems.
4. Italy	4. Her many contributions to world culture.
5. The United Kingdom	5. The development of the idea of democracy and her long rule of a worldwide empire.
6. The U.S.S.R.	6. Her size and diversity; her new forms of government and economy, and the reasons for them.
7. Kenya	7. Building a multiracial society.
8. Israel	8. Judaism and pioneering in a new land; the wise use of human resources.
9. The United Arab Republic (Egypt)	9. The agelong and continuing struggle of sand or land versus water.
10. India	10. The large number of people and where they live; raising living standards for them.
11. Indonesia	11. Her geographical location and resources, and how to develop them.
12. China	12. What constitutes "the Chinese way of life" and how it is changing.
13. Japan	13. A crowded island and its industrialization.
14. Argentina	14. Agriculture and industry in a Latin American nation.
15. Brazil	15. Its size and potential as well as current problems.
16. Mexico	16. A rapidly developing nation economically.

WAYS OF STUDYING A NATION

In studying other countries around the world, the same general pattern can be used which we suggested for the study of the United States. For readers who are using only this chapter, we reproduce that pattern here. A less structured pattern is suggested on page 348.

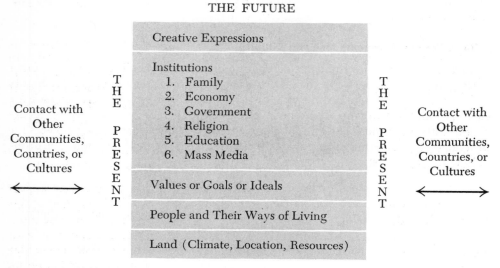

THE FUTURE

		Creative Expressions		
Contact with Other Communities, Countries, or Cultures ⟷	THE PRESENT	Institutions 1. Family 2. Economy 3. Government 4. Religion 5. Education 6. Mass Media Values or Goals or Ideals People and Their Ways of Living Land (Climate, Location, Resources)	THE PRESENT	Contact with Other Communities, Countries, or Cultures ⟷

THE PAST

In using the above chart, it is assumed that the emphasis will be upon the present, with some attention to the past. It is probably best to start with the land base, although in the study of some countries you may want to start with the number of people and where they live — and why they live there. This would be a good approach for India, for example. At some time in the study of every country, all or almost all of the topics suggested in the chart should be examined.

Let us look now at these various topics in relation to the study of other countries around the world.

1. The Land. There are many facets to the topic of land. Each of them helps to explain a different aspect of a nation. One approach is to see whether the country is landlocked, like Paraguay or Afghanistan, or whether it has good harbors and easy access to the oceans.

Another aspect of the geographic base is the type of soil in a nation. Is it a country with a highly productive river valley or several river valleys such as the United States and the Soviet Union have, or is it a nation without such advantages?

Climate and weather are extremely important, too, and help determine the ways of living of the people, their health standards, and sometimes their desire to work hard or to take life easier.

Consideration also should be given to the location of a country. For example, India's position in relation to China and the Soviet Union as well as to Pakistan explains much that India does in international relations. Canada's proximity to both

Based on a chart appearing in *Social Education*, April 1959, p. 161. By permission of the author and the publisher.

the United States and the Union of Soviet Socialist Republics is another example of the location of a country and its effect on a nation.

The resources of a nation and whether they have been developed explain much about a nation, too. Nations with oil in the Middle East, for example, are likely to have a different attitude toward the Western nations than those without oil. The copper in the Congo explains much of the interest of outsiders in that nation. The rice bowl of Southeast Asia is a factor in China's foreign policy. So one could continue to enumerate examples of the effect of resources.

2. *The People and Their Ways of Living.* There may be little diversity in some countries, like Sweden, but even there the Lapps are an important minority. In some nations, however, the diversity of people is of prime importance. How can one explain the Soviet Union or India without taking into account this important factor?

Many aspects of the ways of living of the people of a given country need to be examined. They include food, houses, clothes, transportation, and communication.

As in other places in this book, we want again to emphasize the word "why," here. Boys and girls should discover why people build their houses as they do, why they wear the clothes they wear, and why they have certain kinds of schools.

3. *Values.* Certainly the most difficult but, at the same time, the most important aspect of any group of people is their value system or their value systems. This topic will lead directly into the most simple explanations of religions. In some countries it will lead into a discussion of tribalism and the advantages and disadvantages of that way of life. In Nigeria, for example, there are many tribes. These play a very important role in the politics of that region, as has been evidenced in recent times by the hostility toward the enterprising Ibos of the Eastern Region and the consequent turmoil in that nation.

In every nation the religion or religions need to be examined, at least briefly. In the case of the Soviet Union, communism takes the place of a formal religion to a large extent.

4. *Institutions.* Then come the various institutions of a nation — its families, its economy, its government, its educational system, its mass media, and its religious institutions. Each of these will take some time to develop, especially where they are different from the institutions the pupils know.

There is no easy way to explain to children the differences in institutions in various countries. The basic approach is to try to get them to understand that people disagree on how to organize their lives. They develop different types of schools. They develop different types of government. They develop different religions. Only with such a frame of reference will pupils begin to understand the differences they discover.

Teachers should not try to "sell" their pupils on other kinds of institutions, but should present them as the ways of millions and millions of people. Ways which are different from ours should be presented as having strengths and weaknesses. For example, Buddhism, with its stress on contemplation and acceptance of life as it is, has made for happy people for the most part. But it has also kept millions of them from expecting or demanding changes until quite recently.

5. *Creative Expressions.* Every individual has some creativity in him or in her. Each person loves some types of beauty and produces some things which are beautiful. To the man in the desert, that terrain may be gloriously beautiful and the sky may be his television screen. No wonder the Arabs developed an early interest in

astronomy. To the jungle dweller, the thick forests may be filled with infinite beauty and mystery. Children need to understand that other people have a sense of beauty and awe and wonder, too. Perhaps they will understand this best by hearing the poetry and folk tales of other peoples. Perhaps they will sense this as they see pictures of some of the handicrafts of others. Perhaps they will understand this idea by a trip to a museum where they see the works of art of a given people. Children will grasp some of this when they figure out who wrote the various sayings which Miriam Elting has collected in her book *The Sun Is a Golden Earring*, a superb volume showing how people see the sky in different ways.

Teachers should help children to look for creativity in different forms — in the clothes of people, in their hairdos, in their paddles, in their games, in the music and dances, or in scores of other ways. In this connection, the booklets in the *"Fun and Festival in . . . ."* series of Friendship Press can be helpful. Similar in design are the books in the *Hi Neighbor* series of the United States Committee for UNICEF.

Another approach is to look at the national heroes of a country and discover what qualities the people of that nation saw and see in those persons.

6. The Future of Countries.   It will not be easy to discuss this topic with children, since no one of us knows what the future of any nation will be. But we can surmise from its present status and the direction in which it seems to be moving, or from the resources which it has, what a given nation may be like in the foreseeable future. Children can learn to make intelligent "guesstimates."

7. Contacts with Other Countries.   No nation is an island unto itself in today's world. Each country is dependent on others. Boys and girls need to learn this fact and to realize that it is not only a question of imports and exports, but of a whole host of factors. There are plenty of examples which can be found and discussed — whether it is the relationship of China and Japan, of Israel and her neighbors, or of India and her neighboring giants.

8. The History of a Country.   Although children will usually start with the present of any given nation to be studied, they should learn that they have to delve into its past in order to understand its present. Teachers need not stress the historical roots of any country, but they should certainly examine its past as it relates to the present. Teachers may want to stress the history of a few countries, such as Italy, Turkey, India, China, and the Soviet Union. Such nations seem to warrant more attention to their past than others.

Accentuating the Present

It has been assumed throughout this chapter that, in the study of countries, the emphasis will be on the present. That is because more interest is likely to be aroused in the minds of boys and girls at this stage in their development, if such studies concentrate on the contemporary scene.

It is also because children at this stage cannot grasp easily time concepts. However, the past should not be neglected, as we have pointed out.

Some Other Ways of Looking

Although the topics we have just mentioned seem to be the essential ones to think about and to use in examining any nation, there are many other approaches which can be used, often to advantage. Space precludes the mention of many of them, but here are a few which you might like to consider:

1. What are the forms of greeting and farewell? What do they mean? What are their significance?
2. What holidays does this country celebrate? Why? What does this tell us about this country?
3. In what areas of life does this country seem to excel? Why?
4. What percentage of the national budget is devoted to various items — such as education, defense, health, etc.? What does this tell us about the needs of this country? What does it tell us about the values of the people of this nation — or its rulers?

These are not easy questions, and you may not want to use them except with your more able students. But they are interesting ones to explore.

What Factors Help to Make a Nation?

Some classes of more mature pupils may even want to delve into the question of what makes a nation. You may want to touch lightly upon this topic even with other groups. If you have discussed with them how they "became" Americans, it will be of help in ascertaining how other children became Mexicans or Egyptians or Kenyans or Germans. Here are a few factors which seem to help build a feeling of nationalism among large groups of people. Perhaps you can think of other factors which should be added:

1. A common enemy.
2. A common history or myth of a common history.
3. A common religion.
4. Good transportation and communication.
5. A common language.

6. Common schools.
7. Common symbols, such as songs, flags, holidays, and heroes.
8. Common ways of living.
9. Common laws.
10. A shared or common government.

Now let us turn to some of the difficulties and dangers in teaching about other nations.

Some Difficulties and Dangers in Teaching about Other Countries

Every topic studied in schools is fraught with dangers. This is as true of studying about other countries as it is with other themes. The errors of commission and omission vary from individual to individual, but here are a few of the possible pitfalls in studying other countries:

1. The "Cute" Approach.   Some teachers are intrigued by the "local color" of a country. They think of The Netherlands as a land of wooden shoes, tulips, and windmills, or of Japan as a nation of kimonos, getas, and tea ceremonies. They may do this because they know little about such countries today, or because they consider this good material for art work and dramatics by pupils.

Actually we tend to teach about nations as they were many years ago instead of as they are today. For example, the windmills of The Netherlands are as archaic as our covered bridges in the United States. Windmills still exist, but they are now being placed in outdoor museums. Wooden shoes are rare, too, except where tourists want to take photographs to show back home. The tulips remain! In Japan one finds kimonos and getas, but they are not representative of the New Japan. The tea ceremony is important and beautiful, but not as common as many teachers would lead children to believe.

2. *Stereotyping Countries.*   A second danger is that of stereotyping the people of a nation. We tend to think and talk about *"the* French," *"the* Burmese," or *"the* Brazilians,"* as if all of the people thought and acted alike in those and in other countries. It is well for teachers to bear in mind the differences within a country as well as among countries.

3. *Measuring Nations with an American Yardstick.*   As readers and as travelers, many of us go around the world equipped with American yardsticks. We tend to judge others by our standards and our values. For example, we rate the people of various nations as to how closely they adhere to our standards of cleanliness, or we judge them by our fetish for punctuality.

When studying other nations, we ought to try to understand their values, and at least appreciate the areas of life in which they excel. For example, in many countries, the dance, music, and other art forms are much more important than in our land. We would do well to bear that in mind when studying other nations.

4. *The Transference of Loyalty.*   Occasionally one finds a teacher who teaches about another nation he or she has visited or lived in as if it were the only "civilized" place on earth. Such persons have transferred their loyalty from the United States to another nation, knowingly or not. They may wear the costumes of that country upon every conceivable occasion and revel in the attention they receive. Anglophiles and Francophiles are particularly noticeable in the United States, but this transference can be made to any nation. Teachers need to remain as objective as they can about any country they are studying with their pupils, discovering weaknesses as well as strengths in all of them.

The study of other nations should not diminish loyalty to the United States. One does not have to attack his own nation merely because he is trying to understand the nations of others.

5. *Hating Certain Countries.*   Occasionally one finds teachers who seem to have to hate certain countries. The countries they detest are likely to shift with the political winds of the times. In the 1930s and 1940s they probably hated Germany and Japan. In the 1950s it was the Soviet Union and possibly Egypt and Ghana. Today it may be China. As a result of their own dislike of a leader or a country, such teachers tell their pupils that everything is wrong with that particular country. The people are stupid. Everything they do is immoral.

Such instruction is not teaching. It is indoctrination. Teachers need not like everything that goes on in any nation on the globe, but, as social studies instructors, their job is to help children to see why people act in the ways they do. The social scientist does not have a double standard: one for his political friends and another for his political foes. He tries to view other lands and other peoples as objectively as possible, while still holding to his own values and ways of living.

6. *The Green, Brown, and White Countries Approach.*   For a long time many geographers and textbooks have divided the world into the green, brown, and white countries, or the wet, hot and cold, dry nations. This is an unfortunate division of the world inasmuch as countries do not fall so easily into those rigid categories.

7. *Ignoring Problems in Countries.* In their desire to present countries in the best light, some teachers ignore or minimize the problems inside nations. They gloss over difficulties or do not permit problems to be discussed. This should not be done. Pupils should learn that the people of all countries have problems with which they must deal, and that most countries are wrestling with their worst problems.

8. *Assuming That Knowledge Equals Understanding.* Occasionally teachers assume that if children know enough facts about a country, they will understand it and respect its people. In the section on attitudes in this book, we pointed out that this is not true. Knowledge does not lead necessarily to understanding or respect. Teachers need to bear in mind that boys and girls have to discover their own information to bring about change, and that children's emotions, as well as their minds, need to be educated.

Avoiding all of these obstacles will not be easy. None of us will be completely successful in hurdling the pitfalls mentioned above. Complete objectivity will escape us. Nevertheless, we should attempt to become amateur anthropologists and to train our pupils in the ways of investigating that anthropologists use. That means that we will try to help our pupils walk in the shoes of other people.

Now let us turn to a consideration of some of the methods and resources which are available for studying other countries with boys and girls. At the same time we shall be thinking about ways of strengthening the background of teachers for studying a few carefully selected nations in various parts of our globe. Hopefully you already have begun to think of ways in which you can gain additional background yourself. As you read books written for adults and/or pupils, view films and filmstrips, and talk with people from those countries, you may want to keep notebooks or folders on each country, arranged by the "construct" presented on page 430.

Launching the Study of a Nation

There are a good many ways in thich the study of any country may be started. Those ways depend upon a number of factors — you, your class, and the particular school situation in which you are teaching.

If you are team-teaching, this will place some limitations upon you as a planner, because you will have to go along with the group in planning certain common activities, such as the use of resource personnel, films, and filmstrips.

If you are teaching your own class, you will have much more freedom to decide how to launch a unit. It will help you if you obtain the permission of the principal, supervisor, or person in charge of the curriculum, to study the various countries in a different order from other teachers. Perhaps you will want to work out a master schedule so that each of you can use the resources of the library on one nation while others are working on other countries. Even in the best-equipped schools, a library is strained if several sections of a grade are using the materials at the same time.

Having decided upon the order, at least in a general way, at the beginning of the year, there are several ways you can proceed. Here are a few suggestions to be considered:

1. Plan an exhibit in your room of all the books and visual materials you can find. Let the pupils examine them, with ample time to do so. Discuss the materials

they have seen and encourage them to ask questions. In this way you will motivate an interest in the unit.

2. Show a good film or filmstrip or even two or three such visualizations. This
   will give your pupils a common background and probably will arouse their
   interest in a given country.
3. Use the "word association game" described earlier in the book, and see what
   the pupils already know and what attitudes they already have toward the
   country you plan to study.
4. Use current events to motivate interest in the country to be studied.
5. Use pictures in an opaque projector to launch your study of a nation.
6. Use an outside speaker to arouse the interest of pupils.
7. Utilize the letters of pen pals to launch your study.

At the beginning of the year you may want to present the list of countries to be
studied during the ensuing months and let the pupils decide which they want to
study first.

If you are just starting your teaching, you may not want to give your pupils any
choice in the order of the countries to be studied. It is highly possible that you need
to feel comfortable in those early months of teaching and so should select the country
which you feel you know most about. By doing this you will undoubtedly do a better
job in these initial weeks of teaching. Later on, you may feel free enough to involve
the pupils in some planning.

Some teachers are overly anxious to make a list of questions to which the pupils
will find answers. This is a good idea, but it should not be done until the interest of
the pupils is aroused and they actually do have questions for which they want to
find answers.

METHODS AND MATERIALS FOR STUDYING OTHER NATIONS

There are a great many ways in which you can help boys and girls to understand and respect the people of other nations. Most of the methods mentioned in the
section on methods in this book apply to the study of other countries. But there are a
few approaches which seem particularly desirable when studying other nations.
We will mention them in the pages that follow. As we do so, we will also include,
from time to time, a few resources which will help you as a teacher.

1. *People.*   In teaching about other countries, you should use persons from those
places wherever possible. You can have education about international understanding without such people, but you cannot have education in international understanding without people.

The people available include students from abroad in any nearby college or university. ("Student from abroad," incidentally, is much preferred to that of "foreign
student.") If you use such persons, try to include them in some school activity
besides speaking from a public platform. Your pupils need to become acquainted
with them as human beings off the stage and the visitors need to work with pupils
in classrooms inasmuch as they will be "experts" on American education when
they return to their homelands. Since they are here in the United States to study
and earn a degree, do not pressure them unduly to visit your school. If they do come,
try to arrange some gift for them for their time and effort.

There are often other people in a community who have lived or traveled abroad
and who can be used as resource personnel. Your school system should have a list of

such persons. If not, perhaps you can help prepare such a list. In larger communities, the American Association of University Women or the United Nations Association may have speakers you can use.

In looking for such resource persons, try to discover if they can communicate with children. Not all students from abroad or adults can!

2. Pictures. You certainly should use a great many pictures in the study of other countries. Some of them will be in books. But many of them should be loose pictures which are pasted on cardboard for use by pupils individually and in small groups, or through an opaque projector. You may even want to conduct a "drive" for back copies of *Look, Holiday, Life, Travel,* and the *National Geographic,* cutting out pictures from these magazines and mounting them for future use. A good library should have hundreds of such pictures, filed by topics or nations.

There are many sources of such pictures. Travel agencies can help you. The embassies and information offices of governments are a good source. Several specialized magazines like the *Aramco World, The Lamp,* and *The Arab World* will provide you with good pictures. The large, colored pictures in the UNESCO Geography Series, sold by Unipub in New York City, are especially recommended.

The Bruce Miller Company in Riverside, California sells three small booklets which are helpful in locating sources of pictures and posters. They are "Sources of Free Pictures," "So You Want to Start a Picture File," and "Sources of Free Travel Posters and Geographic Aids." For other ideas, turn to the section in this book on pictures (pages 101–102).

3. Maps. You should have at least one good map of a country as a start. Then other maps can be obtained or made by the pupils. Sometimes it is good to develop a series of overlay maps to use with a basic map of the topography of a nation. These can be made with oaktag as a base and the thin plastic material used to cover suits and dresses at the cleaners for the overlays, done with magic markers, or they can be made for use in overhead projectors. You also may want to develop maps on flannelboards with the pupils figuring out the object to be made. For other suggestions on the use of maps, turn to the section of this book on that topic, on pages 133–140.

4. Realia. Wherever possible, pupils should handle objects from the country they are studying. Such objects may be tools, hats or other forms of clothing, replicas of buildings, or related materials. Sometimes these can be seen and handled in a museum. Often they can be borrowed from a museum. A few such materials are contained in the kits of the International Communications Foundation (870 Monterey Pass Road, Monterey Park, California). Occasionally pupils can make replicas themselves and they can be saved for use by future classes, although the time involved does not always pay.

5. Time-Lines. Since boys and girls in the upper grades of elementary school are beginning to develop a sense of time, simple time-lines should be made and used by them. Often these time-lines can be stretched across the front, back, or side of the room on a string or wire, or posted on the space above the chalkboard as shown in the illustration on page 344 of a time-line of United States history. Time-lines are much better learning devices when they include illustrations done by pupils as is the case in the time-line on our own country just referred to. We cannot emphasize too much that such time-lines should be simple and should include only a very few dates for this age group.

6. Problem-solving. In the study of other countries, as in all aspects of the social studies, a great deal of use should be made of the problem-solving method. Pupils can be "dropped" into a country and then asked to list all the things they would need to know about that nation in order to live there. Or they can plan a trip to a country and list all the things they would need to do in order to make such a trip profitable to them. They can be told the per capita income of the people of a given country and then be asked to discover what the basic resources of that nation are and how they could be used in order to raise the standard of living of all the people. In working on that problem they will learn the importance of education and capital and managerial skills as well as the importance of foreign markets for goods. Another approach might be for them to discover what names of national heroes keep recurring in the books they read. Then they could try to figure out why these particular people are important to that nation — and even, in some cases, what these heroes reveal about the value system of that nation. Still another approach is to have the pupils discover how many people there are in a nation and then ask them to read about the topography of the country and figure out where the people probably would live — and why. If we are intent upon developing boys and girls who can think, then this method of problem-solving should be central in our teaching.

7. Role-playing. This is one of the very best methods to use in the study of any country, as it enables pupils to think in terms of the people of that nation as much as boys and girls of this age possibly can do. You may want to assign to pupils various roles and let them do their reading to discover how those persons might think. For example, they might be people in a village in Thailand or Indonesia. Some of them would be in favor of changes; others would be opposed. Let them figure out how they would feel about proposed changes in their lives. They can be made officials of the government and plan ahead for a few years, presenting their conclusions to the rest of the class as members of the country's parliament. But they should not always be confronted with problems. They can be made agents in a travel agency and try to figure out what they would like to show to visitors to their country. Often such role-playing can be combined with problem-solving, as is evidenced by the foregoing examples. There are few methods which will provoke such good learning as role-playing.

8. Films and Filmstrips. Of course films and filmstrips can be extremely helpful in helping pupils to "see" and "hear" the people in the country they are studying at the moment. Often these aids can create an atmosphere of a nation in a very short time. They also can take pupils to places they have never visited and give them a front seat view of history.

Unfortunately there are not any good, annotated lists of films and filmstrips on all the countries you will want to study. The Asia Society has a list of films for children on that part of the world. Occasionally *Intercom* magazine (Center for War/Peace Studies, 218 East 18th Street, New York, New York 10003) publishes lists of films and filmstrips. Some embassies and information offices produce lists of the films and filmstrips available from them. The best single listing is the *Educational Media Index*, published by McGraw-Hill, but the materials listed there unfortunately are not annotated. This is a field in which we sorely need help. Perhaps you are a person who should view a good many films and filmstrips on a given country and then write a commentary on them for an educational journal so that others may benefit from your experience. Such annotated lists need to include the reactions of pupils to the films and filmstrips which have been used.

Occasionally when using films, you may want to shut off the sound track and let the pupils see the pictures and make their own comments on them.

9. Action Projects.   We have already pointed out that people learn better when they are involved in action projects. This is as true in studying other countries as in other learnings. You and your pupils may want to participate in some action project on a given country. You may want to raise money for Books U.S.A. (Box 1960, Washington, D.C.), for the UNESCO Book Coupon Scheme (UNESCO Office, United Nations, New York), or some similar project. You may want to purchase one of the CARE packages to send abroad or to participate in a local Red Cross project involving some nation you are studying. Any of these undertakings should help to promote learning about a given nation.

If your students can be persuaded to earn the money for such projects, their experience will be a much more meaningful one than if they merely contribute money from their parents for such a project. Projects which involve long-term exchange are even better as they do not perpetuate the "Lady Bountiful" attitude.

10. Music and Dances.   We in the United States do not always think about music and dances in conjunction with the study of other nations. But these two forms of expression are high in the value systems of many countries. Pupils can listen to recordings or tapes or sing the songs of the nations they are studying, examining their music to see what it reveals about that country.

The same thing can be done with the dances of a people. If there is a physical education instructor who can help interpret the dances and teach the children a simple dance or two, this will enhance learning. If you can do this yourself, fine. If there is a parent or someone else in the community who can be used as a resource person, try to corral that person or persons.

Teachers should know about the small, inexpensive pocket songbooks of the Cooperative Recreation Service (Delaware, Ohio) on many nations. They should also have a current catalogue of Folkways Records (165 West 46th Street, New York, New York 10036) and catalogues of other record companies.

There are some books on the music of countries or regions, such as Betty Dietz's volume on *Musical Instruments of Africa* (John Day) and her *Folk Songs of China, Japan, and Korea* (John Day).

Some songs and dances are included in the various *Hi Neighbor* books already referred to, and in the *Fun and Festival* books which already have been mentioned in this chapter.

11. Games.   Playing some of the games played by children in other countries also can be a means of arousing interest in the people of other parts of the world. The best single source on games is Nina Millen's *Children's Games from Many Lands* (Friendship Press), a book which is revised every few years and is available as a paperback. Games also can be found in the *Fun and Festival* and *Hi Neighbor* volumes.

An organization known as World-Wide Games is located in Delaware, Ohio, and a catalogue of their materials should be in every school library. You also may want a copy yourself.

12. Current Events.   While you are studying a given nation, certainly you will want to alert the pupils to the newspapers, magazines, and television programs for current events on that country. This will help them to become interested in that nation and will assist them in understanding it.

13. *Flags.* A great deal of time need not be spent on the study of the flags of different nations, but some time can be spent on this. Sometimes one pupil can report on the flag as a part of the class study of a nation. Some countries have flags which are fascinating and revealing. For example, the flag of Surinam (the former Dutch Guiana) has five stars on a white background. Each star stands for one group of people in that nation. The red star is for the Amerindians, the yellow one for the Asians, the black one for the people of African descent, the brown one for the people of mixed ancestry, and the white one for the whites. "What is the meaning of the flag of the country we are studying?" is a question that can be posed with boys and girls from time to time.

14. *Pen Pals.* Some pupils are ready in the middle and upper grades of school to start corresponding with pupils in other countries. The number of persons they write to should be severely limited and encouragement and suggestions should be given to them to prolong their correspondence over a period of several months. The letters they receive sometimes can be used by a class in the study of a country.

There are several sources for names of pen pals. One of the largest and oldest organizations in the field is the International Friendship League, 40 Mount Vernon Street, Boston, Massachusetts.

15. *All Kinds of Reading.* This approach comes last in this list of methods for studying other countries, but it certainly is one of the most important ways to promote an interest in other nations and to gather information on other countries.

There are many kinds of reading which can be carried on. Novels often give readers the "feel" of a country much better than books of fact. Biographies and autobiographies are often helpful, although there are not too many of them about people in other countries for children in the middle and upper grades. Poetry, plays, and other forms of literature frequently can be used to advantage. Folk tales may appeal to some pupils; often they reveal a great deal about the expectations placed in front of children in a given place.

Pupils need to learn to look at the dates when books were printed. They need to be taught to use the indexes of books when they need information on only one topic. The ought to be helped with the study of pictures. There are many skills to be reinforced or taught as boys and girls increasingly use a wide variety of books.

Encyclopedias often have much valuable information, presented in interesting ways. However, the authors usually have squeezed a good deal of data into a small space. It is almost impossible for boys and girls, therefore, to condense what the authors have said. Therefore, "reports" based on condensations of articles in encyclopedias should not be encouraged.

It is impossible here to list all the books which can be used on the various countries you will probably study. We propose, therefore, to do three things: (1) to suggest certain helpful bibliographies, (2) to list books on thirteen nations which you may want to study, and (3) to list some series of books on nations.

Here, then, are a few bibliographies which may prove helpful to you:

1. "Books on Asia for Children." Asia Society. Single copy free to teachers.
2. McWhirter, Esther. "Books for Friendship." American Friends Service Committee and the Anti-Defamation League.
3. Kenworthy, Leonard S. "Studying Africa in Elementary and Secondary Schools," "Studying the Middle East in Elementary and Secondary Schools," "Studying South America in Elementary and Secondary Schools," and "Studying the U.S.S.R. in Elementary and Secondary Schools." All sold by the Teachers College Press, New York, New York. Include a stamped, self-addressed envelope.

Excellent, annotated reading lists on almost every nation have been compiled by the Information Center on Children's Cultures (331 East 38th Street, New York, New York 10016) and are available free upon request. This is a service of the United States Committee for UNICEF.

Series of Books and Booklets on Individual Countries

Below are two lists. The first is of various series of books and booklets for elementary school pupils on individual nations. The second list is of series of books for better readers.

1. Children's Press.	"Let's Travel to . . ." series.
2. Coward-McCann.	"Getting to Know . . ." and "Challenge" books.
3. The John Day Company.	"Let's Visit . . ." series and "World Neighbors" series.
4. Dodd, Mead.	"Young . . ." series.
5. Dutton.	"Young Traveler" series.
6. Fideler Press.	A series on individual nations.
7. Laidlaw.	"Understanding Your World" series.
8. Lippincott.	"Looking at . . ." series.
9. Nystrom.	Booklets with colored pictures, edited for children, from the "Around the World" series of the American Geographical Society and Nelson Doubleday.
10. Organization of American States.	Booklets on each of the Latin American nations.
11. Pantheon Press.	"My Village in . . ." series.
12. U.S. Committee for UNICEF.	"Hi Neighbor" books (four to six nations in each booklet).
13. Watts.	"First Book of . . ." series.

Most of the books and booklets listed below are for good readers in elementary schools. The Life series is intended for adults but the superb pictures can be used with elementary school pupils.

1. Foreign Policy Association.	Booklets on individual nations.
2. Ginn.	"Studies in Depth" series.
3. Laidlaw and the North Central Association.	Booklets on individual nations, stressing U.S. foreign policy.
4. Lippincott.	"Portraits of the Nations" series on over 100 nations.
5. Life magazine.	Beautifully illustrated volumes on many nations. Text for adults.
6. McCormick-Mathers.	Booklets on several nations.
7. Praeger.	Difficult reading but suitable for a few good readers.
8. World Publishing Company.	Books on several nations, written for junior and senior high school students.

Now let us turn to several nations specifically, with more extensive bibliographies on each of them, including a few references to films, filmstrips, and other resources.

A Checklist for Teachers in Studying Nations

As you prepare for the study of any country, you may want to refer to this checklist to determine whether you have explored every possible resource. Space is provided for three nations. You may want to make copies of this list and place them in your folders for various countries.

	Country of	Country of	Country of
1. People as resources			
2. Materials listed in this book on families of the world			
3. Materials listed in this book on communities of the world			
4. Maps of various kinds			
5. Films			
6. Filmstrips			
7. Records and tapes			
8. Songs and music			
9. Games			
10. Slides			
11. Material from embassies and information bureaus			
12. Special magazines on the country			
13. Several textbooks			
14. Several trade books			
15. Articles in current events newspapers for children			
16. Encyclopedias			

Gaining Background as a Teacher

If you feel that you need background on several countries or even on a few in order to teach "in depth," you have lots of company. Many teachers need more background than they have, or they need to "update" their information.

You may want to work with another teacher or several teachers, each of you doing considerable work on a small number of nations and then sharing your findings with others.

Or you may want to work alone.

No matter what you do, here are some suggestions which you may want to consider in order to gain more background on various nations of the world: Space is provided for you to check yourself on these approaches.

	Am doing this now or have background	Need to tackle
1. Read a book like Vera M. Dean's paperback on *The Nature of the Non-Western World* (Mentor) or Andrew Boyd's *An Atlas of World Affairs* (Praeger).		
2. Read a good newspaper and clip articles on nations. *The New York Times* and the *Christian Science Monitor* are highly recommended.		
3. Read two or more current events magazines, preferably with different points of view. If you read *Time*, *Newsweek*, or *U.S. News and World Report*, read also *The Progressive*, *The New Republic*, or *The Nation*.		
4. Join some organization interested in world affairs, such as the local Foreign Policy Association or the United Nations Association.		
5. Subscribe to such publications as the *Headline* books of the Foreign Policy Association, *Current History*, or *Focus* (The American Geographical Society).		
6. Send for free and inexpensive materials. See items in the Peabody College *Free and Inexpensive Learning Materials* booklet or the Teachers College Press *Free and Inexpensive Materials on World Affairs*.		
7. Start folders on individual nations, filing articles, notes on books, pictures, etc.		
8. Make a survey of children's trade books about the country.		
9. Take a background course in a nearby college.		
10. Outline for yourself a short reading program about the country.		
11. Arrange to meet two or three people from the nation you are studying. Contact them through a nearby college or a local organization.		
12. Preview films and/or filmstrips about the country being studied.		
13. Start a pen pal correspondence with a person in the country you are studying.		
14. Plan to travel to that country in the foreseeable future if possible.		
15. Make a survey of resource units about that country.		
16. Compare textbooks about the country you are studying.		

17. Make a list of the major concepts I want to stress in studying one (or more) nations.

18. Make a list of the generalizations I want to stress on one (or more) nations.

BIBLIOGRAPHY ON SELECTED NATIONS

Brazil

A small and inexpensive packet of materials on Brazil may be purchased from the Pan American Union, Washington, D.C. 20006. Additional material may be obtained free from the Brazilian Government Trade Bureau, 551 Fifth Avenue, New York, New York 10017.

Books

1. BOWEN, DAVID. *Hello Brazil.* New York: Norton (1967), 118 pp. Grades 5–8.
2. BREETVELD, JIM. *Getting to Know Brazil.* New York: Coward-McCann (1960), 64 pp. Grades 4–6.
3. BROWN, ROSE. *The Land and People of Brazil.* Philadelphia: Lippincott (1960), 128 pp. Grades 6–9. In the Portraits of Nations series.
4. CALDWELL, JOHN C. *Let's Visit Brazil.* New York: John Day (1961), 96 pp. Grades 5–8.
5. CALDWELL, JOHN C. and ELSIE F. *Our Neighbors in Brazil.* New York: John Day (1962), 47 pp. Grades 3–5.
6. CAVANNA, BETTY. *Paulo of Brazil.* New York: Watts (1962), 63 pp. Grades 3–5.
7. COVERLEY-PRICE, VICTOR. *Rivers of the World — The Amazon.* New York: Oxford University Press (1960), 32 pp. Grades 5–8.
8. FORMAN, LEONA S. *Bico: A Brazilian Raft Fisherman's Son.* New York: Lothrop, Lee and Shepard (1969), 96 pp. Grades 3–5.
9. GARTLER, MARION and GEORGE C. HALL. *Understanding Brazil.* River Forest, Illinois: Laidlaw (1962), 64 pp. Grades 5–7.
10. GIDAL, SONIA and TIM. *My Village in Brazil.* New York: Pantheon (1968), 80 pp. Grades 5–7.
11. HALL, FREDERICK. *Land of Coffee.* Chicago: Encyclopaedia Britannica (1964), 36 pp. Grades 3–5. With colored illustrations.
12. HERMANNS, RALPH. *River Boy: Adventure on the Amazon.* Chicago: Follett (1965), 44 pp. Grades 3–5.
13. JOY, CHARLES R. *Getting to Know the River Amazon.* New York: Coward-McCann (1963), 64 pp. Grades 4–6.
14. MANNING, JACK. *Young Brazil.* New York: Dodd, Mead (1970), 64 pp. Grades 4–6.
15. MAY, STELLA B. *Brazil.* Grand Rapids, Michigan: Fideler (1966), 144 pp. Grades 5–8. With large black and white photographs.
16. SHEPPARD, SALLY. *The First Book of Brazil.* New York: Watts (1962), 83 pp. Grades 4–6.
17. SPERRY, ARMSTRONG. *Amazon: River Sea of Brazil.* Champaign, Illinois: Garrard (1961), 96 pp. Grades 4–6.
18. SYME, RONALD. *The Man Who Discovered the Amazon.* New York: Morrow (1960), 192 pp. Grades 6–9.
19. WEBB, KEMPTON E. *Brazil.* Lexington, Massachusetts: Ginn (1964), 120 pp. For better readers.

Pictures, Films, and Filmstrips

The Fideler Company has a set of pictures for sale on Brazil.

For films and filmstrips see the *Educational Media Index* or the catalogues of various publishers.

Canada

Teachers' kits and student kits are available free of charge from the Canadian Consulate General, 680 Fifth Avenue, New York, New York 10019 or from the nearest consulate.

Books

1. AMERICAN GEOGRAPHICAL SOCIETY. *Canada.* Garden City, New York: Doubleday (1968), 64 pp. With colored photographs.
2. BOONER, MARY GRAHAM. *Made in Canada.* New York: Knopf (1966), 128 pp. Grades 5–7. On arts and crafts.
3. BRAITHWAITE, MAX. *Land, Water and People: The Story of Canada's Growth.* Princeton, New Jersey: Van Nostrand (1962), 287 pp. Grades 6–9.
4. *Canada.* Columbus, Ohio: Merrill (1961), 32 pp. Grades 4–6.
5. EISENBERG, LARRY. *Fun and Festival from the United States and Canada.* New York: Friendship Press (1956), 48 pp.
6. FIELD, JOHN L. and LLOYD A. DENNIS. *Land of Promise — The Story of Early Canada.* New York: Abelard-Schuman (1963), 265 pp. Grades 5–8.
7. HAIG-BROWN, RODERICK. *The Whole People.* New York: Morrow (1963), 256 pp. Grades 5–8. Canadian Indians.
8. HARE, KENNETH. *Canada.* Chicago: Nystrom (1964), 63 pp. Grades 5–8. With a teacher's guide.
9. HILLS, THEODORE L. and SARAH JANE. *Canada.* Grand Rapids, Michigan: Fideler (1965), 160 pp. Grades 6–9. Large black and white photographs.
10. HOLBROOK, SABRA. *Aluminum from Water: Challenge of Canada's River Power.* New York: Coward-McCann (1960), 128 pp. Grades 5–8.
11. JUDSON, CLARA I. *St. Lawrence Seaway.* Chicago: Follett (1963), 160 pp. Grades 5–8.
12. LAUBER, PATRICIA. *Changing the Face of North America: Challenge of the St. Lawrence Seaway.* New York: Coward-McCann (1959), 128 pp. Grades 5–8.
13. LEITCH, ADELAIDE. *Canada: Young Giant of the North.* New York: Nelson (1964), 223 pp. Grades 6–9.
14. LINEAWATER, CHARLES. *Canada.* New York: Watts (1967), 81 pp. Grades 5–8.
15. MOORE, BRIAN. *Canada.* New York: Time-Life Books (1968), 160 pp. Text for adults but beautiful illustrations can be used to advantage by pupils.
16. PECK, ANNE M. *The Pageant of Canadian History.* New York: McKay (1963), 386 pp. Grades 6–9.
17. ROSS, FRANCES A. *Land and People of Canada.* Philadelphia: Lippincott (1960), 128 pp. Grades 6–9.
18. WOOD, DOROTHY. *Canada.* Chicago: Children's Press (1964), 93 pp. Grades 5–8.

Films and Filmstrips

For free films, contact the Canadian Travel Film Library, 680 Fifth Avenue, New York, New York 10019. Several companies have films and filmstrips on Canada, including a set of films from Bailey on various aspects of that nation.

China

Teachers should be on the alert for new materials on China and examine them with great care as it is difficult at this point in history to obtain objective books on the contemporary scene.

Books

1. AMERICAN GEOGRAPHICAL SOCIETY. *Taiwan.* New York: American Geographical Society (1968), 64 pp. Colored illustrations.
2. BRYAN, DERECK. *China.* New York: Macmillan (1965), 96 pp. Grades 5–8.
3. DIETZ, BETTY and THOMAS CHOONBAI PARK. *Folk Songs of China, Japan and Korea.* New York: John Day (1964), 47 pp.

4. FESSLER, OREN and EDITORS of LIFE. *China.* New York: Life and Time (1963), 176 pp. Text for adults, but the splendid color photographs can be used with pupils.
5. GEIS, DARLENE. *Let's Travel in China.* Chicago: Children's Press (1965), 86 pp. Grades 6–8.
6. HOFF, RHODA. *China: Adventures in Eyewitness History.* New York: Walck (1965), 172 pp. Grades 6–9.
7. JOY, CHARLES R. *Getting to Know the Two Chinas.* New York: Coward-McCann (1960), 64 pp. Grades 4–6.
8. KIMMON, WILLIAM. *The First Book of Communist China.* New York: Watts (1962), 85 pp. Grades 5–7.
9. PINE, TILLIE S. and JOSEPH LEVINE. *The Chinese Knew.* New York: McGraw-Hill (1958). Grades 3–5. Inventions of the Chinese.
10. RAU, MARGARET. *The Yangtze River.* New York: Messner (1969), 96 pp. Grades 6–9.
11. RAU, MARGARET. *The Yellow River.* New York: Messner (1969), 96 pp. Grades 5–7.
12. SEEGER, ELIZABETH. *Pageant of Chinese History.* New York: McKay (1962), 414 pp. Grades 6–9. Revised edition.
13. SPENCER, CORNELIA. *Ancient China.* New York: John Day (1964), 107 pp. Grades 6–9.
14. SPENCER, CORNELIA. *China's Leaders in Ideas and Actions.* Philadelphia: Macrae Smith (1965), 192 pp. Grades 6–9. Twelve persons from Confucius to Chou En-Lai.
15. SPENCER, CORNELIA. *Made in China.* New York: Knopf (1966), 288 pp. Grades 7–10. On arts and crafts. A revised edition.
16. SPENCER, CORNELIA. *The Yangtze: China's River Highway.* Champaign, Illinois: Garrard (1963), 96 pp. Grades 6–9.
17. SWISHER, EARL. *China.* Lexington, Massachusetts: Ginn (1964), 122 pp. Grades 7–10.
18. WALKER, RICHARD L. *Ancient China and Its Influence in Modern Times.* New York: Watts (1969), 86 pp. Grades 5–7.
19. WIENS, HAROLD J. *China.* Grand Rapids, Michigan: Fideler (1965), 192 pp. Grades 5–8.

Pictures, Films and Filmstrips

The Fideler Press has a set of pictures on China for sale. For films and filmstrips consult the *Educational Media Index* or catalogues of publishers.

France

Some free literature is available from the French Embassy — Cultural Services, 972 Fifth Avenue, New York, New York 10021.

Books

1. BISHOP, CLAIRE H. *Here Is France.* New York: Farrar, Straus and Giroux (1969), 215 pp.
2. BRAGDON, LILLIAN J. *The Land and People of France.* Philadelphia: Lippincott (1960), 128 pp. Grades 6–9.
3. BROGAN, DENIS W. *France.* New York: Time and Life (1963), 176 pp. Text is for adults but the pictures can be used with boys and girls.
4. CHURCH, R. J. HARRISON. *Looking at France.* Philadelphia: Lippincott (1969), 64 pp. Grades 5–8.
5. DOLCH, EDWARD. *Stories from France.* Champaign, Illinois: Garrard (1963), 165 pp. Grades 4–6. Folk tales.
6. *France.* Grand Rapids, Michigan: Fideler (1956), 128 pp. Grades 5–8.
7. *France.* Columbus, Ohio: Merrill (1962), 32 pp. Grades 4–6.
8. GEIS, DARLENE (Editor). *Let's Travel in France.* Chicago: Children's Press (1960), 86 pp. Grades 5–8.
9. GIDAL, SONIA and TIM. *My Village in France.* New York: Pantheon (1965), 78 pp. Grades 4–6.

10. GRANT, ROLAND. *French Folk and Fairy Tales.* New York: Putnam (1963), 192 pp. Grades 4–7.
11. HARRIS, LEON A. *Young France.* New York: Dodd, Mead (1964), 64 pp. Grades 5–7. Well illustrated.
12. KEATING, KATE and BERN. *A Young American Looks at France.* New York: Putnam (1963), 120 pp. Grades 4–6.
13. LAUBER, PATRICIA. *Highway to Adventure: The River Rhone of France.* New York: Coward-McCann (1956), 96 pp. Grades 5–8.
14. NEWMAN, BERNARD and JOHN C. CALDWELL. *Let's Visit France.* New York: John Day (1967), 94 pp. Grades 5–7.
15. PASTORE, ARTHUR B. *Dynamite Under the Alps: The Challenge of the Mont Blanc Tunnel.* New York: Coward-McCann (1963), 128 pp. Grades 5–8. A thrilling story of the tunnel between France and Switzerland.
16. SMITH, IRENE. *Paris.* Chicago: Rand McNally (1961), 128 pp. Grades 6–9.
17. WALLACE, JOHN A. *Getting to Know France.* New York: Coward-McCann (1962), 69 pp. Grades 4–6.
18. WEISS, H. *A Week in Daniel's World: France.* New York: Collier-Macmillan (1969), 48 pp. Grades 4–6.
19. WILSON, HAZEL. *The Seine: River of Paris.* Champaign, Illinois: Garrard (1961), 96 pp. Grades 5–8.

Pictures, Films, and Filmstrips

The Fideler Press has a series of pictures on France. Silver Burdett has a set of colored pictures of a family in France.

"Chansons de Notre Chalet" is the title of a small songbook sold by the Cooperative Recreation Service in Delaware, Ohio. Large colored pictures on France are available in the UNESCO Geography Series from Unipub.

Germany

Some help may be derived from the German Information Center, 410 Park Avenue, New York, New York 10022. Also contact the Carl Schurz Memorial Association, 420 Chestnut Street, Philadelphia, Pennsylvania 19106.

Books

1. Editors of Sterling. *Berlin: East and West in Pictures.* New York: Sterling (1970), 64 pp.
2. GIDAL, SONIA and TIM. *My Village in Germany.* New York: Pantheon (1964), 78 pp. Grades 4–6.
3. HOLBROOK, SABRA. *Capital Without a Country.* New York: Coward-McCann (1960), 128 pp. Grades 5–8.
4. HOLBROOK, SABRA. *Germany: East and West.* New York: Meredith (1968), 241 pp. Grades 6–9.
5. KNIGHT, DAVID C. *The First Book of Berlin: Tale of a Divided City.* New York: Watts (1967), 96 pp. Grades 5–7.
6. Life and Time. *Germany.* New York: Time (1962), 176 pp. Text for adults but the pictures can be used to advantage with pupils.
7. MICHAEL, MAURICE and PAMELA. *German Folk and Fairy Tales.* New York: Putnam (1963), 190 pp. Grades 4–6.
8. MOORE, JAMES and JOHN C. CALDWELL. *Let's Visit Germany.* New York: John Day (1970), 95 pp. Grades 4–6.
9. NORRIS, GRACE. *Young Germany: Young Germans at Work and at Play.* New York: Dodd, Mead (1969), 64 pp. Grades 4–6.
10. POUNDS, NORMAN (Editor). *Europe: With Focus on Germany.* Grand Rapids, Michigan: Fideler (1965), 340 pp. Grades 6–9.
11. SAVAGE, KATHERINE. *People and Power: The Story of Three Nations.* New York: Walck (1959), 250 pp. Grades 6–9.

12. SEGAR, GERHART. *Germany*. Grand Rapids, Michigan: Fideler (1959), 160 pp. Grades 5–7. Many black and white photographs.
13. SHIRER, WILLIAM L. *The Rise and Fall of Adolph Hitler*. New York: Random House (1961), 181 pp. Grades 6–9. A children's edition of a famous book.
14. WILSON, BARBARA K. *Fairy Tales of Germany*. New York: Dutton (1960), 48 pp. Grades 5–8.
15. WOHLRABE, RAYMOND and W. KRUSCH. *The Land and People of Germany*. Philadelphia: Lippincott (1957), 128 pp. Grades 6–9.

Films and Filmstrips

1. "Germany: People of the Industrial Ruhr." Encyclopaedia Britannica, color, 17 minutes.
2. "Living in East Germany Today." Society for Visual Education, color, 58 frames.
3. "Modern West Germany." Society for Visual Education, color, 64 frames.
4. "Western Germany: The Land and the People." Coronet, color or black and white, 11 minutes.
5. Jam Handy has a set of six color filmstrips on German history, land and farming, transportation and industry, Berlin, city life, and East Germany.

India

Some help may be obtained from the India Information Services, 2107 Massachusetts Avenue, N.W., Washington, D.C., 20008.

Books

1. BOTHWELL, JEAN. *The Animal World of India*. New York: Watts (1961), 202 pp. Grades 3–5.
2. BOTHWELL, JEAN. *Cobras, Cows and Courage: The Challenge of India's Plains*. New York: Coward-McCann (1956), 96 pp. Grades 5–8.
3. BROWN, JOE DAVID. *India*. New York: Time (1961), 160 pp. Text is too difficult for pupils, but the pictures are unique.
4. BRYCE, WINIFRED. *India: Land of Rivers*. Camden, New Jersey: Nelson (1966), 224 pp. Grades 6–9.
5. CALDWELL, JOHN C. *Let's Visit India*. New York: John Day (1960), 96 pp. Grades 4–6.
6. CALDWELL, JOHN C. and ELSIE F. *Our Neighbors in India*. New York: John Day (1960), 48 pp. Grades 3–5.
7. FERSCH, SEYMOUR. *The Story of India*. Wichita, Kansas: McCormick-Mathers (1965), 188 pp. Grades 6–9.
8. FITCH, FLORENCE MARY. *Their Search for God: Ways of Worship in the Orient*. New York: Lothrop (1950), 144 pp. Grades 5–8.
9. GIDAL, SONIA and TIM. *My Village in India*. New York: Pantheon (1958), 75 pp. Grades 4–6. Fine photographs in black and white.
10. HAMPDEN, JOHN. *A Picture History of India*. New York: Watts (1966), 64 pp. Grades 4–6.
11. *India*. Columbus, Ohio: Merrill (1961), 32 pp. Grades 4–6.
12. JOY, CHARLES R. *Taming Asia's Indus River: The Challenge of the Desert Drought and Flood*. New York: Coward-McCann (1964), 119 pp. Grades 6–9.
13. KINGSBURY, ROBERT C. *India*. Chicago: Nystrom (1964), 64 pp. Grades 6–9. Includes a teacher's edition.
14. LASCHEVER, BARNETT D. *Getting to Know India*. New York: Coward-McCann (1960), 64 pp. Grades 4–7.
15. McDOWELL, BART. "Orissa: Past and Promise in an Indian State." *National Geographic*. October, 1970. Pp. 546–578. With colored illustrations.

16. MODAK, MANORAMA R. *Land and People of India.* Philadelphia: Lippincott (1963), 132 pp. Grades 6–9.
17. POLK, EMILY. *Delhi: Old and New.* Chicago: Rand McNally (1963), 144 pp. Grades 6–9.
18. SHORTER, BANI. *India's Children.* New York: Viking (1960), 175 pp. Grades 5–7. Extremely well written accounts of 12 children in India.
19. SONI, WEETHY. *Getting to Know the River Ganges.* New York: Coward-McCann (1964), 64 pp. Grades 5–8.
20. THAMPI, PARVATHI. *Geeta and the Village School.* Garden City, New York: Doubleday (1960), 64 pp. Grades 3–5.
21. TURNBALL, LUCIA. *Fairy Tales of India.* New York: Criterion (1959), 175 pp. Grades 5–7.
22. U.S. COMMITTEE FOR UNICEF. *Hi Neighbor — Book Four.* New York: Hastings (1961), 64 pp.
23. ZINKIN, TAYA. *India and Her Neighbors.* New York: Watts (1968), 96 pp. Grades 5–8.

Israel

Considerable material is available from the Israel Office of Information, 11 East 70th Street, New York, New York 10021, free of charge.

Books
1. American Geographical Society. *Israel.* New York: American Geographical Society (1970), 64 pp. Text for adults but colored illustrations for pupils.
2. BAKER, RACHEL. *Chaim Weizmann: Builder of a Nation.* New York: Messner (1950), 180 pp. Grades 6–9. Old, but the only account of this famous man for younger pupils.
3. COMAY, JEAN and MOSHE PEARLMAN. *Israel.* New York: Macmillan (1964), 120 pp. Grades 6–9.
4. EDELMAN, LILY. *Israel: New People in an Old Land.* Camden, New Jersey: Nelson (1969), 223 pp. Revised edition. Grades 6–9.
5. GIDAL, SONIA and TIM. *My Village in Israel.* New York: Pantheon (1959), 76 pp. Grades 4–6. Includes many fine photographs.
6. GILLSATER, SVEN and PIA. *Pia's Journey to the Holy Land.* New York: Harcourt Brace Jovanovich (1961), unpaged. K–3. Many color photographs of a journey by a photographer and his young daughter.
7. HAMORI, LASSLO. *Flight to the Promised Land.* New York: Harcourt Brace Jovanovich (1963), 189 pp. Grades 6–9. A boy who emigrates from Yemen to Israel.
8. HOFFMAN, GAIL. *Land and People of Israel.* Philadelphia: Lippincott (1955), 124 pp. Grades 6–9.
9. HOLISHER, DESIDER. *Growing Up in Israel.* New York: Viking (1963), 180 pp. Grades 6–9.
10. JOY, CHARLES R. *Getting to Know Israel.* New York: Coward-McCann (1961), 64 pp. Grades 4–7.
11. KUBIE, NORA B. *Israel.* New York: Watts (1968), 96 pp. Grades 6–9.
12. LOTAN, JOEL (Editor). *A Kibbutz Adventure.* New York: Warne (1963), 64 pp. Grades 4–6.
13. MEEKER, ODEN. *Israel Reborn.* New York: Scribner (1964), 192 pp. Grades 6–9.
14. PINNEY, ROY. *Young Israel.* New York: Dodd, Mead (1963), 64 pp. Grades 4–6. Pictures and text of children at work and at play.
15. RABINOWICZ, RACHEL A. *The Land and People of Israel.* New York: Macmillan (1959), 96 pp. Grades 6–9.
16. ST. JOHN, ROBERT. *Builder of Israel: The Story of Ben-Gurion.* Garden City, New York: Doubleday (1961), 185 pp. Grades 5–8.

17. SAMUELS, GERTRUDE. *Ben-Gurion: Fighter of Goliaths: The Story of David Ben-Gurion.* New York: Crowell (1961), 275 pp. Grades 6–9.
18. SASEK, MIROSLAV. *This Is Israel.* New York: Macmillan (1962), 60 pp. Grades 2–4.
19. SHAMIR, MOSHE. *Great Day in Israel: Why Ziva Cried on the Feast of the First Fruits.* New York: Abelard-Schuman (1960), unpaged. Grades 2–4.
20. TOR, REGINA. *Discovering Israel.* New York: Random (1960), 64 pp. Grades 4–6.

Films and Filmstrips

Several films and filmstrips are available on loan from the Israel Office of Information (see address at the beginning of the bibliography on Israel). For other films and filmstrips see the listings in *Educational Media Index* or catalogues of various companies.

Italy

Some free material may be obtained from the Italian Embassy — Cultural Division, 686 Park Avenue, New York, New York 10021.

Books

1. BARTLETT, VERNON and JOHN C. CALDWELL. *Let's Visit Italy.* New York: John Day (1968), 95 pp. Grades 5–8.
2. CRAZ, AL. *Getting to Know Italy.* New York: Coward-McCann (1961), 64 pp. Grades 5–6.
3. DUGGAN, ALFRED. *The Romans.* New York: World (1964), 125 pp. Better readers.
4. EPSTEIN, SAM and BERYL. *First Book of Italy.* New York: Watts (1958), 68 pp. Grades 4–6.
5. *Italy.* Columbus, Ohio: Merrill (1961), 32 pp. Grades 4–6.
6. JASHEMSKI, WILHELMINA F. *Letters from Pompeii.* Lexington, Massachusetts: Ginn (1963), 155 pp. Grades 6–8.
7. KEATING, KATE and BERN. *A Young American Looks at Italy.* New York: Putnam (1963), 126 pp. Grades 6–9.
8. Life. *Italy.* New York: Life-Time (1964), 160 pp. Text for adults but wonderful pictures can be used with pupils.
9. MARTIN, RUPERT C. *Looking at Italy.* Philadelphia: Lippincott (1966), 64 pp. With colored photographs. Grades 5–7.
10. NOBLE, IRIS. *Leonardo da Vinci.* New York: Norton (1965), 222 pp. Grades 6–9.
11. SAMACHSON, DOROTHY and JOSEPH. *Rome.* Chicago: Rand McNally (1964), 152 pp. Grades 6–9.
12. TOOR, FRANCES. *Made in Italy.* New York: Knopf (1966), 224 pp. Grades 6–9.
13. VANCE, MARGUERITE. *Dark Eminence: Catherine de Medici and Her Children.* New York: Dutton (1961), 160 pp. Grades 6–9.
14. WINWAR, FRANCIS. *Land of the Italian People.* Philadelphia: Lippincott (1961), 128 pp. Grades 6–9.

The *National Geographic* has had several articles in recent years on Italy: November 1961 on the 100th anniversary of a united Italy, June 1963 on the Riviera, July 1967 on Florence, and June 1970 on Rome.

Films

Among the films on Italy are the following:
"Italy: The Land and the People," Coronet, color, 14 minutes.
"Italy—Peninsula of Contrasts," Encyclopaedia Britannica, color, 17 minutes.

Filmstrips
    Among the filmstrips on Italy are the following:
    "Modern Italy," Society for Visual Education. 60 frames, color.
    "Seeing Italy," Coronet, 2 loops.

Japan

Packets of materials on Japan may be purchased from the Japan Society, 333 East 47th Street, New York, New York 10027. Indicate that you are interested in elementary school materials. Some material may be obtained free from the Information Service, Consulate General of Japan, 235 East 42nd Street, New York, New York 10020.

Books
1. ASHBY, GWYNNETH. *Looking at Japan.* Philadelphia: Lippincott (1969), 64 pp. Grades 5–8.
2. BROWN, DELMER, M. *Japan — Today's World in Focus.* Lexington, Massachusetts: Ginn (1968), 122 pp. Grades 7–9.
3. CALDWELL, JOHN C. and ELSIE F. *Our Neighbors in Japan.* New York: John Day (1961), 48 pp. Grades 3–4.
4. CARR, RACHEL. *The Picture Story of Japan.* New York: McKay (1962), 61 pp. Grades 4–6.
5. DEARMIN, JENNIE T. and HELEN E. PECK. *Japan: Home of the Sun.* San Francisco: Harr Wagner (1963), 250 pp. Grades 5–7.
6. DIETZ, BETTY and THOMAS CHOONBAI PARK. *Folk Songs of China, Japan and Korea.* New York: John Day (1964), 47 pp.
7. GIDAL, SONIA and TIM. *My Village in Japan.* New York: Pantheon (1966), 96 pp. Grades 4–6.
8. GLUBOK, SHIRLEY. *The Art of Japan.* New York: Macmillan (1970), 48 pp. Grades 4–6.
9. GULLAIN, ROBERT. *The Japanese Challenge.* Philadelphia: Lippincott (1970), 352 pp. For better readers.
10. HALL, ROBERT B. *Japan.* Garden City, New York: Nelson Doubleday (1966), 64 pp. With colored illustrations. Text for adults but pictures suitable for pupils.
11. LEWIS, RICHARD. *There Are Two Lives: Poems by Children of Japan.* New York: Simon and Schuster (1970), 96 pp.
12. NEWMAN, ROBERT. *The Japanese People: People of the Three Treasures.* New York: Atheneum (1964), 187 pp. Grades 6–9.
13. PETERSON, LORRAINE. *How People Live in Japan.* Chicago: Benefic Press (1963), 96 pp. Grades 5–7.
14. PITTS, FORREST R. *Japan.* Grand Rapids, Michigan: Fideler (1965), 160 pp. Grades 5–8. With large black and white photographs.
15. SAKADE, FLORENCE (Editor). *Japanese Children's Favorite Stories.* Rutland, Vermont: Tuttle (1958), 120 pp. Grades 4–6.
16. SELDENSTICKER, EDWARD. *Japan.* New York: Time-Life (1964), 176 pp. Text for adults but the wonderful pictures can be used with pupils.
17. SHELDON, WALTER J. *The Key to Tokyo.* Philadelphia: Lippincott (1962), 128 pp. Grades 5–8.
18. SHIRAKIGAWA, TOMIKO. *Children of Japan.* New York: Sterling (1969), 96 pp. Grades 6–9.
19. SPENCER, CORNELIA. *Made in Japan.* New York: Knopf (1963), 216 pp. Grades 6–9. On arts and crafts in Japan.
20. YAMAGUCHI, TOHR. *The Golden Crane: A Japanese Folk Tale.* New York: Holt, Rinehart and Winston (1963), unpaged. Grades 3–4.

Mexico

A kit for teachers on Mexico can be purchased from the Pan-American Union, Washington, D.C. 20006, inexpensively.

Books

1. BLACHER, IRWIN R. *Cortez and the Aztec Conquest.* New York: American Heritage (1965), 152 pp. Grades 6–9.
2. BLEEKER, SONIA. *The Aztec Indians of Mexico.* New York: Morrow (1963), 156 pp. Grades 5–8. By an anthropologist who has specialized in writing about Indians for children.
3. BRIGHT, RODERICK. *Mexico.* New York: Macmillan (1958), 90 pp. Grades 6–9.
4. COY, HAROLD. *The Mexicans.* Boston: Little, Brown (1970), 326 pp. Grades 6–9.
5. GLUBOK, SHIRLEY. *The Art of Ancient Mexico.* New York: Harper and Row (1968), 41 pp. Grades 4–7.
6. GOMEZ, BARBARA. *Getting to Know Mexico.* New York: Coward-McCann (1959), 64 pp. Grades 4–7.
7. JORDAN, PHILLIP D. *The Burro Benedicto.* New York: Coward-McCann (1960), 92 pp. Grades 3–5. Folk tales of Mexico.
8. KEATING, BERN. *Life and Death of the Aztec Nation.* New York: Putnam (1964), 153 pp. Grades 6–9.
9. KIRTLAND, G. B. *One Day in Aztec Mexico.* New York: Harcourt Brace Jovanovich (1963), 40 pp. Grades 5–8.
10. LARRALDE, ELSA. *The Land and People of Mexico.* Philadelphia: Lippincott (1964), 160 pp. Grades 6–9.
11. MARX, M. RICHARD. *About Mexico's Children.* Chicago: Melmont (1959), 47 pp. Grades 3–5.
12. ROSS, BETTY. *The Young Traveler in Mexico and Central America.* New York: Dutton (1958), 128 pp. Grades 5–8.
13. ROSS, PATRICIA. *Made in Mexico: The Story of a Country's Arts and Crafts.* New York: Knopf (1966), 324 pp. Grades 5–8.
14. ROSS, PATRICIA. *Mexico.* Grand Rapids, Michigan: Fideler (1966), 160 pp. Grades 5–8. Large black and white photographs enhance this volume.
15. SCHLOAT, G. WARREN, JR. *Conchita and Juan: A Girl and Boy of Mexico.* New York: Knopf (1964), 27 pp. Grades 3–5.
16. TINKLE, LON. *Miracle in Mexico: The Story of Juan Diego.* New York: Hawthorn (1965), 192 pp. Grades 5–8.
17. WOOD, FRANCES E. *Mexico.* Chicago: Children's Press (1964), 93 pp. Grades 5–8.

Pictures, Films, and Filmstrips

The Fideler Press of Grand Rapids, Michigan, has a portfolio of pictures on Mexico for sale.

The International Communications Foundation has a kit of materials on Mexico for sale, including filmstrips.

For films and filmstrips on Mexico, refer to *Educational Media Index* and the catalogues of various companies. There is a wealth of such materials.

Nigeria

Considerable help may be obtained from the Embassy of Nigeria, 1838 16th Street, N.W., Washington, D.C. 20036.

Books

1. BLEEKER, SONIA. *The Ibo of Biafra.* New York: Morrow (1969), 160 pp. Grades 5–8.
2. BUCKLEY, PETER. *Okolo of Nigeria.* New York: Simon and Schuster (1962), 125 pp. Grades 5–8. With many black and white photographs.

3. COURLANDER, HAROLD. *Olode the Hunter and Other Tales from Nigeria.* New York: Harcourt Brace Jovanovich (1968), 153 pp. Grades 5–7.
4. DALY, MAUREEN. *Twelve Around the World: True Accounts of the Lives in Countries of a Dozen Teen-Agers.* New York: Dodd, Mead (1957), 239 pp. Grades 6–9. Chapter 11 on "Ilowu Somnyiwa of Nigeria."
5. FORMAN, BRENDA-LU. *The Land and People of Nigeria.* Philadelphia: Lippincott (1964), 160 pp. Grades 6–9.
6. FREVILLE, NICHOLAS and JOHN C. CALDWELL. *Let's Visit Nigeria.* New York: John Day (1970), 95 pp. Grades 4–6.
7. GEIS, DARLENE (Editor). *Let's Travel in Nigeria and Ghana.* Chicago: Children's Press (1964), 85 pp. Large, colored illustrations enhance the book.
8. KENWORTHY, LEONARD S. *Profile of Nigeria.* Garden City, New York: Doubleday (1960), 96 pp. Grades 5–8. Many black and white photographs.
9. KENWORTHY, LEONARD S. and ERMA FERRARI. *Leaders of New Nations.* Garden City, New York: Doubleday (1967), 373 pp. Grades 7–10. Chapter on Balewa and the winning of independence in Nigeria.
10. NIVEN, C. R. *The Land and People of West Africa.* New York: Macmillan (1958), 84 pp. Grades 6–9. Includes some material on Nigeria.
11. OLDEN, SAM. *Getting to Know Nigeria.* New York: Coward-McCann (1960), 60 pp. Grades 6–9.
12. SCHATZ, LETTA. *Taiwo and Her Twin.* New York: Morrow (1964), 128 pp. Grades 5–7. On the schools of Nigeria.
13. U.S. Committee for UNICEF. *Hi Neighbor — Book Three.* New York: U.S. Committee for UNICEF (1960), 64 pp. Pp. 4–15 on Nigeria.
14. WRIGHT, ROSE H. *Fun and Festival from Africa.* New York: Friendship Press (1959), 48 pp. Includes some material on Nigeria.

Films and Filmstrips

1. "Africa Awakens: Modern Nigeria." Atlantis Productions, black and white, 21 minutes. A film.
2. "Contrasts in Nigeria." Encyclopaedia Britannica, color. 46 frames.
3. "Lagos: Federation of Nigeria." Eye Gate House, color. 51 frames.
4. "Nigeria: Giant in Africa." McGraw-Hill, black and white. 52 minutes.
5. "Nigeria: Important Cities." Eye Gate House, color. 52 frames.
6. "Nigeria: Other Cities." Eye Gate House, color. 50 frames.
7. "Nigeria: The People." Eye Gate House, color. 51 frames.
8. "Nigeria: Land, Transportation and Communication." Eye Gate House, color. 50 frames.

Turkey

A large number of items may be obtained free from the Turkish Information Office, 500 Fifth Avenue, New York, New York 10036.

Books

1. BROCK, RAY, *Ghost on Horseback: The Incredible Ataturk.* Boston: Little, Brown (1954), 408 pp. Grades 8–10.
2. BROCKETT, ELEANOR. *Turkish Fairy Tales.* Chicago: Follett (1968), 199 pp. Grades 4–6.
3. DAVIS, FANNY. *Getting to Know Turkey.* New York: Coward-McCann (1957), 64 pp. Grades 4–6.
4. EKREM, SELMA. *Turkish Fairy Tales.* Princeton, New Jersey: Van Nostrand (1964), 128 pp. Grades 5–7.
5. JUDA, L. *The Wise Old Man: Turkish Tales of Nasteddin Nodja.* New York: Nelson (1964), 113 pp.

6. KRUMGOLD, JOSEPH. *The Most Terrible Turk: A Story of Turkey.* New York: Crowell (1968), 40 pp. A boy and his uncle contrast new and old ways in Turkey.
7. MILLEN, NINA. *Children's Games from Many Lands.* New York: Friendship Press (1962), 240 pp. Pp. 77–78 on Turkey.
8. NORRIS, MARIANNA. *Young Turkey.* New York: Dodd, Mead (1965), 64 pp. Grades 4–7.
9. RIZA, ALI. *The Land and People of Turkey.* New York: Macmillan (1958), 90 pp. Grades 5–8.
10. SPENCER, WILLIAM. *The Land and People of Turkey.* Philadelphia: Lippincott (1958), 128 pp. Grades 6–9.
11. SPIEGELMAN, JUDITH M. *Ali of Turkey.* New York: Messner (1969), 64 pp. Grades 4–6.

Special Visual Materials

The International Communications Foundation in Monterey Park, California has various materials on Turkey, including films, filmstrips, and realia.
The Turkish Information office has several films on loan.

Films and Filmstrips

1. "Living in Turkey." Society for Visual Education, 61 frames.
2. "Modern Turkey." Two filmstrips: "Farmlands" and "Education." Eye Gate House, color. 38 frames each.
3. "Turkey." Eye Gate House, color. 28 frames.
4. "Village and City Life in Turkey." Encyclopaedia Britannica, color. 53 frames.
5. "Turkey." Two filmstrips: "City Life" and "The Art of Asia Minor." International Communications, color.
6. "Turkey: A Strategic Land and Its People." Coronet, 11 minutes, color or black and white.

The United Arab Republic (Egypt)

Write to the Press Department, Embassy of the United Arab Republic, 2310 Decatur Place, N. W., Washington, D. C. 20008, for materials. Books on ancient Egypt are more plentiful than those on the contemporary scene, but there are books representing the past and present in the list below:

Books

1. BERRY, ERICK. *Honey of the Nile.* New York: Viking (1962), 192 pp. Grades 5–8. A historical novel of ancient Egypt.
2. COTTRELL, LEONARD. *Land of the Pharaohs.* Cleveland: World (1960), 128 pp. Grades 6–9.
3. FAIRSERVIS, WALTER A. *Egypt: Gift of the Nile.* New York: Macmillan (1963), 148 pp. Grades 6–9.
4. GARTLER, MARION and others. *Understanding Egypt.* River Forest, Illinois: Laidlaw (1962), 64 pp. Grades 5–8.
5. JOY, CHARLES R. *Island in the Desert: The Challenge of the Nile.* New York: Coward-McCann (1959), 96 pp. Grades 6–9. Emphasizes the contemporary scene.
6. LEACROFT, HELEN and RICHARD. *The Buildings of Ancient Egypt.* New York: Scott (1963), 40 pp. Grades 6–9.
7. MAHMOUD, ZAKI N. *The Land and People of Egypt.* Philadelphia: Lippincott (1965), 128 pp. Grades 6–9. Revised edition.
8. MAYER, JOSEPHINE and TOM PRIDEAUX. *Never to Die: The Egyptians in Their Own Words.* New York: Viking (1961), 234 pp. Grades 7–10. Beautifully illustrated.

9. MEADOWCRAFT, ENID. *The Gift of the River.* New York: Crowell (1965), 235 pp. Grades 4–6. A brief history.
10. MELLERSH, H. E. L. *Finding Out About Ancient Egypt.* New York: Lothrop (1962), 144 pp. Grades 6–9.
11. NEURATH, MARIE. *They Lived Like This in Ancient Egypt.* New York: Watts (1965), 48 pp. Grades 3–5.
12. PINE, TILLIE S. and JOSEPH LEVINE. *The Egyptians Knew.* New York: Whittlesey (1964), 30 pp. Grades 2–4.
13. ROBINSON, CHARLES A. *The First Book of Ancient Egypt.* New York: Watts (1961), 61 pp. Grades 4–6.
14. SONDEGARD, ARENSA. *My First Geography of the Suez Canal.* Boston: Little, Brown (1960), 61 pp. Grades 3–5.
15. U.S. COMMITTEE FOR UNICEF. *Hi Neighbor — Book Five.* New York: Hastings (1962), 64 pp. Pp. 52–63 on Egypt. Games, songs, stories, etc.
16. WALLACE, JOHN A. *Getting to Know Egypt: U.A.R.* New York: Coward-McCann (1961), 64 pp. Grades 5–7.
17. WEINGARTEN, VIOLET. *The Nile: Lifeline of Egypt.* Champaign, Illinois: Garrard (1964), 96 pp. Grades 6–9.

Films and Filmstrips

For films and filmstrips on Egypt, refer to the listings in *Educational Media Index* or to other catalogues.

The United Kingdom

For background material on the United Kingdom, contact the British Information Services, 845 Third Avenue, New York, New York, and for posters write the British Travel Centre, 680 Fifth Avenue, New York, New York.

Books
1. BURTON, ELIZABETH. *Here Is England.* New York: Farrar (1965), 210 pp. Grades 5–8.
2. BUCHANAN, FREDA M. *The Land and People of Scotland.* Philadelphia: Lippincott (1962), 128 pp. Grades 6–9.
3. DE MARE, ERIC. *London's River: The Story of a City.* New York: McGraw-Hill (1965), 126 pp. Grades 5–7.
4. GEIS, DARLENE. *Let's Travel in England.* Chicago: Children's Press (1964), 86 pp. Grades 4–6.
5. HUNTER, LESLIE. *The Boys and Girls Book of the Commonwealth.* New York: Roy (1962), 144 pp. Grades 5–7.
6. JOY, CHARLES R. *Young People of the British Isles.* New York: Duell (1965), 227 pp. Grades 5–8.
7. LAUBER, PATRICIA. *Valiant Scots: People of the Highlands Today.* New York: Coward-McCann (1957), 96 pp. Grades 6–8.
8. MALKUS, ALIDA S. *The Story of Winston Churchill.* New York: Grosset and Dunlap (1957), 181 pp. Grades 5–8.
9. MALSTROM, VINCENT Y. *Life in Europe: The British Isles.* Grand Rapids, Michigan: Fideler (1959), 192 pp. Grades 5–7. Many large black and white photos.
10. MIDDLETON, DREW. *England.* New York: Macmillan (1964), 136 pp. Grades 6–9.
11. MIERS, EARL S. *The Story of Winston Churchill.* New York: Grosset and Dunlap (1965), 48 pp. Grades 5–8.
12. MOORE, MARIAN. *The United Kingdom: A New Britain.* New York: Nelson (1966), 224 pp. Grades 6–9.
13. NATHAN, ADELE G. *Churchill's England.* New York: Grosset and Dunlap (1963), 95 pp. Grades 6–9.

14. SAYWELL, JOHN T. *Commonwealth of Nations.* New York: Scholastic (1966), 160 pp. Grades 6–9.
15. ROSENBAUM, MAURICE. *London.* Chicago: Rand McNally (1963), 128 pp. Grades 5–8.
16. STREATFIELD, NOEL. *The First Book of England.* New York: Watts (1958), 72 pp. Grades 4–6.
17. STREET, ALICIA. *The Land of the English People.* Philadelphia: Lippincott (1953), 130 pp. Grades 6–9.
18. WALL, DAPHNE. *The Story of the Commonwealth.* New York: Watts (1960), 64 pp. Grades 4–7.

Pictures, Films, and Filmstrips

For large, colored pictures see the UNESCO Geography Series, sold by Unipub. For films and filmstrips contact the British Information Services, see the *Educational Media Index*, or examine the catalogues of various publishers.

The U.S.S.R.

Books

1. ALMEDINGEN, E. M. *A Picture History of Russia.* New York: Watts (1964), 63 pp. Grades 5–8. With colored illustrations.
2. ALMEDINGEN, E. M. *Russian Folk and Fairy Tales.* New York: Putnam (1963), 192 pp. Grades 5–8.
3. APPEL, BENJAMIN. *Why the Russians Are the Way They Are.* Boston: Little, Brown (1966), 192 pp. Grades 6–9.
4. DOLCH, EDWARD W. and MARGUERITE P. *Stories from Old Russia.* Champaign, Illinois: Garrard (1964), 168 pp. Grades 4–6.
5. FEUCHTER, CLYDE E. and GRACE E. POTTER. "The U.S.S.R." Columbus, Ohio: Merrill (1960), 32 pp. Grades 4–6. For less able readers.
6. GEIS, DARLENE. *Let's Travel in the Soviet Union.* Chicago: Children's Press (1964), 85 pp. Grades 5–8.
7. GUNTHER, JOHN. *Meet Soviet Russia.* New York: Harper and Row (1962). Grades 5–7. Volume I *Land, People, Sights,* 180 pp. Volume II *Leaders, Politics, Problems,* 180 pp. Abridgments of the book on *Inside the U.S.S.R.*
8. HABBERTON, WILLIAM. *Russia: The Story of a Nation.* Boston: Houghton Mifflin 1965, 282 pp. Grades 6–9.
9. HOFF, RHODA. *Russia: Adventures in Eyewitness History.* New York: Walck (1964), 207 pp. Grades 6–9.
10. JACKSON, W. A. *Soviet Union.* Grand Rapids, Michigan: Fideler (1963), 192 pp. Grades 5–8. Large black and white photographs.
11. KISH, GEORGE. *The Soviet Union.* Chicago: Nystrom (1964), 63 pp. Grades 4–6.
12. LEVIN, DEANA. *Nikolai Lives in Moscow.* New York: Hastings (1968), 48 pp. Grades 3–4. For slower readers. Well illustrated.
13. Life World Library. *Russia.* New York. Time and Life (1963), 176 pp. Text for adults but the excellent pictures can be used with pupils in elementary schools.
14. MASEY, MARY LOU. *The Picture Story of the Soviet Union.* New York: McKay (1971), 64 pp. Grades 4–6.
15. *My Mother Is the Most Beautiful Woman in the World.* New York: Lothrop (1960), 40 pp. Grades 4–6. A Russian folk tale.
16. NAZAROFF, ALEXANDER. *The Land and People of Russia.* Philadelphia: Lippincott (1966), 190 pp. Grades 6–9.

17. PARKER, FAN. *Russian Alphabet Book*. New York: Coward-McCann (1961), 40 pp. Grades 4–6. A clever presentation of geographical information through the letters of the Russian alphabet.
18. POPESCU, JULIAN and JOHN C. CALDWELL. *Let's Visit Russia*. New York: John Day (1968), 96 pp. Grades 5–7.
19. RUBIN, ROSE N. and MICHAEL STILLMAN (Editors). *A Russian Song Book*. New York: Vintage (1962), 197 pp.
20. SALISBURY, HARRISON F. *The Key to Moscow*. Philadelphia: Lippincott (1963), 128 pp. Grades 6–9.
21. VANDERVERT, RITA. *Young Russia*. New York: Dodd, Mead (1960), 62 pp. Grades 5–8.

Large colored pictures of the U.S.S.R. in the UNESCO Geography Series are sold by Unipub. For films and filmstrips see the *Educational Media Index*.

53. Discovering parts of the world through use of a globe

Studying the World

54. Developing map skills
by producing a map of Africa

55. Developing knowledge of the United Nations through a flannelgram

22 Studying the World–General

Studying selected families, communities, and countries of the world is important, but these three approaches certainly do not exhaust the ways in which children should be introduced to the world. Nor do they cover all the aspects of the world to which boys and girls should be exposed in their elementary and junior high school years. Furthermore, these seemingly disparate parts need to be joined together in the minds of pupils so that they see the world as a whole rather than separate parts of it.

It seems wise, therefore, to have at least one segment of the social studies curriculum K–8 in which an overview of the world is undertaken. The earth would be seen as part of the solar system and some attention would be given to outer space and the race to the moon. Pupils would discover the major geographical features of the earth and its major resources. They would find the major concentrations of people and the reason why they live where they do. Some emphasis would be placed upon the interdependence of the people of our planet and the development of regional and international organizations, with particular emphasis upon the United Nations and its specialized agencies.

Such a unit or segment of the social studies curriculum might be used as an introduction to the study of selected countries of the world. That would place it at the beginning of the sixth grade in most school systems. This would enable pupils to see the world as a whole and then to study in depth certain selected parts of it.

It could be used even more advantageously as a summary unit at the end of the sixth grade program. If teachers are pressed for time in that year, such a unit could be an introductory unit in the seventh grade. Pupils could then proceed to discover some of the problems of the people of our planet. Four or five of the major trends in the world could be introduced as a background to the study of problems. They might include such topics as the wise use of resources, industrialization, urbanization, nationalism, and internationalism.

Parts of such a unit might well be approached by committees or individuals, assuming that time is limited and that pupils, by that stage of their work in schools, have learned a good many skills in group work and elementary research.

Four to six weeks should give a teacher and a class enough time to examine in some depth the major topics suggested for this unit. Much more time could be devoted profitably to such a unit, but there is always the factor of other topics in the social studies and of other subject fields.

Such a unit might be labeled "The Earth As the Home of Man," thus tying together the idea of the earth with the people on it. The earth, then, would be viewed as man's workshop, his laboratory, and his playground. The extension of his "home" would then include a brief study of outer space. This should be an important and fascinating unit for teachers and for pupils.

Such a unit might follow the outline below in a general way, with adaptations made by school systems and by teachers to fit the special needs of a locality or a group of pupils.

THE EARTH AS THE HOME OF MAN

1. *The Earth as a Part of the Solar System.*
2. *The Earth Itself.*
   (a) Its shape, size, movements, origin and composition, changes.
   (b) How we use globes and maps to represent the earth.
   (c) Major water areas.
   (d) Major land areas and forms — deserts, mountains, plateaus, plains, ice areas, etc.
3. *The Major Resources of Our Earth.*
   (a) Soil.
   (b) Minerals.
   (c) Conserving or using our resources wisely.
4. *Where the People of the World Live.*
   (a) Major concentrations of people.
   (b) Reasons why people live where they do.
5. *How People Have Changed the Land.*
   (a) Irrigation.
   (b) Reclamation of deserts.
   (c) Improvement of the soil.
   (d) Other changes.
6. *The Interdependence of People on Our Planet Today.*
   (a) Dependence on resources.
   (b) Markets.
   (c) The interchange of ideas and inventions.
7. *The Need for Larger Units of Government in Today's World.*
   (a) Regional organizations.
   (b) International organizations, with emphasis upon the United Nations and its specialized agencies.
8. *Man Extends His World.*
   (a) The Arctic and the Antarctic.
   (b) The reclamation of deserts and other land.
   (c) Exploring the waters of the world — oceanography.
   (d) Outer space.

Such a plan is merely a suggestion. Parts of it will have been treated adequately in many school systems and in many classrooms before children reach the end of the sixth grade year. Therefore it should be used merely as a reference point for teachers developing their own course of study, or by curriculum committees in developing a course of study. Adaptations should be made according to the needs of a particular system, school, or class. You will need to determine whether the time for such a unit will be well spent or whether it would be better if you devoted that time to work on one or more of the nations being studied.

SOME SUGGESTED METHODS

A wide variety of methods can and should be used in developing this unit, as is the case with any unit in the social studies. However, this particular unit lends itself to the extensive and intensive use of globes and maps.

Certainly there should be one large globe in the classroom, preferably a "cradle globe" which can be picked up and used by the teacher and the pupils. Hopefully, there will be several small globes which pupils can use individually and in small groups. The class might want to make a large globe for use by themselves and by future classes, or a small group might work on this project.

A large map of the world certainly should hang in the classroom throughout this unit. It might be a simplified world map like the one prepared by Rand McNally, with a polar projection. Such polar projection or air-age maps certainly should be used even more than the Mercator projection maps, because of the world in which these pupils live and will continue to live.

Overlay maps should be used widely, too. Some of them should be prepared by the pupils themselves. For example, a map of the distribution of the population of the world can be placed on top of a map of the world, showing its major geographical features, and pupils can discover the relation between these two maps. With gifted children, maps of the world's resources and of "trouble spots" in the world can be used to discover how oil in the Middle East, or copper in Katanga and Rhodesia in Africa, or rice in Southeast Asia, relates to rivalries between nations. The overhead projector can be used to great advantage with overlay maps.

A collection of pictures is a "must" for this unit, as for many other units in the social studies. The pupils themselves can prepare many such pictures by thumbing through current and old magazines and cutting out illustrations of the various types of land in the world. These can be mounted on cardboard and used in small groups or in the opaque projector by the entire class. These should be jumbled often, so that pupils can classify them and learn to identify various types of land.

Perhaps a warning needs to be given about the undue emphasis in many elementary schools on longitude and latitude. They are important, but too much time and attention are often given to them in elementary schools.

Maps can also be made, on oaktag or on cardboard, of the major resources of the world, of trade between two nations, and of other examples of interdependence.

Considerable problem-solving can be done, especially with maps of places which do not exist, with the pupils deciding how they would live and earn a living in such places.

Fortunately, there are many films and filmstrips which can be used to advantage in such a unit. Some television programs can be utilized profitably. Bulletin boards can be prepared. And, of course, much reading can be done from the many books cited in this chapter, plus others discovered by the class.

If there are teachers, parents, or other adults in your community who have traveled widely, they should be of great help to you and your pupils. However, you may want to warn them that you are discussing the world rather than individual nations in this unit and ask them to help you with this emphasis.

GAINING BACKGROUND AS A TEACHER

Some teachers are frightened by the prospect of having to teach about the world as a whole. This seems to them like an enormous assignment. They feel that their background is limited and, in some cases, out-of-date.

This is an understandable feeling. Even experts have it. Some people spend all their lives studying one small region of the world or one topic pertaining to the world.

It is not an impossible task, however, to teach about the world as a whole. Teachers already know quite a bit about the earth and a little about space. In general, they are far better informed than their pupils. And they can learn more as they teach, and as the months and years roll by. If we all waited until we were fully qualified to teach, probably we would never teach. Or we would have such an enormous fund of knowledge that we might not be able to communicate with children.

So the job is one of self-education. Teachers need to read as widely as possible in a variety of newspapers and periodicals. They need to attend lectures and in-service courses. They need to watch good television programs on the world or aspects of it. And they need to read.

For teachers who need some general background on the world, Vera M. Dean's little paperback on *The Nature of the Non-Western World* (Mentor) is an excellent little book, covering much of the world, including Latin America.

For teachers whose background in human geography or cultural geography is weak, Emrys Jones's *Human Geography* (Praeger) might be a good volume to read. Or Preston James's *A Geography of Man* (Blaisdell) could be a good textbook or review book.

For those who feel that they lack a point of view, with special emphasis upon anthropology, the pamphlet by Harold Courlander titled "On Recognizing the Human Species" (Anti-Defamation League) might be a good place to start. Or you might want to pick up the highly readable paperback by Clyde Kluckhohn on *Mirror for Man* (Premier-Fawcett).

On the cultural areas of the world, no better book exists for teachers than Preston James's *One World Perspective* (Blaisdell).

There are three publications for teachers which should help them to see how teaching about the world should be carried on. One is the curriculum bulletin of the Board of Education of New York City entitled "Toward Better International Understanding: A Manual for Teachers." Another is Anna L. Rose Hawkes's book on *The World in Their Hands* (Kappa Delta Pi Press). The third is the writer's book on *Introducing Children to the World: In Elementary and Junior High Schools* (Harper). This volume suggests ten themes which might appear at various points in the social studies curriculum of elementary and junior high schools.

Teachers also can learn a great deal by reading books written for boys and girls. A good many of them about people of the world and about the earth as the home of man follow.

BOOKS FOR BOYS AND GIRLS ABOUT THE PEOPLE OF THE WORLD

Unfortunately there are not many books which deal directly with the people of the world. Several deal with the problem of food and people and some with race. Here are a few books which are available:

1. BLOCK, IRVIN. *People.* New York: Watts (1956), 179 pp. Grades 5–8.
2. BOYD, WILLIAM C. and ISAAC ASIMOV. *Races and People.* New York: Abelard-Schuman (1955), 189 pp. Grades 8–10.
3. COHEN, ROBERT. *The Color of Man.* New York: Random House (1968), 114 pp. Grades 5–8.
4. EVANS, EVA KNOX. *People Are Important.* New York: Capitol and Golden (1962), 87 pp. Grades 4–6. Anthropology, with a sense of humor included. Highly recommended.

5. HELFMAN, ELIZABETH S. *This Hungry World*. New York: Lothrop, Lee and Shepard (1970), 160 pp. Grades 5–8.

6. HEY, NIGEL and the EDITORS OF SCIENCE BOOK ASSOCIATES. *How Will We Feed the Hungry Billions?* New York: Messner (1971), 192 pp. Grades 7–9.

7. KENWORTHY, LEONARD S. *Three Billion Neighbors*. Lexington, Massachusetts: Ginn (1965), 160 pp. A "Family of Man" for children, featuring nearly 500 black and white photographs, arranged by the major activities of man.

8. LISITZKY, GENE. *Four Ways of Being Human: An Introduction to Anthropology*. New York: Viking (1956), 303 pp. Grades 6–9.

9. MEAD, MARGARET. *People and Places*. New York: World (1959), 318 pp. Grades 6–9.

10. NEURATH, MARIE. *Living With One Another*. New York: Watts (1965), 128 pp. Grades 5–8. Includes many pictographs of the people of the world.

11. PRINGLE, LAURENCE. *One Earth, Many People: The Challenge of Human Population Growth*. New York: Macmillan (1971), 128 pp. Grades 5–8.

BOOKS FOR BOYS AND GIRLS ABOUT THE EARTH AS THE HOME OF MAN

1. ADLER, IRVING and RUTH. *Irrigation: Changing Deserts into Gardens*. New York: John Day (1964), 48 pp. Grades 3–5.

2. AGUIRRE, A. J. M. and F. GOICO. *A First Look at the Earth*. New York: Watts (1960), 72 pp. Grades 1–3.

3. AGUIRRE, A. J. M. and F. GOICO. *A First Look at the Sea*. New York: Watts (1960), 72 pp. Grades 1–3.

4. AMES, GERALD and ROSE WYLER. *Planet Earth*. New York: Golden (1963), 105 pp. Grades 6–9.

5. ARCHER, SELLERS. *Rain, Rivers and Reservoirs: The Challenge of Running Water*. New York: Coward-McCann (1963), 120 pp. Grades 5–8.

6. BENDICK, JEANNE. *The Shape of the Earth*. Chicago: Rand McNally (1965), 72 pp. Grades 4–6.

7. BRINDZE, RUTH. *The Story of the Trade Winds*. New York: Vanguard (1960), 68 pp. Grades 4–6.

8. CHANDLER, M. H. *Man's Home: The Earth*. Chicago: Rand McNally (1966), 96 pp. Grades 5–7.

9. CULLEN, ALLAN H. *Rivers in Harness: The Story of Dams*. Philadelphia: Chilton (1962), 175 pp. Grades 6–9.

10. DAUGHERTY, CHARLES M. *Searchers of the Sea: Pioneers in Oceanography*. New York: Viking (1961), 160 pp. Grades 6–9.

11. DEMPSEY, MICHAEL W. and ANGELA SHEEHAN. *Water*. New York: World (1970), 30 pp. Grades 2–4.

12. FRANK, R. *Burning Lands and Snow-Capped Mountains*. New York: Crowell (1964), 224 pp. Grades 6–9. Includes 135 photos and maps.

13. FREEMAN, MAE B. *Do You Know About Water?* New York: Random House (1970), 24 pp. Grades 3–4.

14. GALLANT, RAY A. *Exploring the Weather*. Garden City, New York: Doubleday (1969), 64 pp. Grades 5–8.

15. GOETZ, DELIA. *Islands of the Ocean*. New York: Morrow (1964), 64 pp. Grades 3–5.

16. GOETZ, DELIA. *Mountains*. New York: Morrow (1962), 64 pp. Grades 4–6.

17. HICKS, CLIFFORD B. *The World Above*. New York: Holt, Rinehart and Winston (1965), 153 pp. Grades 5–8.

18. KAUFMAN, JOHN. *Winds and Weather*. New York: Morrow (1971), 64 pp. Grades 4–6.

19. KOVALIK, VLADIMIR and NADA. *The Ocean World*. New York: Holiday (1966), 191 pp. Grades 6–9.

20. LAMBERT, ELIZABETH. *The Living Sea.* New York: Coward-McCann (1963), 126 pp. Grades 6–9.
21. LAUBER, PATRICIA. *This Restless Earth.* New York: Random House (1970), 134 pp. Grades 5–8.
22. LEWIS, ALFRED. *This Thirsty World: Water Supply and Problems Ahead.* New York: McGraw-Hill (1964), 96 pp. Grades 6–9.
23. MATTHEWS, WILLIAM H. *The Earth's Crust.* New York: Watts (1971), 92 pp. Grades 5–8.
24. MATTISON, C. W. and JOSEPH ALVAREZ. *Man's Resources in Today's World.* Mankato, Minnesota: Creative Educational Society (1967), 144 pp. Grades 5–8.
25. MAY, JULIAN. *Why the Earth Quakes.* New York: Holiday (1970), 40 pp. Grades 4–6.
26. NADEN, CORINNE J. *Grasslands Around the World.* New York: Watts (1970), 66 pp. Grades 4–6.
27. PILKINGTON, ROGER. *The River.* New York: Walck (1965), 64 pp. Grades 5–8.
28. POND, ALONZO W. *Deserts — Silent Lands of the World.* New York: Norton (1965), 158 pp. Grades 5–8.
29. QUILICI, FALCO. *The Great Deserts.* New York: McGraw-Hill (1969), 128 pp. Grades 6–9.
30. SCHARFF, ROBERT. *The How and Why Wonder Book of Oceanography.* New York: Grosset and Dunlap (1964), 48 pp. Grades 6–9.
31. SCHWARTZ, JULIUS. *The Earth Is Your Spaceship.* New York: McGraw-Hill (1963), 30 pp. Grades 2–4.
32. SHERMAN, DIANE. *You and the Oceans.* Chicago: Children's Press (1965), 62 pp. Grades 4–6.
33. SIMMS, SEYMOUR. *Weather and Climate.* New York: Random House (1969), 134 pp. Grades 5–8.
34. SMITH, FRANCES. *The First Book of Mountains.* New York: Watts (1964), 87 pp. Grades 6–9.
35. TYLER, MARGARET. *Deserts.* New York: John Day (1970), 48 pp. Grades 5–8.
36. WINCHESTER, JAMES. *The Wonders of Water.* New York: Putnam (1963), 123 pp. Grades 5–8.

THE ARCTIC AND THE ANTARCTIC

At some point in their studies in elementary and junior high school, pupils should learn a little about the polar regions of the earth. These are interesting regions and are likely to become increasingly important in the years ahead.

These two regions might well be examined in a very brief unit at almost any time in the middle and upper grades. One of the best ways to handle them is to include them in a unit on The Earth as the Home of Man. If teachers are pressed for time, the Arctic and the Antarctic could be studied by individuals or small committees and the findings reported to the class.

In studying the polar regions, the stress should be in part on the inaccessibility of these areas for a very long period of history, on the feats of the explorers in traversing them, and in their potential. Both are now scientific laboratories and weather stations for the world. The Antarctic is one of the finest examples of international cooperation in scientific research.

The story of the Arctic should include some attention to the variety of people in that region, including the Lapps, the Eskimos, the Aleuts, and several other groups. It should include the land masses that surround the Arctic Ocean. It should also include the importance of this area from a military point of view, in conjunction with airplane flights, and its mineral resources.

The story of the Antarctic is one of the world's greatest icebox, with about 86 per cent of the world's ice. It is also the story of the last large area of the earth to be discovered and explored, with several nations taking part in the race for supremacy. Much more important, the Antarctic illustrates how an icy continent has become an international laboratory, with twelve nations cooperating in research. No one knows what the future holds for this area but we do know that it might become one of the major sources of food some day, as well as one of the largest sources of coal for the world.

A few books for pupils on these two regions are as follows:

1. ANGELL, PAULINE K. *To the Top of the World.* Chicago: Rand McNally (1964), 288 pp. Grades 6–9.
2. BERRY, ERICK. *Mr. Arctic.* New York: McKay (1966), 183 pp. Grades 6–8.
3. BOXBY, WILLIAM. *The Race to the South Pole.* New York: McKay (1961), 215 pp. Grades 6–9.
4. DARBOIS, DOMINIQUE. *Achouna: Boy of the Arctic.* Chicago: Follett (1962), 47 pp. Grades 5–7.
5. ICENHOWER, J. B. *The First Book of the Antarctic.* New York: Watts (1956), 72 pp. Grades 5–7.
6. OGLE, ED. *Getting to Know the Arctic.* New York: Coward-McCann (1961), 64 pp.
7. RINK, PAUL. *Richard E. Byrd: Conquering Antarctica.* Chicago: Encyclopaedia Britannica (1961), 190 pp. Grades 6–9.
8. RONNE, FINN and HOWARD LISS. *The Ronne Expedition to Antarctica.* New York: Messner (1971), 96 pp. Grades 5–8.
9. SULLIVAN, WALTER. *Polar Regions.* New York: Golden (1962), 54 pp. Grades 5–8.

Teachers will find good background for themselves and pictures for pupils in an article in the *National Geographic* for November 1971 on the Antarctic.

INTERDEPENDENCE

Few themes are as important in the social studies as that of interdependence. As Carleton Washburne wrote in *The World's Good: Education for World Mindedness,* "We are living in one world. Yet our thinking, our emotions, and our actions are largely based upon the assumption of at least relative independence of individuals, groups, or religious, economic, political, racial, or national societies. To our failure to realize that the well-being of any individual, group, or society in the world today is interdependent with the well-being of all individuals, groups, and societies is due the confusion and conflict in which we are living." That is a strong statement but not overdrawn.

At many points in elementary and junior high school this theme needs to be struck. In a unit on The Earth as the Home of Man it should be a central concept. Once boys and girls have discovered something about the earth and its people, they should discover how much they depend upon each other in today's world.

We depend upon each other in almost every field of life. We obtain food from each other. We use the raw materials of the world to make finished products and then sell these at home and abroad. We borrow from each other in styles of art and architecture. We share medical and other scientific discoveries. We play the music of the composers of many parts of the world. We read books written by people from all over the world. We belong to religious groups whose founders were largely from other parts of the globe. So one might continue to enumerate the many ways in which we rely on each other. As John Wesley said a good many years ago, "We are debtors to all the world."

Teachers and pupils can find many striking examples of such interdependence. The telephone is one example with at least forty-eight products coming from sixteen nations to make it. An automobile takes three hundred products from fifty-six nations, as a chart issued by the Automobile Manufacturers Association (New Center Building, Detroit 2, Michigan) shows. And the tiny candy bar is a simple example of a product made of ingredients from many parts of the world. Pupils can discover other examples by talking with merchants and manufacturers about the materials that go into their stores and into the products of their factories. A study of the imports and exports of the United States lends itself admirably to this topic, too. Boys and girls who are interested in music can take the records in their home and discover who the composers were and in what part of the world they lived.

Unfortunately, there are almost no books on this theme, written for boys and girls. The few books that do exist are very old, such as Madeline Gekiere's *Who Gave Us?* (Pantheon) and the American Council on Education's *China's Gifts to the West. The Made in. . . .* series of Knopf gives some material on individual countries. Perusal of the author's *Three Billion Neighbors* (Ginn) might give children some leads, too, on this concept. Tillie S. Pine has written several books on this theme, such as *The Chinese Knew, The Africans Knew,* and *The Indians Knew.* In each of these books she shows what one group invented or developed and how we are debtors to them. This series is published by McGraw-Hill.

This is another area of the social studies where we could use many more materials for elementary and middle school boys and girls.

REGIONAL AND INTERNATIONAL ORGANIZATIONS

In the world of the foreseeable future, regional and international organizations will be much more important than they are today. And undoubtedly there will be more regional organizations than exist at the present time. Such federations or confederations will be necessary for the survival of many of today's smaller nations. They simply cannot exist economically with such small populations and therefore small markets. Regional and international planning will be necessary in order to maintain relatively high prices for resources and goods. In the political realm, international control of weapons will be necessary to save the world from global suicide. These larger units of organization are inevitable.

Therefore, social studies teachers should obtain as much background as possible on such organizations today and on suggestions for the strengthening of existing groups as well as the formation of other regional groups. They should know about the Arab League, the European Community, the Organization of American States, and several similar regional groupings. And they should know quite a bit about the United Nations and some of its specialized agencies.

Somewhere in the social studies curriculum, K–8, pupils should begin to learn about these organizations and their importance in the world of today and tomorrow. In the primary grades children might well be introduced to the U. N. through the use of the U. N. flag in their classroom and in their school, along with the use of our own national flag. They might well have a birthday party for the U. N. and learn about it as a "club" of adults which is trying to build a better world in which we can all live. Munro Leaf's *Three Promises to You* and the book and filmstrip on *A Garden We Planted* are the types of materials that are appropriate for this age.

In the middle grades pupils might well learn about three or four of the specialized agencies of the United Nations. Probably the Food and Agriculture Organization, the World Health Organization, the International Civil Aviation Organization, and the Universal Postal Union are the easiest for boys and girls to understand. Middle grade children might well take part in the annual Trick or Treat programs of the United Nations Children's Fund, too, and learn about that organization, with its emphasis upon children around the world.

By the sixth or seventh grades, pupils should be ready to study in some detail the United Nations family. The structure of that organization should be included. Much more learning will take place, however, if they study some of the world's major problems and how the U. N. is trying to help solve them and thereby "build" the structure of the U. N. rather than starting with the structure.

At the upper elementary or junior high school level individual pupils or groups could study one specialized agency in depth, reporting their findings to the entire class. There is no reason for every pupil to study each of the many parts of this international organization. Neither is there time in crowded programs to attempt such broad coverage. Of course the study of the U. N. and its agencies can also be approached through the use of current events.

The most comprehensive account of ways of teaching about the U. N. is the writer's book on *Telling the U. N. Story: New Approaches to Teaching About the United Nations and Its Related Agencies* (UNESCO and the Oceana Press). A few books for children about the U. N. are listed on the next page; there are a good many films and filmstrips which can also be used advantageously.

BOOKS FOR BOYS AND GIRLS ABOUT THE UNITED NATIONS AND ITS AGENCIES

1. BREETVELD, JIM. *Getting to Know How the U. N. Crusaders Keep Us Free: The Human Rights Commission.* New York: Coward-McCann (1961), 64 pp. Grades 5–8.
2. DOBLER, LAVINIA. *Arrow Book of the United Nations.* New York: Scholastic (1965), 80 pp. Grades 5–8.
3. EPSTEIN, EDNA. *The First Book of the U. N.* New York: Watts (1965), 89 pp. Grades 6–9.
4. EPSTEIN, SAM and BERYL. *The First Book of the World Health Organization.* New York: Watts (1964), 82 pp. Grades 4–6.
5. FISHER, ROGER. *International Conflict for Beginners.* New York: Harper and Row (1969), 231 pp. Grades 7–10.
6. GRIFFIN, ELLA. *Getting to Know UNESCO.* New York: Coward-McCann (1962), 64 pp. Grades 4–6.
7. HOKE, HENRY. *The First Book of International Mail: The Story of the Universal Postal Union.* New York: Watts (1963), 41 pp. Grades 4–6.
8. JOY, CHARLES A. *Race Between Food and People: The Challenge of a Hungry World.* New York: Coward-McCann (1961), 121 pp. Grades 6–9.
9. KELEN, EMERY. *The Valley of Trust.* New York: Lothrop (1962), 64 pp. Grades 2–4. Quarreling animals organize, like the U. N.
10. LARSEN, PETER. *The United Nations at Work Throughout the World.* New York: Lothrop, Lee and Shepard (1971), 127 pp. Grades 5–8.
11. LEAF, MUNRO. *Three Promises to You.* Philadelphia: Lippincott (1957), 48 pp. Grades 1–4. The Charter of the U. N. in very simple terms.
12. PIERSON, SHERLEIGH. *What Does a U. N. Soldier Do?* New York: Dodd, Mead (1965), 64 pp. Grades 4–6.
13. RABE, OLIVE. *United Nations Day.* New York: Crowell (1965), 40 pp. Grades 1–3.

14. SASEK, M. *This Is the United Nations.* New York: Macmillan (1968). Grades 4–6.
15. SAVAGE, KATHARINE. *The Story of the United Nations.* New York: Walck (1970), 224 pp. Grades 6–9.
16. SCHLINING, PAULA. *The United Nations and What It Does.* New York: Lothrop (1962), 61 pp. Grades 4–7.
17. SHAPP, MARTHA and CHARLES. *Let's Find Out About the United Nations.* New York: Watts (1962), 48 pp. Grades 1–3.
18. SHIPPEN, KATHERINE B. *The Pool of Knowledge: How the United Nations Share Their Skills.* Garden City, New York: Doubleday (1965), 99 pp. Grades 6–9.
19. SHOTWELL, LOUISA. *Beyond the Sugar Cane Field:* UNICEF *in Asia.* Cleveland: World (1964), Grades 6–9.
20. SMITH, RALPH LEE. *Getting to Know the World Health Organization.* New York: Coward-McCann (1963), 64 pp. Grades 4–6.
21. STERLING, DOROTHY. *United Nations, N. Y.* Garden City, New York: Doubleday (1961), 80 pp. Grades 5–8. Largely a pictorial account of the headquarters building.
22. TOR, REGINA and ELEANOR ROOSEVELT. *Growing Toward Peace.* New York: Random (1960), 86 pp. Grades 5–7.
23. UNITED NATIONS. *A Garden We Planted.* New York: McGraw-Hill (1952), 48 pp. Grades 1–3. Also a filmstrip. The U. N. is like a garden.

SPACE

On October 4, 1957, a new age in the history of man began — The Space Age. Today's children have known no other age. To them, the space age, with its astronauts, rockets, and satellites, seems the only natural and normal world. They are probably more attuned to it than most teachers. In this regard, it may be the teachers who are the "slow learners." They may be the ones who have to do additional "homework" and get some special "tutoring" from their scientifically-minded friends and colleagues. If teachers have been accustomed to teaching about the air-age world, they will now have to jump to aerospace education. They may need to learn along with their pupils — or even from some of them.

Together, then, teachers and pupils need to make some study of the space age and of space. In the primary grades this may mean special attention to the spectacular events shown on television in the race to the moon — and beyond. For older pupils it probably will mean a short unit on space, possibly as a part of a unit on The Earth as the Home of Man.

The study of space provides a wonderful opportunity to correlate science and the social studies. In some classes it is an almost ideal topic for team teaching, with science and social studies teachers cooperating.

Among the several topics which might be studied in such a unit are: (1) The Secrets of the Solar System, (2) The History of Flight and Space Exploration, (3) Rockets and Satellites, (4) The Training of Astronauts, (5) The Costs of Space Exploration, (6) Effects on the U. S. Economy, (7) International Complications of Space Exploration, (8) International Cooperation, (9) Peaceful Uses of Space, and (10) Some Possible Next Steps in Building a Stairway to the Stars. These are merely suggestions from which you may want to build a unit.

The methods you will want to use in such a unit will not vary greatly from those used in other units. However, you certainly will want to make much use of current events and therefore of television programs and spectaculars, newspapers, and magazines. You certainly will want to make use of the airport near your school and of the personnel there, as well as other adults with scientific background. Some boys may want to build small models of airplanes, rockets, and satellites. A time-line is an appropriate experience in connection with the history of space travel. A good many new words will be used, and this means attention to a vocabulary of space.

Following is a list of a few books on space which have been printed in fairly recent years. However, since the topic changes so quickly, you will want to find even more recent books and supplement them with articles from newspapers and magazines. Some of the encyclopedias have very good accounts of space and space travel, with arresting illustrations. Some states and cities have developed curriculum bulletins on space. One of them is the State of California, with its "Aviation Education and the Space Age." Another is Pennsylvania, with its "Earth and Space Guide for Elementary Schools." The best single source of information is the National Aerospace Education Council, 1025 Connecticut Avenue N. W., Washington, D. C. 20036.

BOOKS FOR BOYS AND GIRLS ON SPACE

1. AHNSTROM, D. N. *The Complete Book of Jets and Rockets.* New York: World (1970), Revised Edition. 184 pp. Grades 6–9.
2. ASIMOV, ISAAC. *ABC's of Space.* New York: Walck (1969), 47 pp. Grades 4–6.
3. BENDICK, JANNE. *Space Travel.* New York: Watts (1969), 94 pp. Grades 4–6.
4. BOVA, BENJAMIN. *The Uses of Space.* New York: Holt, Rinehart and Winston (1965), 144 pp. Grades 4–6.
5. CARLISLE, NORMAN. *Satellites: Servants of Man.* Philadelphia: Lippincott (1971), 96 pp. Grades 6–9.
6. DEMPSEY, MICHAEL W. and ANGELA SHESHAN. *Into Space.* New York: World (1970), 31 pp. Grades 3–6.
7. DWIGGINS, DON. *Eagle Has Landed: The Story of Lunar Expeditions.* San Carlos, California: Golden Gate Junior Books (1970), 80 pp. Grades 6–9.
8. DWIGGINS, DON. *Into the Unknown: The Story of Space Shuttles and Space Stations.* San Carlos, California: Golden Gate Junior Books (1971), 96 pp. Grades 5–8.
9. Editors of *The American Heritage. Americans in Space.* New York: American Heritage (1965), 153 pp. Grades 6–9.
10. GALLANT, ROY A. *Man's Reach Into Space.* Garden City, New York: Doubleday (1964), 152 pp. Grades 8–10. Striking illustrations but difficult text.
11. GOODWIN, HAROLD L. *All About Rockets and Space Flight.* New York: Random House (1970), 145 pp. Grades 6–9.
12. GOTTLIEB, WILLIAM P. *Space Flight and How It Works.* Garden City, New York: Doubleday (1964), 61 pp. Grades 5–8. Includes several home experiments.
13. GURNEY, GENE. *Americans to the Moon.* New York: Random House (1970), 147 pp. Grades 5–8. Apollo Mission described.
14. HAGGERTY, JAMES J. *Apollo: Lunar Landing.* New York: Rand (1969), 159 pp. Grades 6–9.
15. HENDRICKSON, WALTER B., JR. *Apollo II: Men to the Moon.* Irvington-on-Hudson, New York: Harvey House (1970), 46 pp. Grades 4–6.
16. HYDE, MARGARET O. *Flight Today and Tomorrow.* New York: McGraw-Hill (1970), Grades 5–8. Third edition of this volume.
17. HYDE, MARGARET O. *Off Into Space!* New York: McGraw-Hill (1969), 64 pp. Grades 5–8.
18. LEY, WILLY. *Our Work in Space.* New York: Macmillan (1964), 142 pp. Grades 6–9. Concentrates on Telstar and communication, Tiros and weather, Transit and ship guidance.

19. NICOLSON, LAIN. *Exploring the Planets*. New York: Grosset and Dunlap (1971), 172 pp. Grades 5–8.
20. OLNEY, ROSS R. *Americans in Space*. New York: Nelson (1971), 188 pp. Revised edition. A history of manned space travel.
21. ROSS, FRANK, JR. *Model Satellites and Spacecraft: Their Stories and How to Make Them*. New York: Lothrop, Lee and Shepard (1969), 159 pp. Grades 5–8.
22. SIMON, TONY. *The Moon Explorers*. New York: Four Winds (1970), 128 pp. Grades 5–8.
23. SMITH, NORMAN F. *Uphill to Mars: Downhill to Venus*. Boston: Little, Brown (1970), 128 pp. Grades 4–6.
24. THARP, EDGAR. *Giants of Space*. New York: Grosset and Dunlap (1970), 128 pp. Grades 6–9.

For films and filmstrips, see the *Educational Media Index* and catalogues of publishers.

23

Studying Personal Problems and Problems of the Local Community, the United States, and the Rest of the World

"We are living in an age when America seems to be bursting with issues and problems. We must secure civil rights for all our citizens. We must end poverty. Our economy must grow in all parts of the country. Automation and technology must create jobs, not more jobless. Old age must be welcomed with serenity and lived in dignity. We must rebuild cities, revitalize rural areas. We must provide wholesome leisure activity and recreation. We must conserve national resources."

In that one paragraph, Hubert Humphrey summarized many of the pressing problems of our times. It is an agenda for action today and tomorrow. Yet he did not mention all of the problems of this period of history. Many more might be added to round out his remarkable list.

These and other significant problems ought to be the stuff of social studies teaching. But are they? Perhaps we devote a little time to some of them in a current events lesson once a week, if the papers have arrived on time. Often in the senior year of high school we have a course on Problems of Democracy, but all too often that is an elective and offered largely for those who are not going on to college. Problems of Democracy are apparently appropriate only for the nonacademic student. The college-bound, some people say, should not be subjected to such simple stuff.

And we protect elementary school and junior high school pupils from the life around them. Our social studies courses are usually a mockery of our contention that we are helping boys and girls to learn about themselves and the world around them in order that they may comprehend themselves and the world better, cope with it more effectively, and contribute as much as possible to the segments of society in which they are members now or will soon be. We wrap early adolescents in cellophane or absorbent cotton or try to hide them from the realities of the world around them. This is just as true of suburban children as of inner-city children, perhaps more so, as Dr. Alice Miel has pointed out so cogently and dramatically in her superb pamphlet on *The Short-Changed Children of Suburbia*.

All too often we require students in the sixth or seventh grades to examine in detail Old World Backgrounds or Life in the American Colonies rather than letting them wrestle with some of the contemporary problems which they see over television, read about in the newspapers and magazines, hear about at the family dinner table or elsewhere, and often experience in the raw in our communities. By the time pupils have reached the sixth or seventh grades, many of them are eager to be a part of the adult world. They are increasingly conscious of the world around them.

Their lives are pivoted around the present rather than the past. Why, then, should chronology rather than contemporaneity always be the center of our social studies curriculum? Why should history, rather than current affairs, always or almost always be the focus of social studies learnings in these grades?

In the concluding chapter of their provocative volume on *Social Studies in the United States: A Critical Appraisal* (Harcourt Brace Jovanovich, 1967), C. Benjamin Cox and Byron G. Massialas state that "The point is that the entire social studies program — the curriculum, the instructional process, textbooks, materials, audio-visual communication media, etc. — must focus on controversial issues and must produce appropriate models for dealing with them in a spirit of inquiry."

Shirley Engle of Indiana University has suggested almost the same approach in an article on "Decision Making — The Heart of Social Studies Instruction," which appeared in *Social Education* in November, 1960. In that article, he stated that:

> The mark of the good citizen is the quality of decisions which he reaches on public and private matters of social concern. The social studies contribute to the process of decision making by supplying reliable facts and principles upon which to base decisions — they do not supply the decisions ready made. The facts are there for all to see but they do not tell us what to do. Decision making requires more than mere knowledge of facts and principles; it requires a weighing in the balance, a synthesizing of all available information and values.

At the present time it is probably impossible to organize all of our social studies offerings upon the basis of controversial issues. This author feels that any one approach, including that of controversial issues, is limiting. But we certainly should do far more than we are now doing to enable boys and girls and young people to be confronted by some of the pressing issues of the world in which they live.

To this writer there seems no excuse for limiting the study of problems primarily to the senior year of high school — and to a limited group of students. Why should there not be at least one year in the upper years of elementary school in which pupils explore personal problems, problems of the local community and region, problems of the United States, and international problems?

Such a course might well be in the sixth, seventh, or eighth grade years, depending upon the rest of the social studies offerings in a given school or school system.

If the general approach suggested by this writer were followed, such a course would be offered in the seventh grade, after a year of studying selected nations of the world. Such a course would precede a year of United States history. Some readers may object to that order, saying that history should come first and then contemporary affairs. That is certainly the logical order, but the psychological order is probably the reverse. Pupils may be better motivated to study history if they have already taken a look at the contemporary scene and have begun to ask why many of today's problems exist. This order also gives boys and girls one more year in which to mature before they tackle a year of history.

Some schools may, however, want to offer United States history in the seventh grade and contemporary problems in the eighth grade. A few may want to substitute a course in personal, local, regional, national, and international problems for the usual ninth grade civics course.

There are several advantages to a course in contemporary problems, placed somewhere in the upper grades of elementary school or in junior high school. Here are a few of them:

1. Such a course would add relevance and reality to social studies instruction or learnings at this point in school.
2. Because of this, boys and girls would be more highly motivated to pursue social studies learnings.
3. Such a course would help to meet the desire of young adolescents to live, at least in part, in the world of adults.
4. This emphasis upon inquiry into current problems would enable teachers and pupils to use a wide variety of materials from current sources, in addition to textbook materials.
5. Such an approach would be problem-centered and would include much attention to the development of skills.
6. With such a frame of reference in the social studies, pupils would be able to draw material and ideas from all of the social science disciplines.
7. Such a course would provide help to many young adolescents in solving some of the personal problems which confront them at this critical juncture of their lives.
8. Through such a course pupils should begin to see the interrelated nature of local, regional, national, and international problems.
9. Such a course should broaden and deepen the concept of civic education, which is so important and which, nevertheless, has been so sterile in recent years.
10. Through such a course, young adolescents could take part in a limited number of action projects.

In the type of course proposed, the emphasis should probably be upon personal, local-regional, and national problems, leaving the problems of the entire world largely to the high school level. The international dimensions of a few problems might be studied so that pupils do not think all problems end at the borders of our nation or that we are the only nation with problems. However, there is certainly enough for a year's program if one attempts to do an adequate job on personal, local and regional, and national problems, without also attempting to do international affairs in the same year.

An unusual approach to the study of personal and group problems has been taken by the social psychologist Ronald Lippitt and two co-authors in a book called *Social Science Resource Book* (Science Research Associates, 1969). In it there are seven units on such topics as "Discovering Differences," "Friendly and Unfriendly Behavior," "Being and Becoming," "Deciding and Doing," and "Influencing Each Other." It is a book based on individual and social psychology.

Another unique approach to a much wider range of problems has been used by Philmore Wass in a book for the junior high school level on *We Are Making Decisions* (Ginn, 1972). By utilizing Life Episodes and Case Studies, the author involves the readers from the very beginning in the decision-making processes. After the pupils have wrestled individually and/or in groups with a problem or a cluster of problems, and have given their tentative answers, they are asked to read on, with background material provided. In the Teachers' Guide to this volume the author refers to this approach as helping pupils "to make sense out of life."

In the next few pages we shall suggest some of the problems which might be considered in such a course. Then we shall propose a few of the major methods which could be utilized. Finally, we shall outline briefly three of the problems with which such a course might deal. Perhaps you would like to take another problem and show how you would develop it.

SOME POSSIBLE TOPICS FOR A COURSE IN CONTEMPORARY PROBLEMS

Obviously there are many problems which might be included in such a course in contemporary problems at the middle school period. Several of them are listed below. From them, teachers or teachers and pupils might select the ones they would like to study in the year. Probably it is best to select one or two early in the fall and decide upon the others later, rather than selecting the entire list at one time. In some instances personal problems will be handled by the guidance people or by home-room teachers. In such cases they could be omitted from the course in social studies. As we have said before, the local and regional and national dimensions of problems should be accented in this course, with some attention to the worldwide implications of the problem, leaving the international aspects largely to the senior year in high school if a course in World Problems is offered there.

In the chart below an x represents the aspects of a problem which might well be accented.

Problem	Personal	Local and/or regional	National	International
1. Who Am I?	x			
2. Learning to Live with My Family	x			
3. Making Friends	x			
4. Developing a Philosophy of Life	x			
5. The World of Work	x	x	x	
6. Poverty		x	x	x
7. Education	x	x	x	
8. Costs and Services of Government		x	x	
9. Safety, Law, and Order		x	x	
10. Air Pollution		x	x	
11. Prejudice	x	x	x	x
12. Labor and Industry		x	x	
13. Wise Use of Resources: Conservation		x	x	
14. Leisure and Recreation	x	x	x	
15. A Peaceful World			x	x
16. Planning	x	x	x	x
17. Consumer Education	x		x	
18. Our Economic System			x	
19. Our Values	x		x	x
20. Coping With Change	x	x	x	

Other Topics I Would Add:

SOME BASES FOR SELECTING PROBLEMS

Obviously, twenty problems are too many for a single year at this point in school. How, then, should a teacher or a school or a school system decide upon the ones to be selected? Here are a few possible criteria:

1. Problems not handled in other courses.
2. Problems of special importance to the pupils in your class.
3. Problems of special current interest, provided that is not the sole criterion.

4. Problems with which the teacher can deal effectively.
5. Problems on which adequate materials for your class (or school) are available or can be assembled.

A school or school system might have a minimum number of problems to be handled in the course of the year, with a further requirement that other problems be selected by the teacher and the pupils through cooperative planning. That would give teachers considerable freedom in the selection of topics. Better learning would result if the pupils were to take part in this selection process, too. Pupils should not be the sole determinants of the topics to consider, but they should have a part either in the selection of a small number of topics or in choosing between alternatives from a list already narrowed by the teacher.

If you are a beginning teacher in such a problems course or initiating such a course in your school, you may want to select as the first problem one in which you are interested and on which you have background, in order to feel more at ease in this initial undertaking. Of course it must be of interest to the pupils, too.

SOME MAJOR IDEAS TO STRESS

Such a course in personal, local and regional, national, and, to some extent, international problems should focus on the problem-solving method. Pupils should assemble information on a problem from a variety of sources. Then they should discover possible alternative solutions and test them out so far as that can be done, arriving at their own conclusions. This should mean that a wide variety of materials is used and that pupils are helped with the research skills they will need. There may be some common reading on each topic, in a textbook or in a pamphlet, but the reading certainly should not be limited to such a source.

Pupils should learn a great deal about solutions to problems. They should discover for themselves that some problems cannot be solved quickly, that other problems can and/or should be postponed, and that some cannot be solved in the foreseeable future, even though temporary or partial solutions need to be found for them.

Through their study, pupils also should learn that decisions about problems are made by different individuals and groups. Some decisions are made by individuals like themselves, alone, or with the help of parents, siblings, relatives, and/or friends. Some decisions are made by people in authority, whether they are officers of the law, mayors, governors, or Presidents, while other decisions are made by legislative bodies. In this connection, pupils should learn something about pressure groups and lobbies and public opinion. This theme of decision making should be struck often during a year spent in examining contemporary problems.

The importance of majority rule would be another idea to stress, together with the importance of minority rights. The important place of compromise should be discovered, too.

Doubtless some attention should be given to some of the propaganda devices used in creating or mobilizing public opinion, including name-calling, glittering generalities, testimonials, the "plain-folks" approach, card-stacking, and the "bandwagon" device.

Through the study of problems, boys and girls also should learn about some of the major groups in the United States, and in their local areas, which help to make decisions, directly or indirectly.

Through the study of a variety of problems, pupils would use concepts and generalizations from most of the social science disciplines. In units on individual and group behavior, they would be drawing from anthropology, sociology, and the related disciplines of psychology and social psychology. Since so many of our decisions now are determined or influenced by governmental decisions, they would be exposed to several aspects of that social science. Economics would play an especially important role in such a course as it vitally affects personal, local, national, and international decisions. Again and again pupils would discover the central idea of unlimited wants and limited resources.

In such a course, pupils will find that they need to examine some history in order to understand the present. That is fine. Teachers should utilize to the full such awareness on the part of pupils. This will make their history functional and meaningful and will prepare the way for later years when history becomes even more important in the curriculum.

What other "big ideas" would you want to stress in such a course?

SOME METHODS TO USE IN SUCH A COURSE

Some of the methods used in a course in contemporary problems will be similar to those already used by pupils in previous courses. They will need to do a great deal of reading. They will need to use films and filmstrips. They will find records and tape recordings useful at many points.

But there are other methods which should be more widely used than before because of the increasing maturity of pupils and/or of the basic themes of such a course. Problem-solving should be at the center of the methodology of such a program, with many types of references used by pupils to gather data. Notetaking of several kinds should be stressed, too. Pupils will now see the need for increased ability in the cluster of skills called note-taking. Some pupils, alone or in small groups, can interview local citizens and report their findings to the entire class, often tape-recording their interviews. Before this is done, considerable practice should be given to the pupils through role-playing in class. Panels, round tables, and debates also can be used more than in previous grades. And radio and television programs should certainly be utilized often in such a course. Often individual reports can be given and a great deal of use should be made of committee work.

Let us examine very quickly three topics or problems which might be included in a course in contemporary problems. Then you may want to prepare a similar statement on a problem in which you are particularly interested.

WHO AM I?

Perhaps the most significant contribution teachers of the social studies can ever make to their pupils is in helping them to understand themselves a little better than they have previously done. This is true of boys and girls at any stage of their development, but it seems to this writer particularly important with young adolescents. That is one of the most crucial periods in life, as they begin to cross over the invisible line that separates childhood from youthhood. For some, this is a relatively easy transition. But for most it is a difficult time. It is an interval of anxiety, a time of tensions, and a period of bewilderment. Much of the time, boys and girls want to be treated as adults, but some of the time they want to be treated as children. They want the best of both worlds, it seems. In the words of A. A. Milne, adapted for this purpose, they are "halfway up the stairs and halfway down."

Perhaps this is the best time in the years in school to introduce a thorough study of people, drawing heavily from the field of psychology, but also borrowing ideas from the other behavioral sciences. Possibly this study would be a lengthy unit on "People," with many subtopics. Possibly it could be a series of units, with such titles as "Who Am I?", "Learning to Live With My Family," "Making Friends," "Developing a Philosophy of Life," and possibly "The World of Work." Different teachers and different schools or school systems might well handle this general approach in similar yet varied ways.

Some readers may say immediately that such a subject is the responsibility of homes. Granted. We agree that it should be. But we also know that many parents cannot handle this topic effectively and others will not. Where they will, there is little loss in another look at some of the problems at school. Where they will not or cannot, the school certainly has a major responsibility to help. It is barely possible that teachers, despite their limitations, are better qualified in most instances to help than parents. They certainly are less involved emotionally than parents, and often are better qualified by training in psychology.

Other readers, or the same ones, may demur by saying that such a subject should be handled by homeroom teachers, by the health and physical education personnel, or by the guidance counselors. In many cases, that is true. If they are better qualified than social studies teachers, let them handle this part of the ongoing responsibility of the schools. It may be that this is one of the best areas for team-teaching, including some of the persons already mentioned in the "team." It might also be a good topic for core programs where they exist.

A number of related tasks need to be done with sensitivity, with adequate knowledge, and with persons equipped with the ability to communicate freely, frankly, and sympathetically with young adolescents. This writer would hope that thousands of social studies teachers would meet those qualifications.

If you agreed with the writer earlier in this volume when he defined the social studies field as primarily centered on an understanding of people, then there should be little doubt that this is an area in which social studies teachers should work.

Let us take one small part of this overall objective and see what it might mean in terms of content. That is the topic of "Who Am I?" Such a unit might well come at the beginning of the seventh grade, especially where boys and girls are entering junior high school for the first time. The teacher could make it an opportunity to get acquainted with his or her pupils — and for them to learn something about their new teacher. Some teachers, however, will feel that it should come later on in the year when an atmosphere of mutual trust has been developed and pupils will be willing to speak frankly and freely about their problems with one another and with the adult in their midst.

What would you handle in such a unit? Here are some topics you might want to consider:

> The Uniqueness of Every Person; What We Mean by Personality.
> How People Learn about People.
>> A Short and Simple History of the Field of Psychology.
> Our Physical Characteristics.
>> What We Inherit — The Place of Heredity.
>> What We Acquire — The Place of Environment.
> Our Basic Needs as Individuals.
>> Our Physical Needs.
>> Our Psychological Needs.

Security, Acceptance and Love, Recognition, Independence and
    Other Needs.
Our Abilities and Aptitudes.
    Our Strengths and How We Can Capitalize Upon Them.
    Our Weaknesses and How We Can Overcome at Least Some of Them.
Our Emotions.
    Ways of Handling Frustrations, Problems and Conflicts.
    The Fighting Approach.
    The Flight Approach.
    Facing Problems.
    Some of the People Who Can Help Us.
Our Futures.

Perhaps you will want to rewrite this brief outline to suit yourself in the light of
what you know about young adolescents and about how you think you might be able
to help them. Fine. The above outline is merely a suggestion, a "starter."

From such a unit you might then be ready to move with a class onto other short
units, closely related to this first one. The next topics might well depend upon your
class. For some, the next topic might be Understanding the Opposite Sex; for
others, it might be Making Friends or Learning to Live with Our Families. Still
other groups might be ready to move into the theme of Developing Our Philosophy of
Life.

Some teachers might like to put such units into a broader frame of reference,
centering them upon the field of anthropology. From that point, references would be
made to the problems of young adolescents in our society.

In handling such a topic, much depends upon the teacher. If possible, this should
be a topic or unit on which no grades are given. Instead, a brief written statement
or checklist might be filled out by the teacher and by each pupil and a conference
held to compare notes. This is the type of unit for which outside speakers should be
used frequently. They might include doctors, guidance people, psychologists or psychiatrists, youth workers, and parents (probably of children in other classes to save
embarrassment to the pupils in the class). Tape recordings can be used effectively.
Films and filmstrips should be useful. Some use should be made of biographies,
ranging from the lives of prominent psychologists like Freud, Jung, and Adler to stories of the Menningers and the Mayo Brothers. Visits might well be made to a
number of local institutions interested in psychology and young adolescents.

Some teachers may want to open such a unit with a question period, based on
anonymous questions placed in a box on the teacher's desk. To protect yourself, you
may want to go over those questions in advance and have some of them saved for
"experts" who will be brought in later in the unit. At the beginning of the unit or
sometime toward the beginning, you might like to administer the Junior Inventory
of Science Research Associates or some similar test. You might like to use a checklist of problems of young adolescents.

A few teachers might like to consider having pupils keep personal diaries
throughout this unit. Comments should be made on them but no grades should be
given. Autobiographies can serve useful purposes, too.

Materials are not as ample for such a unit as one would like them to be. However,
there are some brief, helpful pamphlets and a few books. The Junior Guidance
Series of Science Research Associates includes such titles as "All About You,"
"Your Problems: How to Handle Them," "You and Your Problems," and "Finding

Out About Ourselves." A very simple booklet, filled with situations involving young adolescents is "The Person You Are" in the Turner-Livingston Reading Series published by Follett. A much more mature approach is contained in Neill A. Rosser's booklet on "Personal Guidance" (Holt, Rinehart and Winston). One of the best books available is Dr. Joseph Noshpitz's *Understanding Ourselves: The Challenge of the Human Mind* (Coward-McCann). Teachers undoubtedly will be able to find other books in their school and/or local library.

There are two textbooks which approach this subject from the standpoint of upper grade boys and girls or junior high school pupils. One is the volume by Ronald Lippitt, Robert Fox, and Lucille Schaible, entitled *Social Science Resource Book* (Science Research Associates, 1969). As already pointed out, this book is primarily a text in psychology and social psychology. In simple language and with the use of black and white pictures, this volume confronts young adolescents with themselves and others and helps them to think about living in today's world. The other text which carries out effectively the search for identity is called *We Are Making Decisions* (Ginn, 1972), written by Philmore Wass. Through Life Episodes and Case Studies, it challenges young adolescents to think deeply about many aspects of life today which affect them.

Some films and filmstrips are listed in the *Educational Media Index*. Teachers and school systems would do us all a service by preparing annotated reading lists for such a pertinent unit. Much work needs to be done on this important topic.

CONFLICTS AND CONFLICT RESOLUTION: WAR AND PEACE

As fellow passengers on the spaceship Earth, all the people on our planet are plagued and threatened, directly or indirectly, by a host of problems, including poverty, prejudice, and pollution; diseases, superstitions, and illiteracy. These are some of the major and persistent problems of life on our planet.

To that list some social scientists and some social studies curriculum planners are now adding conflicts and conflict resolution, war and peace. In recent years several research groups have been organized in the United States and in other nations to study intensively conflicts and conflict resolution, war and peace. A few curriculum centers have also been established to study the relation of these topics to curricula.

Broadly interpreted, these may be the central concepts of the social sciences. A good case could be made for such a statement because personal and group fulfillment, national existence and growth, and the survival of man on this earth depend upon adequate resolution of conflicts.

Unfortunately the word conflict usually has a negative connotation. So far we have used it in that sense. But that need not be so. In the best sense of that term there are positive as well as negative connotations, because all or most human creativity emerges from conflicts of many different kinds.

In times of change, conflict takes on added significance. No one can deny that change is an outstanding characteristic of our times. Perhaps the best current summation of that fact appears in Alvin Toffler's *Future Shock* (Random House, 1970). In the light of such varied and radical changes in society, conflict leaps to the top or near the top of any list of significant social science concepts.

Conflict as a Central Concept in the Social Studies Curriculum. If the foregoing statements are true, then conflict and conflict resolution should be given high priority in social studies curricula that are realistic, relevant, and meaningful. In any problem-centered social studies curriculum these concepts are even more pertinent.

There are two major ways to deal with conflict and conflict resolution in the social studies curricula of elementary and middle schools. One is to use these concepts as recurring themes in the social studies symphony. The other is to devote at least one unit, in depth, to this theme, with particular attention to the topic of war and peace. These are complementary rather than contradictory approaches. Both are highly recommended. At the secondary level three other approaches are suggested. One is a semester (or year) course on conflict and conflict resolution, war and peace. Another would be individual projects on some aspects of this theme, especially in schools where such individualized programs are encouraged. The third would be the organization of a co-curricular group interested in certain aspects of these broad themes.

In previous sections of this book we have frequently referred to conflict and conflict resolution as a theme in the study of families in the United States and in other parts of the world, in the study of communities here and in selected parts of our globe, and in the study of our nation and other nations.

Here we want to concentrate for a short time on the inclusion of a unit on war and peace in the upper levels of elementary schools, in middle schools, or in junior high schools.

A Unit on War and Peace.   By the sixth or seventh grades, pupils should be ready for an intensive study of the broad issues involved in preventing war and buttressing peace. By then boys and girls are old enough to understand that war is something more than the glamorous play with sticks in which they engaged as children. They will have studied the history of their own nation and its wars and probably the history of at least a few carefully selected other nations, including their wars. All this means that they will have data with which to work in such a study.

A unit on this theme should extend for at least four weeks. Hopefully it would last for six to eight weeks, in order to explore the many facets of the subject in considerable depth.

Such a unit would be a culmination of the on-going theme of conflict resolution, with applications now in the international field. At this level war would be seen on one end of a continuum as the failure of efforts to resolve conflicts. At the other end of the continuum would be peace and justice as the result of successful efforts to resolve conflicts. The two words "peace" and "justice" are linked here to indicate that peace is something more than the absence of war.

In such a unit, considerable attention would be paid to the causes of international conflict. Three in particular would be examined. One would be the economic causes, including the desire for raw materials, the desire for markets and investments, the need for outlets to the sea or to strengthen a frontier, the desire to protect economic holdings of an ally, and differences in economic systems. A second cause would be political, including intense patriotism, secret diplomacy, the "need" to have a foreign war, the desire for new territory, political independence, and alliances with other nations. The third would be the psychological causes, including propaganda, sympathy for "the oppressed," religious differences, racial differences, fear, belief in the philosophy of force to settle disputes, and ideological disagreements.

Another major aim would be to explore the costs of wars, including indirect as well as direct costs. Pupils should also learn about the "opportunity costs" of such conflicts, realizing what has to be forfeited when limited goods and manpower are diverted into wars.

Probably the most important part of such a unit would be the study of alternatives to war and measures to build peace with justice. In some respects international conflicts are like conflicts between individuals or groups. Pupils can be encouraged to examine how these smaller conflicts are settled and then to see whether the same methods apply in larger disputes. Such methods might include isolation of the parties involved, outside intervention, and talking over points of disagreement. Discussions might well lead to a study of boycotts, blockades, armed intervention, an international peace force, and the place of international conferences and the international court. Longer-term programs for world peace could then be examined, with discussions of such topics as the use of world resources, world trade, the eradication of the causes of war, the place of decision makers, increased understanding of opponents, and outlets for man's aggressions.

Further ideas on this topic can be found in the references listed here.

Some Curriculum Centers Specializing in War and Peace

1. Center for Teaching about Peace and War, Wayne State University, Detroit, Michigan 48202.
2. Center for Teaching International Relations, University of Denver, Denver, Colorado 80210.
3. Center for War/Peace Studies, 218 East 18th Street, New York, New York 10003.
4. Harvard Social Studies Project, Harvard Graduate School of Education, 210 Longfellow Hall, Appian Way, Cambridge, Massachusetts 02130.
5. Minnesota World Affairs Center, 3300 University Avenue, Minneapolis, Minnesota 55418.
6. Program in War/Peace Studies for Teachers. University of Hawaii, Honolulu, Hawaii.
7. United Nations Association, 833 United Nations Plaza, New York, New York 10017.
8. World Law Fund, 11 West 42nd Street, New York, New York 10036.

A BRIEF BIBLIOGRAPHY FOR TEACHERS

1. BOULDING, KENNETH E. *Conflict and Defense*. New York: Harper and Row (1963), 358 pp.
2. KENWORTHY, LEONARD S. *The International Dimension of Education*. Washington: Association for Supervision and Curriculum Development (1970), 120 pp.
3. KING, DAVID C. *International Education for Spaceship Earth*. New York: Foreign Policy Association (1971), 166 pp.
4. MENDLOVITZ, SAUL H. "Teaching War Prevention" in *Bulletin of the Atomic Scientists*. (February, 1964).
5. NESBITT, WILLIAM A. *Teaching About War and War Prevention*. New York: Foreign Policy Association (1971), 166 pp.
6. WILSON, NORMAN H. "The New Social Studies: Understanding the Systems of War and Peace" in *Social Studies*. (March, 1969), pp. 119–124.
7. WRIGHT, QUINCY. *A Study of War*. Chicago: University of Chicago Press (1965). Abbreviated edition.

Teachers will find much helpful material in the magazine *Intercom*, published monthly by the Center for War/Peace Studies (address above).

A BRIEF BIBLIOGRAPHY FOR PUPILS

1. CARR, ALBERT Z. *A Matter of Life and Death: How Wars Get Started — Or Are Prevented.* New York: Viking (1966), 256 pp.
2. FEHRENBACH, T. R. *United Nations in War and Peace.* New York: Random House (1968), 192 pp.
3. FISHER, ROGER. *International Conflict for Beginners.* New York: Harper and Row (1969), 231 pp.
4. GREENE, WADE. *Disarmament: The Challenge of Civilization.* New York: Coward-McCann (1966), 119 pp.
5. LAWSON, DON (Editor). *Youth and War: World War One to Vietnam.* New York: Lothrop, Lee and Shepard (1969), 192 pp.

Several pamphlets issued in the Harvard Public Issues Series and published by the American Education Press (Education Center, Columbus, Ohio 43216) deal with this topic. Among them are "Diplomacy and International Law," "The Limits of War," "Organizations Among Nations," and "Revolution and World Politics."

For simulation games on war and conflict, see the booklet by William A. Nesbitt on *Simulation Games for the Social Studies* (Crowell, 1970).

ENVIRONMENT, ECOLOGY, CONSERVATION

For centuries we Americans have assumed that our resources were limitless. We have stripped our forests, wasted our farmland, and destroyed our animals and birds. We have turned our waterways into giant sewers and our skies into murky smudges. We have clogged our streets with cars and turned many parts of our countryside into junkyards. Today large parts of our nation represent America the Beautiful, but increasingly large sections represent America the Ugly.

We have done even more damage to our environment than that. We have threatened the life support systems of water and air. We have curbed or destroyed plant life and thereby limited our oxygen supply. We have assumed that the resources of the spaceship Earth are infinite when they are in fact finite. We have even upset the interior conditions of the earth's crust.

And still the wantonness and waste continue. For example, in 1900 we were using 8 per cent of our available water; in 1970 we were using 80 per cent of this supply.

And what of the future? By the year 2000, with a population of approximately 300 million persons, we will need more timber, more minerals, more fuel, more water for personal consumption and industrial use, more food, and more space for recreation.

We are not the only plunderers of our planet. But the toll we have taken and are currently taking is the greatest of any group on earth. For instance, the impact of one American on our environment is as great as that of 25 people in India.

Much of this destruction has been done to promote progress or a higher standard of living, but at a terrific cost to our environment and therefore to ourselves and unborn generations.

This is a sad story. But it is one that boys and girls should learn about early. Parts of it can be taught in the primary years. Some of it can be taught as boys and girls learn about United States history.

Of course you will want to stress the optimistic as well as the pessimistic side of this story. For many years there has been a conservation movement in our nation, led by such persons as Audubon, Muir, Pinchot, and Teddy Roosevelt. In recent years there has been a resurgence of interest in conservation, with many concerned citizens organizing to improve the environment locally, in their states, and in the nation. Such groups have carried on educational programs and political action programs. Often they have met with considerable success. Whether this is a temporary flurry of interest in a popular cause or a continuing movement remains to be seen.

The study of the environment, known as ecology, has also developed as a specialized subject, with experts delving deeply into the intricate and seamless web of interrelationships in nature.

Considerable work can be done on the subject of the environment with young children, providing a bridge between science and the social studies. This is another broad theme which should be considered at many grade levels. By the time boys and girls reach the sixth, seventh, or eighth grades, they should certainly have at least one major unit on the environment. In such a unit you will certainly want to concentrate upon a few "big ideas." You will want to determine the extent of our natural resources and how man has exploited the earth. You will want to learn about individuals and groups who have been concerned and are now concerned about the wise use of our resources. You will probably want to trace the development of the national park movement, study the formation of the U.S. Forest Service and learn about the movement for multipurpose dams and the planning of river valley development programs, as well as many other aspects of our contemporary conservation movement.

This is a topic which lends itself to committee work even more than some others. You might well organize the class into committees to study the conservation of (1) water, (2) soil, (3) grasslands, (4) forests, (5) wildlife, (6) minerals, and (7) human resources. An overview of the topic should be gained by the class as a committee-of-the-whole. Then the committees can go to work on the specifics just mentioned, reporting their findings to the group.

The methods you use can be many and varied. High on the list should be direct experiences through field trips and through contacts with the personnel of private and governmental agencies interested in conservation. Biographies can be explored. Pictures, films, and filmstrips abound on this topic and should be widely used. Maps can be made of the areas of destruction, of the Dust Bowl, of government-owned land and of national parks. Bulletin boards can be developed. Interviews with persons interested in conservation can be conducted and reported back to the class.

Some pupils could be encouraged to develop posters or simple cartoons depicting some aspect of the ecology movement of interest and importance to them.

If possible, some action project or projects should be carried on. These can range from the construction of a feeding station for birds to writing letters to legislators about conservation measures, and from the development of nature trails to investigation of some local problem connected with the wise use of resources. Pupils at this stage in their development need to be intimately involved in action, and conservation is a topic about which they can do something — now.

There are a good many organizations interested in conservation in the United States. Many of them have local chapters. Among the best-known of these are the following:

The National Audubon Society, 1130 Fifth Avenue, New York, New York
Sierra Club, 1050 Mills Tower, San Francisco, California
National Wildlife Federation, 1412 Sixteenth Street, N.W., Washington, D.C.

American Forestry Association, 919 Seventeenth Street, N.W., Washington, D.C.
Defenders of Wildlife, 731 DuPont Circle Building, Washington, D.C.
Wilderness Society, 729 Fifteenth Street, N.W., Washington, D.C.
Soil Conservation Society of America, 838 Fifth Avenue, Des Moines, Iowa
Izaak Walton League, 1326 Waukegan Road, Glenview, Illinois
World Wildlife Fund, 1816 Jefferson Place, N.W., Washington, D.C.

Teachers and/or pupils should also contact the local branches of such federal agencies as the U.S. Forest Service, the U.S. Soil Conservation Service, and the U.S. Department of the Interior, as well as state agencies.

There are many films and filmstrips on conservation which may be used to advantage. They can be found in the volumes of the *Educational Media Index* and in other lists. On the next page are a few references for teachers and a number of books for pupils on various aspects of the topic, The Wise Use of Our Resources.

BOOKS FOR PUPILS ON THE ENVIRONMENT

1. ADRIAN, MARY. *Secret Neighbors: Wildlife in a City Lot.* New York: Hastings House (1971), 64 pp. Grades 3–5.
2. BIXBY, WILLIAM. *A World You Can Live In.* New York: McKay (1971), 144 pp. Grades 5–8.
3. COLBY, CARROLL B. *Soil Savers: The Work of the Soil Conservation Service of the United States Department of Agriculture.* New York: Coward-McCann (1957), 48 pp. Grades 5–8.
4. DUFFEY, ERIC. *Conservation of Nature.* New York: McGraw-Hill (1970), 128 pp. Grades 7–10.
5. ELLIOTT, SARAH M. *Our Dirty Air.* New York: Messner (1971), 64 pp. Grades 4–6. Illustrated with several black and white photographs.
6. EVANS, EVA KNOX. *The Dirt Book: An Introduction to Earth Science.* Boston: Little, Brown (1969), 86 pp. Grades 5–7.
7. FABER, DORIS. *Captive Rivers: The Story of Big Dams.* New York: Putnam (1966), 159 pp. Grades 5–8.
8. HARRISON, C. WILLIAM. *Conservation: The Challenge of Reclaiming Our Plundered Land.* New York: Messner (1964), 191 pp. Grades 6–8.
9. HEFFERNAN, HELEN and GEORGE SHAFTEL. *Man Improves His World.* Syracuse, New York: Singer (1964), 100 pp. Grades 4–6. Separate paperbacks on *The Water Story, The Soil Story, The Mineral Story, The Forestry Story, The Fisheries Story, The Energy Story,* and *The Wildlife Story.*
10. HILTON, SUZANNE. *How Do They Cope With It?* Philadelphia: Westminster (1970), 144 pp. Grades 7–10.
11. HIRSH, S. CARL. *Guardians of Tomorrow: Pioneers in Ecology.* New York: Viking (1971), 192 pp. Grades 7–10.
12. HITCH, ALLEN S. and MARIAN SORENSON. *Conservation and You.* New York: Van Nostrand Reinhold (1964), 126 pp. Grades 6–9.
13. HYDE, MARGARET O. *For Pollution Fighters Only.* New York: McGraw-Hill (1971), 157 pp. Grades 6–9.
14. LAYCOCK, GEORGE. *America's Endangered Wildlife.* New York: Norton (1969), 226 pp. Grades 7–10.
15. LEAF, MUNRO. *Who Cares? I Do.* Philadelphia: Lippincott (1971), 40 pp. Grades 1–3. Several pictures as well as his famous drawings.
16. MAY, JULIAN. *Blue River.* New York: Holiday (1971), 40 pp. Grades 2–3. Colored drawings.
17. PRINGLE, LAURENCE. *Basic Ecology.* New York: Macmillan (1971), 152 pp. Grades 6–9.

18. PRINGLE, LAURENCE. *The Only Earth We Have.* New York: Macmillan (1971), 95 pp. Grades 5–8.
19. RANDALL, JANET. *To Save a Tree.* New York: McKay (1971), 144 pp. Grades 4–6. The story of the fight to preserve the West Coast redwoods.
20. SCHLICHTING, HAROLD E., JR. and MARY S. *Ecology: The Study of Environment.* Austin, Texas: Steck-Vaughn (1971), 48 pp. Grades 4–6.
21. SIMON, HILDA. *Ecology in the Back Yard.* New York: Vanguard (1971), 96 pp. Grades 5–8.
22. TALLEY, NAOMI. *To Save the Soil.* New York: Dial (1965), 76 pp. Grades 6–9.
23. WARNER, MATT. *Your World: Your Survival.* New York: Abelard-Schuman (1970), 128 pp. Grades 5–8.

POVERTY

Despite the affluence in the United States today, there are pockets of poverty in every part of our nation. There is hunger in the midst of abundance, ill-health in a nation which has the latest medical know-how, unemployment at a time of astonishing production, and illiteracy in an otherwise literate populace. "There is an ugly smell rising from the basement of the stately American mansion," is the comment of Gunnar Myrdal, the astute Swedish economist and student of American life for many, many years.

In 1964 there were thirty million people in our nation who lived in families earning less than $2000 annually, plus three million single persons who earned less than $1500 a year. If the figure for poverty is placed at $3000 per year per family, one family in every five in the U.S.A. is included.

Shall we ignore these ugly facts or try to hide them from the oncoming generation, telling them that all is well in the United States today? Or shall we encourage them to discover such facts, learn about what is being done now, and what can be done in the future to eradicate or alleviate such conditions, and become concerned about such conditions?

Even if we tried, we could not hide these facts from many pupils. They live in families which are existing at the poverty level. They have eaten of the fruits of poverty and know that they are bitter. They already bear the scars of poverty, physically and psychologically. Others have seen poverty and have had friends who were poor. It is the writer's considered conviction that those who do not know such facts, should discover them. They should not continue to believe that everyone lives in affluence or that the poor are just "dumb clucks" who don't want to do anything about their conditions. Concern for the well-being of every American is certainly one of the values we should all be trying to develop in our teaching.

Very little attention should be given to poverty with primary and intermediate grade children. Little is to be gained from such studies and some harm can be done. But by the time boys and girls have reached the middle school or junior high school years, they should begin to struggle with some of the seamier sides of our national life. They should discover the facts about poverty, explore ways and means of improving conditions for one-fifth of our families, and develop some concern about this problem. For those who have grown up in poverty and now live at that level, there is great value in discovering why such conditions exist and what is being done or can be done about them. More than anything else, they can learn the advantages of education for themselves as they think in terms of their own future.

Various teachers will want to approach this topic in different ways. Some will want to concentrate on poverty in the U.S.A., with only passing references to poverty in the wider world. Others will want to start with the worldwide scene, placing poverty in the United States in that international setting. Both ways have merit, depending upon the class involved.

In looking for ways of reducing poverty, pupils should examine at least three alternatives in the United States. One would be economic growth, with full employment. A second would be the guaranteed minimum annual income or the negative income tax approach. The third would be the welfare system. They should consider these as alternatives, finding the pros and cons of each approach.

One of the most effective methods of making such material live for boys and girls who have no conception of poverty, is to have them prepare annual budgets for families in some part of the world on the basis of $100 or less per year per person, or, in the U.S.A., for a family of five with an annual income of $200 or so per person. This can be a startling experience for many pupils and worth all the time and effort that goes into it.

For teachers desiring background on this topic, here are a few references. They also may be of use with a few of the more able readers in a seventh grade class, who could report to the class their findings or bring the information they obtain to a committee group, with the committee taking notes and preparing graphs and other pictorial mattter based upon these data.

1. BAGDIKIAN, BEN H. *In the Midst of Plenty: A New Report on the Poor in America.* New York: Signet (1964), 160 pp. A paperback.
2. BLAUSTEIN, ARTHUR I. and ROGER R. WOOCK (Editors). *Man Against Poverty: World War III.* New York: Vintage (1968), 456 pp. A paperback.
3. FERMAN, LOUIS and others. *Poverty in America: A Book of Readings.* Ann Arbor, Michigan: University of Michigan Press (1965), 532 pp.
4. HARRINGTON, MICHAEL. *The Other America: Poverty in the United States.* Baltimore, Maryland: Penguin (1962), 203 pp. A paperback.
5. HUMPHREY, HUBERT. *War on Poverty.* New York: McGraw-Hill (1965), 206 pp.
6. LEINWALD, GERALD. *Poverty and the Poor.* New York: Washington Square Press (1970), 96 pp. In the Problems of American Society series.
7. LENS, SIDNEY. *Poverty: America's Enduring Paradox.* New York: Crowell (1969), 341 pp.
8. McIVER, R. M. *The Assault on Poverty.* New York: McGraw-Hill (1965), 206 pp.
9. MAY, EDGAR. *The Wasted Americans.* New York: Harper and Row (1964), 227 pp.
10. ROMASCO, ALBERT U. *The Poverty of Abundance.* New York: Oxford University Press (1965), 282 pp.
11. SHEPPARD, HAROLD (Editor). *Poverty and Wealth in America.* Chicago: Quadrangle (1970), 279 pp.

Unfortunately there is very little material as yet for the junior high school age on this topic. The only textbook with material on poverty in it is Philmore Wass's volume on *We Are Making Decisions* (Ginn, 1972). The Leinwald volume cited above is for high school students, but some junior high school pupils can use it. Some of the pamphlets of the Public Affairs Committee are suitable for good readers in junior high schools. (Write them for their latest titles.)

In developing a unit on poverty in the U.S.A. or around the world, teachers should rely heavily upon current magazine and newspaper articles.

Three films on the topic of poverty which are available are "Superfluous People" (McGraw-Hill Company) — the story of an Illinois dropout, "America's Crises: The Hard Way" (Indiana University), and "Christmas in Appalachia" (Carousel). Teachers may be able to find other films and filmstrips by the time they begin to teach such a unit.

There is not space here to outline an entire unit on poverty in the world and/or in the United States. But it may be helpful to some readers to have a brief outline of some of the main points which might be considered in working on such a unit, or on two units which are closely related:

A. *In the world*

    1. What is the income of most of the people of the world?

    2. What are the results of that income in terms of food, clothing, shelter, health, education, etc.?

    3. Where do most of the people of low income live? What explanations are given for this belt of poverty around the world?

    4. What is being done to improve standards of living?
       By the national governments?
       By the governments of other nations?
       By the United Nations and its agencies?
       By private organizations?

B. *In the United States*

    1. To what extent does poverty exist in the United States?
       (What is meant by the term poverty in U.S. terms?)

    2. In what parts of the United States does poverty exist?
       By regions.
       Rural, urban, and suburban areas.

    3. Who are the people who live at the poverty level?

    4. What are the effects of poverty on health, education, housing, crime, etc?

    5. What seem to be the causes of poverty according to experts?

    6. What is being done today?
       By local and state governments?
       By private organizations?
       By the federal government?

    7. What other plans are suggested by authorities on poverty? What are their merits? Demerits?

Such an outline would need to be modified by teachers in various parts of the United States according to the needs of their own pupils.

An outline like this should not, however, be presented to pupils. This topic is a natural for problem-solving, and a plan like that above should evolve from the class itself, with assistance from the teacher. The pupils should be confronted with some of the basic facts of poverty in the U.S.A. and/or the world through charts, films, filmstrips, an outside speaker, or in some other way. Once interest is aroused, the members of the class should gather as much relevant material as possible. Then they should begin to organize it, probably agreeing upon a general outline for everyone. This might well be in the form of questions, as indicated above.

STUDYING AFRICA IN ELEMENTARY AND MIDDLE SCHOOLS

In the last few years Americans in general and American educators in particular have begun to pay some attention to Africa. After centuries of neglect, we have suddenly realized that there is an entire continent and over 325 million people about whom we know very little and about whom we have taught almost nothing in our schools. Furthermore, what we *have* taught has often been erroneous and misleading.

Many factors have brought about this interest in Africa and Africans. Most important has been the rise of new nations in that part of the world and their increasing role in world affairs. Linked with this has been the growing awareness on the part of blacks in the United States of the need for knowledge about their heritage from that part of the world.

As a result of these and other factors, many schools and many school systems have introduced some study of Africa into the curriculum. Usually this has been in a world cultures course in the 9th and/or 10th grades. Sometimes Africa has been introduced in the elementary and middle schools.

As one of the people who has written and spoken for years on this subject and as one who developed years ago an African Curriculum Center at Brooklyn College, this author welcomes the belated interest in Africa and Africans. They represent a large and important part of our world community. They represent a rich cultural heritage. In their continent are found many of the resources and potential resources of our globe. Today they are wrestling with many of the problems with which we are also wrestling — such as industrialization, urbanization, race relations, and conflicts over values. Furthermore, the people of Africa should be of special interest because they represent the heritage of a large segment of our population. It is high time that we devoted an important part of our curriculum to this part of the world.

However, there are several dangers apparent today in our teaching about Africa and Africans. They vary from school to school and school system to school system.

In some schools and school systems teachers have been pressured to devote all or most of their time in the social studies field to Africa and to the black heritage in the United States. Without any background, or with only a poor background, teachers have suddenly begun to teach these two important aspects of the social studies. Consequently, teachers have taken any materials they could find and tried to use them as adequately as they could. Too often this has resulted in repetition, superficiality of approach, and an undue emphasis on one important aspect of the curriculum. In other places this sudden interest in Africa and Africans has resulted in a perpetuation of the many stereotypes so many of us have of that part of the world. Far too often teachers have lumped the entire continent together and taught as if all Africans were alike. Perhaps our greatest need is to discover the diversity of that part of the world. There are approximately 50 nations and territories, hundreds of tribes, and over 325 million people. It is false and misleading to think of them as essentially alike. In no part of the world is there such diversity.

What we need desperately today is a comprehensive, cumulative, and coherent plan for studying Africa and Africans throughout the school years, K–12.

Many such programs or plans can be devised. One would be to place Africa in perspective in an overall curriculum proposal. Children in the primary grades might be introduced to a small number of carefully selected families in a few parts of Africa. In the middle grades pupils would become acquainted with a small number of carefully selected communities in the continent of Africa. In the sixth grade pupils would study two or three nations in Africa. Then, in the 9th or 10th grades, they would study the continent of Africa or Africa south of the Sahara as a cultural region. In the upper grades of high school they would again study Africa as a part of the international or world community of our day, with emphasis upon some of the problems of our contemporary society.

Is such a plan feasible? Are there adequate materials for such a study? One answer is contained in the new social science textbook series from Ginn and Company. In it children are first introduced to Africa in the second grade by the study of the Donkor family in Ghana. This is an actual family which grows cacao to earn its living. The picture is one of changing village life, with the healthmobile coming into the village weekly as a part of the government's attempts to improve health. In the fourth grade of this series, boys and girls learn about life in Moshi, Tanzania and the area around that town. The people are mostly members of the progressive Chagga or Wachagga tribe who grow coffee and sell it through cooperatives. In addition, they study the changing city of Dar-Es-Salaam, the capital of Tanzania and a seaport town. In the sixth grade pupils learn about Nigeria as the largest nation of the continent and the one with the most viable economy. And they are introduced to Kenya in a supplementary paperback.

This is only one approach, but it is one that is new and exciting and tries to expose American boys and girls to several parts of Africa in considerable depth, rather than trying to expose them to the entire continent in a superficial way. How do you react to this general plan? What are its strengths? What are its weaknesses? What plan would you propose for a K–8 or a K–12 program?

It should be obvious, but perhaps it is important to say that all pupils in American schools need to study about Africa and Africans. This is an important part of the curriculum for blacks because it is a part of their heritage. But it is an important part of the curriculum for whites, too. Their stereotypes need to be shattered. They need to learn about the great empires which have existed in the African continent. They need to learn to appreciate how people in that part of the world have adjusted well to their environment and are now adjusting to a new and changing world.

Above all, boys and girls need to meet Africans. There are thousands of Africans now in the United States, largely as students. They are here to obtain further education, and that comes first in their list of priorities. But many of them would welcome an opportunity to meet with American pupils. Such resource persons should be carefully chosen. Not all of them can communicate well with boys and girls. But a program about Africa without Africans is likely to be limited and "thin."

Up-to-date pictures, films, and filmstrips are also highly important in any study of this part of the world. Too many children (and teachers) think in terms of an Africa which never existed or is rapidly disappearing. Pictures of Africa today should help pupils to learn much about the rapid changes which are occurring in that part of our planet.

Some Resources for Learning About Africa and Africans

It is impossible to go into depth about the study of such a large part of the world in the compass of a few pages in this book. But there are several recent publications wherein you can obtain further leads and the names of several organizations from which you can obtain useful background either for yourself or for your pupils.

Here are a few suggested pamphlets which are devoted to the study of this part of our planet in American schools:

1. *Africa: A List of Printed Materials for Children.* New York: Information Center on Children's Cultures: United States Committee for UNICEF (1968), 76 pp. Excellent annotated lists of books for children.
2. *Africa in the Curriculum.* Special issue of *Social Education* magazine for February, 1971. 103 pp.
3. *Africa: Its Historic Past, Its Turbulent Present, Its Exciting Future.* Special issue of *The Grade Teacher* for October, 1968. 188 pp.

4. *Africa in the Curriculum.* By Beryle Banfield. Published by the Edward W. Blyden Press (Box 621, Manhattanville Station, New York, New York 10027), 1968. 124 pp.
5. *Are You Going to Teach About Africa?* New York: African-American Institute (1970), 85 pp.
6. *Studying Africa in Elementary and Secondary Schools.* By Leonard S. Kenworthy. New York: Teachers College Press (1970), 74 pp.
7. *Teaching About Africa.* Albany, New York: The State Education Department (1970), 185 pp. Intended for teachers of 9th grade classes but of value to teachers in other grades.

Some of the organizations specializing in the field of Africa and African studies include the following:

1. African-American Institute, 866 United Nations Plaza, New York 10017.
2. American Committee on Africa, 164 Madison Avenue, New York 10016.
3. American Society of African Culture, 401 Broadway, New York 10013.

Two of the magazines devoted to the continent of Africa are:

1. *Africa Report.* African-American Institute, 505 DuPont Circle Building, Washington, D.C. 20036.
2. *Africa Today.* American Committee on Africa. Graduate School of International Studies, University of Denver, Denver, Colorado 80210.

For teachers interested in the music of Africa, the Scholastic-Folkways Company is probably the best source of records. Their address is 701 Seventh Avenue, New York, New York 10036. Small pocket handbooks of African songs may be purchased from the Cooperative Recreation Service, Delaware, Ohio 43015. Many teachers will be interested in the book by Betty Dietz and Michael Olatunja on *Musical Instruments of Africa* (John Day, 1965).

On fun and games, teachers should look at the booklet by Rose H. Wright on *Fun and Festival from Africa* (Friendship Press, 1967), Nina Millen's *Children's Games from Many Lands* (Friendship Press, 1965), the various books in the *Hi Neighbor* series (U.S. Committee for UNICEF), and Nina Millen's *Children's Festivals from Many Lands* (Friendship Press, 1964).

PROBLEMS I WOULD LIKE TO TEACH

(*Problem That I Should Like to Teach*)

Focusing on: personal dimensions   _____
             local–regional   _____
             national   _____
             international   _____

A brief statement of the problem as I should like to develop it with pupils:

Some of the major points I would hope to develop:

1.
2.
3.
4.
5.
6.
7.
8.
9.
10.

Possible introductory activities:

Some special methods to stress on this unit or problem:

Background references for teachers:

Some resources for use by pupils:

Addresses of Organizations and Publishers

24

ABELARD-SCHUMAN LIMITED, 257 Park Avenue, New York, New York 10019
ABINGDON PRESS, 201 Eighth Avenue South, Nashville, Tennessee 37202
ADDISON-WESLEY PUBLISHING COMPANY, INC., Reading, Massachusetts 01867
AERO SERVICE CORPORATION (see Nystrom)
ALADDIN BOOKS (see E. P. Dutton and Company, Inc.)
ALLYN AND BACON, INC., 470 Atlantic Avenue, Boston, Massachusetts 02210
AMERICAN BOOK COMPANY, 450 W. 33rd Street, New York, New York 10001
AMERICAN COUNCIL OF LEARNED SOCIETIES, 345 East 46th Street, New York, New York 10016
AMERICAN COUNCIL ON EDUCATION, 1 Dupont Circle N. W., Washington, D. C. 20036
AMERICAN EDUCATION PUBLICATIONS, Education Center, Columbus, Ohio 43216
AMERICAN FRIENDS SERVICE COMMITTEE, 160 North 15th Street, Philadelphia, Pennsylvania 19102
AMERICAN HERITAGE PUBLISHING COMPANY, INC., 551 Fifth Avenue, New York, New York 10017
AMERICAN LIBRARY ASSOCIATION, 50 East Huron Street, Chicago, Illinois 60611
ANTI-DEFAMATION LEAGUE OF B'NAI B'RITH, 315 Lexington Avenue, New York, New York 10016
APPLETON-CENTURY-CROFTS, 440 Park Avenue South, New York, New York 10016
ASIA SOCIETY, 112 East 64th Street, New York, New York 10021
ASSOCIATION FOR CHILDHOOD EDUCATION INTERNATIONAL, 3615 Wisconsin Avenue N. W., Washington, D. C. 20016
ASSOCIATION FOR SUPERVISION AND CURRICULUM DEVELOPMENT, 1201 Sixteenth Street N. W., Washington, D. C. 20036
ATHENEUM PUBLISHERS, 122 E. 42nd Street, New York, New York 10017

BANTAM BOOKS, INC., 666 Fifth Avenue, New York, New York 10017
BARNES & NOBLE, INC., 105 Fifth Avenue, New York, New York 10003
BASIC BOOKS, 404 Park Avenue South, New York, New York 10016
BEACON PRESS, 25 Beacon Street, Boston, Massachusetts 02108
BENEFIC PRESS, 10300 W. Roosevelt Road, Westchester, Illinois 60153
ROBERT BENTLEY INC., 872 Massachusetts Avenue, Cambridge, Massachusetts 02139
BLAISDELL PUBLISHING COMPANY, See Xerox College Publishing
BOBBS-MERRILL COMPANY, 4300 W. 62 Street Indianapolis, Indiana 46206
R. R. BOWKER COMPANY, 1180 Avenue of the Americas, New York, New York 10036
MILTON BRADLEY, 74 Park Street, Springfield, Massachusetts 01101
BUREAU OF PUBLICATIONS T. C. (see Teachers College Press)

CADACO-ELLIS, 1446 Merchandise Mart, Chicago, Illinois 60654
CALIFORNIA STATE EDUCATION DEPARTMENT, Sacramento, California 95800
CAPITOL PUBLISHING COMPANY, 850 Third Avenue, New York, New York 10022
CENTER FOR APPLIED RESEARCH IN EDUCATION, Washington, D. C. 20025
CHANDLER PUBLISHING COMPANY, 124 Spear Street, San Francisco, California 94105
CHANNEL PRESS (see Meredith Publishing Company)
CHILDREN'S PRESS, 1224 W. VanBuren Street, Chicago, Illinois 60607
CHILTON BOOK COMPANY, 401 Walnut Street, Philadelphia, Pennsylvania 19106
E. P. COLLIER and COMPANY, 1000 North Dearborn Avenue, Chicago, Illinois 60610
COMPTON'S PICTURED ENCYCLOPEDIA, F. E. COMPTON AND COMPANY, 1000 North Dearborn
    Street, Chicago, Illinois 60610
CONTRA COSTA DEPARTMENT OF EDUCATION, Contra Costa County, Pleasant Hill,
    California
COOPERATIVE RECREATION SERVICE, Radnor Road, Delaware, Ohio 43015
CORNELL UNIVERSITY PRESS, 124 Roberts Place, Ithaca, New York 14850
COWARD-MCCANN & GEOGHEGAN, INC., 200 Madison Avenue, New York, New York
    10016
GEORGE F. CRAM COMPANY, INC., Box 426, 301 S. LaSalle Street, Indianapolis, Indiana
    46206
CREATIVE EDUCATIONAL SOCIETY, INC., 515 North Front St., Mankato, Minnesota 56001
CRITERION BOOKS, 257 Park Avenue S., New York, New York 10010
THOMAS Y. CROWELL COMPANY, 201 Park Avenue South, New York, New York
    10003
CROWN PUBLISHERS, 419 Park Avenue South, New York, New York 10016

THE JOHN DAY COMPANY, INC., 257 Park Avenue South, New York, New York 10016
DELACORTE PRESS, 750 Third Avenue, New York, New York 10017
DELTA PRESS, 750 Third Avenue, New York, New York 10017
THE DIAL PRESS, INC., 750 Third Avenue, New York, New York 10017
DODD, MEAD AND COMPANY, 79 Madison Avenue, New York, New York 10016
DOUBLEDAY AND COMPANY, 277 Park Avenue South, New York, New York 10017
DOVER PUBLICATIONS, INC., 180 Varick Street, New York, New York 10014
DUELL, SLOAN AND PEARCE, INC., 60 East 42nd Street, New York, New York 10017
E. P. DUTTON AND COMPANY, 201 Park Avenue South, New York, New York 10003

EALING FILMS (see Holt, Rinehart and Winston)
EBONY, 1820 South Michigan Avenue, Chicago, Illinois 60616
EDUCATIONAL DEVELOPMENT CENTER, INC., 15 Mifflin Place, Cambridge, Massachusetts
    02138
EDUCATIONAL PUBLISHING CORPORATION, Darien, Connecticut 06820
EDUCATORS PROGRESS SERVICE, Randolph, Wisconsin 53956
ELEMENTARY ENGLISH JOURNAL, 508 South Sixth Street, Champaign, Illinois 61820
ELK GROVE PRESS, 17420 Ventura Boulevard, Encino, California 91316
ENCYCLOPAEDIA BRITANNICA, INC., 425 North Michigan Avenue, Chicago, Illinois 60611
EXCHANGE TAPES THROUGH WORLD TAPE PALS, Box 9211, Dallas, Texas

FARRAR, STRAUS & GIROUX, 19 Union Square West, New York, New York 10003
FEARON PUBLISHERS, 6 Davis Drive, Belmont, California 94002
FIDELER COMPANY, 31 Ottawa Avenue, N.W., Grand Rapids, Michigan 49502
FIELD EDUCATIONAL PUBLICATIONS, INC., 609 Mission Street, San Francisco, California
    94105
FIELD ENTERPRISES EDUCATIONAL DIVISION, 510 Merchandise Mart Plaza, Chicago, Illinois
    60654
FILENE CENTER, Tufts University, Medford, Massachusetts 02155

FOLLETT PUBLISHING COMPANY, 1010 West Washington Boulevard, Chicago, Illinois 60607

FOUR WINDS PRESS, 50 West 44th Street, New York, New York 10036

FRANKLIN BOOK PROGRAMS, INC., 801 Second Avenue, New York, New York 10017

FRANKLIN PUBLISHING CO., 2047 Locust Street, Philadelphia, Pennsylvania 19103

FREE PRESS OF GLENCOE, 60 Fifth Avenue, New York, New York 10011

FRIENDLY HOUSE, 65 Suffolk Street, New York, New York 10002

FRIENDSHIP PRESS, 475 Riverside Drive, New York, New York 10027

FUNK AND WAGNALLS COMPANY, INC., 53 E. 77 Street, New York, New York 10021

GARDEN CITY BOOKS (DOUBLEDAY AND COMPANY, INC.), 501 Franklin Avenue, Garden City, New York 11531

GARRARD PUBLISHING COMPANY, 1607 North Market Street, Champaign, Illinois 61820

GINN AND COMPANY, 191 Spring Street, Lexington, Massachusetts 02173

GINN AND COMPANY (CANADA), 35 Mobile Drive, Toronto 6, Ontario, Canada

GLOBE BOOK COMPANY, 175 Fifth Avenue, New York, New York 10010

GOLDEN GATE JUNIOR BOOKS, Box 398, San Carlos, California 94070

GOLDEN PRESS, 850 Third Avenue, New York, New York 10022

GOVERNMENT PRINTING OFFICE (see U.S. Government Printing Office)

THE GRADE TEACHER, Darien, Connecticut 06820

GRAPHIC GROUP, 157 East 57th Street, New York, New York 10022

GROLIER EDUCATIONAL CORPORATION, 845 Third Avenue, New York, New York 10022

GROSSET AND DUNLAP INC., PUBLISHERS, 51 Madison Avenue, New York, New York 10010

GROVE PRESS, 53 E. 11th Street, New York, New York 10003

E. M. HALE AND COMPANY, 1201 South Hastings Way, Eau Claire, Wisconsin 54701

C. S. HAMMOND AND COMPANY, 515 Valley Street, Maplewood, New Jersey 07040

HARCOURT BRACE JOVANOVICH, 757 Third Avenue, New York, New York 10017

HARPER AND ROW, 10 East 53rd Street, New York, New York 10022

HARR WAGNER PUBLISHING COMPANY, 600 Mission Street, San Francisco, California 94105

HARVARD UNIVERSITY PRESS, 79 Garden Street, Cambridge, Massachusetts 02138

HARVEY HOUSE PUBLISHERS, Irvington-on-Hudson, New York 10533

HASTINGS HOUSE PUBLISHERS, 10 East 40th Street, New York, New York 10016

HAWTHORN BOOKS, 70 Fifth Avenue, New York, New York 10011

D. C. HEATH AND COMPANY, 125 Spring Street, Lexington, Massachusetts 02173

HILL AND WANG, 72 Fifth Avenue, New York, New York 10011

HOLIDAY HOUSE, INC., 18 East 56th Street, New York, New York 10022

HOLT, RINEHART AND WINSTON, INC., 383 Madison Avenue, New York, New York 10017

HOUGHTON MIFFLIN COMPANY, 2 Park Street, Boston, Massachusetts 02107

THE INSTRUCTOR, Dansville, New York 14437

INTERNATIONAL COMMUNICATIONS FOUNDATION, 870 Monterey Pass Road, Monterey Park, California 91754

JAPAN SOCIETY, 333 East 47th Street, New York, New York 10017

JOHNS HOPKINS PRESS, Baltimore, Maryland 21218

JOINT COUNCIL ON ECONOMIC EDUCATION, 1212 Avenue of the Americas, New York, New York 10036

JOURNAL OF EDUCATION RESEARCH, Box 1605, Madison, Wisconsin 53700

JOURNAL OF GEOGRAPHY, Room 1532, 111 West Washington Street, Chicago, Illinois 60602

ALFRED A. KNOPF, INC., 201 East 50th Street, New York, New York 10022

LAIDLAW BROTHERS, Thatcher and Madison Avenues, River Forest, Illinois 60305
LERNER PUBLICATIONS COMPANY, 241 First Avenue North, Minneapolis, Minnesota 55401
J. B. LIPPINCOTT COMPANY, East Washington Square, Philadelphia, Pennsylvania 19105
LITTLE, BROWN AND COMPANY, 34 Beacon Street, Boston, Massachusetts 02106
LONGMANS GREEN (see David McKay)
LOTHROP, LEE AND SHEPARD COMPANY, INC., 105 Madison Avenue, New York, New York
    10016

MCCORMICK-MATHERS PUBLISHERS, 300 Pike Street, Cincinnati, Ohio 45202
MCGRAW-HILL, INC., 330 West 42nd Street, New York, New York 10036
DAVID MCKAY COMPANY, 750 Third Avenue, New York, New York 10017
MCKINLEY PUBLISHING COMPANY, 112 South New Broadway, Brooklawn, New Jersey
    08030
THE MACMILLAN COMPANY, 866 Third Avenue, New York, New York 10022
MACRAE SMITH COMPANY, 225 South 15 Street, Philadelphia, Pennsylvania 19102
MEDICAL BOOK COMPANY, 336 East 26th Street, New York, New York 10010
MELMONT PUBLISHERS, INC., 1224 W. Van Buren Street, Chicago, Illinois 60607
MENTOR BOOKS, 150 Fifth Avenue, New York, New York 10010
MEREDITH PUBLISHING COMPANY, 1716 Locust Street, Des Moines, Iowa 50309
CHARLES E. MERRILL BOOKS INC., 1300 Alum Creek Drive, Columbus, Ohio 43216
JULIAN MESSNER, INC., 1 West 39th Street, New York, New York 10018
BRUCE MILLER, Box 369, Riverside, California 92502
MINNESOTA WORLD AFFAIRS CENTER, University of Minnesota, Minneapolis, Minnesota
    55455
MODERN EDUCATION PUBLISHERS, Box 651, San Jose, California 95100
WILLIAM MORROW AND COMPANY, INC., 105 Madison Avenue, New York, New York
    10016

NATIONAL AEROSPACE EDUCATION COUNCIL, 806 15th Street, N.W., Washington, D.C.
    20005
NATIONAL CONFERENCE OF CHRISTIANS AND JEWS, 43 West 57th Street, New York, New
    York 10019
NATIONAL COUNCIL FOR THE SOCIAL STUDIES, 1201 Sixteenth Street, N.W., Washington,
    D.C. 20036
NATIONAL COUNCIL OF TEACHERS OF ENGLISH, 1111 Kenyon Road, Urbana, Illinois 61801
NATIONAL EDUCATION ASSOCIATION, 1201 Sixteenth Street, N.W., Washington, D.C.
    20036
NATIONAL ELEMENTARY SCHOOL PRINCIPALS, 1201 Sixteenth Street, N.W., Washington,
    D.C. 20036
NATIONAL GEOGRAPHIC SOCIETY, 17th and M Streets, N.W., Washington, D.C. 20036
NATIONAL PARK SERVICE, Washington, D.C. 20025
NATIONAL SOCIETY FOR THE STUDY OF EDUCATION, University of Chicago Press, Chicago,
    Illinois 60637
THOMAS NELSON AND SONS, Room 1403, 30 East 42nd Street, New York, New York
    10017
NOBLE AND NOBLE PUBLISHERS, INC., 750 Third Avenue, New York, New York 10017
W. W. NORTON AND COMPANY, INC., 55 Fifth Avenue, New York, New York 10003
A. J. NYSTROM AND COMPANY, 3333 Elston Avenue, Chicago, Illinois 60618

OCEANA PRESS, 75 Main Street, Dobbs Ferry, New York 10522
OXFORD BOOK COMPANY, INC., 387 Park Avenue, S., New York, New York 10016

OXFORD UNIVERSITY PRESS, INC., 200 Madison Avenue, New York, New York 10016

PANTHEON BOOKS, INC., 201 East 50th Street, New York, New York 10022
PEABODY COLLEGE, Nashville, Tennessee 37200
PENGUIN BOOKS, INC., 7110 Ambassador Road, Baltimore, Maryland 21207
GEORGE PFLAUM, 38 West Fifth Street, Dayton, Ohio 45402
PHI DELTA KAPPAN, INC., International Headquarters Building, 8th and Union, Bloom-
    ington, Indiana 47401
PITTMAN PUBLISHING CORPORATION (see Delta)
PLAYS, INC., 8 Arlington Street, Boston, Massachusetts 02116
POCKET BOOKS, INC., 630 Fifth Avenue, New York, New York 10020
FREDERICK A. PRAEGER, INC., PUBLISHER, 111 Fourth Avenue, New York, New York 10003
PREMIER-FAWCETT BOOKS, 67 West 44th Street, New York, New York 10036
PRENTICE-HALL, INC., Englewood Cliffs, New Jersey 07632
PRINCETON UNIVERSITY PRESS, Princeton, New Jersey 08540
PUBLIC AFFAIRS COMMITTEE, 381 Park Avenue South, New York, New York 10016
PURDUE UNIVERSITY, West Lafayette, Indiana 47900
G. P. PUTNAM'S SONS, 200 Madison Avenue, New York, New York 10016
PYRAMID BOOKS, 444 Madison Avenue, New York, New York 10022

RAND MCNALLY AND COMPANY, Box 7600, Chicago, Illinois 60680
RANDOM HOUSE, INC., 201 E. 50th Street, New York, New York 10022
REILLY AND LEE COMPANY, 14 East Jackson Boulevard, Chicago, Illinois 60610
REPUBLIC BOOK COMPANY, 104–18 Roosevelt Avenue, Flushing, New York 10068
RINEHART (see Holt, Rinehart and Winston)
THE RONALD PRESS COMPANY, 79 Madison Avenue, New York, New York 10016
ROY PUBLISHERS, 30 East 74th Street, New York, New York 10021

W. H. SADLIER, INC., 11 Park Place, New York, New York 10007
SAINT MARTIN'S PRESS, INC., 175 Fifth Avenue, New York, New York 10010
THE SATURDAY REVIEW, 380 Madison Avenue, New York, New York 10017
SCHOLASTIC MAGAZINES, INC., 50 West 44th Street, New York, New York 10036
SCHOOL PRODUCTS BUREAU, 517 South Jefferson Street, Chicago, Illinois 60680
SCHUMAN (see Abelard-Schuman)
SCIENCE RESEARCH ASSOCIATES, INC., 259 East Erie Street, Chicago, Illinois 60611
WILLIAM R. SCOTT, INC., 333 Avenue of the Americas, New York, New York 10014
SCOTT, FORESMAN AND COMPANY, 1900 E. Lake Avenue, Glenview, Illinois 60025
CHARLES SCRIBNER'S SONS, 597 Fifth Avenue, New York, New York 10017
SEABURY PRESS, 815 Second Avenue, New York, New York 10017
SILVER BURDETT COMPANY, 250 James Street, Morristown, New Jersey 07960
SIMON AND SCHUSTER, INC., 630 Fifth Avenue, New York, New York 10020
L. W. SINGER AND COMPANY, INC., 249 West Erie Boulevard, Syracuse, New York 13202
SOCIAL EDUCATION (see National Council for the Social Studies)
SOCIAL SCIENCE CONSORTIUM, Purdue University, West Lafayette, Indiana 47900
SOCIAL STUDIES, 112 South New Broadway, Brooklawn, New Jersey 08030
SPENCER INTERNATIONAL PRESS, INC., 575 Lexington Avenue, New York, New York 10022
STECK-VAUGHN CO., Box 2028, Austin, Texas 78767
STERLING PUBLISHING COMPANY, INC., 419 Park Avenue South, New York, New York
    10016
HENRY STEWART, 249 Bowen Road, East Aurora, New York 14052
R. H. STONE PRODUCTS, 18279 Livernois, Detroit, Michigan 48221
SYRACUSE UNIVERSITY PRESS, Syracuse, New York 13200

TAYLOR PUBLISHING COMPANY, 1550 West Mockingbird Lane, Dallas, Texas 75221
TEACHERS COLLEGE PRESS, 1234 Amsterdam Avenue, New York, New York 10027
TIME-LIFE BOOKS, Time and Life Building, New York, New York 10036
CHARLES E. TUTTLE, Rutland, Vermont 05701
TWENTIETH CENTURY FUND, 41 East 70th Street, New York, New York 10021

FREDERICK UNGAR PUBLISHING CO., INC., 250 Park Avenue S., New York, New York 10003
UNITED NATIONS ASSOCIATION, 345 East 46th Street, New York, New York 10017
UNIPUB, 650 First Avenue, New York, New York 10016
U.S. COMMITTEE FOR UNICEF, 331 East 38th Street, New York, New York 10017
U.S. GOVERNMENT PRINTING OFFICE, Washington, D.C. 20402
UNIVERSITY OF CALIFORNIA PRESS, 2223 Fulton Street, Berkeley, California 94720
UNIVERSITY OF CHICAGO PRESS, 11030 South Langley, Chicago, Illinois 60637
UNIVERSITY OF MICHIGAN PRESS, 615 East University, Ann Arbor, Michigan 48106
UNIVERSITY OF NORTH CAROLINA PRESS, Chapel Hill, North Carolina 27514
UNIVERSITY OF OKLAHOMA PRESS, 1005 Asp Avenue, Norman, Oklahoma 73069
UNIVERSITY OF UTAH PRESS, Salt Lake City, Utah 84100

THE VANGUARD PRESS, INC., 424 Madison Avenue, New York, New York 10017
VAN NOSTRAND REINHOLD, 450 West 33rd Street, New York, New York 10001
THE VIKING PRESS, 625 Madison Avenue, New York, New York 10022
VINTAGE (see Random House)

WADSWORTH PUBLISHING COMPANY, Belmont, California 94002
HENRY Z. WALCK, INC., 17–19 Union Square West, New York, New York 10003
WALKER AND COMPANY, 720 Fifth Avenue, New York, New York 10019
FREDERICK WARNE AND COMPANY, 101 Fifth Avenue, New York, New York 10003
IVES WASHBURN, 750 Third Avenue, New York, New York 10017
WASHINGTON SQUARE PRESS, 630 Fifth Avenue, New York, New York 10003
FRANKLIN WATTS, INC., 845 Third Avenue, New York, New York 10022
WEBSTER PUBLISHING COMPANY, 1154 Reco Avenue, St. Louis, Missouri 63126
WESTMINSTER, WITHERSPOON BUILDING, Philadelphia, Pennsylvania 19107
ALBERT WHITMAN AND COMPANY, 560 West Lake Street, Chicago, Illinois 60606
WHITTLESEY HOUSE, 330 West 42nd Street, New York, New York 10036
JOHN WILEY AND SONS, INC., 605 Third Avenue, New York, New York 10016
WORLD ENCYCLOPEDIA, FIELD ENTERPRISES EDUCATIONAL DIVISION, 510 Merchandise Mart
    Plaza, Chicago, Illinois 60654
WORLD NEWS OF THE WEEK, 7300 North Linder Avenue, Skokie, Illinois 60076
WORLD PUBLISHING COMPANY, Cleveland, Ohio 44102
WORLD WIDE GAMES, Box 450, Delaware, Ohio 43015

XEROX COLLEGE PUBLISHING, 191 Spring Street, Lexington, Massachusetts 02173

YALE UNIVERSITY PRESS, 149 York Street, New Haven, Connecticut 06511

Index